Modern Chromatography

Modern Chromatography

Edited by
Alex Duncan

C WILLFORD PRESS
www.willfordpress.com

Published by Willford Press,
118-35 Queens Blvd., Suite 400,
Forest Hills, NY 11375, USA

ISBN: 978-1-68285-769-4

Cataloging-in-Publication Data

Modern chromatography / edited by Alex Duncan.
p. cm.
Includes bibliographical references and index.
ISBN 978-1-68285-769-4
1. Chromatographic analysis. 2. Chemistry, Analytic. 3. Chromatograms. 4. Chromatographic detectors.
I. Duncan, Alex.
QP519.9.C47 M63 2020
543.8--dc23

For information on all Willford Press publications
visit our website at www.willfordpress.com

WILLFORD PRESS

Contents

Preface

Chromatography is a technique that is used for the separation of a mixture. The solutes or components of a mixture are separated based on the relative proportion of individual solutes distributed between a moving fluid stream and a contiguous stationary phase. Chromatography can be analytical or preparative. Depending on the mode of operation, geometry of the system, phases involved and the retention mechanism, chromatography can be of two types- column chromatography and planar chromatography. These techniques are of immense use in various chemical and biological fields. This book outlines the processes and applications of chromatography in detail. It includes some of the vital pieces of work being conducted across the world, on various topics related to chromatography. It will prove to be immensely beneficial to students and researchers in this field.

After months of intensive research and writing, this book is the end result of all who devoted their time and efforts in the initiation and progress of this book. It will surely be a source of reference in enhancing the required knowledge of the new developments in the area. During the course of developing this book, certain measures such as accuracy, authenticity and research focused analytical studies were given preference in order to produce a comprehensive book in the area of study.

This book would not have been possible without the efforts of the authors and the publisher. I extend my sincere thanks to them. Secondly, I express my gratitude to my family and well-wishers. And most importantly, I thank my students for constantly expressing their willingness and curiosity in enhancing their knowledge in the field, which encourages me to take up further research projects for the advancement of the area.

Editor

Analysis of Polycyclic Aromatic Hydrocarbons in Ambient Aerosols by using One-Dimensional and Comprehensive Two-Dimensional Gas Chromatography Combined with Mass Spectrometric Method

Yun Gyong Ahn ⓘ,[1] **So Hyeon Jeon,**[1] **Hyung Bae Lim,**[2] **Na Rae Choi,**[3] **Geum-Sook Hwang,**[1] **Yong Pyo Kim,**[4] **and Ji Yi Lee ⓘ**[3]

[1]*Western Seoul Center, Korea Basic Science Institute, Seoul 03759, Republic of Korea*
[2]*Air Quality Research Division, National Institute of Environmental Research, Incheon 22689, Republic of Korea*
[3]*Department of Environmental Science and Engineering, Ewha Womans University, Seoul 03759, Republic of Korea*
[4]*Department of Chemical Engineering and Material Science, Ewha Womans University, Seoul 03760, Republic of Korea*

Correspondence should be addressed to Ji Yi Lee; yijiyi@ewha.ac.kr

Academic Editor: Federica Bianchi

Advanced separation technology paired with mass spectrometry is an ideal method for the analysis of atmospheric samples having complex chemical compositions. Due to the huge variety of both natural and anthropogenic sources of organic compounds, simultaneous quantification and identification of organic compounds in aerosol samples represents a demanding analytical challenge. In this regard, comprehensive two-dimensional gas chromatography with time-of-flight mass spectrometry (GC×GC-TOFMS) has become an effective analytical method. However, verification and validation approaches to quantify these analytes have not been critically evaluated. We compared the performance of gas chromatography with quadrupole mass spectrometry (GC-qMS) and GC×GC-TOFMS for quantitative analysis of eighteen target polycyclic aromatic hydrocarbons (PAHs). The quantitative obtained results such as limits of detection (LODs), limits of quantification (LOQs), and recoveries of target PAHs were approximately equivalent based on both analytical methods. Furthermore, a larger number of analytes were consistently identified from the aerosol samples by GC×GC-TOFMS compared to GC-qMS. Our findings suggest that GC×GC-TOFMS would be widely applicable to the atmospheric and related sciences with simultaneous target and nontarget analysis in a single run.

1. Introduction

Human health research associated with polycyclic aromatic hydrocarbons (PAHs) has raised concerns because certain PAHs are classified as probable human carcinogens [1–4] and have shown tumorigenic activity and endocrine disrupting activity in mammals [5]. The US EPA has included 16 of them in the list of priority pollutants and has established a maximum contaminant level of 0.2 μg/L for benzo[a]pyrene in drinking water [6]. In the European Union (EU), eight PAHs have been identified as priority hazardous substances in the field of water policy [7]. The EPA priority 16 PAHs and two additional PAHs are now being monitored by European agencies, and they have sought to quantify the individual concentrations of benzo[e] pyrene and perylene in environmental samples [6]. PAHs are found in ambient air in the gas phase and as sorbents to aerosols [8]. Thus, air monitoring of PAHs to quantify inhalation exposure and to identify other organic compounds is important for insight into photochemical reactions. The quantification and identification of organic compounds in air samples is an important feature of atmospheric chemistry and represents some demanding analytical challenges [9].

For these reasons, a key issue in current analytical methods is the ability to measure a large number of compounds with quantitative analysis for target analytes. Comprehensive two-dimensional gas chromatography (GC×GC) coupled with mass spectrometry (MS) can screen for nontarget compounds with fast identification of the compounds in an entire sample [10]. Therefore, previous studies applied GC×GC-MS for the identification of numerous compounds present in air samples [11–13]. However, there are limitations on the validation of simultaneous quantification and identification of analytes in air samples. Correspondingly, a validation of simultaneous identification and quantification of PAHs and other compounds in air samples by GC×GC–MS is required. A TOF mass spectrometer was used to acquire sufficient data from a comprehensive two-dimensional chromatographic technique that generated multiple narrow peaks from the short secondary column [14, 15]. Generally, GC coupled with quadrupole MS (GC-qMS) in the selected ion monitoring (SIM) mode has been used for quantitative analysis of PAHs in air samples because of its selective detection for specific target compounds [16, 17]. However, a GC×GC-TOFMS validated method suitable for the quantification of target PAHs in an aerosol sample compared with GC-qMS in the SIM mode has not yet been reported. The aim of this study was to evaluate the effectiveness of GC×GC-TOFMS in the quantitative analysis of target PAHs as well as the fast identification of multiple compounds for aerosol samples. The validity of the quantitative results obtained by both GC×GC-TOFMS and GC-qMS in the SIM mode was demonstrated by several method performance parameters such as linearity, accuracy, and repeatability.

2. Experimental

2.1. Air Sampling. The total suspended particle (TSP) samples were collected at Asan Engineering Building, Ewha Womans University, Seoul, South Korea (37.56°N, 126.94°E, 20 m above ground level), with a PUF sampler (Tisch, TE-1000) on a quartz fiber filter (Quartz fiber filter, QFF, Ø10.16 cm, Whatman, UK). The sampling site is located in the mixed resident area, commercial area, forest area, and nearby roadside. A total of 67 filter samples were obtained during summer (August 12–30, 2013) and winter (January 27–February 16, 2014) and day (9 a.m.~6 p.m.) and night (8 p.m.~6 a.m.). Prior to sampling, the quartz fiber filters were baked for 8 h in an electric oven at 550°C to remove possible organic contaminants. The sampled filters were wrapped in aluminum foils and stored in a freezer at −20°C until analysis.

2.2. Chemicals. All organic solvents were of GC grade and purchased from Burdick and Jackson (Phillipsburg, NJ, USA). Standard solutions of target PAHs (Table 1 for their full chemical names and information) except Per and BeP for quantitative analysis were purchased as a mixture at a concentration of 2000 μg/mL in dichloromethane from Supelco (Bellefonte, PA, USA). Per and BeP standards

(>99%) were purchased from Aldrich (St. Louis, MI, USA), and a standard mixture of eighteen PAHs was prepared at a concentration of 1000 μg/mL. Deuterium-labeled internal standards of seven PAHs were purchased from Aldrich (St. Louis, MI, USA) and Chiron (Trondheim, Norway) and used for the spiking test as listed in Table 1. Working standard solutions (0.01~10 μg/mL) were prepared and then stored at −20°C prior to use.

2.3. Preparation of Samples. Air sampling filters were extracted with a mixture of dichloromethane and methanol (3 : 1, v/v) two times using an accelerated solvent extractor (ASE) (Dionex ASE-200) at 40°C and 1700 psi for 5 min. Prior to the extraction, seven deuterated internal standards (Nap-d8, Ace-d10, Phen-d10, Fla-d10, Chr-d12, Per-d12, and BghiPer-d12) were spiked in the filters to compensate for matrix effects during the extraction procedure. Extracts were blown down to 1 mL using a nitrogen evaporator (TurboVap II, Caliper Life Sciences). GC×GC-TOFMS analysis was carried out using an Agilent GC (Wilmington, Delaware, USA)-Quad-jet thermal modulation Pegasus 4D TOFMS (LECO, St. Joseph, MI, USA). The sample was injected in the splitless mode at 300°C. The GC×GC columns were as follows: DB-5MS (30 m × 0.25 mm ID, film thickness of 0.25 μm) and 1.17 m DB-17MS (0.18 mm OD, 0.18 μm film). The operating conditions of GC-MS and GC×GC-TOFMS are summarized in Table 2.

3. Results and Discussion

3.1. GC-qMS and GC×GC-TOFMS for Characterization of Aerosol Samples. In most studies, separation and quantification of PAHs in aerosol samples have been analyzed using a conventional GC-qMS [18]. Flame ionization detection (FID) has also been widely used for quantification as it features a higher response to PAHs which contain only carbon and hydrogen, while oxygenates and other species that contain heteroatoms tend to have a lower response factor [19]. However, this nonspecific detector may not distinguish inferences, which include a large fraction of aliphatic and aromatic compounds in aerosol samples from alkylated PAH homologues. The coupling of GC with MS is increasingly becoming the analytical tool of choice in this regard because of its superior selectivity and sensitivity. Among the most common analyzers including TOF [20], ion trap, and qMS [21, 22], qMS is the most widely adopted technique for routine analysis of PAHs [23]. GC-qMS data acquisition takes advantages of both a full mass scan range (scan mode) and specific ion masses for target analytes (SIM mode). The sensitivity in the SIM mode is higher than that in the scan mode of GC-qMS due to the increased dwell time on each monitored ion for trace analysis in some matrices such as in atmospheric aerosols [24, 25]. GC-TOFMS has a much faster spectral acquisition rate than GC-qMS does, which is up to 500 full mass scans per second [26]. Consequently, this system is able to widen the application of GC×GC techniques providing very narrow chromatographic peaks, typically 50~600 ms at the baseline with

TABLE 1: Information of target PAHs in the study.

Compound	Abbreviation	CAS number	Molecular formula	MW	Quantitative ion	Qualifier ion	Retention time		
							GC-qMS (min)	GC×GC-TOFMS t_{r1} (min)	t_{r2} (s)
Naphthalene-d$_8$[a]	Nap-d$_8$	1146-65-2	C$_{10}$D$_8$	136.2	136	137	12.25	13.40	1.34
Naphthalene	Nap	91-20-3	C$_{10}$H$_8$	128.2	128	129	12.34	13.47	1.35
Acenaphthylene	Acy	208-96-8	C$_{12}$H$_8$	152.2	152	153	19.08	19.56	1.55
Acenaphthene-d$_{10}$[a]	Ace-d$_{10}$	15067-26-2	C$_{12}$D$_{10}$	164.2	162	164	19.47	20.12	1.52
Acenaphthene	Ace	83-32-9	C$_{12}$H$_{10}$	154.2	153	154	19.61	20.28	1.52
Fluorene	F	86-73-7	C$_{13}$H$_{10}$	166.2	166	165	21.61	22.28	1.53
Phenanthrene-d$_{10}$[a]	Phen-d$_{10}$	1518-22-2	C$_{14}$D$_{10}$	188.2	188	189	25.91	25.96	1.68
Phenanthrene	Phen	85-01-8	C$_{14}$H$_{10}$	178.2	178	179	26.01	26.04	1.71
Anthracene	Ant	120-12-7	C$_{14}$H$_{10}$	178.2	178	179	26.14	26.20	1.68
Fluoranthene-d$_{10}$[a]	Fla-d$_{10}$	93951-69-0	C$_{16}$D$_{10}$	212.2	212	213	30.99	30.68	1.84
Fluoranthene	Fla	206-44-0	C$_{16}$H$_{10}$	202.2	202	203	31.08	30.68	1.87
Pyrene	Pyr	129-00-0	C$_{16}$H$_{10}$	202.2	202	203	32.25	31.56	1.98
Benz[a]anthracene	BaA	56-55-3	C$_{18}$H$_{12}$	228.2	228	226	37.20	36.28	2.21
Chrysene-d$_{12}$[a]	Chr-d$_{12}$	1719-03-5	C$_{18}$D$_{12}$	240.3	240	236	37.42	36.28	2.27
Chrysene	Chr	218-01-9	C$_{18}$H$_{12}$	228.3	228	226	37.54	36.44	2.26
Benzo[b]fluoranthene	BbF	205-99-2	C$_{20}$H$_{12}$	252.3	252	253	41.55	40.12	2.76
Benzo[k]fluoranthene	BkF	207-08-9	C$_{20}$H$_{12}$	252.3	252	253	41.65	40.28	2.74
Benzo[e]pyrene	BeP	192-97-2	C$_{20}$H$_{12}$	252.3	252	253	42.90	41.08	3.16
Benzo[a]pyrene	BaP	50-32-8	C$_{20}$H$_{12}$	252.3	252	253	43.09	41.24	3.23
Perylene-d$_{12}$[a]	Per-d$_{12}$	1520-96-3	C$_{20}$D$_{12}$	264.3	264	260	43.46	41.40	3.36
Perylene	Per	198-55-0	C$_{20}$H$_{12}$	252.3	252	253	43.57	41.48	3.47
Indeno[1,2,3-cd]pyrene	IP	193-39-5	C$_{22}$H$_{12}$	276.3	276	277	48.09	45.48	0.71
Dibenz[a,h]anthracene	DBahAnt	53-70-3	C$_{22}$H$_{14}$	278.3	278	279	48.12	45.64	0.82
Benzo[ghi]perylene-d$_{12}$[a]	BghiPer-d$_{12}$	93951-66-7	C$_{22}$D$_{12}$	288.3	288	284	49.92	46.52	1.54
Benzo[ghi]perylene	BghiPer	191-24-2	C$_{22}$H$_{12}$	276.3	276	277	50.13	46.68	1.78

[a]Internal standard.

TABLE 2: GC-qMS and GC×GC-TOFMS operating conditions.

Parameters	GC-qMS	GC×GC-TOFMS
Injector settings		
Injection volume	1 µL	1 µL
Inlet mode	Splitless	Splitless
Carrier gas	He (99.999%)	He (99.999%)
Carrier gas flow	1.0 mL·min^{-1}	1.3 mL·min^{-1}
Inlet temperature	280°C	300°C
GC oven temperature		
Initial temperature	1 min at 60°C	1 min at 60°C
First rate	6°C/min to 310°C	6°C/min to 300°C
Isothermal pause	15 min at 310°C	15 min at 300°C
2nd oven temperature offset	—	5°C, relative to the 2nd oven temperature
Modulator		
Modulator temperature offset	—	15°C, relative to the 2nd oven temperature
Modulator period	—	4.00 s
Hot pulse time	—	1.00 s
Cool time between stages	—	1.40 s
MS		
Mass range	40~550	40~550
Electron energy	70 eV	70 eV
Ion source temperature	230°C	230°C

sufficient density of data points per chromatographic peak [27]. Environmental samples are generally complex, often with more than hundreds of compounds containing structural isomers and homologues spread over a wide range of concentration and volatility. Accordingly, multidimensional separation is an advanced technique offering the possibility of greatly enhanced selectivity using different separation mechanisms for the analysis of complex environmental samples [28–30]. In this study, a set of columns DB-5×DB-17 ms was applied to increase the resolution and peak capacity. The fast scanning Pegasus 4D TOFMS system was combined to allow efficient processing of data acquisition, handling, peak detection, and deconvolution. In the one-dimensional column, a 30 m-long DB-5 ms (5% diphenyl/95% dimethyl polysiloxane) stationary phase was used to separate analytes based on volatility and combined with a 1.17 m-long DB-17 ms column (50% diphenyl/50% dimethyl polysiloxane) allowing relative polarity-based separation. Figure 1 shows GC×GC-TOFMS chromatograms of aerosol samples collected at day and night during winter in Seoul, South Korea. To compare the identification ability of GC×GC-TOFMS with GC-qMS, analysis with GC-qMS in the scan mode was performed. A comparison of the one-dimensional chromatograms of the same samples obtained by GC-qMS is shown in Figure 2. 2D chromatograms enable the visual classification of chemically related compounds into groups. It was rare to see that the early-eluting

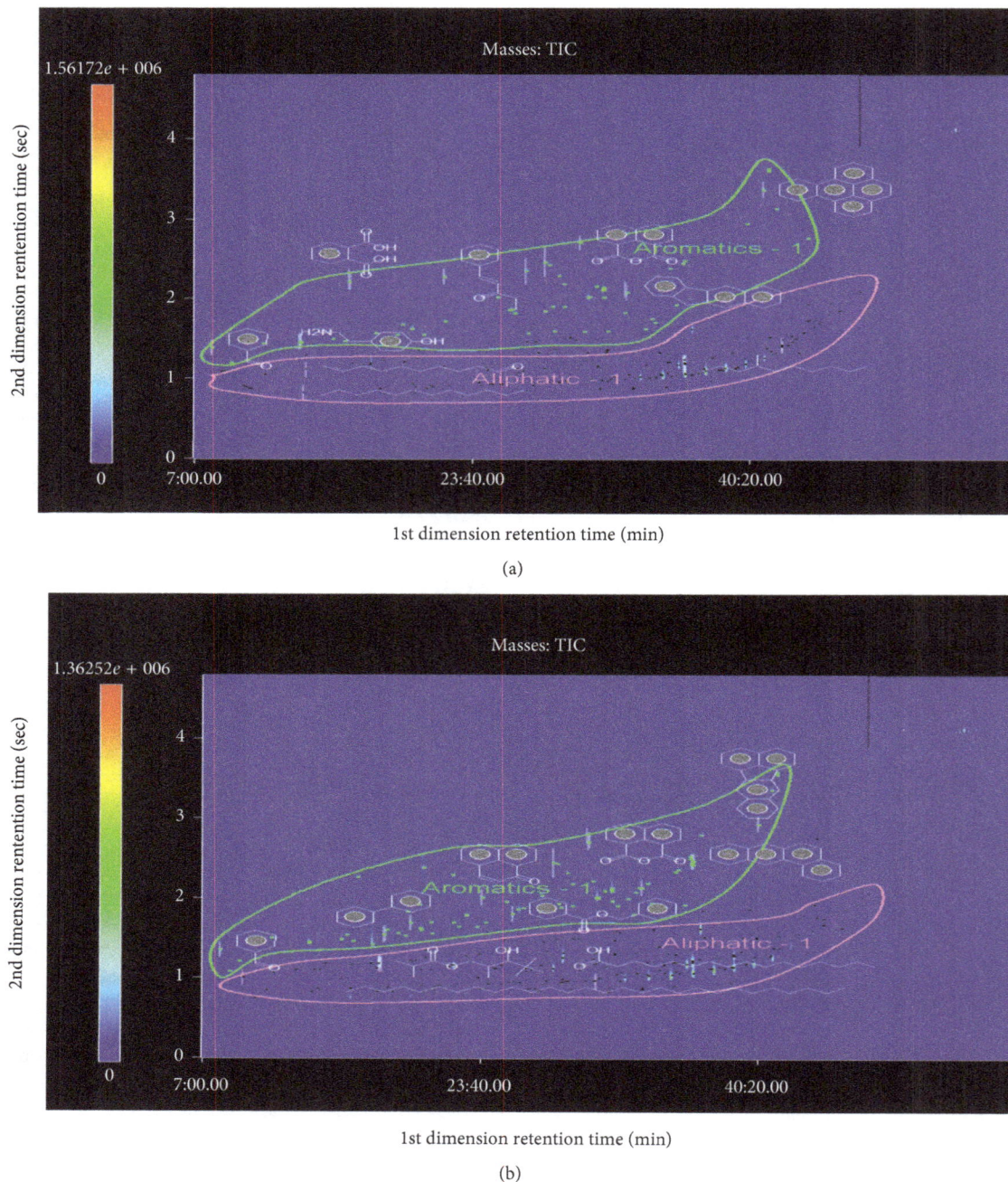

FIGURE 1: GC×GC-TOFMS plots of aerosol samples collected during day (a) and night (b) of winter in Seoul, Korea. A total of 251 and 297 peaks were identified in aerosol samples collected during day (a) and night (b), respectively. Aromatic and aliphatic classes were drawn to divide two regions for ease of viewing.

analytes have an extreme volatility in the chromatogram, as shown in Figure 2. Because of the large losses of these analytes during sample extraction and concentration, particle-associated semivolatile analytes were mainly detected and classified according to their aromatic and aliphatic hydrocarbon groups.

Meanwhile, analytes from the GC-qMS chromatogram were separated based on their vapor pressures or boiling points. The GC×GC technique is rather well suited for group separations, and classifying compounds into chemical-related

groups could be useful for source identification of atmospheric aerosols by means of the large amount of chemical data handling. The combined use with TOFMS provides rapid and reliable identification of analytes using their deconvoluted pure mass spectra. The major limitation of qMS is its limited scan rate; therefore, quantification and identification is seriously compromised because of the mass spectral skew due to the variations in ion abundances at different regions of a chromatographic peak [31, 32]. The numbers of identified chromatographic peaks analyzed by

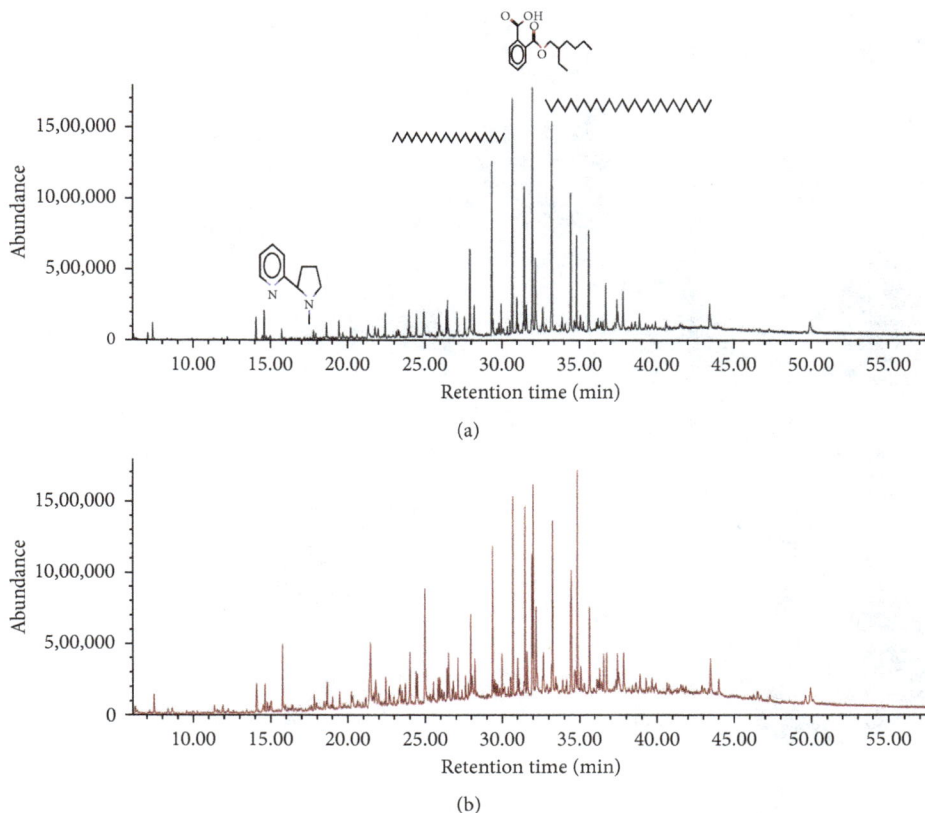

FIGURE 2: Total ion chromatograms of aerosol samples collected in day (a) and night (b) of winter in Seoul, Korea, obtained by GC-qMS. A total of 35 and 64 peaks were identified in aerosol samples collected during day (a) and night (b), respectively. The analytes were separated based on their boiling points.

GC-qMS using the same signal threshold setting from the aerosol samples collected at day and night were 35 and 64, respectively. In the case of results obtained by GC×GC-TOFMS, 251 and 297 peaks from the day- and night-time aerosol samples were, respectively, assigned by individual spectral deconvolution. As a result, phthalic anhydride and 1,2-naphthalic anhydride as the markers of secondary formation for gas-phase PAH reactions were identified in the aerosol sample, as shown in Figure 3. Since the products formed through photochemical reactions are often more toxic than their parent PAHs in atmosphere [17], significant efforts have been expended to identify the photochemical products with PAHs in the fields of atmospheric or environmental sciences. In the case of results obtained using GC-qMS, phthalic anhydride and 1,2-naphthalic anhydride were not detected in the same sample. Limitations of one-dimensional separation have been reported for these photochemical products and complex mixtures of the aerosol sample because of their diverse polarities in a single run [33, 34]. Contrastively, two anhydrides associated with secondary organic aerosol formation were clearly separated and detected by GC×GC-TOFMS. Therefore, it showed advantages for nontarget screening to identify molecular markers or chemical patterns more representative of the aerosol state observed in ambient air.

3.2. Validation of GC-qMS and GC×GC-TOFMS for Quantification of PAHs. GC-qMS and GC×GC-TOFMS were tested individually in order to evaluate their analytical performances. The calibration linearity (regression coefficient, R^2) and relative response factor (RRF) are presented in Table 3. The RRF is the ratio between a signal produced by an individual native analyte and the corresponding isotopically labeled analogue of the analyte (as an internal standard). For calculating RRF, 2 ng of each target PAH and each corresponding deuterated internal standard was spiked, and the relative sensitivity in both the methods was compared. Despite the high-speed scanning performance of GC×GC-TOFMS, the RRFs obtained by this method were approximately equivalent to those obtained by GC-qMS. RRF expresses the sensitivity of a detector for a given substance relative to a standard substance [35, 36]. Thus, it indicated that the sensitivity of GC×GC-TOFMS relative to target PAHs is comparable in quantitative analysis. Calibration curves were generated using the peak area for the 18 PAHs at seven concentrations ranging from 0.01 to 10 μg/mL. The linearity was assessed by calculating the regression equation and the correlation coefficient by the least squares method, as shown in Table 3. The R^2 values were greater than 0.999 for GC-qMS and 0.99 for GC×GC-TOFMS. Although data processing for quantification by GC×GC-TOFMS was derived from the combined peak areas

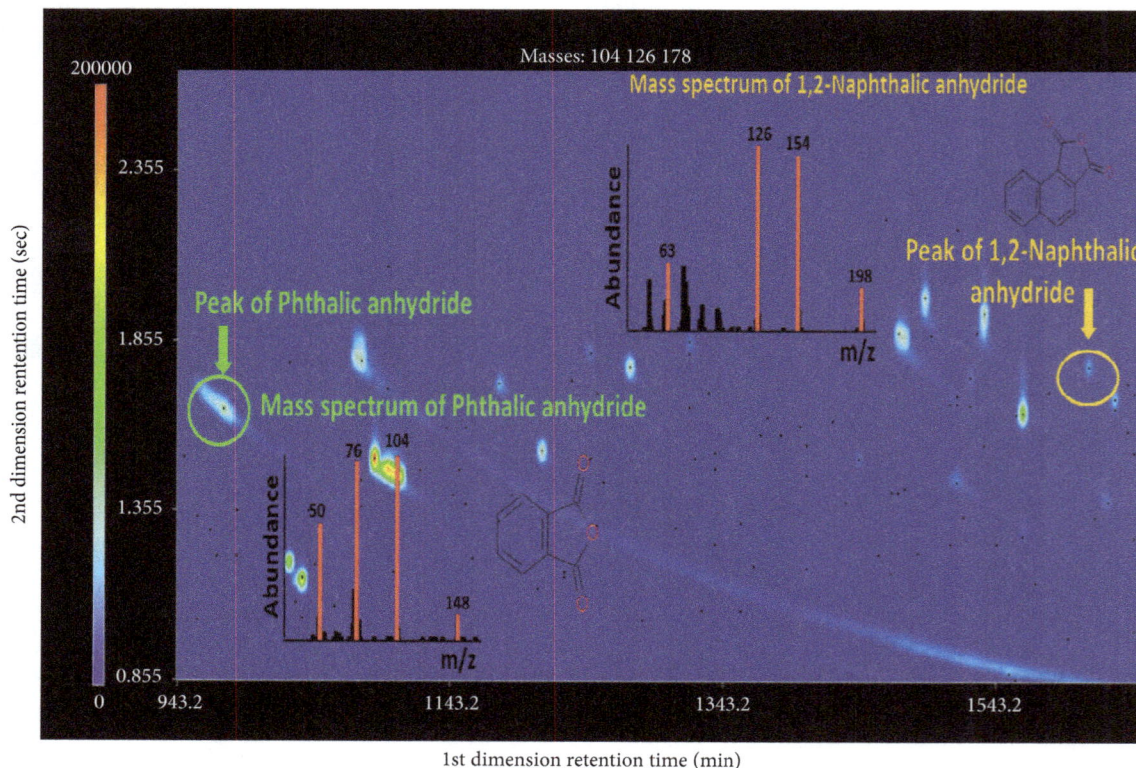

FIGURE 3: GC×GC chromatograms and mass spectrums of phthalic anhydride (marked as green) and 1,2-naphthalic anhydride (marked as yellow) in the aerosol sample. GC×GC chromatograms of phthalic anhydride and 1,2-naphthalic anhydride were certified by molecular ions of m/z 148 and 198, respectively.

TABLE 3: Relative response factors (RRFs) and calibrations of 18 PAHs obtained by the compared methods.

Compound	GC-qMS				GC×GC-TOFMS			
	RRF[a]	Slope	Intercept	R^2	RRF	Slope	Intercept	R^2
Nap	1.04	0.515	−0.003	0.9999	1.69	0.560	0.049	0.9971
Acy	1.57	0.808	−0.004	1.0000	1.95	1.026	−0.012	0.9999
Ace	1.03	0.439	0.009	0.9999	1.16	0.553	−0.006	0.9994
F	1.29	0.667	−0.006	1.0000	1.11	0.609	−0.021	0.9997
Phe	1.17	0.572	−0.004	0.9998	1.47	0.763	−0.039	0.9979
Ant	0.98	0.547	−0.017	0.9992	0.93	0.431	−0.010	0.9982
Fla	1.30	0.678	−0.001	1.0000	1.43	0.802	−0.021	0.9995
Pyr	1.31	0.686	−0.004	0.9999	1.62	0.910	−0.055	0.9972
BaA	0.98	0.575	−0.019	0.9997	1.42	0.574	−0.004	0.9998
Chr	1.06	0.563	−0.003	1.0000	1.26	0.651	−0.005	0.9998
BbF	0.99	0.535	−0.008	0.9999	1.69	0.848	−0.028	0.9993
BkF	1.11	0.576	−0.011	0.9998	0.88	0.327	−0.012	0.9977
BeP	0.91	0.455	−0.007	0.9996	0.90	0.530	−0.014	0.9997
BaP	0.88	0.505	−0.014	0.9997	0.83	0.477	−0.030	0.9985
Per	0.89	0.475	−0.008	0.9998	1.11	0.521	-0.023	0.9979
IP	1.37	0.717	−0.023	0.9995	1.25	0.660	-0.062	0.9922
DBahAnt	1.24	0.629	−0.019	0.9996	1.16	0.490	-0.074	0.9898
BghiPer	1.24	0.594	−0.010	1.000	1.53	0.710	-0.036	0.9991

[a]RRF expresses the sensitivity of a detector for a given analyte relative to its corresponding deuterated internal standards; RRF = $(A_x C_{is})/(A_{is} C_x)$, where A_x is the peak area of a quantifying ion for a given analyte being measured; A_{is} is the peak area of a quantifying ion for its corresponding internal standard; C_x is the concentration of a given analyte; and C_{is} is the concentration of its corresponding internal standard.

for the slices of modulated peaks in contrast to production of the single measured peak by GC-qMS, the results meet the criteria for acceptable linearity within this calibration range.

Naturally, the development of quantitative GC×GC studies based on the quantitative results associated with sophisticated implementation for modulated peaks has been delayed

TABLE 4: Limits of detection and quantification and recoveries of 18 PAHs obtained by the compared methods.

Compound	LOD[a] (ng)		LOQ[b] (ng)		Recovery ± RSD (%)	
	GC-qMS	GC×GC-TOFMS	GC-qMS	GC×GC-TOFMS	GC-qMS	GC×GC-TOFMS
Nap	0.07	0.40	0.21	1.19	94.4 ± 4.2	135 ± 45
Acy	0.17	0.07	0.51	0.22	119 ± 12	116 ± 15
Ace	0.05	0.17	0.16	0.52	105 ± 5.3	105 ± 7.8
F	0.04	0.15	0.13	0.44	158 ± 28	130 ± 29
Phe	0.10	0.34	0.31	1.03	94.5 ± 5.3	86.3 ± 16
Ant	0.19	0.31	0.58	0.92	90.4 ± 4.6	95.1 ± 20
Fla	0.05	0.14	0.16	0.41	90.3 ± 3.9	105 ± 13
Pyr	0.08	0.36	0.25	1.09	97.4 ± 5.3	97.2 ± 13
BaA	0.12	0.09	0.37	0.27	93.4 ± 4.9	86.9 ± 8.2
Chr	0.04	0.08	0.13	0.24	95.8 ± 5.8	101 ± 16
BbF	0.05	0.18	0.15	0.53	96.1 ± 5.7	92.3 ± 10
BkF	0.09	0.35	0.28	1.05	94.2 ± 6.5	105 ± 12
BeP	0.13	0.13	0.40	0.38	92.6 ± 5.8	92.7 ± 5.7
BaP	0.12	0.24	0.37	0.72	93.6 ± 5.3	104 ± 9.0
Per	0.11	0.34	0.32	1.02	93.0 ± 5.5	92.5 ± 8.6
IP	0.15	0.65	0.16	1.94	95.0 ± 5.4	93.9 ± 8.5
DBahAnt	0.13	1.05	0.40	3.14	94.9 ± 5.5	95.8 ± 5.7
BghiPer	0.09	0.22	0.27	0.66	94.6 ± 6.0	87.0 ± 8.5

[a]LOD, smallest amount of analyte that is statistically different from the blank; [b]LOQ, smallest amount of analyte that can be measured with reasonable accuracy.

compared with qualitative reports. Recently, the approach to quantifying multiple analytes at once with comprehensive two-dimensional GC has been extensively studied in accordance with the improvement of data processing for the integration of modulated peaks [37, 38]. In this study, the modulated peaks of each PAH was automatically combined and integrated by the ChromaTOF software based on a similarity of spectra within an allowable time difference between the second dimension peaks in the neighboring slices of the chromatogram. Recovery test was performed by spiking known amounts of the 18 PAH compounds in a prebaked clean filter at a final concentration of 2 µg/mL and analyses of each through all the experiment procedures were compared using the two different methods. Six duplicate tests were performed, and the results of the recovery are shown in Table 4. The average recoveries were in the range of 90.3 to 158% with relative standard deviations (RSDs) ranging from 3.9 to 28% for GC-qMS, while the recoveries were from 86.3 to 135% for GC×GC-TOFMS, with RSDs ranging from 5.7 to 45%. Most of the targeted PAH compounds were afforded acceptable recoveries, excluding F and Nap by using the two analytical methods due to the high volatility of these compounds. Compared with the reproducibility as expressed in %RSDs, the values obtained by GC-qMS were slightly lower than those obtained by GC×GC-TOFMS; however, the %RSD values of the targeted PAHs excluding F and Nap were acceptable (<20% RSD). These observations may vary for the versatile GC×GC technique, since the reproducibility of the modulation phase is dependent on the type of modulator, the stability of the stationary phases, and the chemistry of the analyte, regarding interaction with the stationary phase as presented in several prior studies [39, 40]. The LOD and LOQ were determined based on the standard deviation (SD) of the intersection of the analytical curve (s) and the slope of the

curve (S) as $LOD = 3.3 \times (s/S)$ and $LOQ = 10 \times (s/S)$. The LOD and LOQ for each PAH compound obtained from both the methods are shown in Table 4. The LOD and LOQ values of the 18 PAH compounds obtained by GC-qMS were similar to the results of previous studies [10, 41, 42]. Thus, the suitability of GC×GC-TOFMS for quantification of PAHs was proven by comparing the results with those obtained using GC-qMS.

4. Conclusion

A fast scanning GC×GC-TOFMS was compared to a GC-qMS for the determination of PAHs in aerosol samples. For separation, identification, and characterization, GC×GC-TOFMS was advantageous over GC-qMS owing to the increased peak capacity, and its results showed enhanced detectability and structured chromatograms for nontarget analysis. The qualitative mass separation by TOFMS combined with an automated peak-finding capability provided the resolution of complex mixed mass spectra, resulting from overlapping chromatographic peaks and spectral deconvolution of individual mass spectra for unknown analytes. Furthermore, the obtained quantitative results such as LODs, LOQs, and recoveries of the 18 target PAHs were approximately equivalent for both the analytical methods. Thus, GC×GC-TOFMS had advantages for the simultaneous quantification and qualification of PAHs and other organic compounds in a single run. Because of its high degree of separation and capability of spectral deconvolution of overlapping peaks in highly complex samples, comprehensive GC×GC-TOFMS may become a useful platform in many other fields of research.

Conflicts of Interest

The authors declare that they have no conflicts of interest.

Acknowledgments

This research was supported by the Bio-Synergy Research Project (no. NRF-2017M3A9C4065961) of the Ministry of Science, ICT, and Future Planning through the National Research Foundation and the Korea Basic Science Institute Grant (no. C37705). This research was also supported by the Basic Science Research Program through the National Research Foundation of Korea (NRF) funded by the Ministry of Education (no. NRF-2016R1A2B4015143)

References

[1] IARC, "Polynuclear aromatic compounds, Part 1. Chemical, environmental and experimental data," *IARC Monographs on the Evaluation of the Carcinogenic Risk of Chemicals to Humans*, vol. 32, pp. 1–453, 1983.

[2] World Health Organization, "Environmental health criteria," in *International Programme on Chemical Safety (IPCS)*, vol. 171, WHO, Geneva, Switzerland, 1998.

[3] V. Vestreng, "Emission data reported to UNECE/EMEP: quality assurance and trend analysis and presentation of WebDab: MSC-W status report 2002," Research report, University of Oslo, Oslo, Norway, 2002.

[4] L.-B. Liu, L. Yan, J.-M. Lin, T. Ning, K. Hayakawa, and T. Maeda, "Development of analytical methods for polycyclic aromatic hydrocarbons (PAHs) in airborne particulates: a review," *Journal of Environmental Sciences*, vol. 19, no. 1, pp. 1–11, 2007.

[5] E. Cavalieri, R. Roth, E. Rogan, C. Grandjean, and J. Althoff, "Mechanisms of tumor initiation by polycyclic aromatic hydrocarbons," *Carcinogenesis*, vol. 3, pp. 273–287, 1978.

[6] Z. Zelinkova and T. Wenzl, "The occurrence of 16 EPA PAHs in food—a review," *Polycyclic Aromatic Compounds*, vol. 35, no. 2–4, pp. 248–284, 2015.

[7] T. Wenzl, R. Simon, E. Anklam, and J. Kleiner, "Analytical methods for polycyclic aromatic hydrocarbons (PAHs) in food and the environment needed for new food legislation in the European Union," *Trends in Analytical Chemistry*, vol. 25, pp. 716–725, 2006.

[8] H. I. Abdel-Shafy and M. S. Mansour, "A review on polycyclic aromatic hydrocarbons: source, environmental impact, effect on human health and remediation," *Egyptian Journal of Petroleum*, vol. 25, no. 1, pp. 107–123, 2016.

[9] U. Pöschl, "Atmospheric aerosols: composition, transformation, climate and health effects," *Angewandte Chemie International Edition*, vol. 44, no. 46, pp. 7520–7540, 2005.

[10] M. A. Bari, G. Baumbach, B. Kuch, and G. Scheffknecht, "Particle-phase concentrations of polycyclic aromatic hydrocarbons in ambient air of rural residential areas in southern Germany," *Air Quality, Atmospheric Health*, vol. 3, no. 2, pp. 103–116, 2010.

[11] D. A. Lane, A. Leithead, M. Baroi, J. Y. Lee, and L. A. Graham, "The detection of polycyclic aromatic compounds in air samples by GC×GC-TOFMS," *Polycyclic Aromatic Compounds*, vol. 28, no. 4-5, pp. 545–561, 2008.

[12] D. A. Lane and J. Y. Lee, "Detection of known photochemical decomposition products of PAH in particulate matter from pollution episodes in Seoul, Korea," *Polycyclic Aromatic Compounds*, vol. 30, no. 5, pp. 309–320, 2010.

[13] J. Y. Lee, D. A. Lane, J. B. Heo, S.-M. Yi, and Y. P. Kim, "Quantification and seasonal pattern of atmospheric reaction products of gas phase PAHs in PM2.5," *Atmospheric Environment*, vol. 55, pp. 17–25, 2012.

[14] R. J. Vreuls, J. Dallüge, and U. A. T. Brinkman, "Gas chromatography–time-of-flight mass spectrometry for sensitive determination of organic microcontaminants," *Journal of Microcolumn Separations*, vol. 11, no. 9, pp. 663–675, 1999.

[15] C. Weickhardt, F. Moritz, and J. Grotemeyer, "Time-of-flight mass spectrometry: state-of the-art in chemical analysis and molecular science," *Mass Spectrometry Reviews*, vol. 15, no. 3, pp. 139–162, 1996.

[16] M. X. Xie, F. Xie, Z. W. Deng, and G. S. Zhuang, "Determination of polynuclear aromatic hydrocarbons in aerosol by solid-phase extraction and gas chromatography–mass spectrum," *Talanta*, vol. 60, no. 6, pp. 1245–1257, 2003.

[17] R. Atkinson and J. Arey, "Atmospheric chemistry of gas-phase polycyclic aromatic hydrocarbons: formation of atmospheric mutagens," *Environmental Health Perspectives*, vol. 102, no. 4, pp. 117–126, 1994.

[18] J. D. Pleil, T. L. Vossler, W. A. McClenny, and K. D. Oliver, "Optimizing sensitivity of SIM mode of GC/MS analysis for EPA's TO-14 air toxics method," *Journal of the Air & Waste Management Association*, vol. 41, no. 3, pp. 287–293, 1991.

[19] D. L. Poster, M. M. Schantz, L. C. Sander, and S. A. Wise, "Analysis of polycyclic aromatic hydrocarbons (PAHs) in environmental samples: a critical review of gas chromatographic (GC) methods," *Analytical and Bioanalytical Chemistry*, vol. 386, no. 4, pp. 859–881, 2006.

[20] W. Welthagen, J. Schnelle-Kreis, and R. Zimmermann, "Search criteria and rules for comprehensive two-dimensional gas chromatography–time-of-flight mass spectrometry analysis of airborne particulate matter," *Journal of Chromatography A*, vol. 1019, no. 1-2, pp. 33–249, 2003.

[21] A. Filipkowska, L. Lubecki, and G. Kowalewska, "Polycyclic aromatic hydrocarbon analysis in different matrices of the marine environment," *Analytica Chimica Acta*, vol. 547, no. 2, pp. 243–254, 2005.

[22] K. Ravindra, A. F. L. Godoi, L. Bencs, and R. Van Grieken, "Low-pressure gas chromatography–ion trap mass spectrometry for the fast determination of polycyclic aromatic hydrocarbons in air samples," *Journal of Chromatography A*, vol. 1114, no. 2, pp. 278–281, 2006.

[23] M. Bergknut, K. Frech, P. L. Andersson, P. Haglund, and M. Tysklind, "Characterization and classification of complex PAH samples using GC–qMS and GC–TOFMS," *Chemosphere*, vol. 65, no. 11, pp. 2208–2215, 2006.

[24] T. Tran, *Characterization of Crude Oils and Atmospheric Organic Compounds by Using Comprehensive Two-Dimensional Gas Chromatography Technique (GC×GC)*, Ph.D. thesis, Applied Sciences, RMIT University, Melbourne VIC, Australia, 2009.

[25] Environmental Protection Agency (EPA), *Compendium Method TO-13A, Determination of Polycyclic Aromatic Hydrocarbons (PAHs) in Ambient Air Using Gas Chromatography/Mass Spectrometry (GC/MS)*, EPA, Cincinnati, OH, USA, 1999.

[26] S. H. Jeon, J. H. Shin, Y. P. Kim, and Y. G. Ahn, "Determination of volatile alkylpyrazines in microbial samples using gas chromatography-mass spectrometry coupled with head space-solid phase microextraction," *Journal of Analytical Science and Technology*, vol. 7, no. 1, p. 16, 2016.

[27] A. R. Fernández-Alba, *TOF-MS within Food and Environmental Analysis*, vol. 58, Elsevier, Amsterdam, Netherlands, 2012.

[28] J. H. Winnike, X. Wei, K. J. Knagge, S. D. Colman, S. G. Gregory, and X. Zhang, "Comparison of GC-MS and

GCxGC-MS in the analysis of human serum samples for biomarker discovery," *Journal of Proteome Research*, vol. 14, no. 4, pp. 1810–1817, 2015.

[29] L. I. Osemwengie and G. W. Sovocool, "Evaluation of comprehensive 2D gas chromatography-time-of-flight mass spectrometry for 209 chlorinated biphenyl congeners in two chromatographic runs," *Chromatography Research International*, vol. 2011, Article ID 675920, 14 pages, 2011.

[30] J. Zrostlíková, J. Hajšlová, and T. Čajka, "Evaluation of two-dimensional gas chromatography–time-of-flight mass spectrometry for the determination of multiple pesticide residues in fruit," *Journal of Chromatography A*, vol. 1019, no. 1-2, pp. 173–186, 2003.

[31] P. Antle, C. D. Zeigler, Y. Gankin, and J. A. Robbat, "New spectral deconvolution algorithms for the analysis of polycyclic aromatic hydrocarbons and sulfur heterocycles by comprehensive two-dimensional gas chromatography-quadrupole mass spectrometery," *Analytical Chemistry*, vol. 85, no. 21, pp. 10369–10376, 2013.

[32] S. Samanipour, P. Dimitriou-Christidis, J. Gros, A. Grange, and J. Samuel Arey, "Analyte quantification with comprehensive two-dimensional gas chromatography: assessment of methods for baseline correction, peak delineation, and matrix effect elimination for real samples," *Journal of Chromatography A*, vol. 1375, pp. 123–139, 2015.

[33] P. Mills and W. Guise Jr., "A multidimensional gas chromatographic method for analysis of n-butane oxidation reaction products," *Journal of Chromatographic Science*, vol. 34, no. 10, pp. 431–459, 1996.

[34] R. M. Flores and P. V. Doskey, "Using multidimensional gas chromatography to group secondary organic aerosol species by functionality," *Atmospheric Environment*, vol. 96, pp. 310–321, 2014.

[35] S. Pongpiachan, P. Hirunyatrakul, I. Kittikoon, and C. Khumsup, "Parameters influencing on sensitivities of polycyclic aromatic hydrocarbons measured by Shimadzu GCMS-QP2010 ultra," in *Advanced Gas Chromatography-Progress in Agricultural, Biomedical and Industrial Applications*, M. Ali Mohd, Ed., InTech, Rijeka, Croatia, 2012.

[36] European Pharmacopoeia 7.0, Section 2.2.46, *Chromatographic Separation Techniques*, 2010.

[37] P. Marriott and C. Mühlen, "The modulation ratio in comprehensive two-dimensional gas chromatography: a review of fundamental and practical considerations," *Scientia Chromatographica*, vol. 8, no. 1, pp. 7–23, 2016.

[38] J. Krupcik, P. Majek, R. Gorovenko, J. Blasko, R. Kubinec, and P. Sandra, "Considerations on the determination of the limit of detection and the limit of quantification in one-dimensional and comprehensive two-dimensional gas chromatography," *Journal of Chromatography A*, vol. 1396, no. 117, pp. 117–130, 2015.

[39] P. M. Antle, C. D. Zeigler, N. M. Wilton, and A. Robbat Jr., "A more accurate analysis of alkylated PAH and PASH and its implications in environmental forensics," *International Journal of Environmental Analytical Chemistry*, vol. 94, no. 4, pp. 332–347, 2014.

[40] T. Cajka, "Gas chromatography–time-of-flight mass spectrometry in food and environmental analysis," in *Comprehensive Analytical Chemistry*, I. Ferrer, Ed., pp. 271–302, Elsevier, Amsterdam, Netherlands, 2013.

[41] B. Lazarov, R. Swinnen, M. Spruyt et al., "Optimisation steps of an innovative air sampling method for semi volatile organic compounds," *Atmospheric Environment*, vol. 79, pp. 780–786, 2013.

[42] H. C. Menezes and Z. de Lourdes Cardeal, "Determination of polycyclic aromatic hydrocarbons from ambient air particulate matter using a cold fiber solid phase microextraction gas chromatography–mass spectrometry method," *Journal of Chromatography A*, vol. 1218, no. 21, pp. 3300–3305, 2011.

2

Analysis of the HPLC Fingerprint and QAMS for Sanhuang Gypsum Soup

Yi Peng ⓘ**, Minghui Dong** ⓘ**, Jing Zou** ⓘ**, and Zhihui Liu** ⓘ

Department of Pharmacy, Nanjing University of Traditional Chinese Medicine Affiliated Hospital, Nanjing 210029, China

Correspondence should be addressed to Zhihui Liu; liuzh1008@126.com

Academic Editor: Eduardo Dellacassa

A valid and encyclopaedic evaluation method for assessing the quality of Sanhuang Gypsum Soup (SGS) has been set up based on analysis of high-performance liquid chromatography (HPLC) fingerprint combined with the quantitative analysis of multi-components by single marker (QAMS) method, hierarchical cluster analysis (HCA), and similarity analysis. 20 peaks of the common model were obtained and used for the similarity analysis and HCA analysis. Berberine was selected as an internal reference, and the relative correction factors of mangiferin, geniposide, liquiritin, epiberberine, coptisine, baicalin, palmatine, harpagosid, wogonoside, cinnamic acid, cinnamic aldehyde, baicalein, glycyrrhizic acid, and wogonin were established. The accuracy of quantitative analysis of multicomponents by the single-marker method was verified by comparing the contents of the fourteen components calculated by the external standard method with those of the quantitative analysis of multicomponents by the single-marker method. No significant difference was found in the quantitative results of the established quantitative analysis of multicomponents by a single-marker method and an external standard method. In summary, these methods were applied to evaluate the quality of SGS successfully. As a result, these evaluation methods have great potential to be widely used in the quality control of traditional Chinese medicines (TCM).

1. Introduction

Perspiration is a considerable physiological phenomenon to maintain and control body temperature. Excessive sweat secretion can cause armpit moisture, resulting in unpleasant body odour, embarrassment, and inconvenience [1, 2]. Hyperhidrosis is an excessive sweating disease which can bring severe psychological burden and affect the quality of life of patients negatively. There are a variety of medical treatments and surgery for treating primary hyperhidrosis. However, the side effects of these drugs include thirst, dry eyes, dizziness, drowsiness, constipation, and urinary retention, which limit the scope of their use. Surgical treatment is mainly applied to patients who are not suitable for the abovementioned methods. Surgery is more traumatized and risky than other treatments. Therefore, surgical operation should be as a second or third option [3]. There is still a paucity of effective nonsurgical therapies. With the development of modern society, the elimination of body odour is given more and more people's attention; this article is about introducing Sanhuang gypsum soup (SGS) which is a significant antiperspirant. Through the use of classical Chinese medicine SGS to regulate the internal environment of the human body, it achieves a good antiperspirant effect with small side and remarkable effects, which make up for deficiency of some medicine and surgery. Every year, more than ten thousands of patients have benefited from SGS by reducing excessive perspiration symptoms. SGS is a hospital preparation of Jiangsu Provincial Hospital of Traditional Chinese Medicine, which consists of coptidis rhizoma, phellodendri chinensis cortex, scutellariae radix, scrophulariae radix, anemarrhenae rhizome, gardeniae fructus, cinnamomi cortex, glycyrrhizae radix et rhizoma preparata cum melle, and gypsum fibrosum. These herbs can be used for treating and could exhibit action on excessive perspiration through anti-inflammatory and antipyretic properties. Studies have shown that SGS has many chemical constituents such as mangiferin, geniposide, coptisine, wogonin, wogonoside, baicalein, baicalinin, and cinnamic aldehyde [4–13] in accordance with herbs and preparations known to be beneficial for the treatment of excessive perspiration through anti-inflammatory properties and so on.

Although the SGS is prepared as a prescription with the combination of these herbs in well-defined formulae, no standard quality control method for this product has been reported up to now. Since the effect of SGS might result from the synergy of multiple components, a reliable, sensitive, and uncomplicated quantitative method based on the diverse constituents is need to be developed.

Our findings have established an HPLC method to evaluate the quality of SGS comprehensively. Due to the variety of components of traditional Chinese medicine preparations, any one of the active ingredients cannot reflect the overall curative effect of traditional Chinese medicine. Therefore, a comprehensive macroscopic analysis will become an inevitable trend. Chromatographic fingerprint analysis with integrated, macroscopic, and "fuzzy" nonlinear characteristics is more adapted to the traditional Chinese medicine theory needs. Under the premise of efficacy, toxicology, and clinical trials which have confirmed safety and efficacy of preparation, we can not only verify the authenticity of the preparation but also determine whether the stability of the quality exists or not along with a practical fingerprint. Unlike the content determination, the fingerprint can provide more informative and useful message than the determination of any single component. The US Food and Drug Administration (FDA) allows applicants to provide product chromatographic fingerprinting information in the phytomedical guidance (Draft for Comment). British Herbal Codex, Ayurvedic Codex, the Canadian Society of Medicinal and Aromatic Plants [14], and the German Society of Medicinal Plants [15] also accept chromatographic fingerprint. One of the first measures that China's State Drug Administration has taken to strengthen the supervision of traditional Chinese medicine injections requires the research on the fingerprint of injections, which has taken into account its necessity and feasibility. It is accepted that preparation of acceptable quality can be exerted on its drug efficacy, what really matters is establishing an accurate and easy method.

Previously, our laboratory has researched on the fingerprints of the existing preparations which have been applied in the control of preparation quality. Also, to make up for the limitations of fingerprint that cannot be quantified accurately, a QAMS method using berberine as the standard was developed and validated for the simultaneous quantitative of 14 components [16]. This strategy can not only reduce the cost of the experiment and time of detection but also be independent of the availability of all the target ingredients [15]. To our knowledge, quality control of herb extracts and botanical ingredient by QAMS have been included both in USP 33-NF and in Ch.P.2010 edition (volume I). Our results showed that no significant difference was found in the results between our established QAMS method and the external standard method. No one has yet studied the fingerprints of SGS; this article first established the SGS fingerprinting method and also used the QAMS method to measure the preparation of 14 kinds of pharmacodynamics

components. This method could potentially be applied for the identification of qualitative and quantitative quality of SGS.

This HPLC fingerprint method, therefore, provides a comprehensive platform for quality evaluation of SGS with more chemical information. The combination of HCA and similarity analysis presents the differences and similarities of the HPLC fingerprints. In the meantime, QAMS method was adopted to quantify the main active components by comparing with the external standard method (ESM) in all the SGS samples. Our findings offer a new routine for assessing the quality of TCM.

2. Materials and Methods

2.1. Chemicals and Reagents. Analysis was applied on three different HPLC systems, including (a) Agilent 1100 series with vacuum degasser (G1322A), quaternary pump (G1311A), autosampler (G1316A), and a ChemStation Workstation with VWD detector; (b) Agilent 1260 series with DAD detector and Agilent ChemStation Workstation; and (c) Waters 2695-2996 series with 2998PDA detector and empower workstation. The chromatographic separation was performed on an AmethyC$_{18}$ (4.6 mm × 250 mm, 5 μm) column, Agilent C$_{18}$ (4.6 mm × 250 mm, 5 μm) column, and HedraC$_{18}$ (4.6 mm × 250 mm, 5 μm) column.

The SPSS software (Edition 2.0) was used for conducting cluster analysis.

BP-211D electronic analytical balance (Germany Sartorius Company) was used to weigh the drugs. Sonicator (SK6200H, Shanghai Branch guided ultrasound instrument Co., Ltd.) was used to help dissolve the sample.

2.2. Materials. The batch numbers and origins of eight qualified Chinese herbal pieces of decoction are shown in Table 1. All pieces were purchased from Anhui Concord Pharmaceutical Pieces Co., Ltd. and identified by Professor Zhihui Liu of Nanjing University of Traditional Chinese Medicine.

Fifteen batches of Sanhuang Gypsum Soup were provided by the Department of Pharmacy of Jiangsu Provincial Hospital. Their batch numbers were S1 (1707010), S2 (1704006), S3 (1712019), S4 (1711016), S5 (1704005), S6 (1703004), S7 (1711013), S8 (1711017), S9 (1705007), S10 (1704003), S11 (1704002), S12 (1704001), and S13 (1702015). Each single piece preparations and its negative preparations are made by our laboratory as per the preparation standard process.

2.3. Chemical Reagents and Standards. Mangiferin, geniposide, liquiritin, epiberberine, coptisine, baicalin, palmatine, berberine, harpagosid, wogonoside, cinnamic acid, cinnamic aldehyde, baicalein, glycyrrhizic acid, and wogonin were all supplied by Chengdu Mansi Biotechnology Co., Ltd. The purity of each ingredient was greater than 98% as determined by HPLC. Acetonitrile of HPLC grade and formic acid of analytical grade were purchased from Merck (Darmstadt, Germany) and Roe Scientific Inc. (USA).

TABLE 1: Species and geographical locations of eight Chinese herbal pieces in SGS.

Botanical name	Family	Collection site	Coordinates	Voucher ID
Coptis chinensis Franch.	Ranunculaceae	Sichuan	N30°15′49.6902″ S102°48′19.71″	16121204
Phellodendron chinense Schneid.	Rutaceae	Sichuan	N30°15′49.6902″ S102°48′19.71″	1508120316060500
Scutellaria baicalensis Georgi	Labidae	Heilongjiang	N47°7′17.9292″ S128°44′17.6316″	160401
Scrophularia ningpoensis Hemsl.	Scrophulariaceae	Yunnan	N24°28′31.0254″ S101°20′35.1816″	16060501
Anemarrhena asphodeloides Bge.	Liliaceae	Jiangsu	N33°8′24.6186″ S119°47′20.13″	16032208
Gardenia jasminoides Ellis	Rubiaceae	Jiangxi	N27°5′14.841″ S114°54′15.1956″	16122107
Cinnamomum cassia Presl.	Aceraceae	Sichuan	N30° 15′ 49.6902″ S102° 48′ 19.71″	16112519
Glycyrrhiza uralensis Fisch.	Legume	Neimenggu	N43°22′41.5914″ S115°3′34.1316″	16021710
Gypsum fibrosum	Monoclinic crystal system	Hunan	N27°37′31.0794″ S111°51′24.6924″	16120420

A Milli-Q water (Millipore, Inc., USA) purification system was applied to purify water for the HPLC analysis.

2.4. Preparation of the Sample Solution. The sample solution of SGS was precisely absorbed (5 ml) and immersed in 25 mL volumetric flask with methanol. Additional methanol was added to compensate the weight loss after ultrasonic extraction for 30 min. All solutions were filtered through $0.45\,\mu m$ filter membranes before being injected into the HPLC system precisely.

2.5. Reference Solution Preparation. A mixed stock solution containing reference standards was prepared by dissolving weighed samples of each compound in methanol accurately. Then, the stock solutions were diluted to establish the calibration curves based on six appropriate concentrations with the ranges of 2.80–$88.70\,\mu g \cdot ml^{-1}$ for mangiferin, 14.80–$472.90\,\mu g \cdot ml^{-1}$ for geniposide, 3.20–$101.00\,\mu g \cdot ml^{-1}$ for liquiritin, 1.40–$44.30\,\mu g \cdot ml^{-1}$ for epiberberine, 6.40–$204.10\,\mu g \cdot ml^{-1}$ for coptisine, 19.80–$632.50\,\mu g \cdot ml^{-1}$ for baicalin, 1.80–$56.80\,\mu g \cdot ml^{-1}$ for palmatine, 15.30–$488.40\,\mu g \cdot ml^{-1}$ for berberine, 2.60–$82.60\,\mu g \cdot ml^{-1}$ for harpagosid, 4.60–$145.82\,\mu g \cdot ml^{-1}$ for wogonoside, 0.30–$9.51\,\mu g \cdot ml^{-1}$ for cinnamic acid, 0.20–$6.34\,\mu g \cdot ml^{-1}$ for cinnamic aldehyde, 0.50–$15.85\,\mu g \cdot ml^{-1}$ for baicalein, 1.10–$34.00\,\mu g \cdot ml^{-1}$ for glycyrrhizic acid, and 0.30–$9.51\,\mu g \cdot ml^{-1}$ for wogonin.

2.6. Chromatographic Procedures. Analytes were separated on a reverse phase C_{18} column (Amethyl-ODS-2 C_{18} column, $250\,mm * 4.6\,mm * 5\,\mu m$).

Mobile phase consists of 0.1% phosphoric acid (A)-acetonitrile (B), gradient elution program was as follows: 0~2 min, 12% B; 2~7 min, 12%~20% B; 7~17 min, 20%~25% B; 17~25 min, 25%~32% B; 25~32 min, 32%~35% B; 32~45 min, 35%~44% B; 45~50 min, 44%~45% B; 50~55 min, 45%~50% B;

55~56 min, 50% B; 56~61 min, 12% B. Flow rate: $0.8\,mL \cdot min^{-1}$; column temperature: 35°C; injection volume: $10\,\mu L$; UV detection wavelength: 250 nm. On the basis of chromatographic conditions, all the components had good resolution.

2.7. Data Analysis. The data were analyzed and evaluated by Similarity Evaluation System for chromatographic fingerprint of TCM (Version 2004 A) which was recommended by the SFDA of China for evaluating similarities of chromatographic profiles of TCM. The similarity among different chromatograms was determined by calculating the correlative coefficient or cosine value of the vectorial angle [17–19]. HCA was carried out by calculating Squared Euclidean distance to distinguish preparation of different batches using SPSS. At the same time, we used the external standard method (ESM) and QAMS to calculate the 15 active components in 13 batches of SGS, respectively, to verify the feasibility of QAMS.

3. Results and Discussion

3.1. Chromatograph Optimization. At present, there are no single liquid phase conditions that can divide 15 components of SGS with good resolution. As the ingredients of SGS are very intricate, it is critical to establish a favorable mobile phase system, gradient elution system, and detection wavelength to obtain efficient separation of the numerous target components. The suitable ingredient of the HPLC method was investigated by checking peak resolution and the peak purity of SGS. In this case, some different mobile phases were tested which were acetonitrile-water, methanol-water, methanol-water containing phosphoric acid or formic acid at different concentrations, acetonitrile-water with acetic acid, formic acid, and phosphoric acid at different concentrations. Experimental results show that acetonitrile-water containing 0.2% phosphoric acid system produced

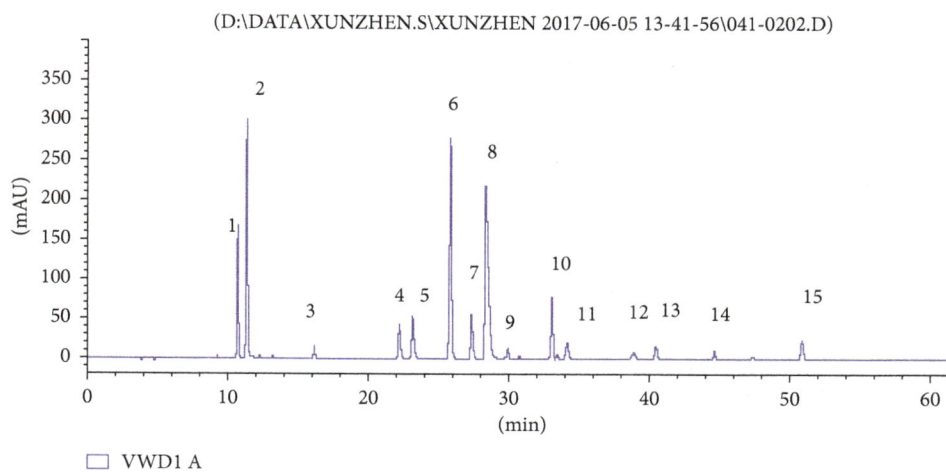

(a)

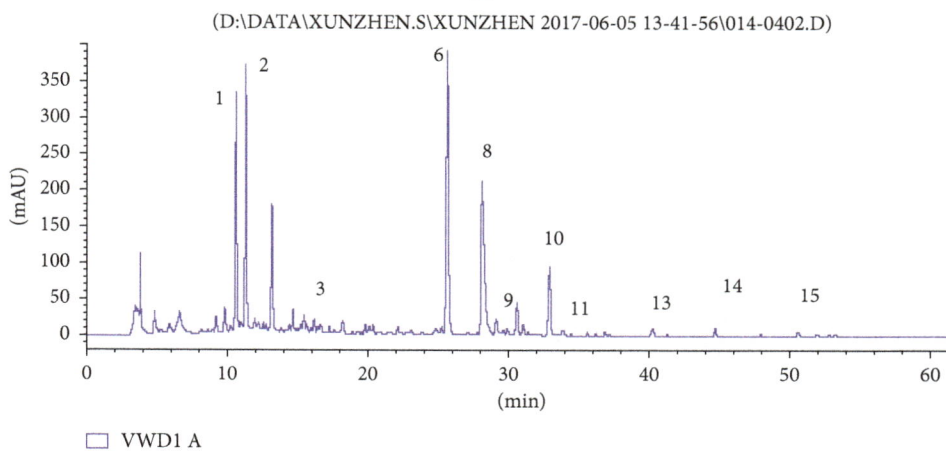

(b)

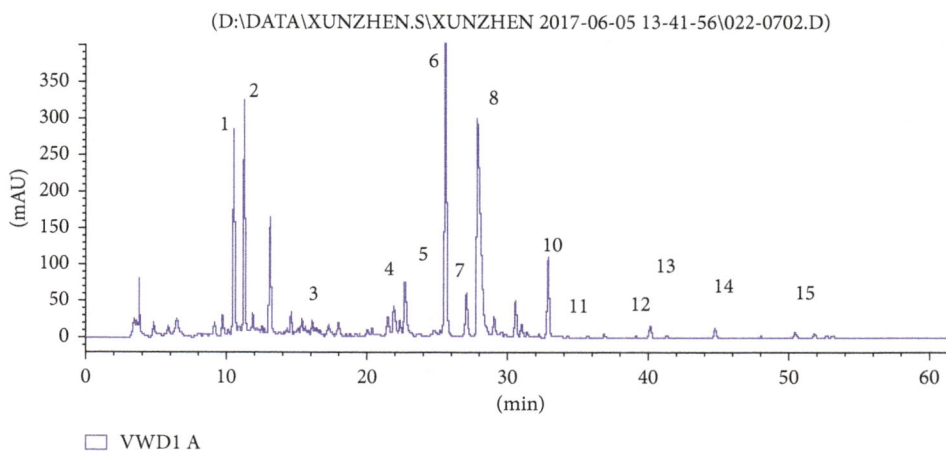

(c)

FIGURE 1: Continued.

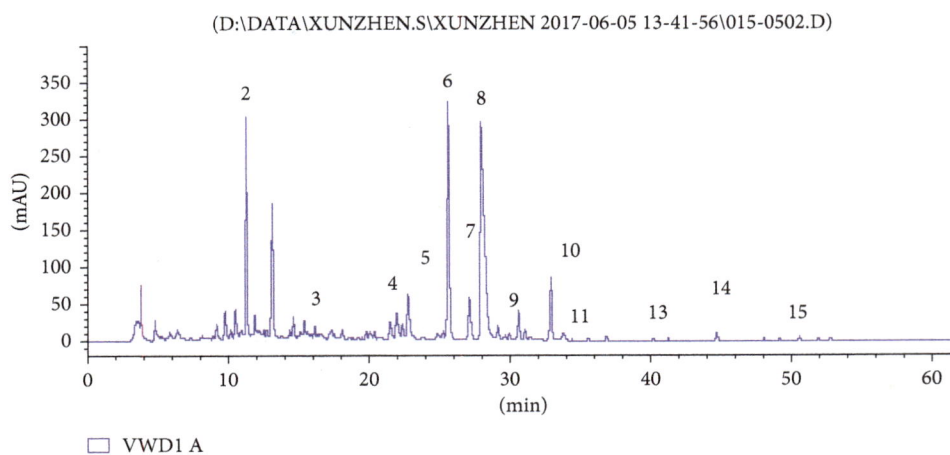

(d)

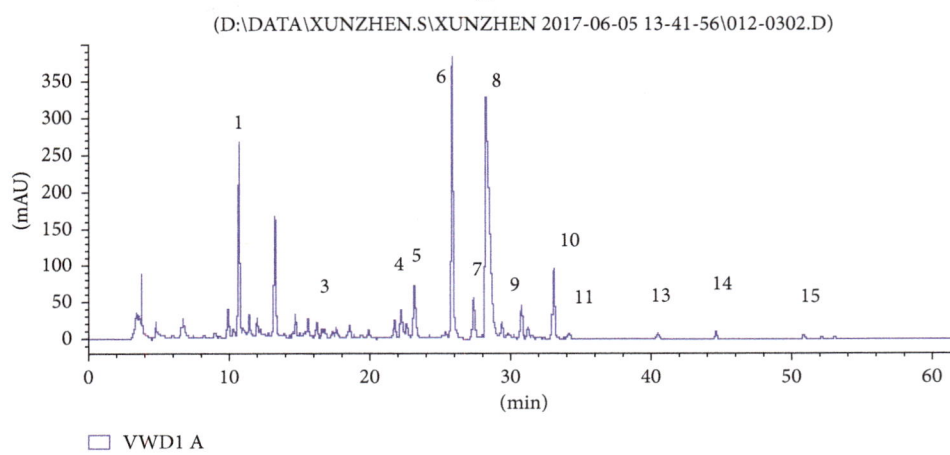

(e)

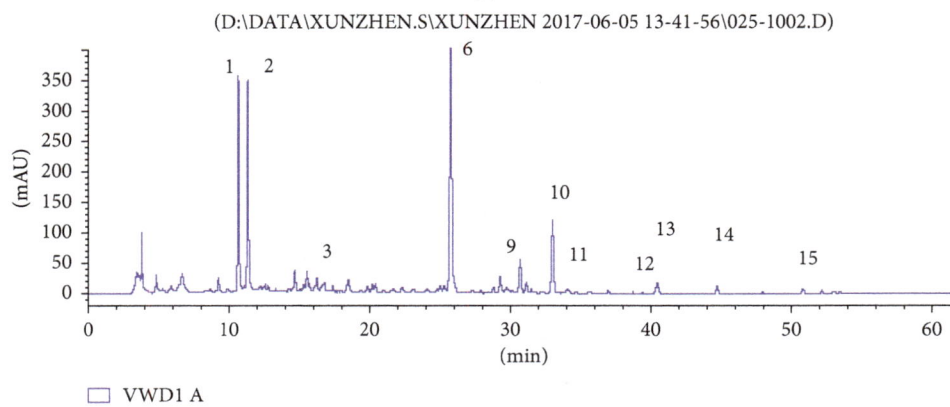

(f)

Figure 1: Continued.

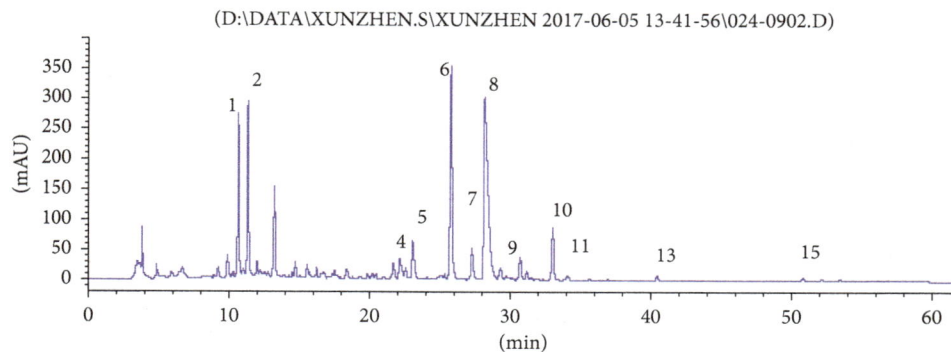

(g)

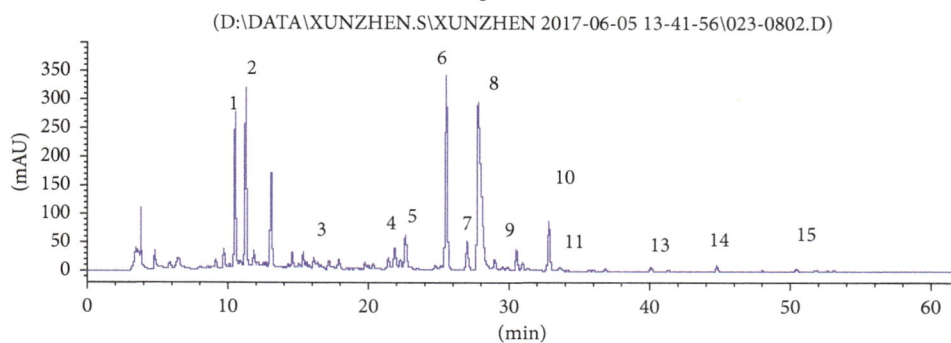

(h)

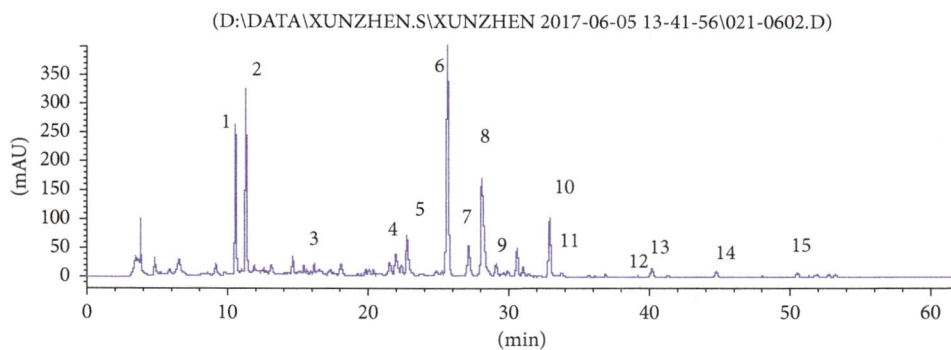

(i)

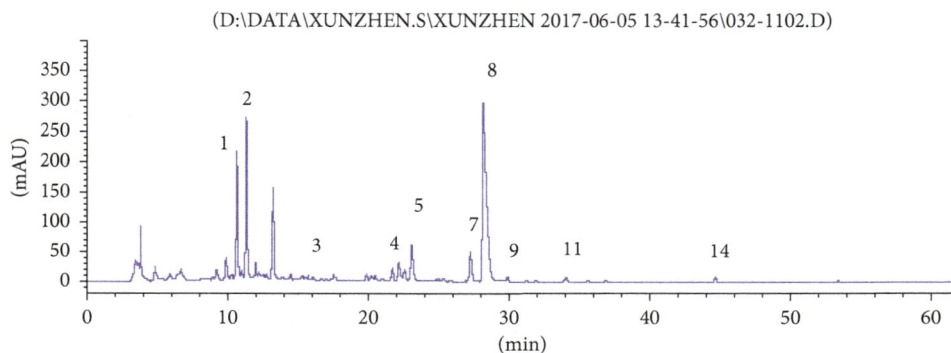

(j)

FIGURE 1: Continued.

(D:\DATA\XUNZHEN.S\XUNZHEN 2017-06-05 13-41-56\011-0103.D)

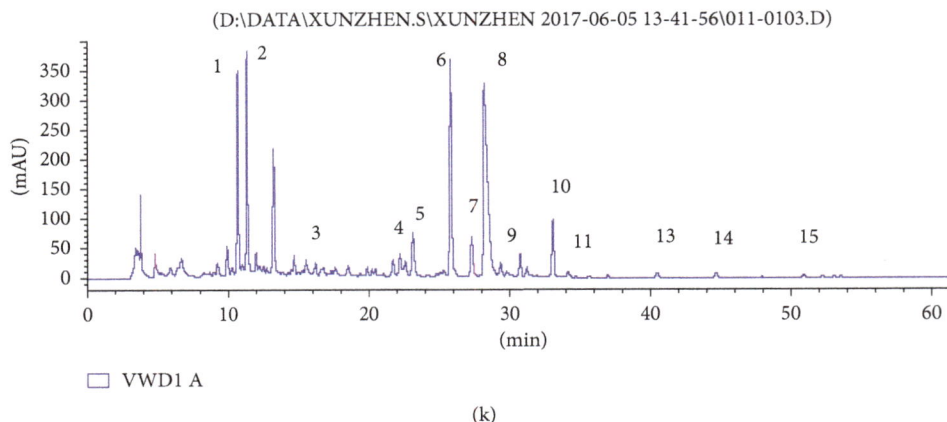

(k)

FIGURE 1: (a) Mixed standard solution, (b) negative sample without coptidis rhizoma, (c) negative sample without scrophulariae radix, (d) negative sample without anemarrhenae rhizome, (e) negative sample without gardeniae fructus, (f) negative sample without coptidis rhizoma and phellodendri chinensis cortex, (g) negative sample without glycyrrhizae radix et rhizoma preparata cum melle, (h) negative sample without cinnamomi cortex, (i) negative sample without phellodendri chinensis cortex, (j) negative sample without scutellariae radix, and (k) SGS sample. 1, mangiferin; 2, geniposide; 3, liquiritin; 4, epiberberine; 5, coptisine; 6, baicalin; 7, palmatine; 8, berberine; 9, harpagosid; 10, wogonoside; 10, cinnamic acid; 11, cinnamic aldehyde; 12, baicalein; 13, glycyrrhizic acid; and 14, wogonin.

TABLE 2: Standard curves of fifteen kinds of reference components.

Compounds	Regression equations	Linear ranges ($\mu g \cdot mL^{-1}$)	R^2
Mangiferin	$y = 29.897x + 30.359$	2.80–88.70	0.9998
Geniposide	$y = 11.306x - 40.123$	14.80–472.90	0.9997
Liquiritin	$y = 6.9865x - 5.2633$	3.20–101.00	1.0000
Epiberberine	$y = 23.477x - 12.998$	1.40–44.30	0.9998
Coptisine	$y = 11.677x - 31.945$	6.40–204.10	0.9998
Baicalin	$y = 17.312x - 189.17$	19.80–632.50	0.9999
Palmatine	$y = 39.570x - 21.105$	1.80–56.80	0.9999
Berberine	$y = 38.357x + 108.6$	15.30–488.40	0.9999
Harpagosid	$y = 9.5715x + 8.4144$	2.60–82.60	0.9998
Wogonoside	$y = 16.462x + 11.105$	4.60–145.82	0.9999
Cinnamic acid	$y = 32.7x + 0.4639$	0.30–9.51	0.9998
Cinnamic aldehyde	$y = 15.523x - 0.1567$	0.20–6.34	0.9998
Baicalein	$y = 25.996x + 3.789$	0.50–15.85	0.9998
Glycyrrhizic acid	$y = 7.5138x + 1.333$	1.10–34.00	0.9999
Wogonin	$y = 21.652x + 0.8509$	0.30–9.51	1.0000

TABLE 3: The results of recovery of fifteen components in samples ($n = 6$).

Compound	Original (mg)	Added amount (mg)	Detected amount (mg)	Recovery (%)	RSD (%)
Mangiferin	0.0241	0.0241	0.0478	93.96	1.18
Geniposide	0.1036	0.1036	0.2032	99.58	2.60
Liquiritin	0.0198	0.0198	0.0380	96.11	3.42
Epiberberine	0.0073	0.0073	0.0142	103.46	3.61
Coptisine	0.0223	0.0223	0.0438	110.19	3.16
Baicalin	0.1226	0.1226	0.2139	81.82	2.61
Palmatine	0.0104	0.0104	0.0200	97.21	2.28
Berberine	0.0748	0.0748	0.1539	101.87	5.31
Harpagosid	0.0212	0.0212	0.0433	100.01	3.56
Wogonoside	0.0307	0.0307	0.0620	99.81	3.33
Cinnamic acid	0.0023	0.0023	0.0048	111.45	2.38
Cinnamic aldehyde	0.0023	0.0023	0.0040	70.08	5.00
Baicalein	0.0048	0.0048	0.0095	79.84	3.73
Glycyrrhizic acid	0.0079	0.0079	0.0155	94.08	2.52
Wogonin	0.0028	0.0028	0.0056	98.34	4.19

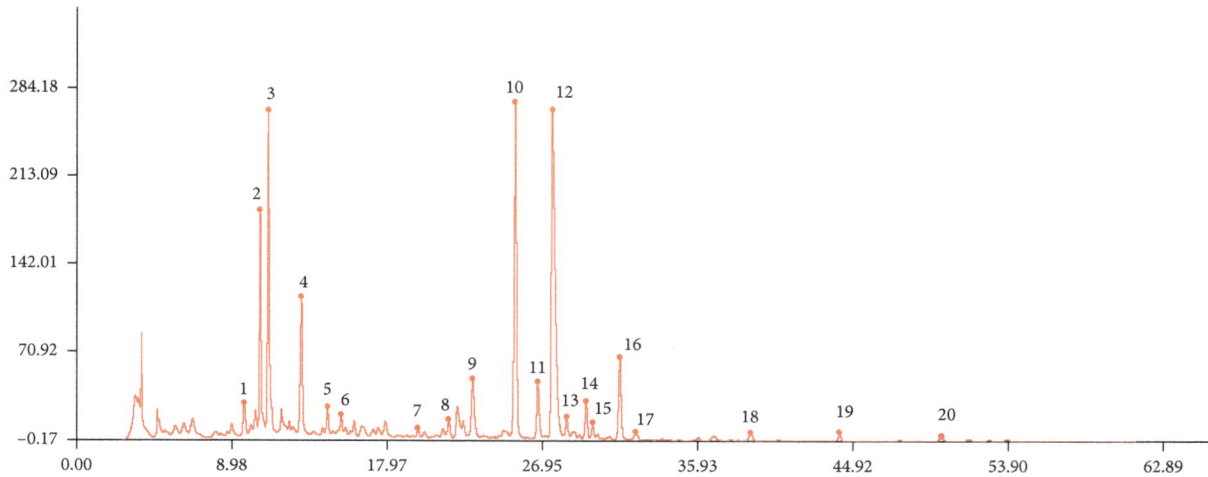

FIGURE 2: Antithesis fingerprint chromatogram of SGS.

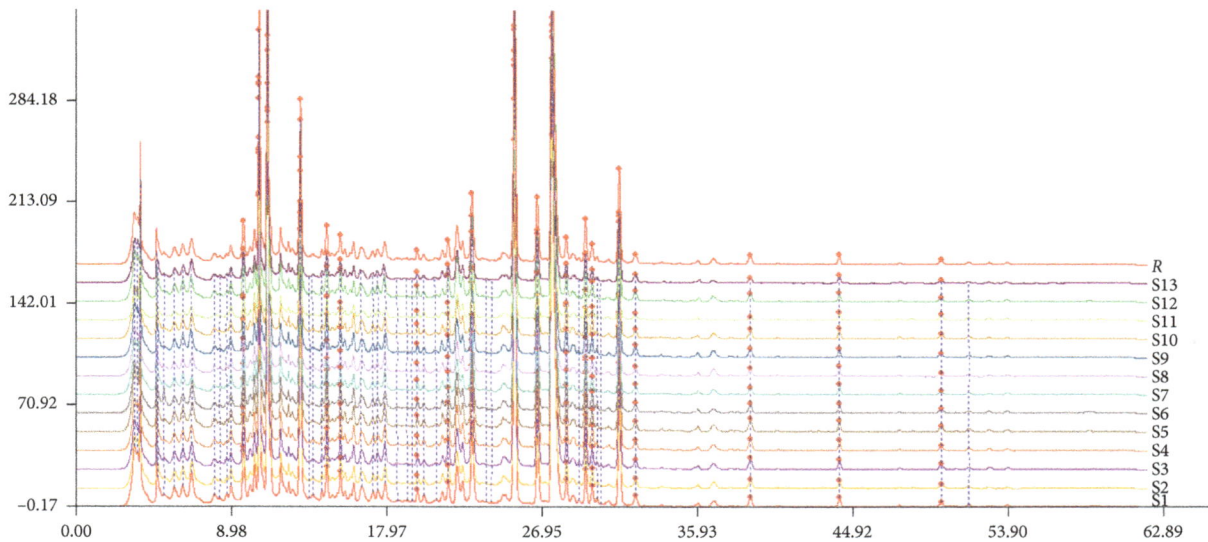

FIGURE 3: Fingerprint chromatograms of 13 batches of SGS.

TABLE 4: Similarities of 13 batches SGS.

	S1	S2	S3	S4	S5	S6	S7	S8	S9	S10	S11	S12	S13	R
S1	1	0.984	0.984	0.982	0.989	0.981	0.991	0.99	0.978	0.984	0.982	0.987	0.978	0.994
S2	0.984	1	0.975	0.978	0.985	0.986	0.985	0.995	0.997	0.987	0.979	0.973	0.994	0.994
S3	0.984	0.975	1	0.991	0.987	0.978	0.974	0.974	0.964	0.983	0.995	0.995	0.959	0.991
S4	0.982	0.978	0.991	1	0.991	0.99	0.98	0.975	0.972	0.99	0.987	0.988	0.96	0.992
S5	0.989	0.985	0.987	0.991	1	0.985	0.992	0.989	0.98	0.985	0.988	0.988	0.975	0.996
S6	0.981	0.986	0.978	0.99	0.985	1	0.983	0.982	0.985	0.993	0.972	0.972	0.974	0.992
S7	0.991	0.985	0.974	0.98	0.992	0.983	1	0.992	0.983	0.98	0.975	0.979	0.983	0.992
S8	0.99	0.995	0.974	0.975	0.989	0.982	0.992	1	0.994	0.982	0.977	0.975	0.993	0.994
S9	0.978	0.997	0.964	0.972	0.98	0.985	0.983	0.994	1	0.985	0.969	0.963	0.994	0.989
S10	0.984	0.987	0.983	0.99	0.985	0.993	0.98	0.982	0.985	1	0.977	0.977	0.973	0.993
S11	0.982	0.979	0.995	0.987	0.988	0.972	0.975	0.977	0.969	0.977	1	0.994	0.966	0.99
S12	0.987	0.973	0.995	0.988	0.988	0.972	0.979	0.975	0.963	0.977	0.994	1	0.962	0.99
S13	0.978	0.994	0.959	0.96	0.975	0.974	0.983	0.993	0.994	0.973	0.966	0.962	1	0.985
R	0.994	0.994	0.991	0.992	0.996	0.992	0.992	0.994	0.989	0.993	0.99	0.99	0.985	1

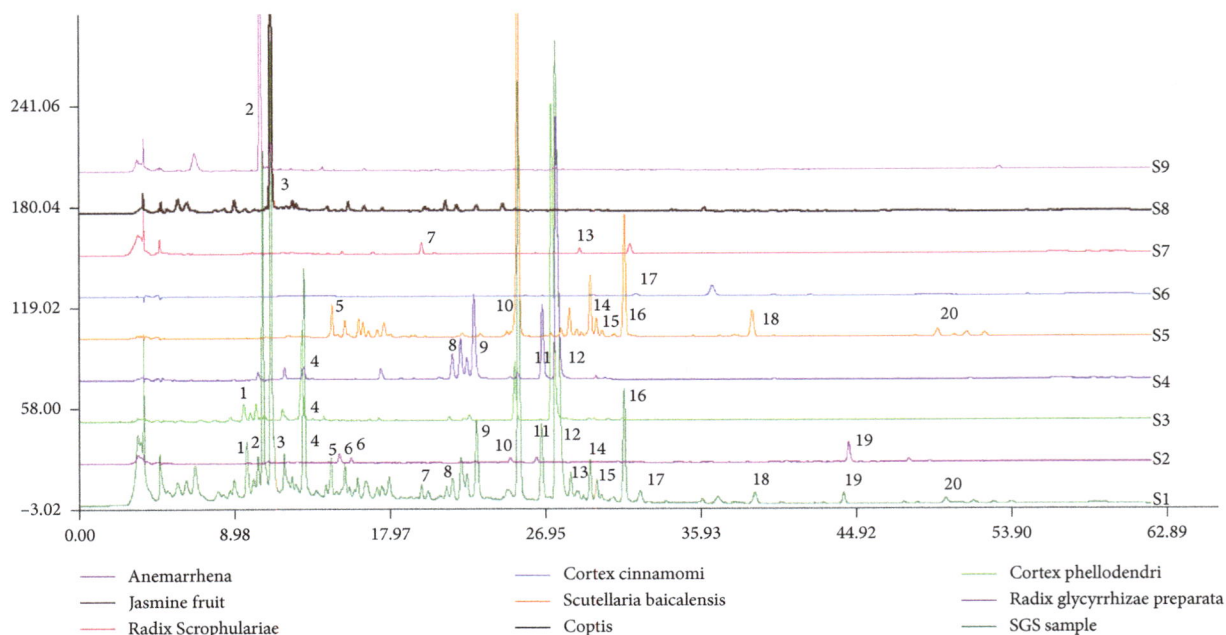

FIGURE 4: Comparison of single TCM pieces and SGS sample.

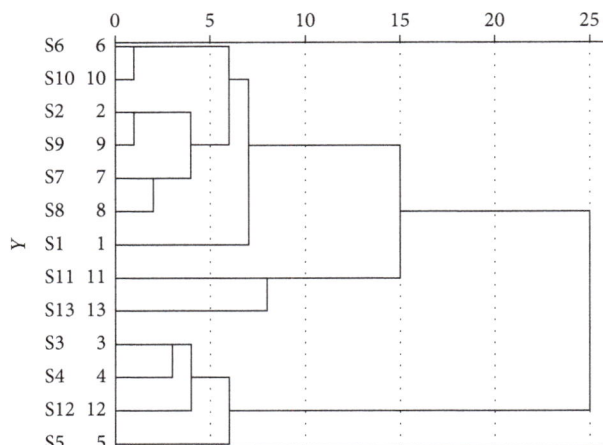

FIGURE 5: Clustering analysis graph of 13 SGS samples.

sharp and symmetrical chromatographic peak shapes, good separation, and prevented the peak tailing. Chromatogram with the maximum number of peaks also relies on best conditions for preparation of sample solution. On the basis of the investigation of different solvent and ultrasonic time, it can be concluded that samples are dissolved in methanol and ultrasound 30 minutes; we can get better resolution and reproducibility of fingerprint chromatograms under the conditions of Section 2.6. Under the above chromatographic conditions, all the components were well separated (Figure 1).

3.2. Method Validation

3.2.1. Linearity. A mixed solution containing all the reference substances were prepared and diluted in series with

methanol to obtain six different concentrations. The different concentration of the mixed solution was used for constructing the reference curve. As shown in Table 2, good calibration curves of 15 compounds were obtained, and high correlation coefficient values ($R^2 > 0.999$) were shown with good linearity at a wide range relatively. In response to sample concentration, the peak area of the analyte is determined by least squares linear regression to obtain a linear equation.

3.2.2. Precision, Stability, Repeatability, and Recovery. The same mixed standard solution of $10 \, \mu l$ was injected for six consecutive times under chromatographic conditions, and their RSDs were calculated. The RSD of mangiferin, geniposide, liquiritin, epiberberine, coptisine, baicalin,

TABLE 5: RCFs of berberine to mangiferin, geniposide, liquiritin, epiberberine, coptisine, baicalin, palmatine, harpagosid, wogonoside, cinnamic acid, cinnamic aldehyde, baicalein, glycyrrhizic acid, and wogonin on different instruments and different columns.

Instrument	Column	Mangiferin	Geniposide	Liquiritin	Epiberberine	Coptisine	Baicalin	Palmatine	Harpagosid	Wogonoside	Cinnamic acid	Cinnamic aldehyde	Baicalein	Glycyrrhizic acid	Wogonin
Waters2695-2998	Amethy	0.7632	0.3112	0.1888	0.6033	0.3144	0.4421	1.0451	0.2561	0.4321	0.8611	0.4043	0.6811	0.1993	0.5532
	Hedra	0.7794	0.3008	0.1791	0.6215	0.3211	0.4351	1.0733	0.2671	0.4411	0.8835	0.4156	0.6632	0.1911	0.5783
	Agilent	0.7794	0.2948	0.1821	0.6121	0.3044	0.4513	1.0316	0.2495	0.4292	0.8525	0.4047	0.6777	0.1959	0.5645
Agilent1100	Amethy	0.7553	0.3138	0.1813	0.5988	0.3198	0.4579	1.0651	0.2355	0.4351	0.8421	0.4255	0.6843	0.2021	0.5547
	Hedra	0.7421	0.3097	0.1923	0.5899	0.3176	0.4621	1.0688	0.2466	0.4284	0.8311	0.4322	0.6731	0.2145	0.5832
	Agilent	0.7342	0.3201	0.1799	0.6031	0.3021	0.4633	1.0803	0.2611	0.4511	0.8941	0.4167	0.6864	0.2145	0.5401
Mean		0.7589	0.3084	0.1839	0.6048	0.3132	0.4520	1.0607	0.2527	0.4362	0.8607	0.4165	0.6776	0.2025	0.5623
RSD (%)		2.48	2.97	2.91	1.80	2.58	2.52	1.75	4.45	1.98	2.81	2.67	1.26	4.56	2.90

TABLE 6: Rentention time (min) of berberine to mangiferin, geniposide, liquiritin, epiberberine, coptisine, baicalin, palmatine, harpagosid, wogonoside, cinnamic acid, cinnamic aldehyde, baicalein, glycyrrhizic acid, and wogonin on different instruments and different columns.

Instrument	Column	Mangiferin	Geniposide	Liquiritin	Epiberberine	Coptisine	Baicalin	Palmatine	Harpagosid	Wogonoside	Cinnamic acid	Cinnamic aldehyde	Baicalein	Glycyrrhizic acid	Wogonin
Waters2695-2998	Amethy	0.38	0.40	0.58	0.80	0.83	0.92	0.97	1.04	1.14	1.17	1.34	1.41	1.60	1.81
	Hedra	0.41	0.43	0.62	0.82	0.84	0.96	0.97	1.06	1.11	1.18	1.28	1.54	1.59	2.02
	Agilent	0.39	0.41	0.60	0.81	0.83	0.94	0.97	1.05	1.13	1.17	1.30	1.48	1.60	1.91
Agilent1100	Amethy	0.36	0.39	0.57	0.75	0.81	0.90	0.95	1.02	1.13	1.16	1.30	1.38	1.59	1.79
	Hedra	0.37	0.41	0.58	0.77	0.83	0.92	0.96	1.05	1.12	1.16	1.33	1.40	1.58	1.78
	Agilent	0.40	0.42	0.60	0.82	0.85	0.94	0.99	1.06	1.16	1.19	1.36	1.43	1.62	1.83
Mean		0.39	0.41	0.59	0.79	0.83	0.93	0.97	1.05	1.13	1.17	1.32	1.44	1.60	1.86
RSD (%)		5.00	3.03	3.38	4.03	2.76	2.66	1.90	1.85	2.32	1.77	2.95	4.33	1.95	5.00

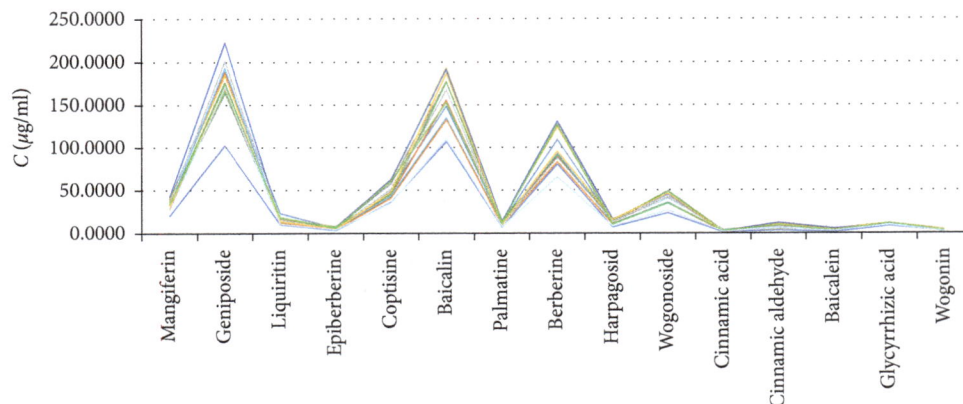

FIGURE 6: The contents of 15 active components in 13 batches of SGS.

palmatine, berberine, harpagosid, wogonoside, cinnamic acid, cinnamic aldehyde, baicalein, glycyrrhizic acid, and wogonin was 1.94%, 0.72%, 0.88%, 0.54%, 0.62%, 0.97%, 0.93%, 1.35%, 0.98%, 1.33%, 0.67%, 1.40%, 1.08%, 0.96%, and 1.49% which indicated that the developed method had a good precision.

The stability of the sample solutions was analyzed at 0, 2, 4, 8, 12, and 24 h at room temperature. It was found that the sample solutions were stable within 24 h (RSD ≤ 5.0%).

To confirm the repeatability of the method, six independently prepared solutions from the same batch were analyzed. The RSD values of the peak area was 0.37%, 0.31%, 1.52%, 0.72%, 1.00%, 0.20%, 0.71%, 0.18%, 0.88%, 0.50%, 3.02%, 3.65%, 2.30%, 1.68%, and 3.23%, respectively. The results indicated the method is reproducible.

The recovery was performed by adding a known amount of individual standards into a certain amount of the SGS sample. The mixture was extracted and analyzed by using the method mentioned above. The average recoveries of 6 samples are shown in Table 3. The results show that the method is accurate. The recoveries of the 15 compounds which are shown in Table 3 ranged from 70.08% to 111.45% with RSDs ≤ 5.0%.

3.3. HPLC Fingerprint and Similarity Analysis. Thirteen batches of samples were prepared according to Section 2.5, and 10 μL was injected into the HPLC system according to the chromatographic conditions given under Section 2.6, and then the chromatograms were recorded and entered into the similarity analysis software. We selected S (1) as the reference chromatogram, the utilization of the average correlation coefficient method of 13 batches of samples for multipoint correction, time window width is set to 0.5, while the establishment of a common model is to generate a control fingerprinting SGS, the antithesis fingerprint chromatogram was shown in Figure 2. Fingerprint chromatograms of 13 batches of SGS can be seen in Figure 3. As compared with the reference fingerprint chromatograms, the similarities of 13 batches of samples shown in Table 4, and the results are all above 0.95. On the basis of these results, we concluded that SGS between different batches are

of good consistency and in line with the relevant requirements of the fingerprints. Palmatine is the main active ingredient of coptidis rhizoma; the corresponding peaks have favorable resolution, and the retention time is stable and moderate. Therefore, we selected palmatine (no. 11 peak) as the reference peak and calculated the relative retention time of the other common peaks. We can see that the retention time of the common peak is stable. According to the retention time of each fingerprint, a total of 20 common peaks were identified while 14 of them were determined. However, it should be pointed out that the chemical property of cinnamic aldehyde is very unstable due to its alkene structure of the molecule which has poor stability when exposed to light and oxygen, so it is not within the category of the common peaks [20]. This can be in accordance with S11 without the peak of cinnamic aldehyde. To gain better understanding of ascription of common peaks, reference standards and single TCM pieces were used. The peaks 2, 3, 6, 8, 9, 10, 11, 12, 13, 16, 17, 18, 19, and 20 were identified as mangiferin, geniposide, liquiritin, epiberberine, coptisine, baicalin, palmatine, berberine, harpagosid, wogonoside, cinnamic acid, cinnamic aldehyde, baicalein, glycyrrhizic acid, and wogonin, respectively (Figure 4). The peak 1 belongs to phellodendri chinensis cortex. The peak 4 belongs to both phellodendri chinensis cortex and coptidis rhizoma. The peaks 5, 14, and 15 belong to scutellariae radix. The peak 7 belongs to scrophulariae radix.

3.4. Hierarchical Cluster Analysis. The 13 * 20 matrices were obtained from 20 common peak areas of fingerprints of 13 batches of SGS. The cluster analysis was performed by using spss 2.0 software. The Euclidean distance was chosen as the measure of the distance between groups. The results are shown in Figure 5. S3, S4, S5, and S12 batches of samples are divided into a category; the remaining batches are divided into another classification, which indicates that there are differences in the content of the components in the samples prepared from different raw material TCM pieces. And it suggested that HCA was a valid method for the identification of the source of TCM pieces.

TABLE 7: Comparison of the results from the ESM and QAMS ($\mu g \cdot ml^{-1}$).

Batch number	Berberine ESM	Mangiferin QAMS	Mangiferin ESM	Geniposide QAMS	Geniposide ESM	Liquiritin QAMS	Liquiritin ESM	Epiberberine QAMS	Epiberberine ESM	Coptisine QAMS	Coptisine ESM	Baicalin QAMS	Baicalin ESM	Palmatine QAMS	Palmatine ESM
1	109.5341	40.2024	40.2261	181.4930	189.7332	23.0219	24.3703	6.3113	7.0281	46.8326	50.7789	134.5759	148.9816	11.4799	12.3100
2	89.2249	37.5741	37.7510	179.5150	188.7602	15.1217	16.3549	5.8212	6.5595	40.4232	44.4416	140.9748	156.3753	10.5081	11.3749
3	129.1133	35.8398	35.6103	157.0895	164.0831	15.6869	16.7843	7.5026	8.2207	54.0515	57.9725	163.0149	177.5168	12.7851	13.5988
4	124.5509	39.7362	39.6241	177.6354	185.2223	16.2344	17.3568	8.0797	8.8171	57.6934	61.7406	178.8141	193.8060	13.3434	14.1801
5	131.7465	42.8953	42.8017	215.0841	223.2552	13.4517	14.4941	8.1313	8.8596	58.7696	62.7683	177.6743	192.4197	13.8791	14.7108
6	91.8581	33.9407	33.9713	167.7467	176.4659	13.4688	14.6373	8.0163	8.8171	41.8712	45.8975	161.3843	177.2857	13.0916	14.0284
7	93.6830	33.7007	33.7037	184.5859	193.7133	12.6430	13.7785	5.6230	6.3466	40.0668	44.0134	121.2770	135.8693	10.4254	11.2738
8	83.4633	28.7922	28.7534	161.0843	170.0976	11.7671	12.9197	4.9025	5.6224	37.9355	41.9581	118.3847	133.3277	9.5326	10.3893
9	87.8692	35.7417	35.8779	191.4261	201.1430	15.2532	16.4980	5.3645	6.0910	47.1242	51.3784	151.7083	167.5237	9.7197	10.5662
10	96.1076	33.3683	33.3358	163.7588	172.1319	13.7647	14.9235	5.8754	6.6021	44.4223	48.4666	171.5295	187.5098	10.4331	11.2738
11	81.0647	21.1692	20.8931	96.3176	103.2304	10.2344	11.3452	3.7042	4.3872	33.2648	37.1624	93.3769	107.5653	7.3989	8.1907
12	127.4969	35.4701	35.2424	160.0752	167.1788	18.2031	19.3607	6.7088	7.4414	56.9690	60.9699	139.2935	153.3139	11.8422	12.6385
13	66.2826	26.8813	27.0141	161.2528	171.6896	10.7070	11.9177	4.4526	5.1965	32.0308	36.1347	95.5600	110.5690	7.9011	8.7719
p value	1.0000	1.0000		0.9990		1.0000		1.0000		1.0000		1.0000		1.0000	
SMD	-4.00%	-0.14%		4.84%		5.00%		4.40%		3.88%		4.64%		5.00%	

Batch number	Harpagosid QAMS	Harpagosid ESM	Wogonoside QAMS	Wogonoside ESM	Cinnamic acid QAMS	Cinnamic acid ESM	Cinnamic aldehyde QAMS	Cinnamic aldehyde ESM	Baicalein QAMS	Baicalein ESM	Glycyrrhizic acid QAMS	Glycyrrhizic acid ESM	Wogonin QAMS	Wogonin ESM
1	14.2582	13.7492	41.5099	41.9147	4.3225	4.4343	10.1104	10.3781	3.8998	3.8544	10.7680	10.8733	2.8363	2.8704
2	14.3794	13.9581	41.7443	42.4007	3.3197	3.4251	8.1171	8.3811	4.2131	4.2006	10.3196	10.4741	3.3573	3.4246
3	13.6995	13.1223	42.3230	42.5829	3.0224	3.0887	8.5102	8.7032	5.2322	5.2008	11.9814	12.0711	3.6155	3.6556
4	14.3017	13.7492	48.7639	49.2042	3.6181	3.7003	10.8341	11.0868	6.0932	6.0856	12.1021	12.2042	4.6062	4.6716
5	17.3874	16.8835	45.4335	45.7417	3.2033	3.2722	12.8653	13.1482	6.2136	6.2010	10.8139	10.8733	4.1144	4.1636
6	15.7097	15.3163	48.3223	49.1435	3.4117	3.5168	6.2494	6.4485	4.9259	4.9315	9.4250	9.5424	3.8532	3.9327
7	12.4737	11.9730	36.3218	36.7513	2.9684	3.0581	7.2536	7.4792	3.2858	3.2390	11.1098	11.2726	2.7346	2.7780
8	11.7217	11.2417	35.2517	35.7794	2.6324	2.7217	9.3460	9.6695	3.3857	3.3544	9.5254	9.6755	2.8142	2.8704
9	14.9799	14.5850	41.7833	42.4614	3.9107	4.0367	7.6140	7.8657	4.4720	4.4699	10.9594	11.1395	4.2506	4.3483
10	16.8469	16.4655	46.1440	46.8351	2.4359	2.5076	5.3190	5.4822	4.5214	4.5084	12.1523	12.3373	4.0377	4.1174
11	8.5808	8.0029	24.3588	24.5414	1.9207	1.9878	4.0460	4.1938	2.0815	2.0080	8.7446	8.8770	3.2131	3.2861
12	11.9582	11.3462	35.5964	35.7186	3.8293	3.9144	9.8943	10.1205	3.9137	3.8544	12.1083	12.2042	3.5242	3.5632
13	9.5187	9.0477	27.3810	27.8824	2.6982	2.8135	5.8692	6.1264	3.3202	3.3159	10.0833	10.3410	2.9233	3.0090
p value	1.0000		1.0000		1.0000		1.0000		1.0000		0.9990		0.9990	
SMD	-4.00%		1.15%		2.84%		2.92%		-0.80%		1.28%		1.75%	

3.5. Quantitative Analysis of Multicomponents by a Single Marker (QAMS). It is well established that many variations of experimental conditions, such as concentrations of standard, detector, and peak measurement parameters, would extremely influence the RCFs. Accordingly, the accuracy of RCFs may affect the final analysis results. Nevertheless, RCFs, which was calculated by linear-regression equation in the experiment, was considered to be accurate and stable [21, 22]. The RCFs were calculated using the calibration curves as follows:

$$F_{K/S} = \frac{a_k}{a_s}. \tag{1}$$

The content of the measured component was calculated as follows:

$$C_k = \frac{A_K}{(A_S * F_{K/S})}. \tag{2}$$

a_s is the ratio of the slope of internal standard reference calibration equations; a_k is the ratio of the slope of measured component calibration equations; A_K is the peak area of the measured component; A_S is the peak area of the internal standard reference.

We investigated the influence of different instruments and different columns on the RCF values, and results are shown in Table 5 which illustrated RCF values had good repeatability on different chromatographic systems and different columns.

In this paper, we selected cheap, readily available, and chemically stable berberine as an internal reference standard for the quantitative determination of other active components. In addition to that, berberine is the main active ingredient of phellodendri chinensis cortex and coptidis rhizoma, so this study eventually takes it as an internal reference standard. The relative retention time has been used to locate target chromatographic peaks.

$$t_{k/s} = \frac{t_k}{t_s}, \tag{3}$$

where t_s is the retention time of internal standard reference and t_k is the retention time of measured component.

The internal reference is berberine; the relative retentions between the other target peaks and berberine were obtained in different columns and HPLC instruments. Results are shown in Table 6. The results showed that their RSDs ≤ 5% and no interference with other components; the relative retention time can be applied to locate the peak component of the analytes.

We measured the multicomponent content of 13 batches SGS (Figure 6); the results showed that there were significant differences in some contents of 15 ingredients, such as cinnamic aldehyde and baicalein, which indicated that only a few ingredients of the standard determination of content could not control the quality of SGS effectively. It is necessary to use multiple active ingredients as index components to control the quality of TCM preparations more comprehensively.

To validate the difference between ESM and QAMS method using RCFs, 13 SGS samples were analyzed for their active ingredients. The calculated results are shown in Table 7. Standard method difference (SMD) is calculated according to the following equation:

$$\text{SMD} = \frac{(C_{ES} - C_{QAMS})}{C_{ES}} * 100\%, \tag{4}$$

where C_{ES} and C_{QAMS} represent the concentrations of an analyte assayed by the external standard method and QAMS method, respectively [23]. All the values of standard deviation (SMD < 0.05) revealed that there were no significant differences between ESM and QAMS methods of all SGS samples.

4. Conclusion

On the basis of these results, we concluded that HPLC fingerprint method based on chemical constituents profiling was an effective and stable tool, and QAMS method was feasible to quantify the active compounds by RCFs for evaluating the quality of SGS. Along with similarity analysis and HCA of synthesis, the quality of SGS would be evaluated and better identified comprehensively. This method could potentially be applied in the quality control of TCM.

Conflicts of Interest

The authors declare no conflicts of interest.

Acknowledgments

This work was supported by the advantages discipline of Universities in Jiangsu Province (PATA2014).

References

[1] S. Jing, "Progress in antiperspirant research," *Detergent and Cosmetics*, vol. 29, no. 10, pp. 23–25, 2006.

[2] K. Sato, N. Kane, and G. Soos, "The eccrine sweat gland: Basic science and disorders of eccrine sweating," *Progress in Dermatology*, vol. 29, pp. 1–11, 1995.

[3] J.-R. Yang and L.-H. Zhou, "Therapeutic status and progress of primary hyperhidrosis," *Journal of Practical Medicine*, vol. 31, no. 3, pp. 493–495, 2015.

[4] C. Zhang, C. Li, F. Sui et al., "Cinnamaldehyde decreases interleukin-1β induced PGE2 production by down-regulation of m PGES-1 and COX-2 expression in mouse macrophage RAW264.7 cells," *China Journal of Chinese Materia Medica*, vol. 9, no. 37, pp. 1274–1278, 2012.

[5] X.-Y. Ma, C.-H. Li, L.-F. Li et al., "Experimental study on antipyretic, analgesic and anti-inflammatory actions of cinnamaldehyde," *Chinese Journal of Clinical Pharmacology and Therapeutics*, vol. 12, no. 11, pp. 1336–1339, 2006.

[6] S. Saha, P. Sadhukhan, and P. Sil, "Mangiferin: A xanthonoid with multipotent anti-inflammatory potential," *BioFactors*, vol. 42, no. 5, pp. 459–474, 2016.

[7] Y.-P. Yu, Y. Zhang, H. Li et al., "Pharmacokinetic-pharmacodynamics study on antipyretic effects of baicalin

on carrageenan-induced pyrexia of rats," *Chinese Traditional and Herbal Drugs*, vol. 4, no. 45, pp. 527–531, 2014.

[8] L. Wang, Y.-F. Hu, D. Tong et al., "Pharmacokinetic-pharmacodynamics study on antipyretic effects of coptisine on endotoxin-induced pyrexia of rats," *Chinese Pharmacological Bulletin*, vol. 21, no. 3, pp. 552–556, 2017.

[9] J. Wu, H. Zhang, B. Hu et al., "Coptisine from *Coptis chinensis* inhibits production of inflammatory mediators in lipopolysaccharide-stimulated RAW 264.7 murine macrophage cells," *European Journal of Pharmacology*, vol. 780, pp. 106–114, 2016.

[10] L. B. Lopes, J. L. Lopes, D. C. Oliveira et al., "Liquid crystalline phase s of monoolein and water for topical delivery of cyclosporine A: characterization and study of in vitro and in vivo delivery," *European Journal of Pharmaceutics*, vol. 63, no. 2, pp. 146–155, 2006.

[11] X. Song, M. Guo, T. Wang, W. Wang, Y. Cao, and N. Zhang, "Geniposide inhibited lipopolysaccharide-induced apoptosis by modulating TLR4 and apoptosis-related factors in mouse mammary glands," *Life Sciences*, vol. 119, no. 1-2, pp. 9–17, 2014.

[12] Y. S. Chi, B. S. Cheon, and H. P. Kim, "Effect of wogonin, a plant flavone from Scutellaria radix, on the suppression of cyclooxygenase-2 and the induction of inducible nitric oxide synthase in lipopolysaccharide-treated RAW 264.7 cells," *Biochemical Pharmacology*, vol. 61, no. 10, pp. 1195–1203, 2001.

[13] H. S. Kim, J. H. Song, U. J. Youn et al., "Inhibition of UVB-induced wrinkle formation and MMP-9 expression by mangiferin isolated from Anemarrhena asphodeloides," *European Journal of Pharmacology*, vol. 689, no. 1–3, pp. 38–44, 2012.

[14] B.-S. Xie, "On the feasibility of application of chromatographic fingerprint identification to herbal medication," *Chinese Traditional Patent Medicine*, vol. 22, no. 6, pp. 391–395, 2000.

[15] R. Bauer, "Quality criteria and phytopharmaceuticals: can acceptable drug standards be achieved?," *Drug Information Journal*, vol. 31, no. 1, pp. 101–110, 1998.

[16] Z.-M. Wang, H.-M. Gao, X.-T. Fu, and W.-H. Wang, "Multicomponents quantitation by one marker new method for quality evaluation of Chinese herbal medicine," *China Journal of Chinese Materia Medica*, vol. 23, no. 31, pp. 1925–1928, 2006.

[17] D.-W. Li, M. Zhu, Y.-D. Shao et al., "Determination and quality evaluation of green tea extracts through qualitative and quantitative analysis of multi-components by single marker (QAMS)," *Food Chemistry*, vol. 197, pp. 1112–1120, 2015.

[18] H.-L. Zhai, B.-Q. Li, Y.-L. Tian, P. Z. Li, and X. Y. Zhang, "An application of wavelet moments to the similarity analysis of three-dimensional fingerprint spectra obtained by high-performance liquid chromatography coupled with diode array detector," *Food Chemistry*, vol. 145, no. 7, p. 625, 2014.

[19] L.-W. Yang, D.-H. Wu, X. Tang et al., "Fingerprint quality control of Tianjihuang by high-performance liquid chromatography—photo diode array detection," *Journal of Chromatography A*, vol. 1070, no. 1-2, pp. 35–42, 2005.

[20] L.-Q. Zhang, Z.-G. Zhang, F.-U. Yan et al., "Research Progress of cinnamic aldehyde in pharmacology," *China Journal of Chinese Materia Medica*, vol. 40, no. 23, pp. 4568–4572, 2015.

[21] X.-Y. Gao, Y. Jiang, J.-Q. Lu, and P. F. Tu, "One single standard substance for the determination of multiple anthraquinone derivatives in rhubarb using high-performance liquid chromatography diode-array detection," *Journal of Chromatography A*, vol. 1216, no. 11, pp. 2118–2123, 2009.

[22] J.-J. Hou, W.-Y. Wu, J. Da et al., "Ruggedness and robustness of conversion factors in method of simultaneous determination of multi-components with single reference standard," *Journal of Chromatography A*, vol. 1218, no. 33, pp. 5618–5627, 2011.

[23] C.-Q. Wang, X.-H. Jia, S. Zhu et al., "A systematic study on the influencing parameters and improvement of quantitative analysis of multi-component with single marker method using notoginseng as research subject," *Talanta*, vol. 134, pp. 587–595, 2015.

3

Establishing Analytical Performance Criteria for the Global Reconnaissance of Antibiotics and Other Pharmaceutical Residues in the Aquatic Environment using Liquid Chromatography-Tandem Mass Spectrometry

Luisa F. Angeles ⓘ **and Diana S. Aga** ⓘ

Department of Chemistry, The State University of New York, Buffalo, NY 14260, USA

Correspondence should be addressed to Diana S. Aga; dianaaga@buffalo.edu

Academic Editor: Veronica Termopoli

The occurrence of antibiotics in the environment from discharges of wastewater treatment plants (WWTPs) and from the land application of antibiotic-laden manure from animal agriculture is a critical global issue because these residues have been associated with the increased emergence of antibiotic resistance in the environment. In addition, other classes of pharmaceuticals and personal care products (PPCPs) have been found in effluents of municipal WWTPs, many of which persist in the receiving environments. Analysis of antibiotics by liquid chromatography-tandem mass spectrometry (LC-MS/MS) in samples from different countries presents unique challenges that should be considered, from ion suppression due to matrix effects, to lack of available stable isotopically labeled standards for accurate quantification. Understanding the caveats of LC-MS/MS is important for assessing samples with varying matrix complexity. Ion ratios between quantifying and qualifying ions have been used for quality assurance purposes; however, there is limited information regarding the significance of setting criteria for acceptable variabilities in their values in the literature. Upon investigation of 30 pharmaceuticals in WWTP influent and effluent samples, and in receiving surface water samples downstream and upstream of the WWTP, it was found that ion ratios have higher variabilities at lower concentrations in highly complex matrices, and the extent of variability may be exacerbated by the physicochemical properties of the analytes. In setting the acceptable ion ratio criterion, the overall mean, which was obtained by taking the average of the ion ratios at all concentrations (1.56 to 100 ppb), was used. Then, for many of the target analytes included in this study, the tolerance range was set at 40% for WWTP influent samples and 30% for WWTP effluent, upstream, and downstream samples. A separate tolerance range of 80% was set for tetracyclines and quinolones, which showed higher variations in the ion ratios compared to the other analytes.

1. Introduction

In recent years, studies have reported the occurrence of pharmaceuticals and personal care products (PPCPs), including antibiotics and selective serotonin reuptake inhibitors (SSRIs), in the environment [1–6]. These drugs are being released through different routes, such as discharges from wastewater treatment plant (WWTP) effluents to surface water, where hospitals and private households contribute a large volume of antibiotics and other pharmaceuticals [2, 7–9]. The presence of PPCPs in effluents of WWTPs in different geographical regions has been documented, with concentrations reported as high as about 125 μg/L [10]. In Germany, the environmental concentration in municipal sewage that comes from the discharge of antibiotics from hospitals and households is predicted to be about 71 mg/L annually [9]. The presence of high levels of pharmaceuticals in the environment has a wide range of ecological effects; for instance, antibiotics may contribute to the development of antibiotic resistance in bacteria due to selective pressure, which is a threat to global health [9, 11, 12].

Analysis of PPCPs in environmental samples is typically performed using liquid chromatography-tandem mass spectrometry (LC-MS/MS) to quantify pharmaceutical concentrations based on triple quadrupole MS [3, 6, 13]. The high selectivity and sensitivity obtained using triple quadrupole MS is achieved when performing selected reaction monitoring (SRM), where a precursor ion is isolated from the first quadrupole and fragmented in the collision cell, followed by isolating selected product ions in the third quadrupole. However, despite this high selectivity, there is still a possibility that a compound other than the target analyte will produce a signal that has a similar m/z value to either the qualifying ion or the quantifying ion at the same retention time [14], resulting in a significant deviation in the expected ion ratio for the selected fragment ions being monitored by the two SRM transitions.

In order to confirm the presence of a compound, the chromatographic peak must have both the quantitative and qualitative ion transitions with retention times matching those of the standard analyte. In addition, the ion ratio of the two SRM transitions has been used as an additional confirmation criterion, as stated in some legal documents from different organizations such as the European Union (EU) and the US Food and Drug Administration (US FDA), which provide guidelines for the analysis of official samples [15–19]. Having this additional criterion is important since LC-MS/MS has now become the mandatory technique for the analysis of official samples that are used for establishing legal policies [15–19]. Monitoring the ion ratios will provide improved confidence in reporting analyte concentrations, avoiding false positives and false negatives, which have been reported in the literature [14].

Different legal guidelines are currently available from the United Nations (UN), the EU, and the United States of America (USA). The UN set the ion ratio tolerance to be ±20% [16] for the testing of illicit drugs in seized materials and biological specimens. The European Commission Decision (2002/657/EC) requires a tolerance of ±20% to ±50% for the ion ratio, depending on the ion intensities [17], for analytical methods that are used for the testing of official samples in control laboratories. The European Workplace Drug Testing Society sets it at ±20% [19]. The US Department of Agriculture requires a ±20% tolerance in the ratio of the ion transitions [18], while the US FDA sets an ion ratio tolerance of ±20% and ±30% if 2 and 3 diagnostic ions are being monitored, respectively [15]. The weakness of these guidelines, however, is that they are not based on experimental data and are arbitrarily assigned.

Recent studies [20, 21] have been published on performance criteria for the analyses of pesticides in fruits and vegetables and veterinary drugs in biological matrices. For pesticides, a tolerance range of ±20% was established for all compounds at all concentrations, except when one or both product ions have an S/N of 3–15, in which case, a range of ±45% was set. For veterinary drugs, a fixed tolerance range of ±50% for all the compounds at all concentrations was set after evaluation of the ion ratios in different matrices such as muscle, urine, milk, and liver [20, 21]. However, these tolerance values cannot be used for PPCPs because the variability of ion ratios differs per compound and the nature of the sample matrix. This variability is due to differences in the ionization behavior of analytes and the extent of matrix effects. It is not unexpected to observe different effects on the ion ratios of the analytes in wastewater and surface water matrices because the composition of the interferences in environmental samples is different relative to biological samples.

Establishing performance criteria is important because it minimizes the occurrence of false-positive and false-negative detections. In fact, a doubling of false-positive detections was reported without the application of the ion ratio criterion in the analysis of veterinary drugs in the muscle, urine, milk, and liver [14]. Most published and existing methods do not mention the use of any ion ratio criteria [3, 6, 13]. In the US Environmental Protection Agency (EPA) Method 1694 for the determination of PPCPs in environmental samples by LC-MS/MS, the presence of a compound in a sample extract is confirmed when the signal-to-noise ratio (S/N) of the fragment ion of the compound is greater than or equal to 2.5 and its retention time is within ±15 seconds of the calibration verification standard. If these criteria are not met, then an experienced analyst must confirm the presence or absence of a compound [22]. Additionally, in the EPA Method 542, which is for the analysis of PPCPs in drinking water, the acceptable retention time window for the compounds in a sample is within 3 standard deviations for a series of injections. Quality control for this method involves the confirmation of the presence of the quantifying ion of the internal standard and requires that it must be within ±50% of the average area measured in the initial calibration [23]. No criteria regarding the ion ratios have been mentioned in both EPA methods. The absence of quality control measures in published methods may be due to the lack of suitable guidelines in the literature. In order to determine an appropriate tolerance value for the ion ratios, variabilities resulting from the physicochemical nature of the analytes should be investigated at high and low concentrations. The variability in the signal intensities of the qualifier ions is expected to be more significant than that of the quantifier ions because of the relatively lower signals for the qualifier ions.

The aim of this study is to validate and provide guidelines on the use of ion ratios as a criterion for quality control in reporting concentrations of PPCPs in wastewater and surface water samples with varying complexity. To achieve this goal, the ion ratios of 30 PPCPs in different matrices were determined at different concentrations in order to determine a tolerance value that is sufficient to eliminate false positives and false negatives. The matrices studied were WWTP influent and effluent samples and surface water samples from upstream and downstream of the WWTP discharge point collected from the US, Sweden, Switzerland, Hong Kong, and the Philippines, allowing the set tolerance levels to be robust, given that the composition of water samples varies significantly in different parts of the world. The data obtained from these analyses were the basis for the construction of a more accurate and reliable ion ratio criterion which takes into account the differences in the properties of compounds at different concentrations.

2. Materials and Methods

Acetaminophen (ACT), acetylsulfamethoxazole (ASMX), azithromycin (AZI), caffeine (CAF), carbamazepine (CBZ), clarithromycin (CLA), enrofloxacin (ENRO), erythromycin (ERY), iopamidol (IOPA), norfloxacin (NOR), oxytetracycline (OTC), sarafloxacin (SARA), sulfachloropyridazine (SCP), sulfadiazine (SPD), sulfadimethoxine (SDM), sulfamerazine (SMR), sulfameter (SMT), sulfamethazine (SMZ), sulfamethizole (SMI), sulfamethoxazole (SMX), tetracycline (TC), and trimethoprim (TMP) were purchased from Sigma-Aldrich. Ciprofloxacin (CIP) and diclofenac (DIC) were obtained from Cambridge Isotope Laboratories, Inc. (Tewksbury, MA). Sulfathiazole was purchased from ICN Biomedicals, Inc. (Irvine, CA). Carbamazepine-d10 (d10-CBZ) was purchased from CDN Isotopes (Quebec, Canada). Chlortetracycline (CTC) was obtained from Acros Organics (VWR International, Westchester, PA). Paroxetine maleate (PRX) and venlafaxine (VEN) were obtained from Cerilliant (Sigma-Aldrich, St Louis, MO). The Barnstead NANOpure™ DIamond (Waltham, MA) purification system was used to obtain 18.2 MΩ water. LC-MS grade methanol and acetonitrile were obtained from EMD Millipore Corporation (Billerica, MA), and formic acid (88%) was purchased from Fisher Chemical (Pittsburgh, PA). Oasis™ HLB solid-phase extraction (SPE) cartridges were purchased from Waters (Milford, MA).

2.1. Sample Preparation. Wastewater and surface water samples (0.5 L) were collected in amber glass bottles which were pre-rinsed with 10% nitric acid. The samples were acidified to about pH 2.5 using 40% phosphoric acid and then passed through 0.45 μm glass microfiber filters to remove microorganisms and particulate matter. Then, 2 mL of Na_2EDTA (5% *w/v* in water) was added to each sample. The samples were then spiked with surrogate standards (50 μL of 1000 μg/L surrogate mix solution).

The samples were passed through Oasis HLB SPE cartridges (500 mg, 6 cc) for cleanup and concentration. The SPE cartridges were first conditioned with 6 mL acetonitrile, followed by 6 mL NANOpure water, before the water samples were loaded at a rate of approximately 3–5 mL/min. After loading, the cartridges were dried by keeping them on the SPE manifold with the vacuum on. Then, the SPE cartridges were wrapped in aluminum foil, stored in Ziploc® bags, and shipped with ice to the University at Buffalo for elution and LC-MS/MS analysis. Once received, the samples were eluted using 8 mL of acetonitrile and then dried under N_2 gas at 35°C. The samples were then spiked with 100 ppb of the internal standard, carbamazepine-d10, in order to account for possible differences in measurements in-between injections due to variations caused by the instrument.

2.2. LC-MS/MS Analysis. A Waters Cortecs™ C18$^+$ column (Milford, MA) with dimensions 2.1 × 150 mm and 2.7 μm particle size was used for the separation of the 30 PPCPs. Analysis was performed using an Agilent 1200 LC system (Palo Alto, CA) and a Thermo Scientific TSQ Quantum Ultra triple quadrupole MS (Waltham, MA) equipped with

TABLE 1: Target pharmaceuticals used for establishing the ion ratio criterion.

Class	Compound
Antibiotics	
Macrolides	Anhydroerythromycin
	Azithromycin
	Clarithromycin
Quinolones	Ciprofloxacin
	Enrofloxacin
	Norfloxacin
	Sarafloxacin
Sulfonamides	Acetylsulfamethoxazole
	Sulfachloropyridazine
	Sulfadiazine
	Sulfadimethoxine
	Sulfamerazine
	Sulfamethazine
	Sulfamethizole
	Sulfamethoxazole
	Sulfamethoxydiazine
	Sulfathiazole
Tetracyclines	Chlortetracycline
	Oxytetracycline
	Tetracycline
Other PPCPs	
	Acetaminophen
	Caffeine
	Carbamazepine
	Diclofenac
	Iopamidol
	Trimethoprim
	Bupropion
	Paroxetine
	Sertraline
	Venlafaxine

a heated electrospray ionization (HESI) probe, operated under positive ionization mode. Timed-SRM mode transition was performed, and the SRM transitions used for the compounds are shown in Table S1.

The mobile phase used for the separation consisted of aqueous 0.3% formic acid (A), and 75% methanol and 25% acetonitrile (B). The gradient began with 90% A and 10% B for three minutes and is ramped up linearly to 100% B for 22 min; this condition was kept for 5 min before it was switched back to 90% A, where it was maintained for 14 min to allow for column equilibration. The flow rate was set at 0.2 mL/min, and the total run time was 45 min.

The spray setting used for the MS was as follows: spray voltage 3000 V, ion sweep gas pressure 0 arbitrary units, vaporizer temperature 350°C, sheath gas pressure 40 arbitrary units (N_2), auxiliary gas pressure 35 arbitrary units (N_2), capillary temperature 325°C, collision gas pressure 1.5 mTorr (Ar), cycle time 0.300 s, and Q1 peak width 0.70 FWHM.

2.3. Design of the Study

2.3.1. Assessment of Ion Ratio Behavior of PPCPs across Varying Concentrations. A total of 30 PPCPs were studied for the development of an ion ratio criterion (Table 1). A mixture of all the native PPCP standards was prepared using the starting mobile phase of the LC-MS/MS method as the solvent. An initial solution of 100 ppb (μg/L) was made, and then it was serially diluted to obtain mixtures with concentrations of 50, 25, 12.5, 6.25, 3.13, and 1.56 ppb. These standards were analyzed by LC-MS/MS, with nine replicates for each concentration, to obtain the areas of both the quantifying and qualifying ions. The ion ratios were calculated by dividing the area of the quantifying ion by the area of the qualifying ion for each analyte. The average ion ratio and the deviations from the average value for the ion ratios were calculated at all concentrations for each compound.

2.3.2. Assessment of Ion Ratio Behavior of Pharmaceuticals in the Matrix.

Samples from WWTP influents and effluents and from receiving surface waters upstream and downstream of the WWTPs were collected from selected sites in five countries: Central, Hong Kong; Manila, Philippines; Vastergotland, Sweden; Zurich, Switzerland; and Virginia, US. The samples from the Philippines were collected in December 2016, while the others were collected in June or July 2016. The exact names and locations of the WWTPs cannot be disclosed as part of the agreement with the WWTP operations. A total of 19 samples were each spiked with 1.56 ppb, 12.5 ppb, 25 ppb, and 100 ppb of the native standard mix and were analyzed by LC-MS/MS to determine the mean ion ratios and standard deviations from the mean for each compound. The variabilities of the ion ratios in the different sample matrices were then evaluated and compared with the values observed in the standards.

2.3.3. Optimization of the Ion Ratio Criterion.

The proposed formula to be used in order to optimize the appropriate ion ratio tolerance that will give the least false negative is a mean ion ratio and a tolerance range that will account for variations in the sample matrix. This tolerance range should not be too wide so as to avoid having false positives. To determine the optimum tolerance level, different values were tested for all compounds; the same test was also used to determine whether a single tolerance range would be used for all the matrices or if different ones should be used for wastewater and for surface water. To check the appropriateness of the selected tolerance values, the number of false negatives will be determined using the water extracts spiked with known amounts of standards.

3. Results and Discussion

3.1. Ion Ratio Variability at Different Concentrations.

A total of 30 PPCPs, which include 23 antibiotics, were studied for the development of an ion ratio criterion (Table 1). The classes of antibiotics that were included in this study were sulfonamides, macrolides, quinolones, and tetracyclines.

First, the ion ratio behavior of the compounds was studied at different concentrations by analyzing nine replicates of the standard solutions of 1.56, 3.13, 6.25, 12.5, 25.0,

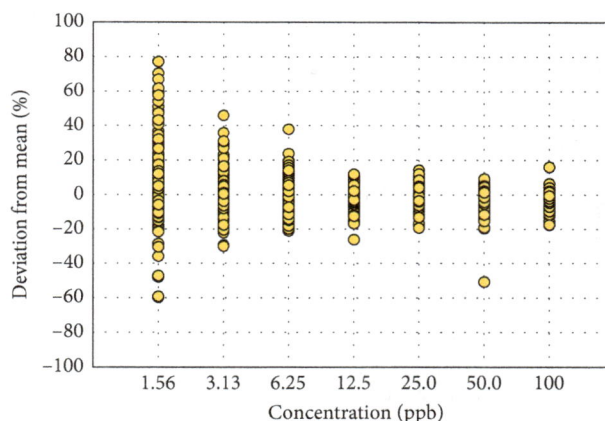

FIGURE 1: Deviation of ion ratios from the overall mean across different concentrations of 23 PPCPs without tetracyclines and quinolones.

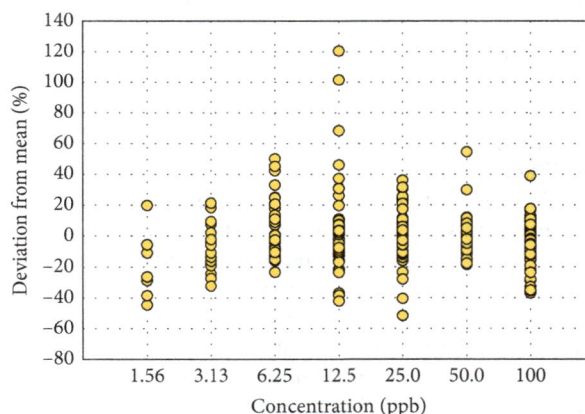

FIGURE 2: Deviation of ion ratios of tetracyclines and quinolones from the overall mean across different concentrations.

50.0, and 100 ppb in the LC-MS/MS. The overall mean, which is the average ion ratio of all nine replicates at all concentrations, was obtained for each compound (Table S2). The relative percent deviation was then calculated by subtracting the overall mean from each of the data points and then dividing by the overall mean. These values were then plotted against the seven concentrations to see how the ion ratios at each concentration vary from the overall mean of each compound, as shown in Figure 1. The trend for all the compounds is that the variation is highest at the lowest concentration. The average relative standard deviation for all the compounds at 1.56 ppb was 18%, while that for compounds at 100 ppb was only 4%. These results indicate that the differences in the ion ratios at different concentrations should be taken into account because if only one tolerance limit is applied across all concentrations, it is likely that false-negative results will occur at low concentrations, especially at concentrations between 1.56 to 12.5 ppb.

The general trend in the ion ratios for 23 PPCPs is shown in Figure 1; a separate plot for tetracyclines and quinolones was prepared (Figure 2) because the variabilities in the ion

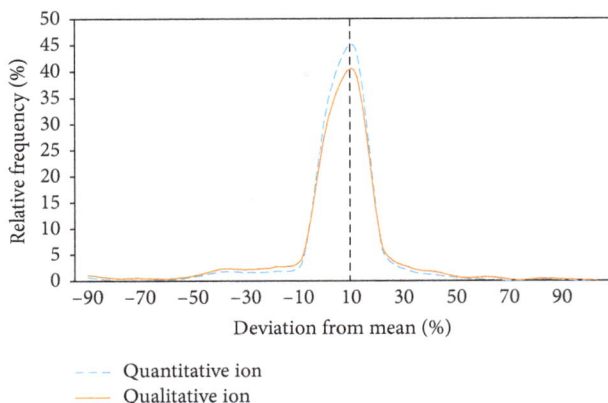

FIGURE 3: Distribution of the deviation of the quantitative and qualitative ion areas from the mean.

ratios were notably higher in these classes of antibiotics than the rest of the PPCPs. Data points at lower concentrations of some compounds were removed in cases where the qualitative ion was not detectable. For example, oxytetracycline was not detected below 25 ppb, chlortetracycline and tetracycline were not detected below 12.5 ppb, sarafloxacin and norfloxacin were not detected below 6.25 ppb, and enrofloxacin was not detected at 1.56 ppb. Therefore, a separate chart (Figure 2) was created for tetracyclines and quinolones since they do not follow the same behavior as the other pharmaceuticals. Based on these data, a separate ion ratio tolerance range is needed for tetracyclines and quinolones in order to capture the wide variations, without affecting the other compounds. It can be observed that the variations in Figure 2 are lower at 1.56 ppb compared to the higher concentrations, but this is because most of these compounds were no longer detected at 1.56 ppb, and these data points were removed in the chart. The deviation from the mean reaches up to 120% at 12.5 ppb for tetracyclines and up to 50% for quinolones at 6.25 ppb. At the highest concentration of 100 ppb, the deviations from the mean in both tetracyclines and quinolones are at 40%, while those for the other PPCPs are only 20%.

The areas of both the quantitative and qualitative ions were investigated separately in order to identify which of the two ions causes high variations. The deviations in the areas of these ions from the mean were calculated and compared with each other. Since the distribution of variation of both the qualitative and quantitative areas is similar, as seen in Figure 3, this means that both of them contribute equally to the variations, and the ion ratio deviations cannot be attributed to just the quantitative or qualitative ion alone.

3.2. Ion Ratio Variability in Wastewater and Surface Water Matrices. A total of 19 different samples were spiked with the pharmaceutical standards at 4 concentrations: 1.56, 12.5, 25.0, and 100 ppb, in order to determine how the differences in the nature of the matrices influence the ion ratios.

Figure 4 shows how the ion ratios change in the influent, effluent, upstream, and downstream water samples in comparison with the clean standard. It can be seen that the

variations are higher in the wastewater as compared to those of the surface water samples, with relative standard deviation values of 13%, 11%, 10%, and 9%, for the influent, effluent, downstream, and upstream samples, respectively. This trend was expected since the upstream and downstream samples are less-complex matrices (lower organic matter content than wastewater). It can be seen in the lowest concentration studied (1.56 ppb) that the standards in the clean matrix varied more than the ones spiked in the samples, with relative standard deviations of 18% for the standards and 20%, 14%, 15%, and 12% for the influent, effluent, downstream, and upstream samples, respectively. This is due to the removal of 22 data points at 1.56 ppb because the qualitative ions were no longer detected in the samples.

3.3. Optimization of a Tolerance Range for the Ion Ratio Criterion. The formula for the ion ratio criterion that was used is a tolerance range from the mean of each standard compound in the clean matrix. This tolerance range should account for the deviations because of differences in concentrations and the matrix being analyzed. The goal in setting this range is to have the least number of false positives and false negatives. False negatives will occur when the ion ratio of analytes in spiked environmental samples does not meet the tolerance criteria such that the analyte in question will be considered "nondetect." On the contrary, one cannot set a tolerance range too wide that will likely result in a significant number of false positives. Therefore, a range that will still capture all the variations at the 95% confidence level in both matrices and at different concentrations is needed.

In order to provide an appropriate criterion, the overall mean, which is obtained by taking the average of the ion ratios at all concentrations (1.56 to 100 ppb), will be used. This way, the variations of the ion ratios from low to high concentrations will be taken into account. The tolerance range must then be optimized for the spiked matrices. Tolerance ranges from 10% to 50% were tested to see which one will give the least number of false negatives for each of the water matrices (Table 2). The overall false-negative rate is the weighted average of the percent false negatives for each matrix type.

Since the tetracyclines and quinolones were found to have greater variations than the rest of the PPCPs as seen in Figure 2, a test was performed to check if tetracyclines and quinolones should use a different tolerance level than what is used for the other classes of PPCPs. Table 3 shows the percentage of false negatives in the samples when the tetracyclines and quinolones were removed. If tetracyclines and quinolones were included, a tolerance range of 50% would give a false-negative result of ≤5%. If removed, a tolerance range of 30% would be enough to give the same value of ≤5% for false negatives.

It is important to have a separate tolerance range for tetracyclines and quinolones because as seen in Table 2, a tolerance range of 50% is needed in order to capture them at the 95% confidence level. This value, however, would be too high for the other PPCPs, where only 30% is required to

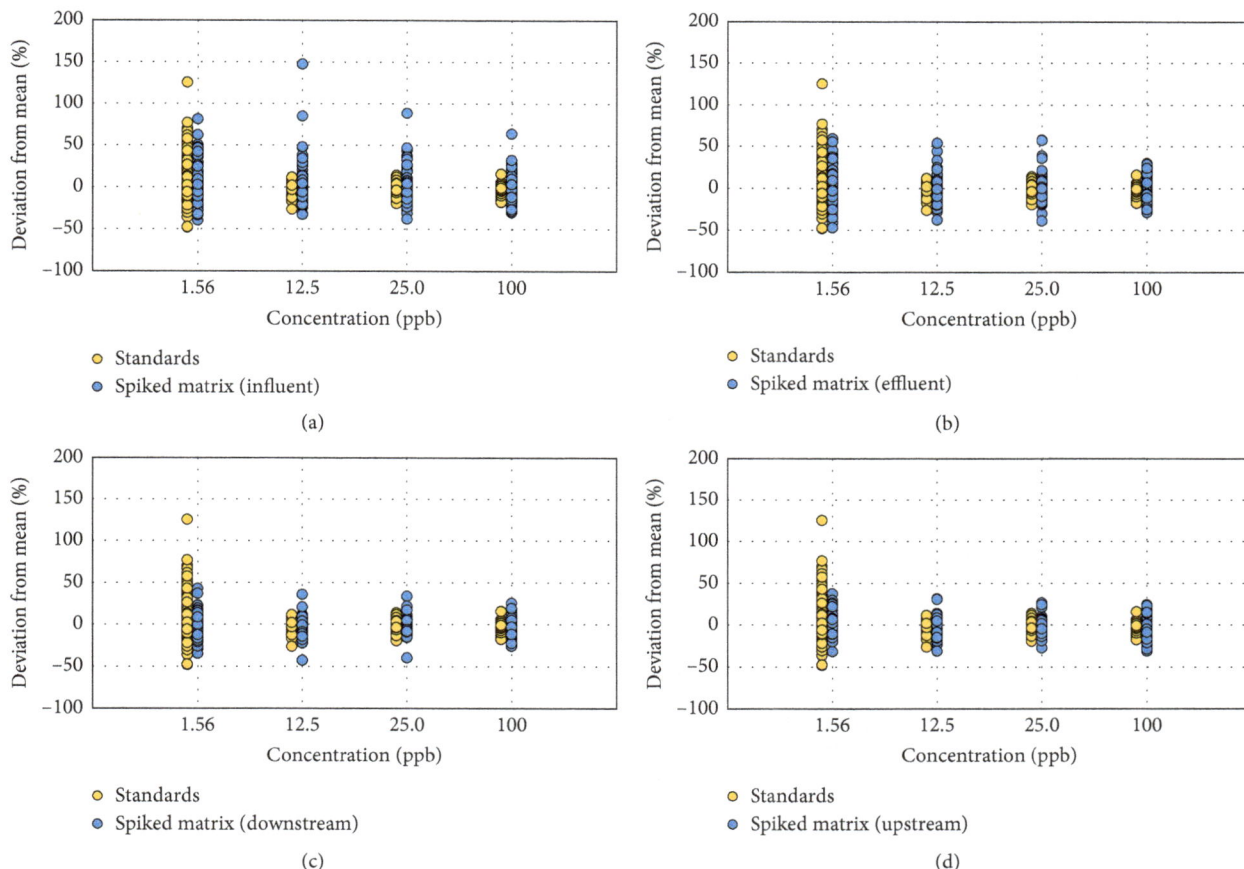

FIGURE 4: Comparison of ion ratios in spiked matrices and in clean standards. (a) WWTP influent samples; (b) WWTP effluent samples; (c) upstream surface water samples; (d) downstream surface water samples.

TABLE 2: Percent false negatives in spiked environmental matrices at different tolerance ranges for all 30 PPCPs.

Tolerance range	Matrix				Overall false negatives
	Influent	Effluent	Downstream	Upstream	
±10%	54%	49%	47%	42%	38%
±20%	26%	20%	16%	14%	16%
±30%	13%	9%	7%	6%	8%
±40%	8%	6%	4%	3%	6%
±50%	5%	4%	2%	2%	4%

TABLE 3: Percent false negatives in spiked environmental matrices at different tolerance ranges for 23 compounds without tetracyclines and quinolones.

Tolerance range	Matrix				Overall false negatives
	Influent	Effluent	Downstream	Upstream	
±10%	48%	42%	41%	35%	31%
±20%	20%	13%	10%	8%	10%
±30%	8%	4%	2%	2%	4%
±40%	4%	2%	0%	0%	2%
±50%	2%	1%	0%	0%	1%

have the same confidence level (Table 3). Therefore, a fixed tolerance range of 50% for all compounds could potentially result in high false negatives for tetracyclines and quinolones and high false positives for the other PPCPs. Also, at the 50% tolerance range, even if the overall false negatives were already below 5% for many PPCPs, as seen in Table 2, it was observed that the tetracyclines and quinolones still had very high values, with ciprofloxacin having 96% false negatives (Table 4).

For chlortetracycline, ciprofloxacin, enrofloxacin, norfloxacin, and tetracycline, the range needed to be from 70%

TABLE 4: Percent false negatives for tetracyclines and quinolones at the 50% tolerance range in spiked environmental matrices.

Compounds	Matrix				Overall false negatives
	Influent	Effluent	Downstream	Upstream	
Chlortetracycline	10%	0%	0%	0%	2%
Ciprofloxacin	96%	71%	33%	13%	61%
Enrofloxacin	5%	0%	0%	0%	1%
Norfloxacin	0%	4%	0%	0%	1%
Oxytetracycline	0%	30%	29%	22%	21%
Sarafloxacin	12%	16%	0%	20%	13%
Tetracycline	11%	0%	0%	0%	3%

TABLE 5: Percent false negatives for tetracyclines and quinolones in spiked environmental matrices at the 80% tolerance range.

Compounds	Matrix				Overall false negatives
	Influent	Effluent	Downstream	Upstream	
Chlortetracycline	10%	0%	0%	0%	2%
Ciprofloxacin	0%	0%	0%	0%	0%
Enrofloxacin	0%	0%	0%	0%	0%
Norfloxacin	0%	4%	0%	0%	1%
Oxytetracycline	0%	0%	29%	0%	6%
Sarafloxacin	6%	11%	0%	20%	10%
Tetracycline	0%	0%	0%	0%	0%

TABLE 6: Results of the application of the ion ratio criterion in real wastewater and surface water samples.

Matrix	Tolerance range	Total no. of detections	No. of detections outside the range
Influent	±40%	102	0
Effluent	±30%	90	2
Downstream	±30%	37	2
Upstream	±30%	39	1

A number of detections outside the range are data points that were considered positive detections but had ion ratios outside the set tolerance range.

to 85% in order to have a false-negative rate of ≤5%. Oxytetracycline and sarafloxacin, on the contrary, still have false negatives of up to 29% (downstream) and 20% (upstream), respectively, at a tolerance range of 80% (Table 5). However, setting a wider range may result in greater probability of false positives. Therefore, a tolerance range of 80% was set for the tetracyclines and quinolones, but it is recommended that other criteria such as retention time, peak areas, and the number of points per peak be investigated more carefully in the confirmation of these compounds.

Once the acceptable tolerance range for the mean ion ratio for each analyte was established based on spiked environmental samples, the ion ratio in each sample matrix was also assessed in order to adjust this range accordingly for the influent, effluent, upstream, and downstream samples. The tolerance range that would give ≤5% false negatives was recorded for each matrix type. These values were 40% for the influent samples and 30% for the effluent, upstream, and downstream samples (Table 3). It is expected that the compounds would have higher variations in more complex matrices such as the influent. Since the tolerance range for the influent differed by 10%, it is recommended to establish a different tolerance limit for influent samples to avoid a high false-negative rate in this matrix. If a fixed range of ±20% is used as the tolerance value (Table 2) for all types of matrices, the number of false negatives would be much

higher, 26%, 20%, 16%, and 14%, for the influent, effluent, downstream, and upstream samples, respectively. Therefore, it is important to have a separate tolerance range for certain compounds in different environmental matrices.

3.4. Applying the Optimized Ion Ratio Criterion in Real Water Samples from around the World. The optimized ion ratio criterion for each of the target PPCPs was applied to real environmental samples that were not spiked with standards. These samples were wastewater influents and effluents and receiving surface waters which are located upstream and downstream of the respective WWTPs, collected from 5 different countries. For influent samples, an analyte is said to be positively detected in the sample if its ion ratio is within the mean ± 40% of the reference standard. For effluent, upstream, and downstream samples, analytes with the mean ion ratio within ±30% of the standards are considered positive detection. Note that the results in Table 6 do not include detections for tetracyclines and quinolones, for which a different tolerance range was set.

The compounds that were detected outside the range were acetylsulfamethoxazole in the effluent samples, azithromycin in the effluent and upstream samples, and clarithromycin in the downstream samples, with the details shown in Table 7.

TABLE 7: Compounds with ion ratios detected outside the set tolerance range of ±40% for WWTP influent samples and 30% for WWTP effluent, upstream, and downstream samples.

Matrix	Compound	Tolerance range	Calculated ion ratio
Effluent	Acetylsulfamethoxazole	0.88–1.64	1.74
	Azithromycin	1.89–3-5	1.86
Downstream	Clarithromycin	1.28–2.38	2.41
	Clarithromycin	1.28–2.38	2.42
Upstream	Azithromycin	1.89–3-5	1.74

The compounds with ion ratios that fell outside the set tolerance range were investigated individually to confirm if these were real detections or not by checking the presence of both the quantitative and qualitative ions and if the shift in retention time is not more than 0.5 min. It was found that all of them had both ions, and their retention times were within the acceptable range. Since their calculated ion ratios are still close to the limits of the range, these were still considered as positive detections. In cases like this where the calculated ion ratios are close to the limits of the range and retention times are within the acceptable shift, it is recommended that the qualitative ion be checked to make sure that its signal is at least 3 times that of the noise (Table 6).

A total number of 37 detections for ciprofloxacin, norfloxacin, and tetracycline were found for tetracyclines and quinolones in the samples. The ion ratios of all 37 peaks in all matrices were within the set tolerance range of 80%, and they passed other criteria for peak confirmation.

An example of a false-positive detection that was found through the use of the ion ratio is diclofenac. The quantitative and qualitative ion transitions of diclofenac are $296 \rightarrow 214$ and $296 \rightarrow 250$ and its retention time is at 27.5 min. Figure 5 shows a comparison of the two chromatograms, both of which have peaks at 27 min for both SRM transitions.

When the ion ratios were calculated, an influent sample (Figure 5) gave an ion ratio of 0.15, which falls outside the range for diclofenac in the influent which is from 1.57 to 3.67. Furthermore, it can be observed that, for WWTP A, the retention time of the qualitative ion, which is at 27.16 min, is slightly different from that of the quantitative one at 27.78 min, further proving that this is a false-positive detection since the retention times of both ions should be the same. Upon removal of 7 false-positive diclofenac peaks that did not match the ion ratio criterion and retention times for both the quantitative and qualitative ions, the total number of detections was reduced from 312 to 305.

4. Conclusions

An ion ratio criterion has been optimized for six classes of pharmaceuticals in wastewater and surface water using LC-MS/MS. For 23 PPCPs, values for mean ± tolerance for the ion ratios in the different types of environmental matrices were established based on the variabilities of the ion ratios in spiked samples. The variabilities of the ion ratios of the compounds were found to increase at lower concentrations from 4% at 100 ppb to 18% at 1.56 ppb. Therefore, the mean ion ratio that was used in the formula is the average of the

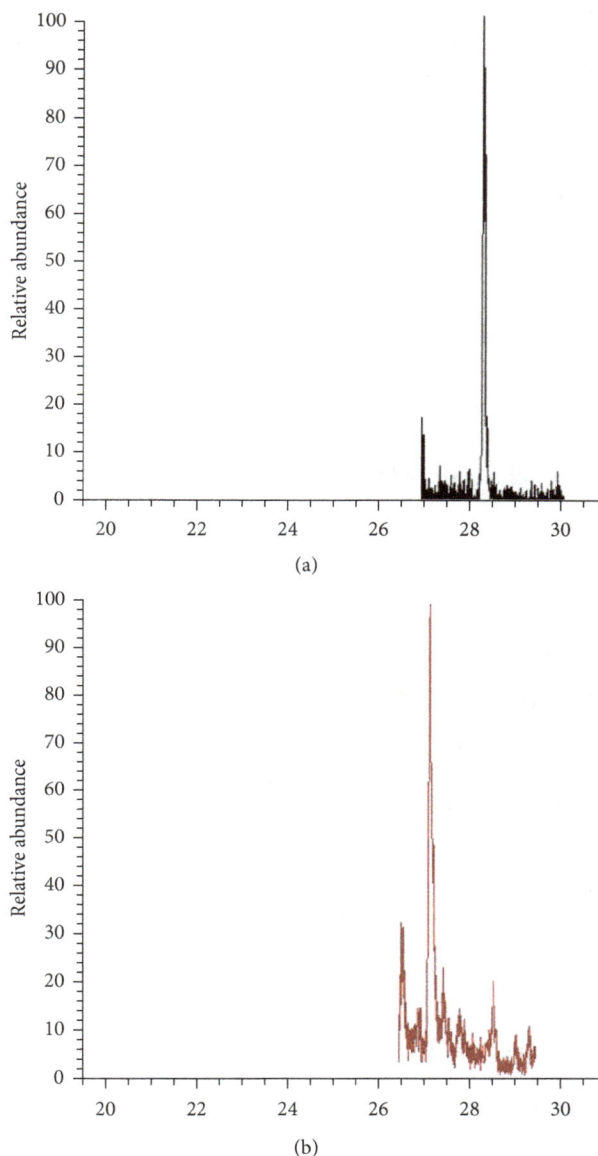

(a)

(b)

FIGURE 5: False-positive detection of diclofenac in wastewater. The calculated ion ratio, 0.15, falls outside the ion ratio tolerance range of 1.57 to 3.67. The chromatograms show (a) the peak for the quantitative ion with a transition of $296 \rightarrow 214$ and (b) the peak for the qualitative ion with a transition of $296 \rightarrow 250$.

ion ratios from 1.56, 3.13, 6.25, 12.5, 25, 50, and 100 ppb so that it can capture the variations at different concentrations. The ion ratios for tetracyclines and quinolones were found to have higher variations, which are twice that of the other PPCPs; therefore, these two classes of compounds were analyzed separately so as not to increase the possibility of false positives for the other compound classes. For tetracyclines and quinolones, the tolerance range was set to 80%, but it is recommended that other criteria such as retention time, peak areas, and the number of points per peak be investigated carefully before reporting their detections.

For the sulfonamides, macrolides, SSRIs, and other PPCPs, the ion ratios were studied in the different environmental matrices. It was found that the variations also increase with the complexity of the matrix. The optimized tolerance range that would give <5% false negatives was 40% for the influent and 30% for the effluent, upstream, and downstream. This optimized ion ratio criterion was then applied to real wastewater and surface water samples that were not spiked with standards and resulted in the reduction of the total number of detections from 312 to 305, after false positives were eliminated.

Conflicts of Interest

The authors declare that they have no conflicts of interest.

Acknowledgments

The authors would like to acknowledge support from the National Science Foundation (PIRE-HEARD, Award no. 1545756).

References

[1] P. Arnnok, R. R. Singh, R. Burakham, A. Pérez-Fuentetaja, and D. S. Aga, "Selective uptsake and bioaccumulation of antidepressants in fish from effluent-impacted Niagara River," *Environmental Science and Technology*, vol. 51, no. 18, pp. 10652–10662, 2017.

[2] K. Kummerer, "Antibiotics in the aquatic environment–a review–part I," *Chemosphere*, vol. 75, no. 4, pp. 417–434, 2009.

[3] I. Senta, I. Krizman-Matasic, S. Terzic, and M. Ahel, "Comprehensive determination of macrolide antibiotics, their synthesis intermediates and transformation products in wastewater effluents and ambient waters by liquid chromatography-tandem mass spectrometry," *Journal of Chromatography A*, vol. 1509, pp. 60–68, 2017.

[4] I. Senta, S. Terzić, and M. Ahel, "Simultaneous determination of sulfonamides, fluoroquinolones, macrolides and trimethoprim in wastewater and river water by LC-tandem-MS," *Chromatographia*, vol. 68, no. 9-10, pp. 747–758, 2008.

[5] L. J. Zhou, G. G. Ying, S. Liu et al., "Occurrence and fate of eleven classes of antibiotics in two typical wastewater treatment plants in South China," *Science of the Total Environment*, vol. 452-453, pp. 365–376, 2013.

[6] M. Pedrouzo, F. Borrull, R. M. Marce, and E. Pocurull, "Ultra-high-performance liquid chromatography-tandem mass spectrometry for determining the presence of eleven personal care products in surface and wastewaters," *Journal of Chromatography A*, vol. 1216, no. 42, pp. 6994–7000, 2009.

[7] E. Kristiansson, J. Fick, A. Janzon et al., "Pyrosequencing of antibiotic-contaminated river sediments reveals high levels of resistance and gene transfer elements," *PLoS one*, vol. 6, no. 2, article e17038, 2011.

[8] L. Rizzo, C. Manaia, C. Merlin et al., "Urban wastewater treatment plants as hotspots for antibiotic resistant bacteria and genes spread into the environment: a review," *Science of the Total Environment*, vol. 447, pp. 345–360, 2013.

[9] K. Kümmerer and A. Henninger, "Promoting resistance by the emission of antibiotics from hospitals and households into effluent," *Clinical Microbiology and Infection*, vol. 9, no. 12, pp. 1203–1214, 2003.

[10] N. H. Tran, M. Reinhard, and K. Y. H. Gin, "Occurrence and fate of emerging contaminants in municipal wastewater treatment plants from different geographical regions-a review," *Water Research*, vol. 133, pp. 182–207, 2018.

[11] WHO, *Antimicrobial Resistance*, WHO, Geneva, Switzerland, 2017, http://www.who.int/mediacentre/factsheets/fs194/en/.

[12] H. L. Schoenfuss, E. T. Furlong, P. J. Phillips et al., "Complex mixtures, complex responses: assessing pharmaceutical mixtures using field and laboratory approaches," *Environmental Toxicology and Chemistry*, vol. 35, no. 4, pp. 953–965, 2016.

[13] N. H. Tran, H. Chen, M. Reinhard, F. Mao, and K. Y. Gin, "Occurrence and removal of multiple classes of antibiotics and antimicrobial agents in biological wastewater treatment processes," *Water Research*, vol. 104, pp. 461–472, 2016.

[14] B. J. Berendsen, T. Meijer, H. G. Mol, L. van Ginkel, and M. W. Nielen, "A global inter-laboratory study to assess acquisition modes for multi-compound confirmatory analysis of veterinary drugs using liquid chromatography coupled to triple quadrupole, time of flight and orbitrap mass spectrometry," *Analytica Chimica Acta*, vol. 962, pp. 60–72, 2017.

[15] FDA, *Guidance for Industry 118 Confirmation of Identity of Animal Drug Residues*, Food and Drug Administration, Silver Spring, MD, USA, 2003.

[16] UNODC, *Guidance for the Validation of Analytical Methodology and Calibration of Equipment Used for Testing of Illicit Drugs in Seized Materials and Biological Specimens*, UNODC, New York, NY, USA, 2009.

[17] The European Communities, *Commission Decision 2002/657/ EC Implementing Council Directive 96/23/EC Concerning the Performance of Analytical Methods and the Interpretation of Results*, The European Communities, 2002.

[18] USDA, *Data and Instrumentation Revision 5. Agricultural Marketing Service*, United States Department of Agriculture, Washington, DC, USA, 2017.

[19] EWDTS, *European Laboratory Guidelines for Legally Defensible Workplace Drug Testing (EWDTS)*, CRC Press, BocaRaton, FL, USA, 2002, http://www.eapinstitute.com/documents/EWDTS-Guidelines.pdf.

[20] H. G. Mol, P. Zomer, M. Garcia Lopez et al., "Identification in residue analysis based on liquid chromatography with tandem mass spectrometry: experimental evidence to update performance criteria," *Analytica Chimica Acta*, vol. 873, pp. 1–13, 2015.

[21] B. J. Berendsen, T. Meijer, R. Wegh et al., "A critical assessment of the performance criteria in confirmatory analysis for veterinary drug residue analysis using mass spectrometric detection in selected reaction monitoring mode," *Drug Testing and Analysis*, vol. 8, no. 5-6, pp. 477–490, 2016.

[22] USEPA, *Method 1694: Pharmaceuticals and Personal Care Products in Water, Soil, Sediment, and Biosolids by HPLC/MS/ MS*, U.S Envorinmrntal Protection Agency, Washington, DC, USA, 2007.

[23] USEPA, *Method 542: Determination of Pharmaceuticals and Personal Care Products in Drinking Water by Solid Phase Extraction and Liquid Chromatography Electrospray Ionization-Tandem Mass Spectrometry (LC/ESI-MS/MS)*, U.S Envorinmrntal Protection Agency, Washington, DC, USA, 2016.

4

A Stability Indicating HPLC Assay Method for Analysis of Rivastigmine Hydrogen Tartrate in Dual-Ligand Nanoparticle Formulation Matrices and Cell Transport Medium

Naz Hasan Huda ⓘ, Bhawna Gauri ⓘ, Heather A. E. Benson, and Yan Chen ⓘ

School of Pharmacy and Biomedical Sciences, Curtin Health Innovation Research Institute, Curtin University, Perth, WA 6845, Australia

Correspondence should be addressed to Yan Chen; y.chen@curtin.edu.au

Academic Editor: Bengi Uslu

The objective of this study was to develop and validate a method for quantitative analysis of rivastigmine hydrogen tartrate (RHT) in dual-ligand polymeric nanoparticle formulation matrices, drug release medium, and cellular transport medium. An isocratic HPLC analysis method using a reverse phase C_{18} column and a simple mobile phase without buffer was developed, optimised, and fully validated. Analyses were carried out at a flow rate of 1.5 mL/min at 50°C and monitored at 214 nm. This HPLC method exhibited good linearity, accuracy, and selectivity. The recovery (accuracy) of RHT from all matrices was greater than 99.2%. The RHT peak detected in the samples of a forced degradation study, drug loading study, release study, and cellular transport study was pure and free of matrix interference. The limit of detection (LOD) and limit of quantification (LOQ) of the assay were 60 ng/mL and 201 ng/mL, respectively. The method was rugged with good intra- and interday precision. This stability indicating HPLC method was selective, accurate, and precise for analysing RHT loading and its stability in nanoparticle formulation, RHT release, and cell transport medium.

1. Introduction

There is a worldwide increase in the prevalence of brain diseases such as Alzheimer's disease (AD), Parkinson's disease, and stroke due to the increase of the aging population. Consequently, there is an increase in demand for effective treatments for these diseases. Rivastigmine hydrogen tartrate (RHT) was approved by the US FDA in 2000 for the treatment of mild-to-moderate dementia of either Alzheimer's type or related to Parkinson's disease [1]. RHT is a cholinesterase inhibitor that inhibits both acetylcholinesterase (AChE) and butyrylcholinesterase (BuChE) enzymes responsible for the degradation of acetylcholine (ACh) into nonfunctional metabolites. Among the many chemical changes that the brain encounters during AD, depletion of ACh is one of the earliest and biggest changes. RHT increases the central cholinergic function by enhancing the ACh level in mild-to-moderate AD patients and inhibits

deposition of amyloid plaques in the brain, slowing down the mental decline [2–6].

RHT has been reported to improve or maintain patients' cognitive function, global function, behaviour, and day-to-day activities [7, 8]. It is commercially available as capsules, oral solution, and patches. However, the current therapeutic regimen of RHT demands frequent dosing, and cholinergic side effects are common. Like other CNS drugs, treatment efficacy of RHT is primarily restricted, not by the drug's inherent potency but by its ability to cross the blood-brain barrier (BBB) into the brain due to its hydrophilic nature [9].

To overcome the challenge of RHT transport into the brain, it has been formulated into nanoparticles (NPs) since 2008 [9]. NPs act as a drug carrier to provide targeted delivery of a concentrated payload to, and sustained release at, the target site for therapeutic action. The targeting ability of NPs is influenced by surface ligands that interact

TABLE 1: Summary of published HPLC conditions for RHT determinations.

Column	Sample matrix	Mobile phase	Flow rate	Detection technique	Analysis time	Detection limits	Reference
Waters Spherisorb silica	Human plasma	Acetonitrile-50 mM aqueous sodium dihydrogen phosphate (17 : 83 v/v, pH 3.1)	1.3 mL/min	UV: 200 nm	6 minutes	LOD: 0.2 ng/mL LOQ: 0.5 ng/mL	[26]
Inertsil ODS-3V C_{18}	Rat plasma and brain	Ammonium acetate buffer (20 mM, pH 4.5) and acetonitrile 74 : 26 (v/v)	1.0 mL/min	Fluorescence detector, Ex/Em wavelength: 220/293 nm	16 minutes	LOD: not given LOQ: 10 ng/mL	[16]
XTerra RP18 C_{18}	Raw material	10 mM sodium-1-heptane sulphonate (pH 3.0) and acetonitrile 72 : 28 (v/v)	1.0 mL/min	UV: 217 nm	13 minutes	LOD: 100 ng/mL LOQ: 300 ng/mL	[27]
Monomeric C_{18}	Rat plasma	Acetonitrile and 20 mmol/L phosphate buffer, pH 3.0 (25 : 75)	1.0 mL/min	Fluorescence detector, Ex/Em wavelength: 220/293 nm	20 minutes	LOD: not given LOQ: 25 ng/mL	[28]
$5C_{18}$-MS	Capsule	Methanol and water (90 : 10)	1.0 mL/min	UV: 217 nm	Not given	LOD: not given LOQ: 10.9 μg/mL	[29]
Kromasil C_8	Liposomes	20 mmol·L^{-1} phosphate buffer (pH 3.0) and acetonitrile (75 : 25%, v/v)	1.0 mL/min	UV: 210 nm	20 minutes	LOD: not given LOQ: 10 ng/mL	[17]
C_{18}	Solid lipid nanoparticles	Acetonitrile and potassium dihydrogen orthophosphate buffer (pH 6.0) (20 : 80 v/v)	1.0 mL/min	UV: 215 nm	Not given	LOD: not given LOQ: 1 μg/mL	[14]
ODS C_{18}	Liposomes	Acetonitrile : water (20 mM $NaH_2PO_4 \cdot 2H_2O$, 10 mM $Na_2HPO_4 \cdot 12H_2O$) (25 : 75, v/v)	1.0 mL/min	UV: 218 nm	Not given	LOD: not given LOQ: not given	[15]

specifically at receptors on the target site. This targeting approach is receiving increased attention for the development of highly efficacious therapeutics with minimal side effects that can be used in a wide range of diseases where ideal pharmaceutical options are currently limited. Accurate and efficient analytical methods that can facilitate the formulation development and evaluation process are essential. Our group recently developed dual-targeting ligand NPs for the brain delivery of RHT [10, 11]. This was facilitated by the development of an analytical method for quantitation of RHT loading in a complex formulation and relevant cellular and stability studies.

In the past, the quantitative determination of RHT has been reported using several analytical techniques such as spectrophotometry [12, 13], HPLC [14–17], gas chromatography-mass spectrometry (GC-MS) [18–20], and liquid chromatography-mass spectrometry (LC-MS) [21–25]. For example, Fazil et al. [12] determined drug loading and encapsulation efficiency by measuring the amount of free RHT in the NPs supernatant using a UV spectrophotometer. The group used the same technique to quantify RHT in phosphate-buffered saline (PBS) in their in vitro permeability studies. Nagpal et al. [13] employed

spectrophotometry to measure drug loading, entrapment efficiency, and RHT release from NPs in PBS. However, spectrophotometry is not selective and cannot separate formulation excipients, impurities, or degradation products from the drug itself.

HPLC methods developed for RHT commonly employ a buffer in the mobile phase and special columns, such as Kromasil C_8, XTerra RP18, and $5C_{18}$-MS, with either UV or fluorescence detection for the analysis of RHT in different samples [14–17, 26–29]. Table 1 summarises the HPLC methods reported to date for separation and quantification of RHT in different sample matrices.

Although HPLC coupled with mass spectrophotometry (MS) is suitable or desirable for separation and quantification of RHT in biological samples (rat, canine, and human plasma, and rat brain and urine) [30], for initial formulation development and evaluation, a fast, economic, and simple yet selective and accurate HPLC method is preferred. Furthermore, it is advantageous if the developed method is also directly applicable for HPLC-MS analysis. Therefore, the objective of the current study was to develop and validate a simple, fast, sensitive, selective, and accurate HPLC method for the quantitative analysis of RHT loading in a dual-ligand NP formulation, its release and cellular transport, and its stability profile under different stressed conditions. Development criteria were that the chromatography should be achieved on a commonly used C_{18} column and that the mobile phase was without a buffer, thus allowing direct translation to HPLC-MS.

2. Experimental

RHT (purity ≥99.2%) was purchased from Innochem Technology Co., Ltd. (Beijing, China). HPLC grade acetonitrile (ACN) (purity ≥99.9%) was obtained from Thermo Fisher Scientific (Scoresby, Australia). Hanks' balanced salt solution (HBSS), 4-2-hydroxyethylpiperazine-1-ethanesulfonic acid (HEPES) (purity ≥ 99.5%), trifluoroacetic acid (TFA) (purity ≥ 99.0%), and phosphate-buffered saline (PBS) pouches were purchased from Sigma-Aldrich (Castle Hill, Australia). D-glucose anhydrous, HCl (32% w/v), and hydrogen peroxide (H_2O_2, 30% w/v) were obtained from Ajax Finechem Pty Ltd. (Taren Point, Australia). NaOH (purity ≥ 98.0) was obtained from BDH Laboratory Supplies (Poole, England). The transport buffer (HBSS-P) was prepared from HBSS containing 10 mM HEPES and 20 mM glucose. Ultrapure (type 1) water was generated using a Milli-Q System (Merck Millipore, Bayswater, Australia).

2.1. Nanoparticle Preparation and Characterisation. The dual-ligand PLGA-based NPs were prepared by a double emulsion solvent evaporation technique [31–33]. RHT was loaded in the NPs as a model drug. The NPs formulation was optimised to achieve optimum particle size for brain drug delivery and maximal drug loading in the NPs. The optimised NPs formulation was evaluated for in vitro characteristics including particle morphology, size, zeta potential, drug loading efficiency, release profile, and stability studies. The developed HPLC method was used for quantitative analysis of RHT loading in the NPs formulation, release and cellular transport, and stability profile under different stressed conditions.

2.2. Chromatographic Conditions. The HPLC system was an Agilent® 1200 instrument (Agilent Technologies, Mulgrave, Australia) with a degasser (G1379B), a binary pump (G1312A), and an autosampler (G1329A) with thermocontrol unit (G1330B) and VWD (G1314B), Waters® 1122/WTC-120 external column heater (Waters Australia Pty Ltd, Rydalmere, Australia), and a diode-array detector (DAD, G1315B). Data acquisition and processing were carried out with Agilent ChemStation® software version B.04.03 SP1.

Samples were maintained at 4°C in the autosampler prior to analysis. An Apollo C_{18} column, 5 μm particle size, 150 mm × 4.6 mm (Grace Davison Discovery Sciences, Baulkham Hills, Australia), was maintained at 50°C. All analyses were conducted with an isocratic mode with a 1.5 mL/min flow rate of the mobile phase (20% v/v ACN in water containing 0.1% TFA, prefiltered through a 0.2 μm

hydrophilic nylon filter: Merck Millipore, Bayswater, Australia) and injection volume of 50 μL. The detection of RHT was monitored at an UV wavelength of 214 nm.

2.3. Forced Degradation Studies. Forced degradation studies were conducted according to published protocols to confirm the selectivity of the developed assay method [34–36]. RHT (25 μg/mL) was used for all degradation studies. For acid decomposition (hydrolysis) studies, RHT solution was prepared in 2 N HCl and incubated for 48 h at 37° and 60°C. For base hydrolysis studies, RHT solution was prepared in 0.5 N NaOH and incubated at 37°C and 60°C up to 48 h. Both the acid and alkaline samples were cooled to RT and neutralised before analysis by HPLC.

The stability of RHT in water was assessed with RHT solution (25 μg/mL) incubated at 37° and 60°C for 48 h, whereas the effect of RHT oxidation was determined by incubating RHT for 48 h at 37° and 60°C in 30% H_2O_2.

All degraded samples were analysed by HPLC, and RHT peak purity was evaluated using a diode-array detector by obtaining five UV spectra across the peak. The similarity among these five spectra was determined and reported using ChemStation software to determine the peak purity. Coelution of any degraded product with the drug peak would make the peak impure, resulting in dissimilar UV spectra. The software also reported whether the peak purity in each spectrum was within the automatically set threshold limit.

2.4. Method Validation. The developed HPLC method was validated with respect to selectivity, linearity, precision, accuracy, limit of detection (LOD), and limit of quantification (LOQ) in accordance with the International Council for Harmonisation (ICH) Guidelines for Validation of Analytical Procedures, Q2B [37] and the United States Pharmacopeia and the National Formulary (USP 37-NF 32) [38].

2.4.1. Selectivity. Forced degradation study samples were used to assess the selectivity of the method. Supernatants of both blank NPs (dual-ligand NPs without any loaded drug) and RHT-loaded dual-ligand NPs were diluted 200 times in the mobile phase and injected into the HPLC to study whether any interfering peaks coeluted at or near the drug peak. Similarly, the matrix interference was also investigated using (i) PBS medium collected from release study control (dual-ligand NPs without any loaded drug) and (ii) uptake/transport study buffer (HBSS-P) comprising HBSS containing 10 mM HEPES and 20 mM glucose (collectively the "experimental media").

2.4.2. Linearity. The linearity of the developed assay method was assessed in two different media: 0.3% vitamin E-TPGS (solvent for dispersion of dual-ligand nanoparticle) and PBS (release study medium). A range of concentrations of RHT solutions (0.1 to 2 mg/mL) was prepared in 0.3% vitamin E-TPGS from a stock solution (2 mg/mL RHT in 0.3% vitamin E-TPGS). Each solution was then diluted 200 times (50 μL into 10 mL) with the mobile phase to obtain the final

RHT standard concentrations of 0.5, 1, 2, 3, 4, 5, 6, 7, 8, 9, and 10 μg/mL. These standards were injected into the HPLC column, in duplicate. Another set of standards with the same concentration range was prepared by diluting an RHT stock solution (1 mg/mL) in type 1 water with 10 mM PBS (pH 7.4). Again, these standards were injected into the HPLC column in duplicate. Average peak area data were plotted against corresponding standard concentrations using Microsoft® Excel 2016 to construct the standard calibration curve. The linearity was established by calculating the R^2 value.

2.4.3. Precision. The precision of the proposed method was determined by injecting four RHT concentrations (1, 4, 6, and 10 μg/mL) in the experimental media, six times into the HPLC. The relative standard deviation (RSD) values were calculated for all concentrations.

2.4.4. Limit of Detection (LOD). The LOD was determined as the drug concentration that produced a signal three times greater than the baseline noise level. Two blank solvents, namely, (1) 0.3% vitamin E-TPGS diluted 200 times in the mobile phase and (2) 10 mM PBS, pH 7.4, were injected six times to determine the average noise levels. Standard RHT solutions prepared in the mobile phase were analysed and calibration curves constructed by plotting average peak heights against the corresponding concentrations. The LOD was calculated by the following formula:

$$\text{LOD} = 3 \times \frac{\text{Peak height of noise}}{\text{Slope of calibration curve constructed by peak height versus conc}}. \tag{1}$$

2.4.5. Limit of Quantification (LOQ). Using the same data, the LOQ was determined as the concentration with the signal at ten times greater than the baseline noise level. The LOQ was calculated by the following formula:

$$\text{LOQ} = 10 \times \frac{\text{Peak height of noise}}{\text{Slope of curve constructed by peak height versus conc}}. \tag{2}$$

2.4.6. Intra- and Interday Repeatability (Ruggedness). Three standard concentrations of RHT (low, medium, and high) in each solvent (0.3% vitamin E-TPGS diluted 200 times in the mobile phase and 10 mM PBS, pH 7.4) within the calibration curve were selected. The intra- and interday repeatability of the method was assessed by analysing these 1, 5, and 10 μg/mL RHT standards, in triplicate, at different time points in the same day and on two different days. In addition, the ruggedness study was conducted by analysing another set of RHT standards of the same concentrations by a second analyst on a different day. The RSD was calculated for each analysed concentration and compared with the nominal limit to evaluate the intra- and interday repeatability of the method and ruggedness.

2.4.7. Accuracy. The accuracy of the developed method was determined in two media to assess the interference of the dual-ligand NPs formulation matrices and solvents. Firstly, a batch of drug-free dual-ligand NPs (blank NPs) was prepared, and the supernatant was collected during the last step of the preparation. The supernatant was then spiked with RHT to obtain a solution of RHT (2 mg/mL). This stock solution was diluted with the same supernatant medium to prepare five RHT solutions with concentrations of 0.1, 0.4, 0.8, 1.2, and 1.6 mg/mL. These solutions were diluted 200 times with the mobile phase to obtain the final drug concentrations of 0.5, 2, 4, 6, 8, and 10 μg/mL and injected into the HPLC column in triplicate. This procedure mimics the method of sample preparation for the determination of RHT loading in dual-ligand NPs.

Secondly, an in vitro release study of the blank dual-ligand NPs was carried out at pH 7.4 in PBS (mimicking the NP matrix in the release medium). 3 mL of blank dual-ligand NPs suspension was loaded in a dialysis tube (MWCO 12000), sealed, and placed in a 60 g amber glass jar containing 50 mL prewarmed PBS at 37°C. The setup was placed on an orbital shaker at 37°C and horizontally shaken at 100 rpm. Release medium from outside the dialysis bag was collected after 24 h, spiked with RHT standard solution (200 μg/mL) prepared in type 1 water to get the final RHT concentrations of 2, 4, 6, 8, and 10 μg/mL, and injected into the HPLC column in triplicate. The concentration of RHT in all samples was determined against RHT standards prepared in the mobile phase.

The method accuracy was determined by calculating the percentage of recovery (measured concentration over the added concentration) in each case.

3. Results and Discussion

The NP formulation was optimised to achieve the particle size between 70 nm and 200 nm as well as to achieve the highest possible drug loading in the NPs. The optimised RHT-loaded dual-ligand NPs were found to have negative zeta potential (-24.3 ± 2.5 mV) and narrow size distribution (139.5 ± 3.9 nm) ideal for targeting the BBB.

3.1. Development and Optimisation of HPLC Method. The HPLC method for separation and quantification of RHT in PBS (pH 7.4) and NP supernatant (containing 0.3% vitamin E-TPGS) was developed and validated. An appropriate combination of the column type, column temperature, mobile phase composition and flow rate, injection volume, and detection system was studied to produce a simple, fast, economic, and yet selective and accurate assay method. We validated 50 μL injection volume as the maximum injection volume for future application in analysis of biological samples. The UV wavelength of 214 nm was selected for the detection of the compound based on the UV spectrum of RHT. A lower wavelength (209 nm) produced a much stronger drug signal, but higher background noise made this approach impractical. The mobile phase composition was developed based on the solubility and pKa of RHT (8.85) [39]. An acidic condition (pH 2.6) was considered necessary to keep all RHT molecules ionised (Figure 1). At this lower

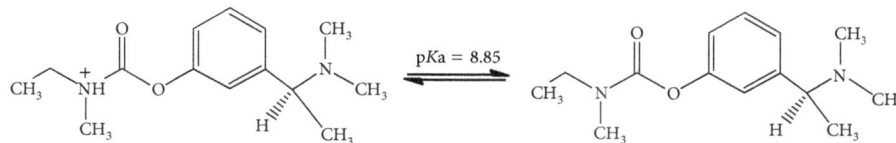

FIGURE 1: Ionisation of rivastigmine tartrate.

TABLE 2: Summary of findings in RHT-forced degradation studies.

Forced degradation condition	Temp. (°C)	Incubation duration (hrs)	Remaining percentage
Acid hydrolysis: RHT in 2 N HCl	60	48	87.4
	37	48	97.8
Base hydrolysis: RHT in 0.5 N NaOH	60	2	80.9
	60	48	0.0
	37	48	29.7
Hydrolysis: RHT in water	60	48	99.2
	37	48	99.7
Oxidation: RHT in 30% (w/v) H$_2$O$_2$	60	48	0.0
	37	48	79.6

pH, the silane groups of the C$_{18}$ column were also fully protonated, leading to weak interaction with RHT, thereby shortening the elution time. TFA was used to provide a good peak shape and avoid the use of buffer salts that may precipitate due to an interaction with formulation excipients. In addition, eliminating a buffer allows the method to be easily adapted for LC-MS analysis of RHT in the future. The initial trial mobile phase composed of ACN and water (50 : 50 v/v) containing 0.1% TFA at a flow rate of 1 mL/min resulted in the RHT eluting with the solvent front. Consequently, the organic phase was optimised at a ratio of 20 : 80 (v/v) for ACN : water to produce the best peak shape and separation. The flow rate was increased to 1.5 mL/min, and column temperature was maintained at 50°C to facilitate separation, sharpen the peaks, and reduce the retention time to 6.8 min. Mullangi et al. [16] also employed a similar mobile phase composed of ACN and water (26 : 74 v/v) and acidic pH of 4.5 for the analysis of RHT. However, our developed method utilises a less-expensive shorter column and a shorter run time, thereby providing economic benefits.

3.2. Forced Degradation of RHT and Selectivity. The main aim of the forced degradation studies of RHT was to assess the selectivity of the analytical method. According to the MSDS supplied by the manufacturer, RHT is chemically stable under normal conditions but incompatible with strong acids, bases, and oxidising agents. No light sensitivity data were provided in the MSDS; however, RHT was reported to be stable when exposed to light for at least ten days [27]. In our investigation, various stress conditions were employed to simulate any possible degradation that might occur during the NPs preparation and in vitro characterisation experiments. RHT was subjected to hydrolysis (acidic, alkaline, and neutral pH) and oxidation. Our results (Table 2) showed a similar degradation pattern for RHT as per the published literature [27]. We found that RHT was most prone to base degradation, showing maximum degradation after 48 h incubation at 37°C, compared to acidic, oxidative, and hydrolysis in water conditions. The drug demonstrated excellent stability against hydrolysis conditions at neutral pH under both test temperatures, but oxidised easily at 37°C, with complete degradation after 48 h incubation at 60°C. RHT stability in 2 N HCl was relatively good with 2.2% degradation at 37°C over 48 h.

Figure 2 illustrates that our developed HPLC assay method is capable of separating RHT from all degradation products and that the RHT peak obtained at 6.8 min is pure. Peak purity analysis was conducted using the default settings of the ChemStation software without any manual data entry. The peak purity of degraded products was not checked because we were only interested in assessing the method's capability of resolving the pure RHT peak. The eluted RHT peak was well separated from the degraded products (retention time < 6 min). Thus, the developed method was selective and can be used as a stability indicating method for the analysis of RHT concentration in various samples including stability samples.

As our RHT samples of interest are from dual-ligand nanoparticle formulation, release, and cellular transport studies, it is important that the matrices present in those samples do not interfere with the RHT quantitation. Therefore, further selectivity studies were carried out to confirm that the developed HPLC method has the capability to generate "true results" that are free from matrix or medium interference. The HPLC spectra in Figure 3 indicate that there was no peak around the RHT retention time (6.8–6.9 min) in any of the experimental media: (i) supernatant of dual-ligand NPs without any loaded drug (after 200 times dilution with the mobile phase), (ii) PBS medium following release study of empty dual-ligand NPs, and (iii) cellular transport study medium (HBSS-P). This

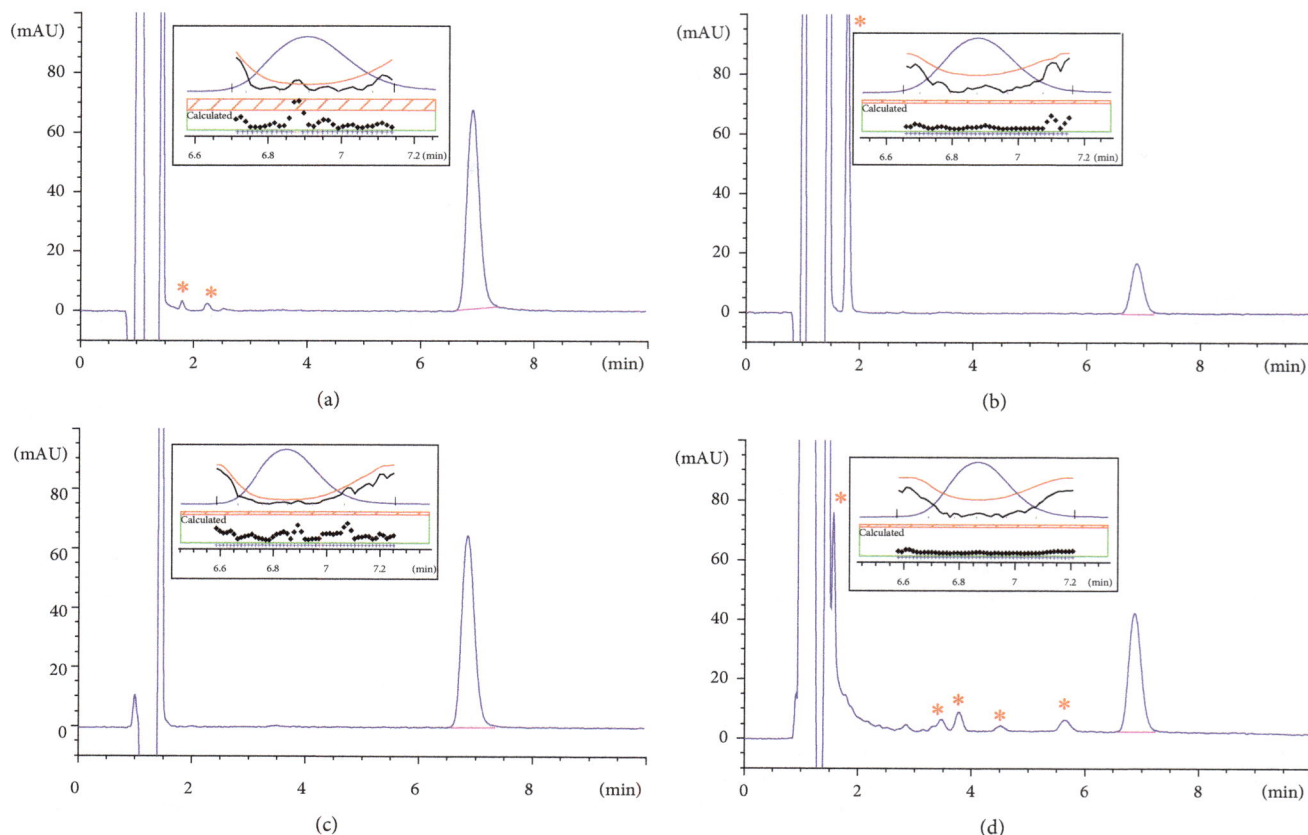

FIGURE 2: HPLC chromatogram of RHT under various stress conditions conducted at 37°C for 48 hours. (a) Acid degradation, (b) alkali degradation, (c) hydrolysis, and (d) oxidation. Analysis of RHT was not interfered by the degradation products (∗). Peak purity reports are shown in the insets, confirming that the RHT peaks are pure and the purity factors are within the calculated threshold limit.

demonstrates that there was no matrix interference and the method is selective or specific for analysis of RHT under various conditions.

3.3. *Linearity.* The detector response to various concentrations of RHT in two media produced a linear relationship. For 50 μL injections of RHT in vitamin E-TPGS (following 200 times dilution with the mobile phase), the regression plot demonstrated a nearly perfect linear relationship (coefficient of variance was 0.9999) over the concentration range of 0.5–10 μg/mL that covered the concentrations encountered in the RHT loading analysis. The same concentration range of RHT in PBS also demonstrated a good linear relationship with a coefficient of variance of 0.9998.

3.4. *Precision.* The precision study was also conducted with the three media used in the selectivity study (Table 3). All RSD values were well below the nominally acceptable level of ≤2% [40]. Even at the low concentrations of RHT (1 μg/mL), the RSD of 1.04–1.28% was achieved, demonstrating that the method is precise.

3.5. *LOD and LOQ.* The LOD of an analytical procedure is the lowest detectable amount of an analyte in a sample but not necessarily a quantifiable value. For the current method, the lowest detectable concentration of RHT in both solvent systems was 60 ng/mL.

The LOQ is the lowest amount of the drug in the sample that can be confidently quantified using the method. For the current method, the lowest quantifiable concentration of RHT in both solvent systems was 201 ng/mL. The LOD and LOQ is in a comparable range or even better than other published methods [14, 27], and this method can also meet the analytical requirements of dual-ligand NP formulation development and evaluation.

3.6. *Intra- and Interday Repeatability.* The intra- and interday repeatability data are shown in Table 4. All RSD values of repeated analysis were within the acceptable limit of ≤2% [40]. The ruggedness study performed by different analysts also demonstrated similar trend with RSD values below 2% (data not shown). These results suggest that the developed method is rugged.

3.7. *Accuracy.* The accuracy of a method demonstrates that the assay can accurately quantify the molecule(s) of interest in the presence of other possible interfering

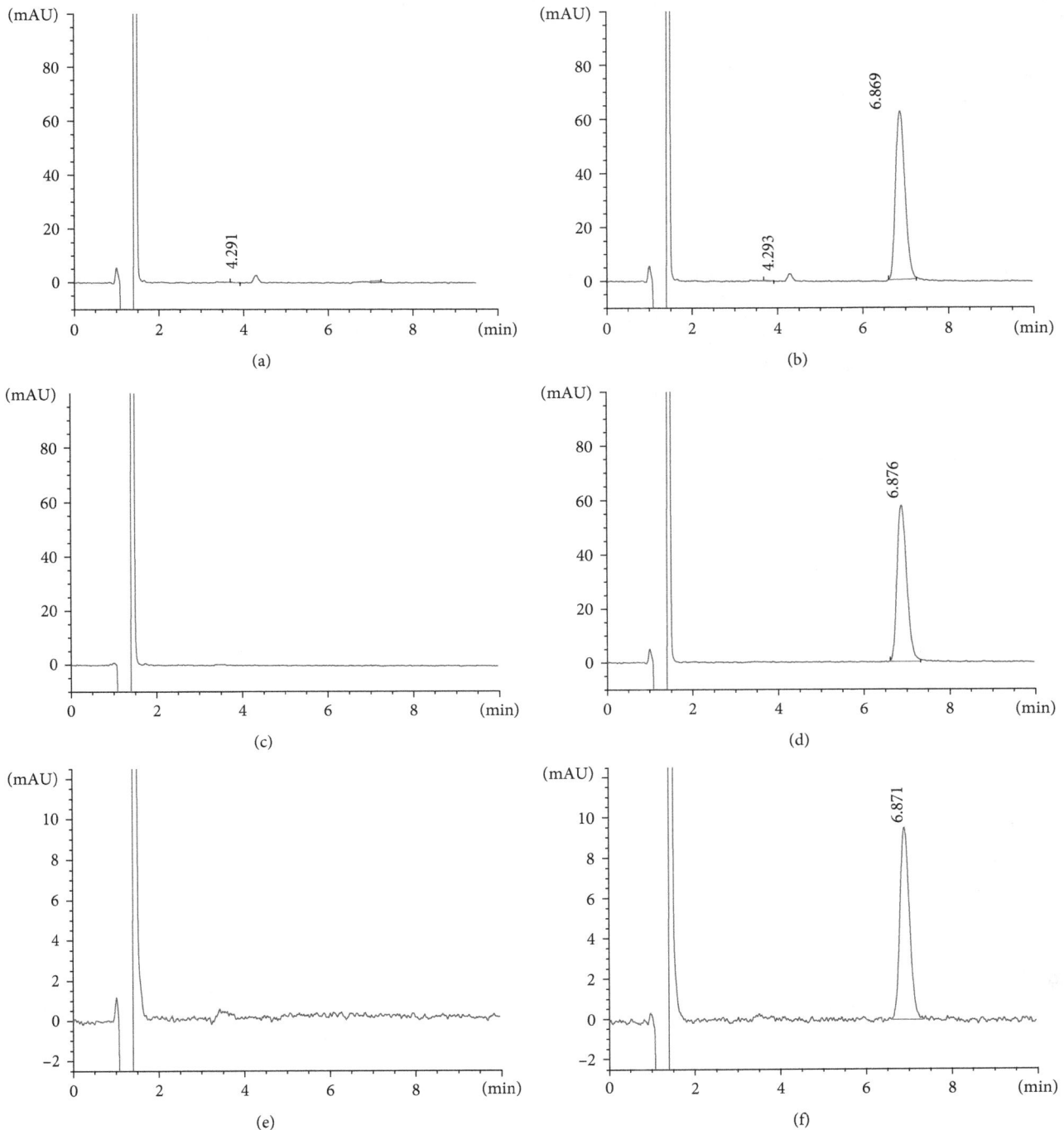

FIGURE 3: HPLC chromatograms illustrating absence of any matrix interfering peak around the RHT retention time (6.9 minutes). Chromatograms of (a) empty dual-ligand NPs matrix, (b) RHT-loaded dual-ligand NPs matrix, (c) release medium after 24-hour release study of empty dual-ligand NPs, (d) release medium after 24-hour release study of RHT-loaded dual-ligand NPs, (e) cell transport medium, and (f) released RHT from NPs in cell transport medium after the transport study.

components such as excipients, reactions components, release medium, and cellular transport medium. The accuracy of the proposed method was calculated as percentage recovery from the six concentrations covering the entire RHT concentration range within the calibration curve. RHT was successfully recovered from all samples in three experimental media (Table 5) with an accuracy of 99.5 ± 1%, which is within the acceptable range [40]. This suggests that none of the matrices in the dual-ligand nanoparticle formulation or medium/buffer is interfering

TABLE 3: The precision of the HPLC method for determination of RHT.

RHT conc. (µg/mL)	RHT in NPs matrix		RHT in release medium		RHT in cell transport medium	
	Average RHT peak area (mAU × sec)	RSD (%)	Average RHT peak area (mAU × sec)	RSD (%)	Average RHT peak area (mAU × sec)	RSD (%)
1	36.50	1.04	36.22	1.28	36.02	1.06
4	145.97	0.59	145.82	0.67	144.24	0.66
6	219.73	0.49	220.98	0.28	218.64	0.48
10	370.20	0.17	369.30	0.19	369.15	0.23

TABLE 4: Intra- and interday repeatability of RHT analysis in NPs matrix, release medium, and cell transport medium.

RHT concentration[a] (µg/mL)	RHT in NPs matrix (200x diluted in mobile phase)		RHT in release medium		RHT in cell transport medium	
	Intraday RSD[b]	Interday RSD[c]	Intraday RSD[b]	Interday RSD[c]	Intraday RSD[b]	Interday RSD[c]
1	0.69	1.32	0.73	0.64	0.94	1.24
5	0.92	0.92	0.65	1.11	0.51	1.11
10	0.49	1.19	0.35	1.02	0.15	0.95

[a]Each concentration was analysed in triplicate ($n = 3$); [b]the analyses were carried out at 0, 3, and 8 hrs on the same day, and all data were included in the calculation; [c]the analyses were carried out at days 1 and 2, and all data were included in the calculation.

TABLE 5: Accuracy data for RHT in NPs matrix, release medium, and cell transport medium.

Prepared RHT concentration (µg/mL)	RHT in NPs matrix		RHT in release medium		RHT in cell transport medium	
	Measured concentration (µg/mL)	Recovery (%)	Measured concentration (µg/mL)	Recovery (%)	Measured concentration (µg/mL)	Recovery (%)
0.50	0.49 ± 0.13	98.0 ± 0.2	0.49 ± 0.11	98.4 ± 0.1	0.49 ± 0.16	98.6 ± 0.2
2.00	1.99 ± 0.16	99.5 ± 0.3	2.01 ± 0.10	100.5 ± 0.3	2.00 ± 0.12	99.8 ± 0.2
4.00	3.96 ± 0.19	99.0 ± 0.3	4.04 ± 0.24	101.0 ± 0.2	3.95 ± 0.31	98.7 ± 0.3
6.00	5.97 ± 0.11	99.5 ± 0.1	5.95 ± 0.25	99.2 ± 0.2	5.93 ± 0.15	98.8 ± 0.2
8.00	8.03 ± 0.16	100.4 ± 0.3	7.98 ± 0.19	99.8 ± 0.6	7.97 ± 0.16	99.6 ± 0.2
10.00	9.99 ± 0.37	99.9 ± 0.2	9.99 ± 0.15	99.9 ± 0.2	9.98 ± 0.24	99.8 ± 0.1
	Mean ± SD = 99.4 ± 0.8		Mean ± SD = 99.8 ± 0.9		Mean ± SD = 99.2 ± 0.9	

with the assay of RHT. It can be concluded that the developed HPLC assay method can be used to produce accurate data.

4. Conclusion

To assist the development of dual-ligand NP formulations for brain drug delivery, we have developed a simple, fast, accurate, and reliable HPLC method for RHT analysis during the formulation development and evaluation. This HPLC method has been validated for analysis of RHT loading in dual-ligand NPs preparations, in vitro drug release, and cellular transport studies. The chromatographic separation was achieved using a C_{18} column maintained at 50°C and an isocratic mobile phase consisting of TFA containing ACN and water with a flow rate of 1.5 mL/min. The method exhibited good linearity over the assayed concentration range and good intra- and interday precision. The developed HPLC method is accurate, selective, and rugged for RHT

analysis with good detection and quantification limits and is suitable for its intended use. This stability indicating analytical method can be adapted easily to analyse RHT in pharmaceutical formulations and biological matrices and for the future use in HPLC-MS analysis.

Conflicts of Interest

The authors declare that there are no conflicts of interest regarding the publication of this article.

Acknowledgments

The authors wish to thank Dr. Andrew Crowe, Mr. Giuseppe Luna, Mr. Michael Boddy, Ms. Yan-Jing Ng, and Ms. Yeakuty Marzan Jhanker for their technical assistance. Naz Hasan Huda and Bhawna Gauri wish to acknowledge the support of Curtin Strategic International Research Scholarship (CSIRS).

References

[1] J. Cherian and K. Gohil, "Cautious optimism for growth in Alzheimer's disease treatments," *Pharmacy and Therapeutics*, vol. 40, no. 4, pp. 288-289, 2015.

[2] N. H. Greig, T. Utsuki, D. K. Ingram et al., "Selective butyrylcholinesterase inhibition elevates brain acetylcholine, augments learning and lowers Alzheimer β-amyloid peptide in rodent," *Proceedings of the National Academy of Sciences of the United States of America*, vol. 102, no. 47, pp. 17213–17218, 2005.

[3] G. T. Grossberg, "Cholinesterase inhibitors for the treatment of Alzheimer's disease: getting on and staying on," *Current Therapeutic Research, Clinical and Experimental*, vol. 64, no. 4, pp. 216–235, 2003.

[4] M. Pohanka, "Inhibitors of acetylcholinesterase and butyrylcholinesterase meet immunity," *International Journal of Molecular Sciences*, vol. 15, no. 6, pp. 9809–9825, 2014.

[5] A. Kumar, A. Singh, and Ekavali, "A review on Alzheimer's disease pathophysiology and its management: an update," *Pharmacological Reports*, vol. 67, no. 2, pp. 195–203, 2015.

[6] F. Zemek, L. Drtinova, E. Nepovimova et al., "Outcomes of Alzheimer's disease therapy with acetylcholinesterase inhibitors and memantine," *Expert Opinion on Drug Safety*, vol. 13, no. 6, pp. 759–774, 2014.

[7] M. R. Farlow, G. T. Grossberg, C. H. Sadowsky, X. Meng, and D. M. Velting, "A 24-week, open-label extension study to investigate the long-term safety, tolerability, and efficacy of 13.3 mg/24 h rivastigmine patch in patients with severe Alzheimer disease," *Alzheimer Disease & Associated Disorders*, vol. 29, no. 2, pp. 110–116, 2015.

[8] J. S. Birks, L. Y. Chong, and J. Grimley Evans, "Rivastigmine for Alzheimer's disease," *Cochrane Database of Systematic Reviews*, no. 9, pp. 1–198, 2015.

[9] B. Wilson, M. K. Samanta, K. Santhi, K. P. S. Kumar, N. Paramakrishnan, and B. Suresh, "Poly(n-butyl cyanoacrylate) nanoparticles coated with polysorbate 80 for the targeted delivery of rivastigmine into the brain to treat Alzheimer's disease," *Brain Research*, vol. 1200, pp. 159–168, 2008.

[10] N. H. Huda, "Development of a novel nanoparticulate carrier system for enhancement of bioactive molecule delivery into the brain," Ph.D. thesis, School of Pharmacy, Curtin University, Perth, Australia, 2017.

[11] B. Gauri, "Development and evaluation of an intranasal nanoparticulate formulation for enhanced transport of rivastigmine into the brain," Ph.D. thesis, School of Pharmacy, Curtin University, Perth, Australia, 2017.

[12] M. Fazil, S. Md, S. Haque et al., "Development and evaluation of rivastigmine loaded chitosan nanoparticles for brain targeting," *European Journal of Pharmaceutical Sciences*, vol. 47, no. 1, pp. 6–15, 2012.

[13] K. Nagpal, S. K. Singh, and D. N. Mishra, "Optimization of brain targeted chitosan nanoparticles of rivastigmine for improved efficacy and safety," *International Journal of Biological Macromolecules*, vol. 59, pp. 72–83, 2013.

[14] B. Shah, D. Khunt, H. Bhatt, M. Misra, and H. Padh, "Application of quality by design approach for intranasal delivery of rivastigmine loaded solid lipid nanoparticles: effect on formulation and characterization parameters," *European Journal of Pharmaceutical Sciences*, vol. 78, pp. 54–66, 2015.

[15] Z.-Z. Yang, Y.-Q. Zhang, Z.-Z. Wang, K. Wu, J.-N. Lou, and X.-R. Qi, "Enhanced brain distribution and pharmacodynamics of rivastigmine by liposomes following intranasal administration," *International Journal of Pharmaceutics*, vol. 452, no. 1-2, pp. 344–354, 2013.

[16] R. Mullangi, A. Ranjithkumar, K. Arumugam et al., "High performance liquid chromatographic fluorescence detection method for the quantification of rivastigmine in rat plasma and brain: application to preclinical pharmacokinetic studies in rats," *Journal of Young Pharmacists*, vol. 3, no. 4, pp. 315–321, 2011.

[17] K. Arumugam, G. Subramanian, S. Mallayasamy, R. Averineni, M. Reddy, and N. Udupa, "A study of rivastigmine liposomes for delivery into the brain through intranasal route," *Acta Pharmaceutica*, vol. 58, no. 3, pp. 287–297, 2008.

[18] L. Lee, M. Hossain, Y. Wang, and G. Sedek, "Absorption of rivastigmine from different regions of the gastrointestinal tract in humans," *Journal of Clinical Pharmacology*, vol. 44, no. 6, pp. 599–604, 2004.

[19] M. Hossain, S. S. Jhee, T. Shiovitz et al., "Estimation of the absolute bioavailability of rivastigmine in patients with mild to moderate dementia of the Alzheimer's type," *Clinical Pharmacokinetics*, vol. 41, no. 3, pp. 225–234, 2002.

[20] Y. Sha, C. Deng, Z. Liu, T. Huang, B. Yang, and G. Duan, "Headspace solid-phase microextraction and capillary gas chromatographic-mass spectrometric determination of rivastigmine in canine plasma samples," *Journal of Chromatography B*, vol. 806, no. 2, pp. 271–276, 2004.

[21] K. Arumugam, M. R. Chamallamudi, R. R. Gilibili et al., "Development and validation of a HPLC method for quantification of rivastigmine in rat urine and identification of a novel metabolite in urine by LC-MS/MS," *Biomedical Chromatography*, vol. 25, no. 3, pp. 353–361, 2011.

[22] F. Pommier and R. Frigola, "Quantitative determination of rivastigmine and its major metabolite in human plasma by liquid chromatography with atmospheric pressure chemical ionization tandem mass spectrometry," *Journal of Chromatography B*, vol. 784, no. 2, pp. 301–313, 2003.

[23] J. Bhatt, G. Subbaiah, S. Kambli et al., "A rapid and sensitive liquid chromatography-tandem mass spectrometry (LC-MS/MS) method for the estimation of rivastigmine in human plasma," *Journal of Chromatography B*, vol. 852, no. 1-2, pp. 115–121, 2007.

[24] S. V. Frankfort, M. Ouwehand, M. J. van Maanen, H. Rosing, L. R. Tulner, and J. H. Beijnen, "A simple and sensitive assay for the quantitative analysis of rivastigmine and its metabolite NAP 226-90 in human EDTA plasma using coupled liquid chromatography and tandem mass spectrometry," *Rapid Communications in Mass Spectrometry*, vol. 20, no. 22, pp. 3330–3336, 2006.

[25] A. Enz, A. Chappuis, and A. Dattler, "A simple, rapid and sensitive method for simultaneous determination of rivastigmine and its major metabolite NAP 226-90 in rat brain and plasma by reversed-phase liquid chromatography coupled to electrospray ionization mass spectrometry," *Biomedical Chromatography*, vol. 18, no. 3, pp. 160–166, 2004.

[26] H. Amini and A. Ahmadiani, "High-performance liquid chromatographic determination of rivastigmine in human plasma for application in pharmacokinetic studies," *Iranian Journal of Pharmaceutical Research*, vol. 9, no. 2, pp. 115–121, 2010.

[27] B. M. Rao, M. K. Srinivasu, K. Praveen Kumar et al., "A stability indicating LC method for rivastigmine hydrogen tartrate," *Journal of Pharmaceutical and Biomedical Analysis*, vol. 37, no. 1, pp. 57–63, 2005.

[28] A. Karthik, G. Subramanian, M. Surulivelrajan, A. Ranjithkumar, and S. Kamat, "Fluorimetric determination of rivastigmine in rat plasma by a reverse phase–high performance liquid chromatographic method. Application to a pharmacokinetic study," *Arzneimittelforschung*, vol. 58, no. 5, pp. 205–210, 2008.

[29] C. P. Li, L. Zheng, M. Mao, G. Rao, and W. Shan, "HPLC determination of rivastigmine hydrogen tartrate capsules," *Chinese Journal of Pharmaceutical Analysis*, vol. 31, no. 6, pp. 1123–1125, 2011.

[30] S. P. Sulochana, K. Sharma, R. Mullangi, and S. K. Sukumaran, "Review of the validated HPLC and LC-MS/MS methods for determination of drugs used in clinical practice for Alzheimer's disease," *Biomedical Chromatography*, vol. 28, no. 11, pp. 1431–1490, 2014.

[31] M. J. Ramalho and M. C. Pereira, "Preparation and characterization of polymeric nanoparticles: an interdisciplinary experiment," *Journal of Chemical Education*, vol. 93, no. 8, pp. 1446–1451, 2016.

[32] F. T. Meng, G. H. Ma, Y. D. Liu, W. Qiu, and Z. G. Su, "Microencapsulation of bovine hemoglobin with high bioactivity and high entrapment efficiency using a W/O/W double emulsion technique," *Colloids and Surfaces B: Biointerfaces*, vol. 33, no. 3-4, pp. 177–183, 2004.

[33] M. Ben David-Naim, E. Grad, G. Aizik et al., "Polymeric nanoparticles of siRNA prepared by a double-emulsion solvent-diffusion technique: physicochemical properties, toxicity, biodistribution and efficacy in a mammary carcinoma mice model," *Biomaterials*, vol. 145, pp. 154–167, 2017.

[34] T. S. Raju, L. Kalyanaraman, V. Venkat Reddy, and P. Yadagiri Swamy, "Development and validation of an UPLC method for the rapid separation of positional isomers and potential impurities of rivastigmine hydrogen tartrate in drug substance and drug product," *Journal of Liquid Chromatography & Related Technologies*, vol. 35, no. 7, pp. 896–911, 2012.

[35] P. P. Dandekar and V. B. Patravale, "Development and validation of a stability-indicating LC method for curcumin," *Chromatographia*, vol. 69, no. 9-10, pp. 871–877, 2009.

[36] ICH, "Stability testing of new drug substances and products (Q1AR)," 2000.

[37] ICH, "Q2B validation of analytical procedures: methodology," 1996.

[38] USP-NF, "Validation of compendial procedures," in *The United States Pharmacopeia 37/National Formulary 32, 2014*, The United States Pharmacopeia Convention, Inc., Rockville, MD, USA, 2006.

[39] D. J. Canney, "Cholinomimetic drugs," in *Remington: The Science and Practice of Pharmacy*, R. Hendrickson, Ed., p. 1397, Wolters Kluwer Health/Lippincott, Williams & Wilkins, Baltimore, MD, USA, 2005.

[40] M. E. Swartz and I. S. Krull, "Method validation basics," in *Handbook of Analytical Validation*, pp. 61–80, CRC Press, Boca Raton, FL, USA, 2012.

5

Rapid Screening and Quantitative Determination of Active Components in Qing-Hua-Yu-Re-Formula using UHPLC-Q-TOF/MS and HPLC-UV

Xin Shao,[1,2] **Jie Zhao,**[3] **Xu Wang**[1] **and Yi Tao**[4]

[1]*The First Clinical Medical College, Nanjing University of Chinese Medicine, Nanjing 210023, China*
[2]*Department of Endocrinology, Nanjing Hospital of Traditional Chinese Medicine, Nanjing 210001, China*
[3]*Pharmaceutical Animal Experimental Center, China Pharmaceutical University, Nanjing 210009, China*
[4]*School of Pharmacy, Nanjing University of Chinese Medicine, Nanjing 210023, China*

Correspondence should be addressed to Xu Wang; wangxunjzy@163.com

Academic Editor: Eduardo Dellacassa

Qing-Hua-Yu-Re-Formula (QHYRF), a new herbal preparation, has been extensively used for treating diabetic cardiomyopathy. However, the chemical constituents of QHYRF remain uninvestigated. In the present study, rapid ultrahigh-performance liquid chromatography coupled with quadrupole-time-of-flight mass spectrometry (UHPLC-Q-TOF/MS) was used to qualitatively analyze the components of QHYRF. Qualitative detection was performed on a Kromasil C_{18} column through the gradient elution mode, using acetonitrile-water containing 0.1% formic acid. Twenty-seven compounds were identified or tentatively characterized, including 12 phenolic acids, nine monoterpene glycosides, two flavonoids, three iridoids, and one unknown compound. Among these, six compounds were confirmed by comparing with standards. A high-performance liquid chromatography (HPLC) method was developed to simultaneously determine the following six active components in QHYRF: danshensu, paeoniflorin, acteoside, lithospermic acid, salvianolic acid B, and salvianolic acid C. These HPLC chromatograms were monitored at 254, 280, and 320 nm. The method was well validated with respect to specificity, linearity, limit of detection, limit of quantification, precision, stability, and recovery. The HPLC-UV method was successfully applied to 10 batches of QHYRF.

1. Introduction

An increasing number of people have suffered from diabetes in recent years, and cardiovascular complications secondary to diabetes have become the main cause of death in diabetic patients. The incidence of cardiovascular disease in patients with diabetes is 2-3 times higher than that of nondiabetic patients [1]. Diabetic cardiomyopathy (DCM), a specific cardiomyopathy and one of the major cardiac complications in diabetic patients, was found in diabetic patients without significant coronary artery atherosclerosis by Rubler et al. in 1972 [2]. Few clinical symptoms were observed in early DCM. However, with the further development of the disease, patients have become more susceptible to heart failure due to myocardial microvascular and metabolic disorders, which lead to changes in myocardial cell dysfunction and structure.

Moreover, DCM is closely related to the high incidence of cardiovascular disease and high mortality in patients with diabetes. It was reported that the prevalence of DCM in patients with type 2 diabetes is approximately 12% [3].

At present, controlling blood sugar and improving heart failure are the main ways to overcome DCM. However, treatment with Western medicine remains unsatisfactory. Therefore, the diagnosis and therapy for DCM have presently become pressing problems. Traditional Chinese herbal formulation (TCMF) has been widely used in clinic due to its well-proven efficacy and few side effects. The Qing-Hua-Yu-Re-Formula (QHYRF) is a new herbal preparation for treating DCM, which was developed by Professor Wang Xu, according to the clinic experience of Chinese Medicine Master Professor Zhou Zhongying. The recipe of QHYRF comprises six herbal medicines: radix rehmanniae recen,

Salvia miltiorrhiza Bge, cortex moutan, *Rhizoma coptidis*, radix paeoniae rubra, and *Euonymus alatus*. Emerging evidences have indicated that the six major active components (danshensu, paeoniflorin, acteoside, lithospermic acid, salvianolic acid B, and salvianolic acid C) in QHYRF protect cardiomyocytes by antioxidation, anti-inflammation, anti-apoptosis, and decreasing calcium overload [4–7]. Moreover, clinical studies have shown that QHYRF could effectively improve symptoms in patients with DCM, including blood sugar, blood rheology, and left ventricular structure and function. According to the known effective components, we speculate that the effect of myocardial protection of QHYRF could be achieved through the following methods: (1) increase myocardial glucose transporter gene expression to improve glucose and lipid metabolism; (2) decrease the level of inflammatory cytokine hypersensitive C-reactive protein and tumor necrosis factor-alpha, and increase serum adiponectin levels to reduce inflammation; (3) reduce glucose toxicity; (4) increase insulin sensitivity and improve insulin resistance; (5) reduce the apoptosis of cardiac muscle cells and decrease the PAI-1 level of the fibrinolysis system; (6) improve blood stasis [8, 9].

It has been widely accepted that the chemical composition of traditional Chinese medicine (TCM) is complex, and its improper use may cause toxic effects [10]. Therefore, the quality control of TCM is extremely important. However, few researches have been carried on the chemical composition of QHYRF. Furthermore, the specific content of its main ingredients remains unknown. It should be noted that the constituents and contents of the main active components of QHYRF may be influenced by harvest time, plant origin, and manufacturing procedures [10, 11]. Thus, there is an urgent need to develop an effective method for QHYRF quality control, in order to guarantee its pharmacological efficacy.

In the present study, both qualitative and quantitative approaches were developed for the comprehensive quality control of QHYRF. Twenty-seven compounds were identified or tentatively characterized by ultrahigh-performance liquid chromatography coupled with quadrupole-time-of-flight mass spectrometry (UHPLC-Q-TOF/MS), including 12 phenolic acids, nine monoterpene glycosides, two flavonoids, three iridoids, and one unknown compound. Moreover, a simple, reliable, and sensitive analytical method, the high-performance liquid chromatography with ultraviolet detection (HPLC-UV) method, was used to determine the quantity of the six major active components (danshensu, paeoniflorin, acteoside, lithospermic acid, salvianolic acid B, and salvianolic acid C) of QHYRF. The potential application of the present study not only supports the quality control of QHYRF but also provides a theoretical research basis for the further research of QHYRF in clinic.

2. Experimental

2.1. Chemicals and Materials. Six crude herbs (cortex mouta, *Rhizoma coptidis*, *Euonymus alatus*, *Rehmannia glutinosa*, *Salvia miltiorrhiza*, and *Paeonia lactiflora*) were obtained from the pharmacy in Nanjing Hospital of Traditional Chinese Medicine, according to the Chinese Pharmacopoeia (2010 edition). The standard compounds (purity > 98%; danshensu, paeoniflorin, acteoside, lithospermic acid, salvianolic acid B, and salvianolic acid C) were purchased from Sichuan Victor Biotechnology Co., Ltd. HPLC-grade acetonitrile was obtained from Merck Co. All solutions and dilutions were prepared with ultrapure water obtained from a Milli-Q water purification system.

The extraction procedures for QHYRF total fraction were as follows: first, 15 g of radix rehmanniae recen, 15 g of *Salvia miltiorrhiza* Bge, 10 g of paeonol, 5 g of *Rhizoma coptidis*, 10 g of radix paeoniae rubra, and 10 g of *Euonymus alatus* were extracted twice with water (×8) to yield a crude extract. Second, the crude extract was concentrated and spray-dried to obtain the final product.

2.2. Sample Preparation. The samples were prepared as follows: 20 mg of QHYRF total fraction was dissolved in 1 mL of methanol and ultrasonicated for 15 minutes. Then, the sample solution was centrifuged at 12,000 rpm for 10 minutes, transferred to vials, and subjected to UHPLC-Q-TOF/MS and HPLC-UV analysis.

2.3. UHPLC-Q-TOF/MS Analysis. The analytical included a Shimadzu UHPLC system and a Q-TOF 5600-plus mass spectrometer equipped with Turbo V source and a TurboIonSpray interface (AB Sciex) and an Agilent 1100 LC-UV system with ChemStation (Agilent). The chromatographic conditions were as follows [12]: Kromasil C_{18} column (4.6×150 mm, 5μm) at 35°C, sample injection volume, 10μL; mobile phases, water containing 0.1% formic acid (solvent A) and acetonitrile (solvent B); gradient program was employed according to the following programmer: 0–5 minutes, 5–15% B; 5–10 minutes, linear increase to 20% B; 10–20 minutes, linear increase to 25% B; 20–30 minutes, linear increase to 100% B; 30–35 minutes, hold on 100% B.

The operating parameters for Q-TOF/MS were set as follows [12]: ion spray voltage, −4.5 kV; collision energy, −35 eV; nebulizer gas (gas 1), 55 psi; declustering potential, −60 V; heater gas (gas 2), 55 psi; turbo spray temperature, 550°C; curtain gas, 35 psi; resolution, 20,000. Full-scan data acquisition was performed from m/z 100 to 1500 in the negative mode.

2.4. HPLC-UV Analysis. The analysis was performed on a Shimadzu HPLC system. Chromatographic separation was achieved on an XBridge C_{18} column (4.6×150 mm, 5μm). The temperature of the column oven was set at 30°C, and the flow rate was 1.0 mL/min. The sample injection volume was 10μL. The mobile phases were a mixture of water with 0.1% formic acid (solvent A) and acetonitrile (solvent B). A gradient program was employed according to the following profile: 0–10 minutes, 10–15% B; 10–25 minutes, linear increase to 30% B; 25–60 minutes, linear increase to 100% B. The UV wavelength was set at 254 nm (paeoniflorin, lithospermic acid, and salvianolic acid B), 280 nm (danshensu and salvianolic acid C), and 320 nm (acteoside).

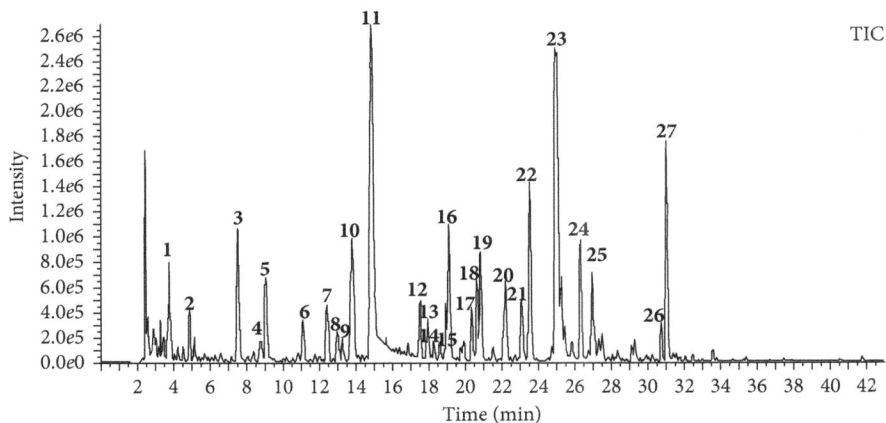

FIGURE 1: TIC chromatogram of QHYRF using UHPLC-Q-TOF/MS.

2.5. Validation of the Established HPLC-UV Approach. The stock solution that contained the six reference compounds (0.5 mg/mL danshensu, 25 mg/mL paeoniflorin, 1 mg/mL acteoside, 1 mg/mL lithospermic acid, 10 mg/mL salvianolic acid B, and 1 mg/mL salvianolic acid C) was prepared in methanol and stored at 4°C. In order to establish the calibration curves, the stock solution was diluted to appropriate concentrations. Solutions that contained different concentrations of the six standard compounds were injected in triplicate. The calibration curves were peak areas versus the concentration for each compound [12].

The limit of detection (LOD) was determined as a signal-to-noise ratio equal to 3, and the limit of quantification (LOQ) was based on 10 times of the signal-to-noise ratio value.

The precision was evaluated by intraday and interday variability. Intraday reproducibility was carried out by analyzing the individual sample solution six times within one day. For interday variability, six samples were determined six times in three consecutive days.

A stability study was performed with a sample solution checked at 0, 4, 8, 12, and 24 hours. Variations were expressed by relative standard deviations (RSDs).

Recovery studies were carried out by spiking known amounts of the reference compounds at low, medium, and high concentration in the samples. Then, the spiked samples were thoroughly mixed, extracted, and analyzed in accordance with the methods mentioned above. Recovery (%) = (amount found – original amount)/amount spiked × 100% [12].

2.6. Application to Different Batches of QHYRF. The established HPLC-UV method was subsequently employed for the simultaneous quantification of danshensu, paeoniflorin, acteoside, lithospermic acid, salvianolic acid B, and salvianolic acid C in QHYRF from 10 different batches.

3. Results and Discussion

3.1. UHPLC-Q-TOF-MS Analysis. The UHPLC-Q-TOF/MS approach was employed for the separation and identification of compounds in QHYRF. The total ion current chromatogram is shown in Figure 1. A total of 27 compounds were identified, including 12 phenolic acids, nine monoterpene glycosides, two flavonoids, three iridoids, and one unknown compound. The detailed information is displayed in Table 1. The retention time and fragmentation information of compounds **2**, **11**, **16**, **22**, **23**, and **26** were compared with that of the standard compounds.

Phenolic acids were the major components of QHYRF. A total of 12 phenolic acids were identified and tentatively characterized based on their mass data and reports in literature. These always exhibited a unique fragmentation behavior. The characteristic mass behaviors of these phenolic compounds were the losses of danshensu (198 Da) and caffeic acid (180 Da) molecules. Peak **2** revealed the $[M-H]^-$ ion at m/z 197.0467 and gave the element composition of $C_9H_{10}O_5$. In the MS^2 spectra, the ions at m/z 135 and 179 were observed, which was consistent with that of danshensu. In comparison with the reference, compound **2** was unambiguously identified as danshensu (Figure 2(a)). Peaks **20**, **23**, and **24** displayed the same $[M-H]^-$ ion at m/z 717.1472 ($C_{36}H_{30}O_{16}$). In comparison with standard compound, compound **23** was unambiguously assigned as salvianolic acid B (Figure 2(e)). The product ion at m/z 339 was formed due to the disconnection of another caffeic acid (180 Da) molecule from the deprotonated daughter ion $[M-H-198]^-$ at m/z 519. Further loss of H_2O led to the yield of product ion at m/z 321. Peaks **20** and **24** shared the same MS^2 ions with that of compound **23**. Compared with literatures [13, 14], compounds **20** and **24** was plausibly deduced as salvianolic acid E and isosalvianolic acid B. Peaks **14**, **18**, and **22** were a group of isomeric compounds, and these displayed $[M-H]^-$ ions at m/z 537.1041 ($C_{27}H_{22}O_{12}$). These had similar fragmentation pathways to those of salvianolic acid A and generated similar fragment ions at m/z 313, 295, and 185. Peak **22** was unequivocally identified as lithospermic acid by comparison with the standard (Figure 2(d)). According to the literature [14], peaks **18** and **14** were tentatively identified as salvianolic acid H or salvianolic acid A. Peak **19** revealed the $[M-H]^-$ ion at m/z 417.0826 and gave the element composition of $C_{20}H_{18}O_{10}$. This generated a series of fragment ions at m/z 373, 197, 179, and 175. Compared with the

TABLE 1: MS^1 and MS^2 information of the QHYRF.

Number	t_R (min)	MS^1	MS^2	Formula	Error	Identification	Source
1	3.7	719.2063	**359.0976**, 197.0432, 179.0326	$C_{48}H_{32}O_7$	−1.7	—	—
2*	4.9	197.0467	179.0340, **135.0448**, 123.0452, 72.9960	$C_9H_{10}O_5$	5.9	Danshensu	Salvia miltiorrhiza
3	7.5	495.1498	465.1405, 345.1118, 137.0228	$C_{23}H_{28}O_{12}$	−2.0	Oxypaeoniflorin	Paeonia lactiflora
4	8.7	525.1612	495.1546, **167.0338**	$C_{24}H_{30}O_{13}$	−0.3	Mudanpioside E	Salvia miltiorrhiza
5	9.0	367.1033	193.0490, 134.0362	$C_{17}H_{20}O_9$	−0.4	5-O-Feruloylquinic acid	Salvia miltiorrhiza
6	11.1	505.1562	459.1522, 293.0870, **165.0543**, 150.0308	$C_{20}H_{28}O_{12}$	3.0	Paeonolide	Paeonia lactiflora
7	12.4	505.1563	459.1526, 293.0873, 233.0656, **165.0547**, 150.0310	$C_{20}H_{28}O_{12}$	0.0	Apiopaeonoside	Salvia miltiorrhiza
8	13.0	647.1615	629.1577, 509.1336, 491.1226, 399.0946, 313.0565, **271.0453**	$C_{30}H_{32}O_{16}$	−0.4	6-O-Galloyloxypaeoniflorin	Paeonia lactiflora
9	13.3	785.252	**623.2223**, 477.1619, 179.0345, 161.0227	$C_{35}H_{46}O_{20}$	1.3	Echinacoside	Rehmannia glutinosa
10	13.8	367.1036	193.0487, 191.0543, **173.0439**, 134.0362	$C_{17}H_{20}O_9$	0.4	3-O-Feruloylquinic acid	Salvia miltiorrhiza
11*	14.8	525.1608	479.1554, 449.1453, 327.1069, 165.0539, **121.0288**	$C_{24}H_{30}O_{13}$	−1.1	Paeoniflorin	Paeonia lactiflora
12	17.5	611.1621	**445.0995**, 343.0670, 169.0124	$C_{27}H_{32}O_{16}$	0.6	Hydroxysafflor yellow A	—
13	17.9	611.1621	**445.0998**, 169.0131	$C_{27}H_{32}O_{16}$	0.6	Hydroxysafflor yellow A isomer	—
14	18.3	537.1041	493.1154, 313.0515, **295.0600**, 185.0230	$C_{27}H_{22}O_{12}$	0.5	Salvianolic acid A	Salvia miltiorrhiza
15	18.6	525.16	479.1589, 449.1472, 317.1086, 283.0818, **121.0293**	$C_{24}H_{30}O_{13}$	−0.1	Paeoniflorin isomer	Paeonia lactiflora
16*	18.9	623.1979	461.1680, **161.0237**	$C_{29}H_{36}O_{15}$	−0.4	Acteoside	Rehmannia glutinosa
17	20.3	623.1983	461.1683, **161.0234**	$C_{29}H_{36}O_{15}$	0.2	Forsythoside A	Rehmannia glutinosa
18	20.6	537.1034	493.1155, 313.0502, **295.0594**, 185.0226	$C_{27}H_{22}O_{12}$	−0.8	Salvianolic acid H	Salvia miltiorrhiza
19	20.8	417.0826	373.0918, 197.0432, 179.0328, **175.0379**, 152.0279	$C_{20}H_{18}O_{10}$	−0.3	Salvianolic acid D	Salvia miltiorrhiza
20	22.2	717.1472	519.0950, 339.0500, **321.0395**, 295.0597	$C_{36}H_{30}O_{16}$	1.5	Salvianolic acid E	Salvia miltiorrhiza
21	23.1	359.0775	197.0442, 179.0334, **161.0233**, 133.0287	$C_{18}H_{16}O_8$	0.7	Rosmarinic acid	Salvia miltiorrhiza
22*	23.5	537.1037	493.1172, 313.0509, **295.0608**, 185.0236	$C_{27}H_{22}O_{12}$	−0.3	Lithospermic acid	Salvia miltiorrhiza
23*	24.9	717.1456	519.0939, 339.0489, **321.0386**, 295.0590, 279.0277	$C_{36}H_{30}O_{16}$	−0.7	Salvianolic acid B	Salvia miltiorrhiza
24	26.3	717.1468	519.0936, 339.0487, **321.0385**, 295.0588, 279.0275	$C_{36}H_{30}O_{16}$	1.0	Isosalvianolic acid B	Salvia miltiorrhiza
25	27.0	599.1769	569.1709, 477.1430, **137.0234**	$C_{30}H_{32}O_{13}$	−0.2	Benzoyloxypaeoniflorin	Paeonia lactiflora
26*	30.7	491.0983	311.0556, **293.0440**, 265.0492, 197.0440, 135.0442	$C_{26}H_{20}O_{10}$	−0.1	Salvianolic acid C	Salvia miltiorrhiza
27	31.0	629.1872	583.1838, 553.1730, 431.1342, 165.0541, **121.0287**	$C_{30}H_{32}O_{12}$	−0.6	Benzoylpaeoniflorin	Paeonia lactiflora

*Indicated compared with standards. The bold numbers represent the most abundant ions.

literature [15], compound **19** was tentatively deduced as salvianolic acid D. Peak **21** displayed the $[M\text{-}H]^-$ ion at m/z 359.0775 and produced fragment ions at m/z 197, 179, and 161. In comparison with the reference [15], compound **21** was plausibly assigned as rosmarinic acid. Peak **26** gave an

$[M\text{-}H]^-$ ion at m/z 491.0983 with the molecular composition $C_{26}H_{20}O_{10}$. Compared with the standard, peak **26** was unequivocally identified as salvianolic acid C (Figure 2(f)). Peaks **5** and **10** displayed the same $[M\text{-}H]^-$ ion at m/z 367.1033 with the molecular composition $C_{17}H_{20}O_9$. These

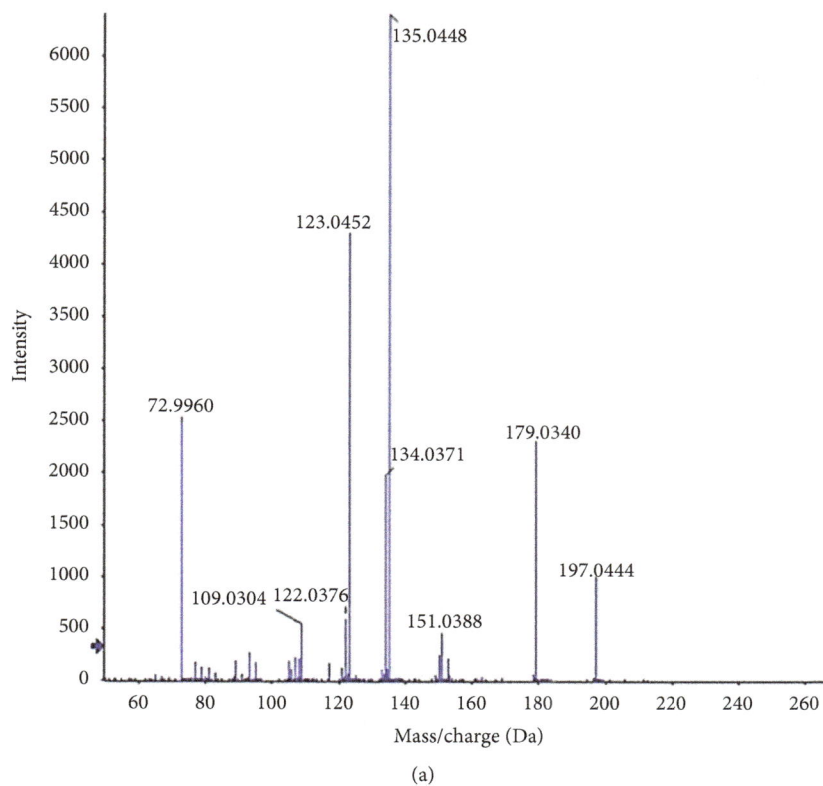

(a)

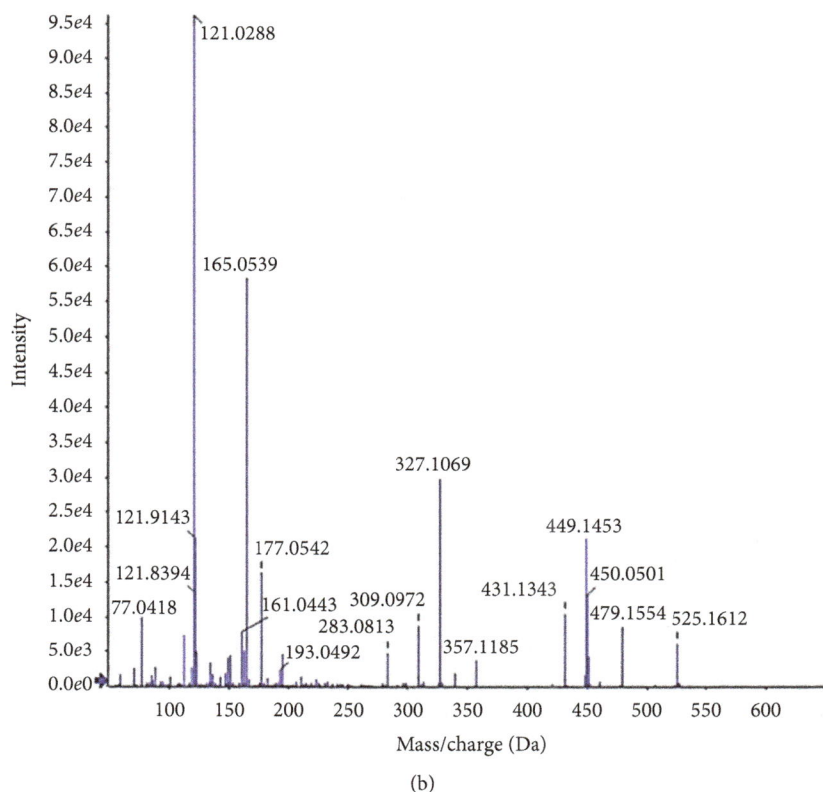

(b)

FIGURE 2: Continued.

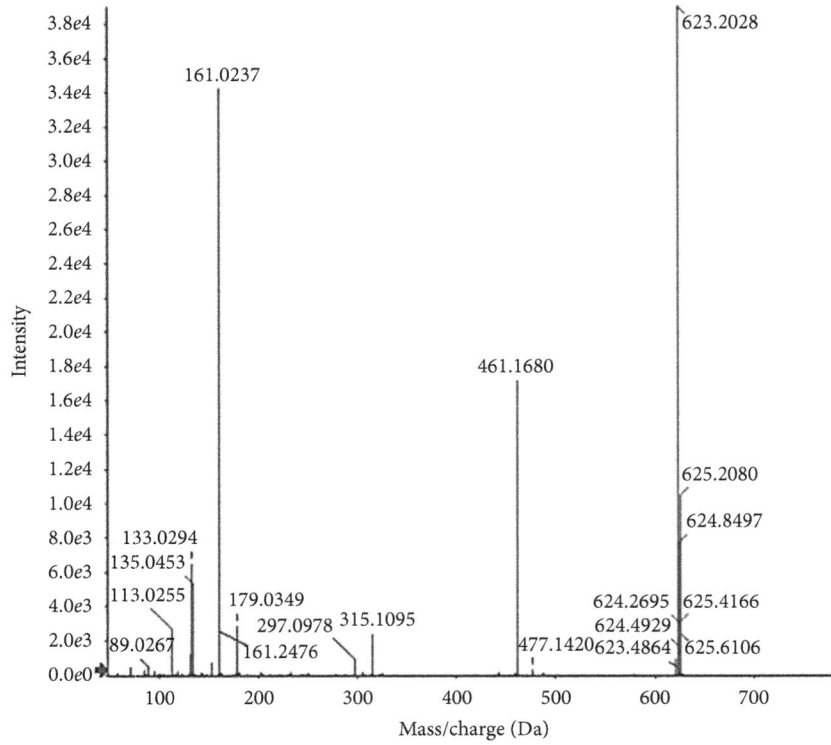

(c)

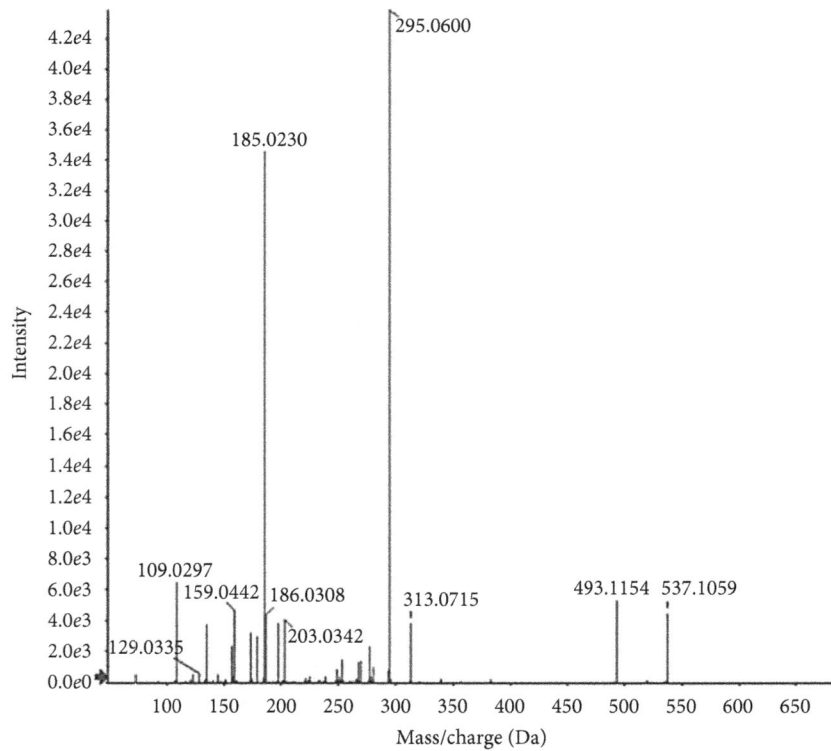

(d)

Figure 2: Continued.

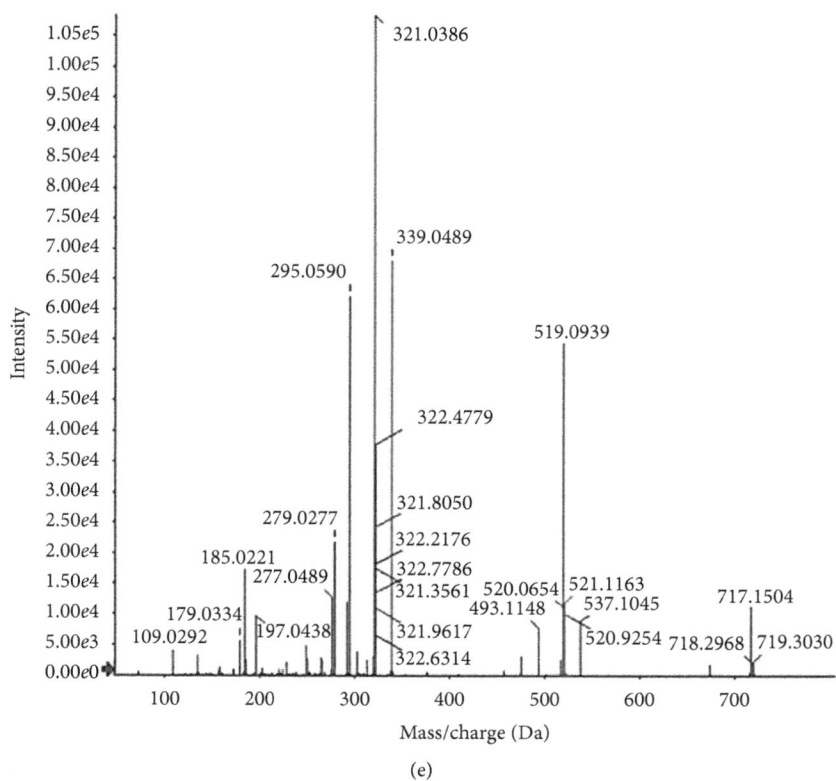

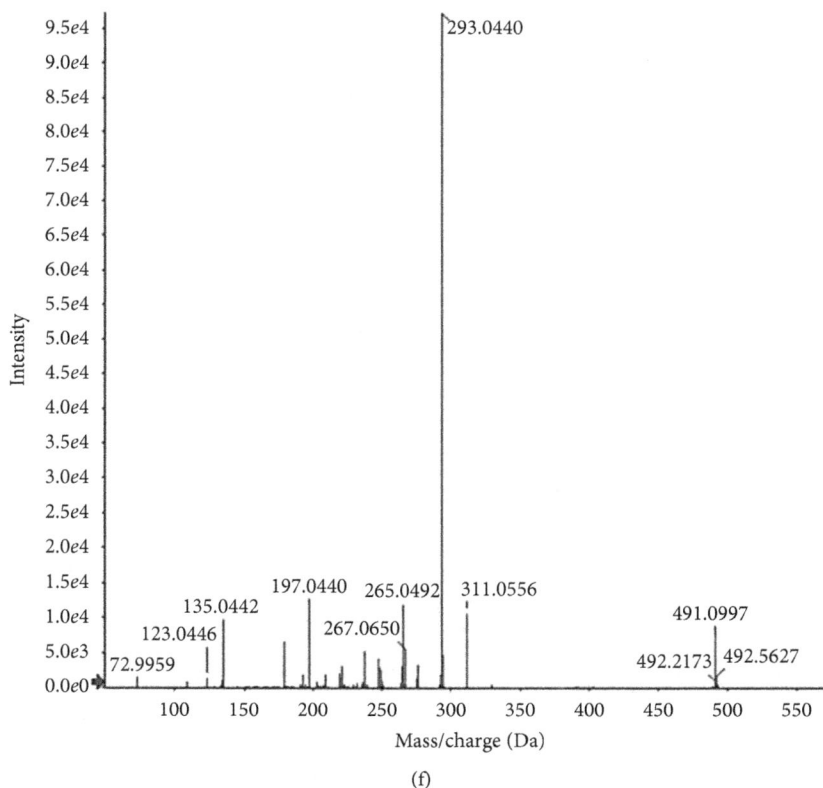

FIGURE 2: MS^2 spectra of peak 2 (a), peak 11 (b), peak 16 (c), peak 22 (d), peak 23 (e), and peak 26 (f) in the QHYRF.

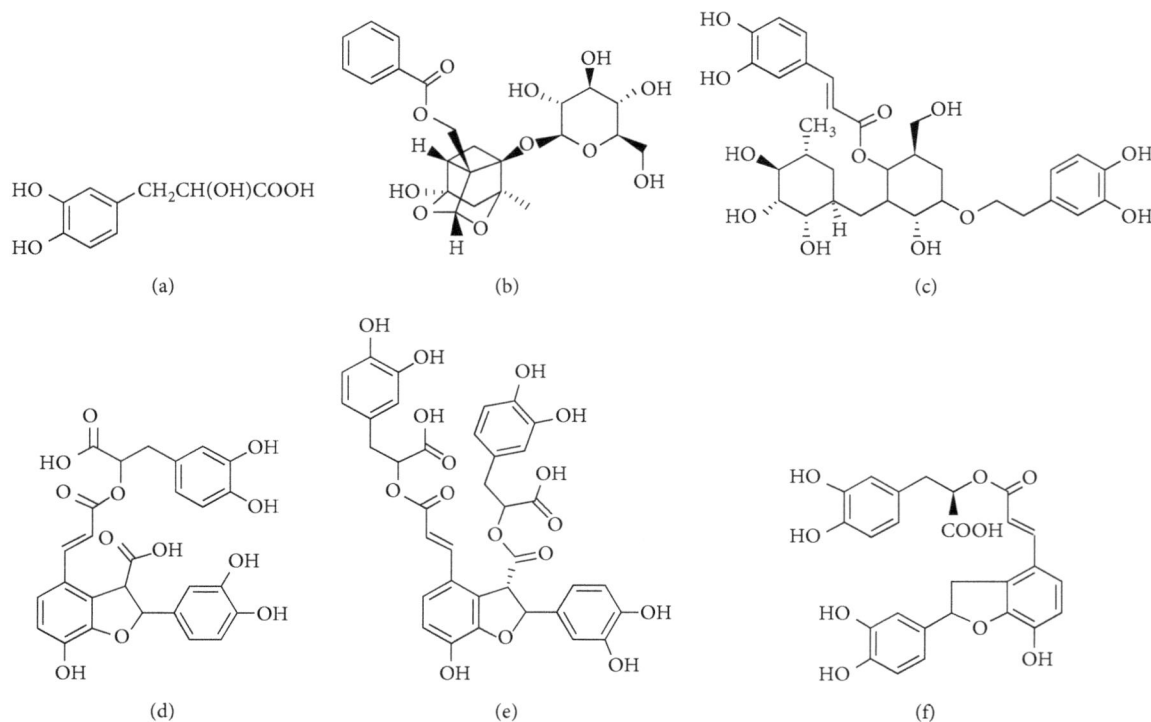

FIGURE 3: Chemical structure of danshensu (a), paeoniflorin (b), acteoside (c), lithospermic acid (d), salvianolic acid B (e), and salvianolic acid C (f).

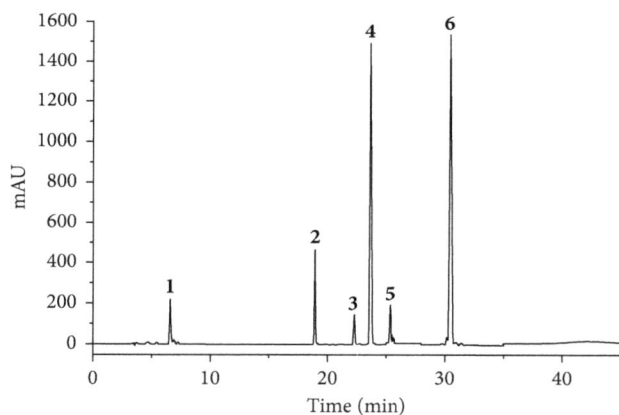

FIGURE 4: HPLC chromatogram of QHYRF using UV detector. Danshensu (1), paeoniflorin (2), acteoside (3), lithospermic acid (4), salvianolic acid B (5), and salvianolic acid C (6).

two compounds yielded similar product ions at m/z 193 and 134. Compared with literatures [16, 17], compounds 5 and 10 were tentatively assigned as 5-O-feruloylquinic acid and 3-O-feruloylquinic acid, respectively.

Monoterpene glycosides exhibited quasi-molecular ions $[M-H]^-$ and $[M-H+HCOO]^-$ in the negative ion mode. The chemical structure of aglycones is a "cage-like" pinane skeleton, which is very unusual among natural products. The monoterpene glycosides in QHYRF were usually esterified with an aromatic acid such as benzoic acid, vanillic acid, methoxybenzoic acid, and gallic acid. A total of nine

monoterpene glycosides were identified. Peaks 4, 11, and 15 displayed the same $[M-H]^-$ ion at m/z 525 with the molecular composition $C_{24}H_{30}O_{13}$. Peaks 11 and 15 produced the same product ions at m/z 479, 449, and 121. In comparison with the standard, peak 11 was unambiguously identified as paeoniflorin (Figure 2(b)). Peak 15 was plausibly assigned as the paeoniflorin isomer. Peak 4 yielded the product ions at m/z 495 and 167. In comparison with the literature [18], peak 4 was deduced as mudanpioside E. Peak 3 gave the $[M-H]^-$ ions at m/z 495.1498 ($C_{23}H_{28}O_{12}$) and was 16 Da (O) more than that of paeoniflorin. It produced fragment ions at m/z 465.1437 ($[M-H-HCOH]^-$), indicating that peak 3 may be oxypaeoniflorin based on a previously reported literature data [13]. Peaks 6 and 7 displayed the same $[M-H]^-$ ions at m/z 505.1562 ($C_{20}H_{28}O_{12}$) and generated similar product ions at m/z 459, 293, and 150. Compared with literatures [19], peaks 6 and 7 were tentatively deduced as paeonolide and apiopaeonoside, respectively. Peak 8 revealed the $[M-H]^-$ ion at m/z 647.1615, which was 152 Da (galloyl group, $C_7H_4O_4$) more than that of peak 3. This produced fragment ions at m/z 629, 509, and 491 by a series of losses of one H_2O molecule, one benzoic acid molecule, and one H_2O molecule, respectively. Compared with the literature [20], peak 8 was tentatively characterized as 6-O-galloyloxypaeoniflorin. Peak 25 gave the $[M-H]^-$ ions at m/z 599.1769 ($C_{30}H_{32}O_{13}$) and was 104 Da more than that of peak 3. This produced fragment ions at m/z 569 and 477, indicating that peak 25 may be benzoyloxypaeoniflorin, based on a previously reported literature data [19]. Peak 27 revealed the $[M-H]^-$ ions at m/z 629.1872

TABLE 2: Linear regression data, LODs, and LOQs of six compounds.

Components	Regression equations	R^2	Linear range (μg/mL)	LODs (μg/mL)	LOQs (μg/mL)
Danshensu	$y = 5508.1x + 5.404$	0.9999	2.6–500	0.7	2.6
Paeoniflorin	$y = 419.45x - 6.7219$	0.9994	20.8–25,000	5.0	20.8
Acteoside	$y = 3283.3x - 11.072$	0.9997	5.2–1000	1.0	5.2
Lithospermic acid	$y = 12900x - 123.54$	0.9991	10.4–1000	0.6	10.4
Salvianolic acid B	$y = 1616.6x - 9.4854$	0.9993	5.2–10,000	2.0	5.2
Salvianolic acid C	$y = 19071x + 88.12$	1.0000	10.4–1000	0.8	10.4

$y = Ax + B$; y is the peak area; x is the concentration of the analytes; R^2 is the correlation coefficient of the equation.

TABLE 3: Precision and stability of six compounds in QHYRF ($n = 6$).

Components	Precision		Stability RSD (%)
	Intraday RSD (%)	Interday RSD (%)	
Danshensu	3.8	2.2	4.7
Paeoniflorin	1.6	1.7	4.2
Acteoside	4.3	1.2	2.8
Lithospermic acid	4.1	1.6	2.8
Salvianolic acid B	4.8	2.1	2.4
Salvianolic acid C	4.7	1.2	3.9

and was 104 Da more than that of paeoniflorin. Compared with the literature [16], peak **27** was tentatively characterized as benzoylpaeoniflorin.

Iridoids were also observed in the formula. Peaks **16** and **17** revealed the same [M-H]$^-$ at m/z 623 and produced the same fragments at m/z 461 and 161. Compared with the standard, compound **16** was unambiguously identified as acteoside (Figure 2(c)). The retention time of peak **17** was a little later than that of peak **16**. Compared with the literature, compound **17** was tentatively deduced as forsythoside A. Peak **9** displayed the [M-H]$^-$ ion at m/z 785.2520 with the molecular composition $C_{35}H_{46}O_{20}$. This produced product ions at m/z 623 [M-Glc]$^-$, 477 [M-Glc-Ara]$^-$, and 179. In comparison with the literature, compound **9** was tentatively deduced as echinacoside. Peaks **12** and **13** revealed the same [M-H]$^-$ ions at m/z 611.1621 and yielded product ions at m/z 445 and 169. Compared with the reference [21], compounds **12** and **13** were tentatively characterized as hydroxysafflor yellow A and its isomer.

3.2. Development and Validation of the HPLC-UV Approach.
The HPLC separation conditions were optimized including the mobile phase system, column temperature, and UV detection wavelength. In order to obtain chromatograms with better resolution of the adjacent peaks and shorter time, methanol and acetonitrile were compared in the experiment. The result indicated that acetonitrile was better than methanol due to shorter analysis time and better peak shape. In addition, the different column temperatures were also optimized. Finally, 30°C was chosen as the best column temperature. Under present HPLC conditions, the samples of QHYRF and standard solutions were analyzed. The UV absorption and detection wavelength of each compound were confirmed as follows: paeoniflorin, lithospermic acid and salvianolic acid B (254 nm), danshensu and salvianolic acid C (280 nm), and acteoside (320 nm) (chemical structure is shown in Figure 3).

According to the representative chromatograms of the standard solution, sample solution, and negative control samples solution, it could be found that the determination of these six compounds does not interfere with each other (Figure 4).

The calibration curves of danshensu, paeoniflorin, acteoside, lithospermic acid, salvianolic acid B, and salvianolic acid C were linear within the range of 2.6–500, 20.8–25,000, 5.2–1000, 10.4–1000, 5.2–10,000, and 10.4–1000 μg/mL, respectively. The correlation coefficient (R^2) of all six standards revealed a good linearity of >0.991 in the aforementioned ranges. The LODs and LOQs of these six compounds were 0.6–5.0 and 2.6–20.8 μg/mL, respectively (Table 2).

Intraday and interday variations were chosen to evaluate the precision of the method. These results are shown in Table 2. It was revealed that the intraday and interday RSDs were within the range of 1.6%–4.8% and 1.2%–2.2% ($n = 6$). The stability of six major compounds ranged from 2.8% to 4.7%, which met the requirements of the analytical approach (Table 3).

The recovery data are shown in Table 4, which represents the accuracy of the method and was sufficient for the analysis. The measured data indicated that the recovery of these six components ranged from 99.0% to 104.7%.

3.3. Application.
The established HPLC-UV approach was employed for simultaneously quantifying the six active components in 10 batches of QHYRF. These results are presented in Table 5. Furthermore, these results revealed that there were small differences among the contents of the six major constituents in different batches of QHYRF.

Optimized chromatographic conditions for the good resolution of adjacent peaks were achieved after several trials with elution systems of acetonitrile-water, methanol-water, acetonitrile-acid, and methanol-acid in different proportions.

TABLE 4: Recovery of six compounds in QHYRF ($n = 3$).

Components	Contents (mg/mL)	Quantity added (mg/mL)	Theoretical amount (mg/mL)	Recorded amount (mg/mL)	Recovery (%)	RSD (%)
Danshensu	0.22	0.05	0.27	0.27	100.6	1.7
		0.11	0.33	0.33	100.0	3.3
		0.2	0.42	0.42	100.2	1.8
Paeoniflorin	12.32	2.45	14.77	14.78	100.4	0.4
		4.95	17.27	17.27	100.0	2
		7.85	20.17	20.17	100.0	5.4
Acteoside	0.07	0.15	0.22	0.22	103.1	3.7
		0.22	0.29	0.29	99.0	5.5
		0.42	0.49	0.49	100.6	5.3
Lithospermic acid	0.1	0.08	0.18	0.18	102.6	4.3
		0.15	0.25	0.25	103.3	5.4
		0.27	0.37	0.38	102.2	1.7
Salvianolic acid B	4.96	1	5.96	5.96	99.9	3.7
		2.23	7.19	7.19	100.0	3.5
		3.41	8.37	8.37	100.1	2.2
Salvianolic acid C	0.01	0.09	0.1	0.10	103.6	3
		0.21	0.22	0.22	99.0	2.3
		0.35	0.36	0.38	104.7	4.8

TABLE 5: Content of six components in 10 batches of the QHYRF.

Components	Content (mg/g)									
	1	2	3	4	5	6	7	8	9	10
Danshensu	9.8	10.0	9.8	10.0	10.0	9.8	10.1	10.2	10.0	9.7
Paeoniflorin	582.6	591.2	612.4	579.8	617.3	591.4	609.3	582.4	582.0	601.4
Acteoside	3.4	3.4	3.3	3.4	3.3	3.3	3.5	3.2	3.4	3.4
Lithospermic acid	4.5	4.7	4.6	4.4	4.5	4.6	4.6	4.5	4.6	4.6
Salvianolic acid B	265.9	266.4	260.5	260.7	263.4	262.6	272.8	263.5	264.8	270.7
Salvianolic acid C	0.5	0.5	0.5	0.5	0.5	0.5	0.5	0.5	0.5	0.5

It was found that the presence of 0.1% fomic acid resulted in the significant improvement on the retention behavior of the different components. The optimal mobile phase, which consists of acetonitrile-0.1% fomic acid, was finally employed. This led to its high resolution and symmetrical peak shape.

According to the theory of TCM, QHYRF was prepared by six herbal medicines to cure the disease of diabetic cardiomyopathy. The presence of the researched active compounds in QHYRF was proved, making it possible to suggest a therapeutic effect in clinical applications. However, it enhances the complexity of the constituents and preparation procedures, which make it difficult to ensure the batch-to-batch uniformity of QHYRF. Thus, the quantitative measurement of its bioactive components is extremely necessary during the preparation and application of this prescription.

In recent years, methods for multicomponent analysis have become a credible solution for the analysis of a complex system in TCM. At present, we developed a sensitive HPLC-UV method to simultaneously quantify six active components in 10 different batches of QHYRF. A slight variation was observed in the different batches of QHYRF during the determination of the six major constituents, which reveal that the present preparation of QHYRF has good reproducibility. Furthermore, good reproducibility ensures the safety and effects of QHYRF in clinical application.

4. Conclusion

In summary, a simple and sensitive UHPLC-Q-TOF/MS method was established to qualitatively analyze the chemical components of QHYRF. Twenty-seven compounds were identified or tentatively characterized based on retention time, and MS[1] and MS[2] data, or by comparing with standards and literatures. These compounds include 12 phenolic acids, nine monoterpene glycosides, two flavonoids, three iridoids, and one unknown compound. A HPLC-UV method for the simultaneous determination of six active components from QHYRF was developed and well validated, showing its excellent precision, stability, and recovery. In addition, these results indicate a slight variation in the contents of the six major constituents among the different batches of QHYRF. This is the first comprehensive study that qualitatively and quantitatively determined the chemical constituents of QHYRF using UHPLC-Q-TOF/MS and HPLC-UV. Thus, these findings could serve as a foundation for the quality control or further research of QHYRF.

Conflicts of Interest

The authors declare that they have no conflicts of interest regarding the publication of this study.

Authors' Contributions

Xin Shao and Jie Zhao contributed equally to this work.

Acknowledgments

This study was financially supported by the Natural Science Foundation of Jiangsu Province (no. BK20151572), the Natural Science Foundation for the Youth of Jiangsu Province (no. BK20140963), the Nutritional Science Foundation of By-Health Co. (no. TY0141103), and the Priority Academic Program Development of Jiangsu Higher Education Institution (PAPD).

References

[1] J. R. Sowers, M. Epstein, and E. D. Frohlich, "Diabetes, hypertension, and cardiovascular disease: an update," *Hypertension*, vol. 37, no. 4, pp. 1053–1059, 2001.

[2] S. Rubler, J. Dlugash, Y. Z. Yuceoglu, T. Kumral, A. W. Branwood, and A. Grishman, "New type of cardiomyopathy associated with diabetic glomerulosclerosis," *American Journal of Cardiology*, vol. 30, no. 6, pp. 595–602, 1972.

[3] A. G. Bertoni, W. G. Hundley, M. W. Massing, D. E. Bonds, G. L. Burke, and D. C. Goff Jr., "Heart failure prevalence, incidence, and mortality in the elderly with diabetes," *Diabetes Care*, vol. 27, no. 3, pp. 699–703, 2004.

[4] X. Zhou, S. W. Chan, H. L. Tseng et al., "Danshensu is the major maker for the antioxidant and vasorelaxation effects of Danshen (*Salvia miltiorrhiza*) water-extracts produced by different heat water-extractions," *Phytomedicine*, vol. 19, no. 14, pp. 1263–1269, 2012.

[5] Y. Sun, J. Zhang, R. Huo et al., "Paeoniflorin inhibits skin lesions in imiquimod-induced psoriasis-like mice by downregulating inflammation," *International Immunopharmacology*, vol. 24, no. 2, pp. 392–399, 2015.

[6] K. K. Au-Yeung, D. Y. Zhu, O. Karmin, and Y. L. Siow, "Inhibition of stress-activated protein kinase in the ischemic used heart: role of magnesium tanshinoate B in preventing apoptosis," *Biochemical Pharmacology*, vol. 62, no. 4, pp. 483–493, 2001.

[7] Y. Ren, S. Tao, S. Zheng et al., "Salvianolic acid B improves vascular endothelial function in diabetic rats with blood glucose fluctuations via supression of endothelial cell apoptosis," *European Journal of Pharmacology*, vol. 791, pp. 308–315, 2016.

[8] H. Bing, *Effects of Qing-Hua-Yu-Re Formula on Diabetic Cardiomyopathy in Rats and Possible Mechanisms*, Ph.D. thesis, Nanjing University of Chinese Medicine Department, Nanjing, China, 2010.

[9] W. Xu, B. Hong, Z. XuePing, S. Xin, and Z. Li, "Clinical research of Qing-Hua-Yu-Re formula in treating diabetic cardiomyopathy," *Journal of Nanjing University of Traditional Chinese Medicine*, vol. 26, pp. 412–414, 2010.

[10] K. An, G. Jin-Rui, Z. Zhen, and W. Xiao-Long, "Simultaneous quantification of ten active components in traditional Chinese formula sijunzi decoction using a UPLC-PDA method," *Journal of Analytical Methods in Chemistry*, vol. 2014, Article ID 570359, 8 pages, 2014.

[11] K. Zhong, "Current analysis of the factors affecting the efficacy of traditional Chinese medicine," *Journal of North Pharmacy*, vol. 8, pp. 67-68, 2011.

[12] Y. Tao, Y. Jiang, W. Li, and B. Cai, "Rapid characterization and determination of isoflavones and triterpenoid saponins in Fu-Zhu-Jiang-Tang tablets using UHPLC-Q-TOF/MS and HPLC-UV," *Analytical Methods*, vol. 8, no. 21, pp. 4211–4219, 2016.

[13] H. Chen, Q. Zhang, X. Wang, J. Yang, and Q. Wang, "Qualitative analysis and simultaneous quantification of phenolic compounds in the aerial parts of *Salvia miltiorrhiza* by HPLC-DAD and ESI/MS(n)," *Phytochemical Analysis*, vol. 22, no. 3, pp. 247–257, 2011.

[14] M. Liu, S. Zhao, Y. Wang et al., "Identification of multiple constituents in Chinese medicinal prescription Shensong Yangxin capsule by ultra-fast liquid chromatography combined with quadrupole time-of-flight mass spectrometry," *Journal of Chromatographic Science*, vol. 53, no. 2, pp. 240–252, 2015.

[15] X. Li, F. Du, W. Jia et al., "Simultaneous determination of eight Danshen polyphenols in rat plasma and its application to a comparative pharmacokinetic study of DanHong injection and Danshen injection," *Journal of Separation Science*, vol. 40, no. 7, pp. 1470–1481, 2017.

[16] D. Gao, B. Wang, Z. Huo et al., "Analysis of chemical constituents in an herbal formula Jitong Ning tablet," *Journal of Pharmaceutical and Biomedical Analysis*, vol. 140, pp. 301–312, 2017.

[17] Y. Zhao, M. X. Chen, K. T. Kongstad, A. K. Jäger, and D. Staerk, "Potential of polygonum cuspidatum root as an antidiabetic food: dual high-resolution α-glucosidase and PTP1B inhibition profiling combined with HPLC-HRMS and NMR for identification of antidiabetic constituents," *Journal of Agricultural and Food Chemistry*, vol. 65, no. 22, pp. 4421–4427, 2017.

[18] Y.-H. Shi, S. Zhu, Y.-W. Ge et al., "Characterization and quantification of monoterpenoids in different types of peony root and the related Paeonia species by liquid chromatography coupled with ion trap and time-of-flight mass spectrometry," *Journal of Pharmaceutical and Biomedical Analysis*, vol. 129, pp. 581–592, 2016.

[19] P. Li, W. Su, C. Xie, X. Zeng, W. Peng, and M. Liu, "Rapid identification and simultaneous quantification of multiple constituents in Nao-Shuan-Tong capsule by ultra-fast liquid chromatography/diode-array detector/quadrupole time-of-flight tandem mass spectrometry," *Journal of Chromatographic Science*, vol. 53, no. 6, pp. 886–897, 2015.

[20] Q. Wang, Z. Liang, Y. Peng et al., "Whole transverse section and specific-tissue analysis of secondary metabolites in seven different grades of root of *Paeonia lactiflora* using laser microdissection and liquid chromatography-quadrupole/time of flight-mass spectrometry," *Journal of Pharmaceutical and Biomedical Analysis*, vol. 103, pp. 7–16, 2015.

[21] L. Zuo, Z. Sun, Y. Hu et al., "Rapid determination of 30 bioactive constituents in XueBiJing injection using ultra high performance liquid chromatography-high resolution hybrid quadrupole-orbitrap mass spectrometry coupled with principal component analysis," *Journal of Pharmaceutical and Biomedical Analysis*, vol. 137, pp. 220–228, 2017.

6

Analysis of the High-Performance Liquid Chromatography Fingerprints and Quantitative Analysis of Multicomponents by Single Marker of Products of Fermented *Cordyceps sinensis*

Li-hua Chen ⓘ,[1] Yao Wu ⓘ,[1] Yong-mei Guan,[1] Chen Jin,[1] Wei-feng Zhu,[1] and Ming Yang[2]

[1]*Key Laboratory of Modern Preparation of TCM, Ministry of Education, Jiangxi University of Traditional Chinese Medicine, No. 18 Yun Wan Road, Nanchang 330004, China*
[2]*Jiangxi Sinopharm Co. Ltd., No. 888 National Medicine Road, Nanchang 330004, China*

Correspondence should be addressed to Li-hua Chen; chlly98@163.com

Academic Editor: Mohamed Abdel-Rehim

Fermented *Cordyceps sinensis*, the succedaneum of *Cordyceps sinensis* which is extracted and separated from *Cordyceps sinensis* by artificial fermentation, is commonly used in eastern Asia in clinical treatments due to its health benefit. In this paper, a new strategy for differentiating and comprehensively evaluating the quality of products of fermented *Cordyceps sinensis* has been established, based on high-performance liquid chromatography (HPLC) fingerprint analysis combined with similar analysis (SA), hierarchical cluster analysis (HCA), and the quantitative analysis of multicomponents by single marker (QAMS). Ten common peaks were collected and analysed using SA, HCA, and QAMS. These methods indicated that 30 fermented *Cordyceps sinensis* samples could be categorized into two groups by HCA. Five peaks were identified as uracil, uridine, adenine, guanosine, and adenosine, and according to the results from the diode array detector, which can be used to confirm peak purity, the purities of these compounds were greater than 990. Adenosine was chosen as the internal reference substance. The relative correction factors (RCF) between adenosine and the other four nucleosides were calculated and investigated using the QAMS method. Meanwhile, the accuracy of the QAMS method was confirmed by comparing the results of that method with those of an external standard method with cosines of the angles between the groups. No significant difference between the two methods was observed. In conclusion, the method established herein was efficient, successful in identifying the products of fermented *Cordyceps sinensis*, and scientifically valid to be applicable in the systematic quality control of fermented *Cordyceps sinensis* products.

1. Introduction

In recent years, fermented *Cordyceps sinensis*, as a substitute for natural *Cordyceps sinensis*, has been attracting increasing attention from across the globe. Some of the major products prepared from fermented *Cordyceps sinensis* include Jinshuibao capsules, Jinshuibao tablets, and Bailing capsules. Modern pharmacological studies have demonstrated that this material has antioxidative, anticancer [1], antihyperglycaemic [2], and antifatigue [3] properties, and it has a similar chemical composition and pharmacological activities to *Cordyceps sinensis*. Moreover, modern pharmacological and clinical studies have also revealed that fermented *Cordyceps sinensis* can have a major impact on the treatment of chronic

hepatitis and kidney failure. These curative effects could be attributed to their active chemical components, including nucleosides, polysaccharides, sterols, amino acids, flavonoids, and metal complexes [4–8]. The type and quantity of the bioactive ingredients in fermented *Cordyceps sinensis* vary with the strain and the condition of the material [9, 10]. Among these components, nucleosides support the regulation and modulation of various physiological processes in the body [11], and they have been recognized as the main bioactive components [12] in fermented *Cordyceps sinensis*. Their pharmacological activities include antitumour, antileukaemic [13], antiviral [14], and anti-HIV [15] activities. The major nucleosides in fermented *Cordyceps sinensis* are guanosine, uridine, adenosine, uracil, and adenine.

Currently, the products of fermented *Cordyceps sinensis* have no uniform quality standards, which have resulted in the flooding of the market with products of uneven quality. Only Jinshuibao capsules, Jinshuibao tablets, and Bailing capsules, which are prepared by fermenting with *Paecilomyces hepiali* chen and *Hirsutella sinensis* Liu, GUO, Yu-et Zeng, have been recorded in Chinese Pharmacopoeia and have relatively comprehensive quality standards. The quality of the three fermented *Cordyceps sinensis* products is higher overall and is easier to control since they are produced exclusively by one company. The quality standards for fermented *Cordyceps sinensis* products vary from one manufacture to another. For instance, in the Pharmacopoeia of the People's Republic of China (Edition 2015) [16], the quality standards for Jinshuibao capsules included the identification of five required nucleosides, amino acids, and mannitol as well as required contents of adenosine and ergosterol. The standards for Bailing capsules include the identification of two required nucleosides, amino acids, and ergosterol as well as required contents of mannitol, adenosine, and total amino acid. Thus, the development of a method to control the quality of these products more systematically and effectively is an urgent issue. Since the overall chemical composition is contained within the chromatographic fingerprint, this type of fingerprinting may provide the foundation for such method. Meanwhile, chromatographic fingerprinting has been regarded as a feasible and rational approach to the quality evaluation of crude drug materials in recent years [17]. In addition, methods based on this technique have been accepted by the WHO, the FDA, and the State Food and Drug Administration (SFDA) of China [18]. However, chemical fingerprinting is a type of qualitative analysis, and thus, it can only reflect the general characteristics of the contents of herbs and serve as an indicator of their quality, consistency, and stability. Thus, the cataloguing and evaluation of chromatographic fingerprints are of great importance to evaluating and distinguishing medicinal materials and related preparations.

Multicomponent quantitative analysis was developed for applications in comprehensive quality control. Unfortunately, the use of multicomponent quantitative analysis is limited by the high price, limited availability, and poor stability of some of the necessary standards. Thus, a QAMS method was carried out. A predominant advantage of this strategy is that the content of each target component can be determined independently since their concentrations can be calculated based on standard materials. In recent years, the QAMS method has been widely used to evaluate a large number of Chinese herbal medicines [19–21] and was reported in the Chinese Pharmacopoeia to be applicable to the determination of the contents of TCM herbs, including *Coptis chinensis*, *Salvia miltiorrhiza*, and *Ganoderma lucidum* [16].

However, the current method was not sufficient to fully assess the quality of products of fermented *Cordyceps sinensis*. In this work, a partial method involving analysis of HPLC fingerprint chromatograms by SA and HCA combined with QAMS was established for the first time. Adenosine was chosen as the internal reference substance, and the relative correction factors (RCF) between the internal marker and test specimen were used to calculate the

other four components. These steps in combination with HPLC fingerprint analysis based on SA and HCA can be used to obtain more general information on the products of fermented *Cordyceps sinensis*. The rapid, practical, and effective combinatorial approach can be used to comprehensively assess the quality of products prepared from fermented *Cordyceps sinensis*.

TABLE 1: Gradient elution program.

T (min)	A (%)	B (%)
0	99	1
15	99	1
30	85	15

2. Materials and Methods

2.1. Reagents and Chemicals. Four reference substances (guanosine, adenosine, uracil, and adenine) were purchased from the National Institutes for Food and Drug Control (China), and uridine was purchased from the Nanchang Beta Biotechnology Co. Ltd. 10 Jinshuibao capsules (strain: *Paecilomyces hepiali* chen, batch number: 1,60,40,185, 1,50,50,347, 1,60,50,229, 1,60,20,075, 1,60,30,133, 1,60,90,394, 1,51,10,701, 1,60,80,393, 1,60,10,105, and 1,60,70,345) and 10 Jinshuibao tablets (strain: *Paecilomyces hepiali* chen, batch number: 1,60,703, 1,60,601, 1,60,401, 1,60,701, 1,60,501, 1,60,702, 1,61,202, 1,61,203, 1,61,204, and 1,60,502) were purchased from Jiangxi Jimingkexin Jinshuibao Co. Ltd., and 10 Bailing capsules (strain: *Hirsutella sinensis* Liu, GUO, Yu-et Zeng, batch number: 16,04,172, 1,6,04,101, 16,04,137, 16,05,171, 16,03,200, 16,03,102, 16,05,172, 16,06,124, 16,06,203, and 16,06,106) were purchased from Hangzhou Huadong Pharmaceutical Co. Ltd. The purity of each of the 5 reference substance was more than 98%. HPLC grade methanol was purchased from TEDIA reagent company (USA).

2.2. Instruments and Chromatographic Conditions. Chromatographic separations were performed on an Agilent 1260 HPLC coupled with a DAD, an Agilent 1100 HPLC coupled with a UV detector, and a Waters 2695 HPLC coupled with a PDA detector. The separations of the analytes were achieved on a reversed-phase C18 column (Agilent ZORBAX-RP, 250×4.6 mm, $5 \mu m$; Phenomenex OOG-4435-EO, 250×4.6 mm, $5 \mu m$). A sample volume of $10 \mu L$ was injected by an autosampler, and the column temperature was maintained at 30°C. The mobile phase was composed of ultrapure water (A) and methanol (B) with a gradient elution system as follows: 0–15 min, 1% B; and 15–30 min, 1%–15% B. As show in Table 1, the flow rate was 1 mL/min, and the wavelength of the DAD was set at 260 nm. Under the chromatographic conditions described above, all five components could be baseline separated within 30 min (Figure 1).

2.3. Preparation of Standard Solutions. Five stock solutions were prepared by dissolving the requisite components in ultrapure water. A working solution that contains the five standard solutions was prepared prior to analysis.

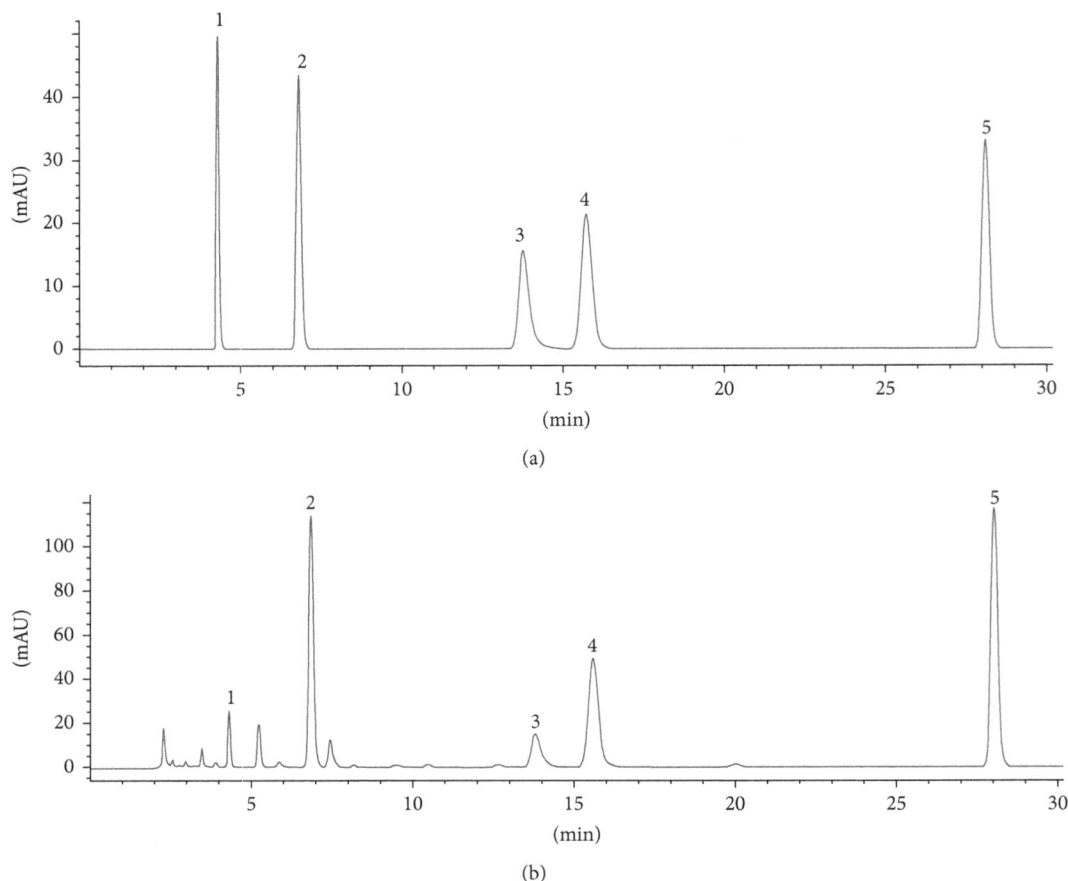

FIGURE 1: A chromatogram of the mixture of the standard compounds (a), and a chromatogram of fermented *Cordyceps sinensis* in Jinshuibao capsules (b). Peak 1: uracil; peak 2: uridine; peak 3: adenine; peak 4: guanosine; and peak 5: adenosine.

The concentrations of the components in the mixed solution were as follows: uracil $33.8\,\mu g/mL$, uridine $93.6\,\mu g/mL$, adenine $30.1\,\mu g/mL$, guanosine $120.8\,\mu g/mL$, and adenosine $93.6\,\mu g/mL$.

2.4. Preparation of Sample Solution.
A $1.00\,g$ sample of the fermented *Cordyceps sinensis* was precisely weighed and immersed in $40\,mL$ of ultrapure water in a $50\,mL$ volumetric flask. The total mass of the flask and sample was determined, and the flask was placed in an ultrasonic bath for 30 min. The procedure was repeated twice. After sonication, additional ultrapure water was added to return the sample to its original weight. The mixture was separated by centrifugation, and the extracted solution was cooled to room temperature. Next, the diluted extract was filtrated through a $0.22\,\mu m$ membrane filter into an HPLC vial to prepare it for HPLC analysis.

2.5. Similar Analysis (SA).
The similarity evaluation method for the chromatographic fingerprints of TCM (Version 2004A) recommended by the SFDA of China was used to obtain standardized fingerprint chromatograms of three kinds of fermented *Cordyceps sinensis*. Acquiring the standardized fingerprint chromatograms involved the calibration and normalization of the retention times of all the common peaks of the three forms of fermented *Cordyceps sinensis*.

2.6. Hierarchical Cluster Analysis (HCA).
The similarity or dissimilarity of each sample was visually displayed with a dendrogram in HCA. In this paper, samples of different products prepared from fermented *Cordyceps sinensis* were compared using SPSS 21.0 software (IBM company, New York, America) based on the between-groups linkage method and the cosine of the angular distance.

2.7. Calculation of Relative Conversion Using the QAMS Method.
In a certain linear range, the concentration of a component is proportional to the response of the detector. Adenosine was selected as the internal reference analyte because it is stable, readily available, inexpensive, and present in a high concentration in Jinshuibao capsules. The RCFs between adenosine and the other analytes were determined using (1). The contents of other five components can be calculated using (2) and (3).

TABLE 2: Calibration curves for the five target compounds.

Peak number	Analyte	Calibration curve	R^2	Linear range (μg/mL)	LOD (μg/mL)	LOQ (μg/mL)
1	Uracil	$Y = 41.219X + 5.2065$	0.9997	0.338–33.8	0.005	0.025
2	Uridine	$Y = 23.568X - 1.2283$	1.0000	0.936–93.6	0.016	0.049
3	Adenine	$Y = 62.462X + 4.2509$	0.9998	0.301–30.1	0.001	0.005
4	Guanosine	$Y = 22.077X - 27.477$	0.9992	1.208–120.8	0.038	0.151
5	Adenosine	$Y = 28.112X + 2.0196$	1.0000	0.936–93.6	0.024	0.073

TABLE 3: Precision, repeatability, stability, and recovery of makers in Jinshuibao capsules ($n = 6$).

peak number	Analyte	Intraday RSD (%)	Interday RSD (%)	Repeatability RSD (%)	Stability RSD (%)	Recovery (%)
1	Uracil	0.16	0.10	0.55	0.19	97.5
2	Uridine	0.42	0.05	0.57	0.19	96.7
3	Adenine	1.10	0.78	1.36	0.48	102.1
4	Guanosine	1.61	0.31	0.61	0.13	95.9
5	Adenosine	0.14	0.13	0.23	0.22	96.0

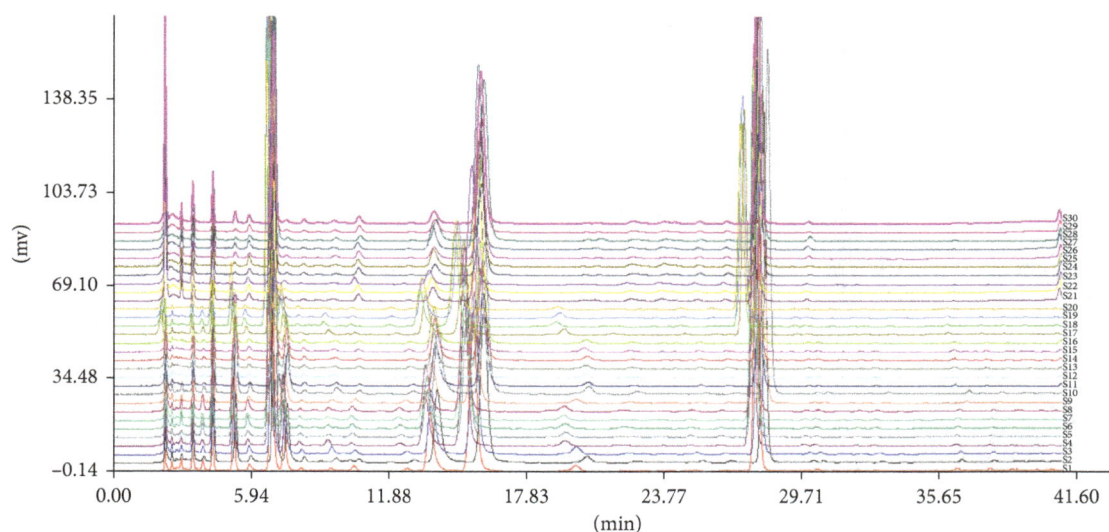

FIGURE 2: S1–S10: Jinshuibao capsule; S11–S20: Jinshuibao tablet; and S21–S30: Bailing capsule.

$$F_{i/s} = \frac{f_i}{f_s} = \frac{C_i}{C_s} \times \frac{A_s}{A_i}, \qquad (1)$$

$$C_i = \frac{f_i}{f_s} \times \frac{A_i}{A_s} \times C_s, \qquad (2)$$

$$m_i = C_i \times V_i, \qquad (3)$$

where A_i is the peak area of the analyte, A_s is the peak area of the standard solution, C_i is the concentration of the analyte, C_s is the concentration of the standard, and m_i is the mass of the analyte in the capsule or tablet.

3. Results and Discussion

3.1. Method Validation

Calibration Curves, Limits of Detection, and Limits of Quantity. The linearity of the calibration curves were determined by plotting the results from six standard solutions at different concentrations. The standard curve and linear range were obtained as shown in Table 2. There was a good linear relationship over a wide concentration range. The LODs and the LOQs of the five analytes were determined at the noise-signal ratios of $3:1$ and $10:1$, within the ranges of 0.005–0.038 μg/mL and 0.025–0.151 μg/mL, respectively.

3.2. Precision Stability, Repeatability, and Recovery. The precision of the method was validated by their intraday and interday variability. The intraday precision was evaluated by injecting the same standard solution for six times. The interday variability was evaluated on 3 successive days using the same standard solution. The stability of the sample solutions was analysis at 0, 2, 4, 6, 8, 10, and 24 h. The sample solutions were found to be stable for 24 h (RSD $\leq 0.48\%$). To validate the repeatability of the method, five different sample solutions from the same sample were injected into the HPLC system, and the RSD values were found to be 0.23–1.36%.

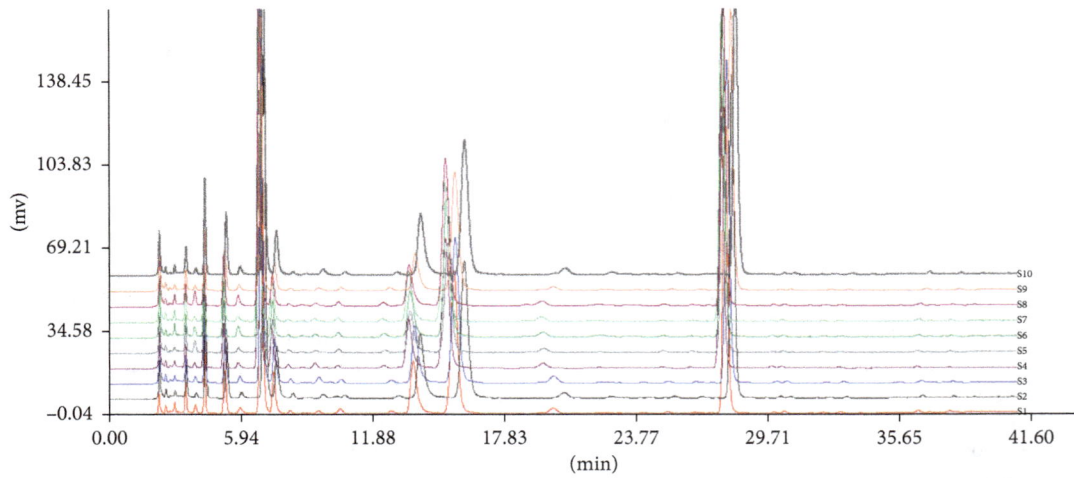

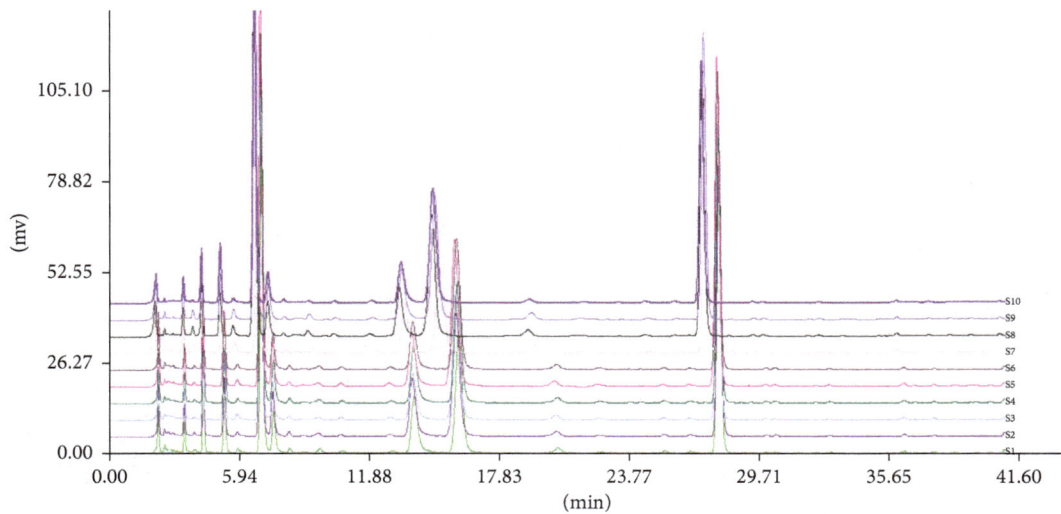

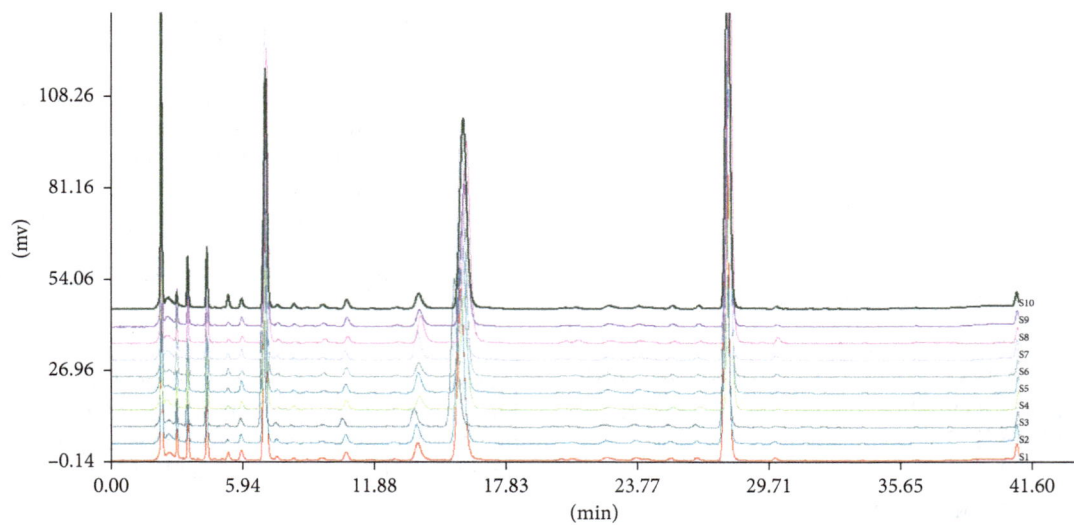

FIGURE 3: HPLC fingerprints for 30 fermented *Cordyceps sinensis* samples: 10 Jinshuibao capsule samples (a), 10 Jinshuibao tablet samples (b), and 10 Bailing capsule samples (c).

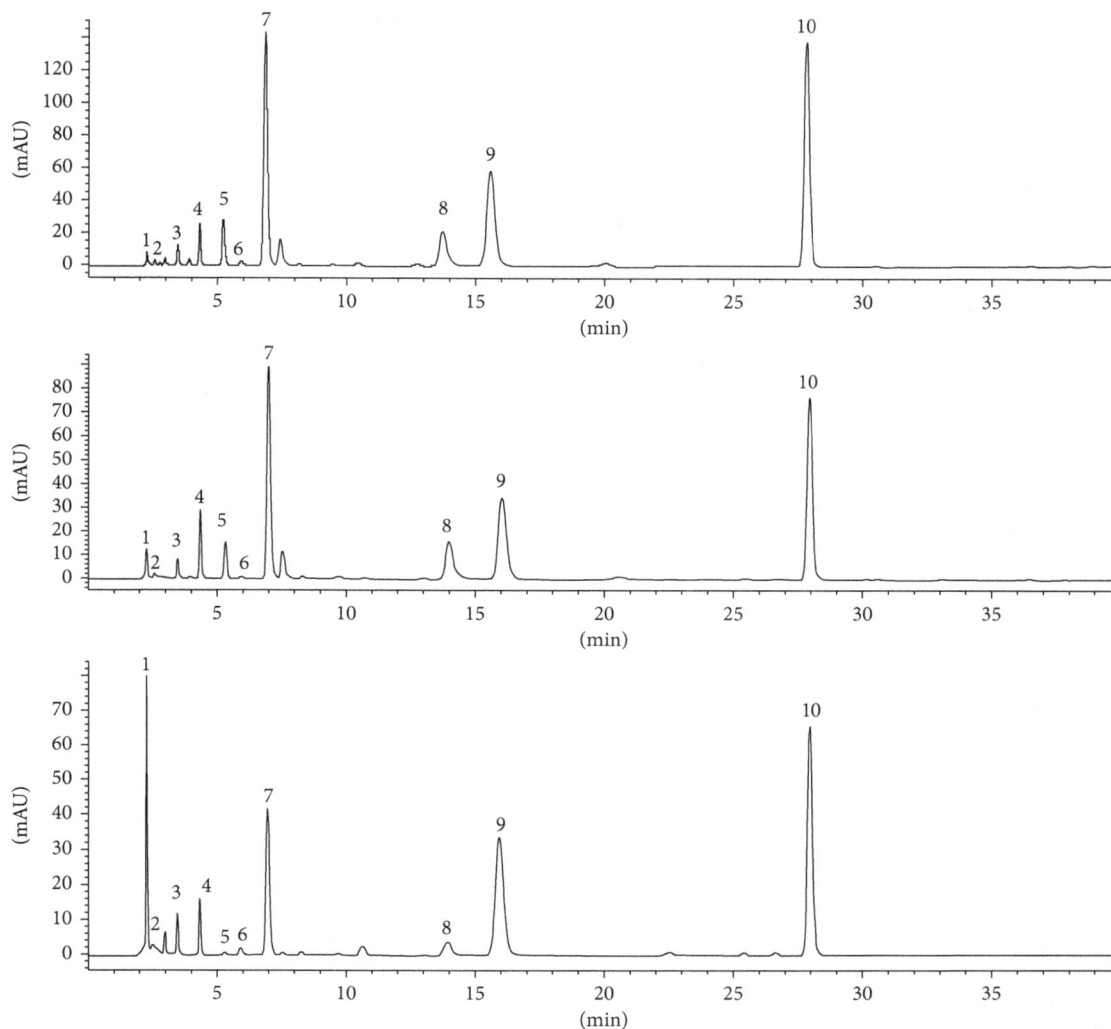

FIGURE 4: Common model fingerprints of Jinshuibao capsules (S1), Jinshuibao tablets (S2), and Bailing capsules (S3).

Recovery was evaluated by adding the same amount of an individual standard into a certain amount (0.2 g) of fermented *Cordyceps sinensis* sample that was found to have a concentration in the middle of the linear range (this procedure was repeated six times). In addition, the data from the DAD on the purities of the peaks indicate that the concentrations of the five nucleosides were not beyond the range of the threshold value, and the peak purity index of each compound was more than 990. This result confirmed that the purities of the chromatographic peaks were good, and the method is sufficiently accurate. All the results are showed in Table 3 and suggest that these methods are reliable and effective.

3.3. HPLC Fingerprint Analysis.

The fingerprint chromatograms of all three fermented *Cordyceps sinensis* products (Figure 2) and the fingerprint chromatograms of each sample (Figure 3) were established. The common patterns of each fermented *Cordyceps sinensis* were established according to the relative times and peak areas (RRTs and

RPAs, resp.) of 10 characteristic peaks (Figure 4). Furthermore, peaks 4, 7, 8, 9, and 10 were identified as uracil, uridine, adenine, guanosine, and adenosine, respectively. Bailing capsules had the lowest contents of the compounds with retention times of 5.3 min, 13.5 min, and 20 min (almost no absorption), and Jinshuibao tablets had the lowest content of the compound with a retention time of 3.0 min (Figure 3). The similarity values of the Jinshuibao capsules, Jinshuibao tablets, and Bailing capsules were 0.988, 0.990, and 0.994, respectively. The correlation coefficients between each chromatogram of the fermented *Cordyceps sinensis* samples and the simulated mean chromatogram were 0.998, 0.996, 0.996, 0.997, 0.998, 0.999, 0.998, 0.998, 0.999, 0.989, 0.993, 0.993, 0.995, 0.993, 0.994, 0.997,0.981, 0.980, 0.981, 0.994, 0.978, 0.983, 0.984, 0.977, 0.975, 0.981, 0.985, 0.986, 0.979, and 0.974. These coefficients indicate that the samples from different companies are all of uniformity good quality since they are produced by the same method. In addition, the similarities of the samples fermented using the same strain and samples from the same company were fairly close.

Dendrogram using average linkage (between-groups)

Rescaled distance cluster combaine

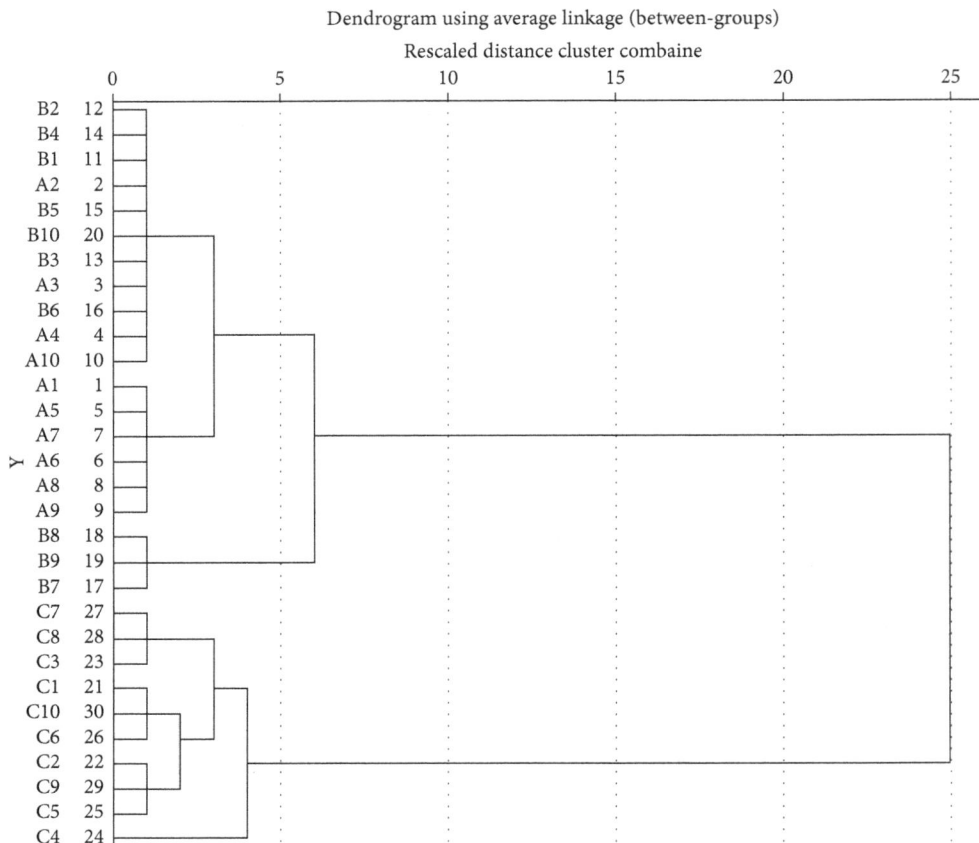

FIGURE 5: Clustering analysis graph of 30 fermented *Cordyceps sinensis* samples.

TABLE 4: The values of RCF based on different injection volumes ($n = 3$).

Injection volume	f_1	f_2	f_3	f_4
2	0.702	1.227	0.482	1.384
4	0.706	1.235	0.474	1.359
6	0.709	1.240	0.486	1.398
8	0.711	1.245	0.474	1.387
10	0.712	1.249	0.471	1.385
12	0.719	1.267	0.480	1.410
15	0.717	1.254	0.478	1.398
Mean	0.711	1.245	0.478	1.389
RSD (%)	0.83	1.05	1.09	1.15

f_1: $f_{\text{uracil/adenosine}}$; f_2: $f_{\text{uridine/adenosine}}$; f_3: $f_{\text{adenine/adenosine}}$; and f_4: $f_{\text{guanosine/adenosine}}$.

TABLE 5: The values of RCF based on different columns and instrument conditions ($n = 3$).

Instrument	Column	f_1	f_2	f_3	f_4
Agilent 1100	Phenomenex	0.674	1.193	0.501	1.281
Agilent 1100	Agilent	0.676	1.189	0.461	1.300
Agilent 1260	Phenomenex	0.685	1.200	0.444	1.32
Agilent 1260	Agilent	0.712	1.249	0.471	1.385
Waters 2695	Phenomenex	0.676	1.203	0.432	1.269
Waters 2695	Agilent	0.693	1.199	0.451	1.306
Mean		0.686	1.206	0.460	1.310
RSD (%)		1.46	2.19	2.42	4.09

f_1: $f_{\text{uracil/adenosine}}$; f_2: $f_{\text{uridine/adenosine}}$; f_3: $f_{\text{adenine/adenosine}}$; and f_4: $f_{\text{guanosine/adenosine}}$.

TABLE 6: RCFs values based on different flow rates ($n = 3$).

Flow rate	f_1	f_2	f_3	f_4
0.9 mL/min	0.674	1.189	0.443	1.300
1.0 mL/min	0.693	1.199	0.451	1.306
1.1 mL/min	0.676	1.193	0.443	1.306
Mean	0.681	0.194	0.446	1.304
RSD (%)	1.04	0.50	0.46	0.35

f_1: $f_{\text{uracil/adenosine}}$; f_2: $f_{\text{uridine/adenosine}}$; f_3: $f_{\text{adenine/adenosine}}$; and f_4: $f_{\text{guanosine/adenosine}}$.

TABLE 7: RCFs based on different column temperatures ($n = 3$).

Column temperature	f_1	f_2	f_3	f_4
25°C	0.677	1.195	0.445	1.310
30°C	0.693	1.199	0.451	1.306
35°C	0.676	1.194	0.443	1.290
Mean	0.682	1.196	0.446	1.302
RSD (%)	0.95	0.26	0.42	1.06

f_1: $f_{\text{uracil/adenosine}}$; f_2: $f_{\text{uridine/adenosine}}$; f_3: $f_{\text{adenine/adenosine}}$; and f_4: $f_{\text{guanosine/adenosine}}$.

3.4. *Hierarchical Cluster Analysis (HCA)*. HCA on the 30 samples was carried out using SPSS 21.0 software based on the between-groups linkage method and the cosine of the angular distance. Jinshuibao capsules, Jinshuibao tablets, and Bailing capsules are referred to A, B, and C, respectively, for simplicity. The results are shown in Figure 5. The 30 samples were

TABLE 8: Comparison of the results from the QAMS and ESM of fermented *Cordyceps sinensis* products (mg·g^{-1}).

Sample number	Uracil		Uridine		Adenine		Guanosine		Adenosine
	QAMS	ESM	QAMS	ESM	QAMS	ESM	QAMS	ESM	ESM
1	0.1393	0.1341	2.504	2.398	0.3191	0.3016	2.593	2.404	2.670
2	0.2357	0.2269	2.540	2.432	0.3891	0.3677	2.565	2.379	2.784
3	0.2220	0.2138	2.810	2.694	0.3530	0.3336	2.695	2.466	2.851
4	0.2235	0.2151	2.368	2.267	0.2805	0.2652	2.287	2.121	2.476
5	0.1123	0.1081	1.899	1.189	0.2468	0.2332	1.950	1.808	2.275
6	0.1275	0.1228	2.535	2.427	0.2694	0.2417	2.653	2.460	2.982
7	0.1326	0.1276	2.517	2.409	0.2913	0.2753	2.464	2.284	2.740
8	0.1100	0.1058	2.582	2.518	0.2439	0.2305	2.620	2.430	3.009
9	0.1404	0.1351	1.997	1.192	0.2219	0.2097	2.180	2.021	2.518
10	0.2235	0.2151	3.024	2.896	0.3798	0.3590	2.574	2.387	2.714
11	0.1599	0.1539	1.573	1.506	0.2369	0.2264	1.547	1.434	1.583
12	0.1646	0.1585	1.624	1.555	0.2479	0.2343	1.591	1.475	1.629
13	0.0999	0.0961	1.775	1.670	0.1984	0.1875	1.650	1.530	1.745
14	0.1617	0.1557	1.593	1.525	0.2394	0.2262	1.558	1.444	1.595
15	0.1149	0.1106	2.039	1.925	0.2301	0.2174	1.861	1.726	1.957
16	0.1278	0.1231	1.718	1.645	0.2090	0.1975	1.674	1.552	1.752
17	0.1101	0.1065	2.335	2.236	0.2277	0.2152	1.698	1.575	1.858
18	0.1099	0.1058	2.295	2.197	0.2091	0.1977	1.665	1.544	1.828
19	0.1143	0.1100	2.385	2.284	0.2406	0.2273	1.737	1.611	1.896
20	0.1077	0.1031	1.850	1.772	0.2059	0.1946	1.762	1.634	1.830
21	0.1098	0.1057	1.158	1.109	0.0785	0.0742	2.241	2.078	2.269
22	0.1068	0.1028	1.283	1.229	0.0834	0.0789	2.325	2.156	2.207
23	0.1023	0.0985	1.067	1.022	0.0756	0.0714	1.946	1.804	1.684
24	0.0914	0.0879	0.737	0.707	0.0597	0.0564	1.524	1.413	1.370
25	0.0897	0.0864	0.974	0.932	0.0925	0.0875	2.111	1.958	1.901
26	0.1028	0.0989	1.249	1.197	0.0636	0.0599	2.251	2.088	2.257
27	0.1569	0.1510	1.697	1.625	0.1281	0.1208	3.048	2.827	2.755
28	0.1401	0.1348	1.532	1.467	0.1211	0.1145	2.689	2.493	2.538
29	0.0851	0.0819	0.969	0.928	0.0741	0.0701	1.893	1.756	1.816
30	0.1007	0.0969	1.245	1.192	0.0711	0.0672	2.526	2.343	2.595
Correlation coefficient	0.9999		0.9998		0.9997		0.9999		—

categorized into two groups: one containing 20 samples (Group 1) and the other containing 10 samples (Group 2). Group 1 was further divided into Subgroups 1a (A2, A3, A4, A10, B1, B2, B3, B4, B5, B6, and B10), 1b (A1, A5, A6, A7, A8, and A9), and 1c (B17, B18, and B19). Subgroup 1a contained 4 Jinshuibao capsules and 7 Jinshuibao tablet samples, Subgroup 1b contained 6 Jinshuibao capsules, and Subgroup 1c contained 3 Jinshuibao tablets. All Bailing capsule samples were grouped together. The 11 samples with cosine and correlation coefficient values between 0.989 and 0.998 are in Subgroup 1a, the 6 samples with values greater than 0.998 are in Subgroup 1b, and the 3 samples with values less than 0.986 are in Subgroup 1c. These groupings implied that HCA was effective for identifying *Cordyceps sinensis* samples fermented with different bacterial strains.

3.5. The Quantitative Analysis of Multicomponents by Single Marker (QAMS). The RCF values of the four analytes were calculated at different injection volumes (2, 4, 6, 8, 10, 12, and 15 μL) and are shown in Table 4. Many factors impacted the RCF values. In this work, the flow rate, column temperature, type of reversed-phase column, and chromatographic system were optimized to investigate the robustness of the RCFs, and the results are shown in Tables 5–7. The values were found to be accurate and stable.

3.6. Comparison of the Result of QAMS with Those of the External Standard Method (ESM). To assess the differences in the results of the QAMS and ESM methods, 30 samples, including 10 Jinshuibao capsules, 10 Jinshuibao tablets, and 10 Bailing capsules, were analysed. Vector angles were used to determine the consistency of the results. The larger the cosine of the angle, the higher the similarity between the two samples. The results are shown in Table 8, and the values of the cosine of all the angles were more than 0.999, which indicated there were no significant differences between the results of QAMS and ESM. From these results, we can see that all samples contained all five active ingredients, but their contents are different in samples fermented with different strains; for example, for the Jinshuibao capsules (adenosine > uridine > guanosine > adenine > uracil), Jinshuibao tablets (adenosine > uridine > guanosine > adenine > uracil), and Bailing capsule (adenosine > guanosine > uridine > uracil > adenine). As a whole, the contents of adenosine, uridine, and guanosine were relatively high and met the standard given by the Pharmacopoeia of China.

4. Conclusion

The results presented herein suggest that fingerprint chromatography based on HPLC is an effective method for assessing the quality of fermented *Cordyceps sinensis* products, and the QAMS method is convenient and reliable for qualitifying the five active components in fermented *Cordyceps sinensis* by RCF. When HPLC fingerprint analysis was combined with the SA, HCA, and QAMS methods, it was a practical, feasible, and effective approach to better identifying and comprehensively evaluating the quality of fermented *Cordyceps sinensis* products. Hence, the method established herein is a promising and broadly applicable strategy for the comprehensive quality control of TCMs.

Conflicts of Interest

All authors have announced that there are no conflicts of interest in the publication of this study.

Acknowledgments

This work was supported by the National Traditional Chinese Medicine Standardization Project (ZYBZH-C-JX-39), The State "Major New Drug Creation" Science and Technology Major Project (2011ZX09201-201-30), and the Jiangxi High School Science and Technology Project (12th "Five-Year" renovation project of large variety of major new drug creation (KJLD13061)).

References

[1] Y. S. Kim, E. K. Kim, Y. J. Tang et al., "Antioxidant and anticancer effects of extracts from fermented *Haliotis discus hannai* with *Cordyceps militaris* mycelia," *Food Science and Biotechnology*, vol. 25, no. 6, pp. 1775–1782, 2016.

[2] H. C. Lo, T. H. Hsu, S. T. Tu, and K. C. Lin, "Antihyperglycemic activity of natural and fermented *Cordyceps sinensis* in rats with diabetes induced by nicotinamide and streptozotocin," *American Journal of Chinese Medicine*, vol. 34, no. 5, pp. 819–832, 2006.

[3] T. Y. Zheng, Z. M. Wu, C. S. Zhao et al., "Enhanced muscle contractility and anti-fatigue proprieties of CordyMax5," *FASEB Journal*, vol. 21, no. 6, p. A939, 2007.

[4] P. Zhang, X. Y. Xiao, Y. K. Li, and R. Lin, "Chemical analysis of nucleosides and alkalines components in five preparations of fermental Cordyceps," *Chinese Journal of Pharmaceutical Analysis*, vol. 29, no. 6, pp. 889–893, 2009.

[5] X. L. Li and D. Li, "Enhancing antioxidant activity of soluble polysaccharide from the submerged fermentation product of *Cordyceps sinensis* by using cellulase," *Advanced Materials Research*, vol. 641-642, no. 1, pp. 975–978, 2013.

[6] J. Sun, X. E. Zhao, J. Dang et al., "Rapid and sensitive determination of phytosterols in functional foods and medicinal herbs by using UHPLC-MS/MS with microwave-assisted derivatization combined with dual ultrasound-assisted dispersive liquid-liquid microextraction," *Journal of Separation Science*, vol. 40, no. 3, pp. 597–820, 2016.

[7] X. Wei, N. Xu, D. Wu, and Y. He, "Determination of branched-amino acid content in fermented *Cordyceps sinensis* mycelium by using FT-NIR spectroscopy technique," *Food and Bioprocess Technology*, vol. 7, no. 1, pp. 184–190, 2014.

[8] H. Y. Ahn, K. R. Park, K. H. Yoon, J. Y. Lee, and Y. S. Cho, "Biological activity and chemical characteristics of *Cordyceps militaris* powder fermented by several microscopic organisms," *Journal of Life Science*, vol. 25, no. 2, pp. 197–205, 2015.

[9] S. Lin, Z. Q. Liu, P. J. Baker et al., "Enhancement of cordyceps polysaccharide production via biosynthetic pathway analysis in *Hirsutella sinensis*," *International Journal of Biological Macromolecules*, vol. 92, pp. 872–880, 2016.

[10] S. K. Sharma, N. Gautam, and N. A. Singh, "Optimized extraction, composition, antioxidant and antimicrobial activities of exo and intracellular polysaccharides from submerged culture of *Cordyceps cicadae*," *BMC Complementary and Alternative Medicine*, vol. 15, no. 1, pp. 446–456, 2015.

[11] S. Ahmad, M. Khan, R. Parveen, K. Mishra, and R. Tulsawani, "Determination of nucleosides in *Cordyceps sinensis* and *Ganoderma lucidum* by high performance liquid chromatography method," *Journal of Pharmacy & Bioallied Sciences*, vol. 7, no. 4, pp. 264–266, 2015.

[12] S. Y. Zong, H. Han, B. Wang et al., "Fast simultaneous determination of 13 nucleosides and nucleobases in *Cordyceps sinensis* by UHPLC-ESI-MS/MS," *Molecules*, vol. 20, no. 12, pp. 21816–21825, 2015.

[13] T. S. Bozhok, E. N. Kalinichenko, B. B. Kuz'mitskii, and M. B. Golubeva, "Synthesis, hydrolytic stability, and antileukemic activity of azacytidine nucleoside analogs," *Pharmaceutical Chemistry Journal*, vol. 49, no. 12, pp. 804–809, 2016.

[14] E. Kim and J. H. Hong, "Synthesis and antiviral activity evaluation of $2',5',5'$-trifluoro-apiosyl nucleoside phosphonic acid analogs, nucleosides," *Nucleotides and Nucleic Acids*, vol. 35, no. 3, pp. 130–146, 2016.

[15] N. Khatri, V. Lather, and A. K. Madan, "Diverse models for anti-HIV activity of purine nucleoside analogs," *Chemistry Central Journal*, vol. 9, no. 1, pp. 1–10, 2015.

[16] Chinese Pharmacopoeia Commission, *Pharmacopoeia of the People's Republic of China*, Chemical Industry Press, Beijing, China, 2015th edition, 2015.

[17] R. M. Yu, B. Ye, C. Y. Yan et al., "Fingerprint analysis of fruiting bodies of cultured *Cordyceps militaris* by high-performance liquid chromatography–photodiode array detection," *Journal of Pharmaceutical & Biomedical Analysis*, vol. 44, no. 3, pp. 818–823, 2007.

[18] L. L. Cui, Y. Y. Zhang, W. Shao, and D. Gao, "Analyis of the HPLC fingerprint and QAMS from *Pyrrosia* species," *Industrial Crops and Products*, vol. 85, pp. 29–37, 2016.

[19] A. Stavrianidi, E. Stekolshchikova, A. Porotova, I. Rodin, and O. Shpigun, "Combination of HPLC–MS and QAMS as a new analytical approach for determination of saponins in ginseng containing products," *Journal of Pharmaceutical and Biomedical Analysis*, vol. 132, pp. 87–92, 2017.

[20] Y. H. Jiang, H. Chen, L. L. Wang, Z. Liu, J. Zou, and X. Zheng, "Quality evaluation of polar and active components in crude and processed fructus corni by quantitative analysis of multicomponents with single marker," *Journal of Analytical Methods in Chemistry*, vol. 12, pp. 1–13, 2016.

[21] C. P. Yan, Y. Wu, Z. B. Weng et al., "Development of an HPLC method for absolute quantification and QAMS of flavonoids components in *Psoralea corylifolia* L," *Journal of Analytical Methods in Chemistry*, vol. 23, pp. 1–7, 2015.

7

Optimization and Validation of Thermal Desorption Gas Chromatography-Mass Spectrometry for the Determination of Polycyclic Aromatic Hydrocarbons in Ambient Air

Iñaki Elorduy [ID], Nieves Durana, José Antonio García, María Carmen Gómez, and Lucio Alonso

Chemical and Environmental Engineering Department, School of Engineering, University of the Basque Country, Alameda de Urquijo s/n, 48013 Bilbao, Spain

Correspondence should be addressed to Iñaki Elorduy; inaki.elorduy@ehu.eus

Academic Editor: Verónica Pino

Thermal desorption (TD) coupled with gas chromatography/mass spectrometry (TD-GC/MS) is a simple alternative that overcomes the main drawbacks of the solvent extraction-based method: long extraction times, high sample manipulation, and large amounts of solvent waste. This work describes the optimization of TD-GC/MS for the measurement of airborne polycyclic aromatic hydrocarbons (PAHs) in particulate phase. The performance of the method was tested by Standard Reference Material (SRM) 1649b urban dust and compared with the conventional method (Soxhlet extraction-GC/MS), showing a better recovery (mean of 97%), precision (mean of 12%), and accuracy (±25%) for the determination of 14 EPA PAHs. Furthermore, other 15 nonpriority PAHs were identified and quantified using their relative response factors (RRFs). Finally, the proposed method was successfully applied for the quantification of PAHs in real 8 h-samples (PM$_{10}$), demonstrating its capability for determination of these compounds in short-term monitoring.

1. Introduction

Polycyclic aromatic hydrocarbons (PAHs) are a class of persistent organic pollutants (POPs) comprising hundreds of individual substances. These compounds contain two or more fused aromatic rings (made up of carbon and hydrogen atoms) in linear, angular, or cluster arrangements [1]. They are semivolatile organic compounds (SVOC); thus, they are present in the atmosphere in both the gas and the particulate phases as well as dissolved or suspended in precipitation (fog or rain) [2].

PAHs are as by-products of incomplete combustion processes of organic matter [3], and primarily emitted from anthropogenic sources [4], being the mobile and major sources in urban areas [5, 6]. Their harmful health effects and persistence pose an environmental concern. Thus, these compounds were among the first atmospheric pollutants identified as suspected carcinogens [7]. Moreover, PAHs belong to the group of POPs included in the list of 16 POPs specified by the UNECE Convention on Long-range Transboundary Air Pollution

Protocol on Persistent Organic Pollutants [8, 9]. Due to these features, the United States Environmental Protection Agency (US-EPA) has listed 16 of them as priority pollutants (16 EPA PAHs) [10]. The most toxic PAHs (5 and 6 rings) are linked to the particulate matter [11, 12]. Accordingly, many air pollution studies have been focused on PAHs bound to particulate matter, particularly PM$_{10}$ and PM$_{2.5}$ in order to assess their concentration, distribution, and sources.

Air monitoring for PAHs in urban areas is an important issue because the risk associated with human exposure is higher considering the population density [13, 14]. However, PAH data in urban air show large spatial and temporal uncertainties because of the complex sampling and analytical procedures required.

Sampling of particulate PAHs is mostly done by the collection of them on a filter (quartz or glass fiber), using high- or low-volume samplers [15–17]. Once the PAHs have been collected, they have to be extracted for the final determination. The extraction of PAHs from multiple matrices is a difficult step. PAHs are found in the environment in very low concentrations; consequently, an effective extraction method, able to quantitatively separate the analytes from the matrix, is required. The widely used method is solvent desorption of sampling media (Soxhlet extraction, accelerated solvent extraction, microwave-assisted extraction, and ultrasonic-assisted extraction) followed by analysis of the compounds of interest by GC-MS (gas chromatography coupled to mass chromatography) or high-performance liquid chromatography coupled with florescence detection (HPLC-FLD) [18, 19], where the detection methods allow cutting most of analytical interferences.

The use of toxic organic solvents in the solvent-based extraction methods causes added difficulties with sample handling and generates large amounts of solvent waste, which is costly and can generate additional environmental problems. Furthermore, the sensitivity of the current analytical procedures limits time resolution of measurements; thus, most of the urban pollution studies rarely achieved temporal resolution measurements better than 24 h. Since the PAH composition of aerosols can vary according to the diurnal changes in the sources, meteorological conditions and atmospheric reactivity [20], the time resolution of 24 h seems not sufficient to comprehend their variability, fate, and behavior in the atmosphere [21]. For these reasons, the development of simpler and sensitive methods or the improvement of the existing ones is of great interest, for the detection, determination, and monitoring of PAHs.

In recent years, alternative analytical procedures for PAHs based on the use of solvent-free extraction methods have been studied [22, 23]. Thermal desorption (TD) involves heating sample materials or sorbents in a flow of inert carrier gas, so that retained organic volatiles and semivolatiles are released and transferred or injected into the analytical system (e.g., into the carrier gas stream of the GC column).

The power and potential of TD allow configuring the technique in multiple adsorption-desorption stages, thus enhancing the concentration of the compounds of interest and detection limits. This higher sensitivity may provide shorter sampling times or lower sampling volumes. Another benefit of TD is that it is often possible to quantitatively retain target compounds during one or more of the trapping stages, while unwanted, for example, water and/or permanent gases, is selectively purged to vent. This avoids the entrance of unwanted compounds into the analytical system that could generate interferences during the analysis and/or damage to the equipment.

This work has tested and optimized different TD-GC/MS operation conditions in order to develop the best method that is able to sample and analyze airborne PAHs in particulate phase. The TD-GC/MS method was later validated by using a Standard Reference Material (SRM) 1649b urban dust and comparing with the conventional method based on solvent extraction (Soxhlet extraction-GC/MS). Moreover, the method was applied to measure PAH levels of 8 h PM_{10} samples in ambient air.

2. Materials and Methods

2.1. Reagents and Materials.
A liquid certified mixture of 16 EPA PAHs ($2000\,\mu g\cdot mL^{-1}$, SV Calibration Mix 5, Restek Corporation, USA) and a liquid deuterated mixture

($200\,\mu g\cdot mL^{-1}$, predeuterated internal standard PAH Mixture 6, Chiron AS, Norway) were used during the study. In Soxhlet extraction, decafluorobiphenyl, 4,4'-dibromooctafluorobiphenyl, 4,4'-dibromobiphenyl. (Restek, $2000\,\mu g\cdot mL^{-1}$), and indeno [1,2,3-cd]pyrene-d12 (Chiron, $100\,\mu g\cdot mL^{-1}$) were used as recovery standard for the assessment of extraction efficiency. Solutions were prepared by appropriate dilution in methanol, HPLC grade (99.9%, Lab-Scan Analytical Sciences, Poland).

The method was validated using the Standard Reference Material (SRM) 1649b urban dust, obtained from the National Institute of Standards and Technology (NIST, Gaithersburg, MD, USA).

2.2. TD-GC/MS Method.
Sampling tubes (stainless steel tube of 5 mm outer diameter × 90 mm length) packed with filter were analyzed by using TD-GC/MS. The 16 EPA PAHs and deuterated PAHs were spiked on two one-eighth parts of a 47 mm quartz fiber filter (Whatman International Ltd., UK). These portions, suitably folded, were introduced into the sampling tubes.

Prior to use, the packed sampling tubes were conditioned by thermal cleaning under a helium flow rate of $100\,mL\cdot min^{-1}$ at 350°C for 30 min.

The NIST Standard Reference Material 1649b was handled in a similar way. Samples of the urban dust (10 mg) were weighed and placed on a one-eighth section of a 47 mm quartz fiber filter which was rolled and put into the sampling tube. Silanized glass wool (Supelco Inc., Bellefonte, USA) was introduced at the end and at the head of the desorption tubes in order to prevent system contamination.

Prior to use, filters and glass wool plugs were heated in a muffle furnace at 500°C for 24 h to remove trace organic compounds.

PAHs analysis was carried out using an automatic thermal desorber unit (Turbomatrix 150 ATD, Perkin Elmer S.L., USA) coupled by a fused silica capillary transfer line (5 m length × 0.32 mm I.D.) to a GC/MS detector (Clarus 500, Perkin Elmer S.L., USA). The chromatographic separation of PAHs was conducted on a Meta. X5 (silphenylene phase) capillary column: 30 m length × 0.25 mm I.D. × 0.25 mm film thickness (Teknokroma, Spain).

The helium gas carrier pressure employed in the GC/MS system was 145 kPa, and the column temperature was programmed as follows: initial temperature 100°C for 3 min, ramp of $10°C\cdot min^{-1}$ until 250°C, ramp of $5°C\cdot min^{-1}$ until 320°C, and finally temperature held at 320°C for 10 min. The total analysis time was 42 min per sample. The temperature of both the transfer lines (from TD to GC and from GC to MS) was held at 280°C, whereas the source temperature was 250°C. Simultaneous full scan (SCAN) and selective ion monitoring (SIM) modes were used for the identification and quantification of PAHs. Table 1 shows, according to their elution order, the PAHs determined in this study with their quantification ions. Supplementry Figures S1–S3 show the representative SCAN chromatograms of the 16 EPA and deuterated PAHs.

2.3. Soxhlet Extraction-GC/MS Method.
Between 300 and 500 mg of the NIST SRM 1649b urban dust was weighted and placed on one-eighth of a 150 mm prebaked (at 500°C

Table 1: Abbreviations and quantification ions of PAHs determined by the TD-GC/MS method.

PAH	Abbreviation	Ion (m/z)
Naphthalene-d$_8$[b]	Naph-d$_8$	136
Naphthalene[a]	Naph	128
Biphenyl-d$_{10}$[b]	Bph-d$_{10}$	164
Acenaphthylene[a]	Acy	152
Acenaphthene[a]	Ace	154
Fluorene[a]	FL	166
Phenanthrene-d$_{10}$[b]	Phe-d$_{10}$	188
Phenanthrene[a]	Phe	178
Anthracene[a]	Ant	178
Fluoranthene[a]	Ft	202
Pyrene-d$_{10}$[b]	Pyr-d$_{10}$	212
Pyrene[a]	Pyr	202
Benzo[ghi]fluoranthene[c]	BghiFt	226
Benzo[c]phenanthrene[c]	BcP	228
Cyclopenta[cd]pyrene[c]	CPP	226
Benzo[a]anthracene-d$_{12}$[b]	BaA-d$_{12}$	240
Benzo[a]anthracene[a]	BaA	228
Triphenylene[c]	Triph	228
Chrysene[a]	Chry	228
Retene[c]	Ret	234
Benzo[b]fluoranthene[a]	BbFt	252
Benzo[j]fluoranthene[c]	BjFt	252
Benzo[k]fluoranthene[a]	BkFt	252
Benzo[a]fluoranthene[c]	BaFt	252
Benzo[e]pyrene[c]	BeP	252
Benzo[a]pyrene-d$_{10}$[b]	BaP-d$_{10}$	264
Benzo[a]pyrene[a]	BaP	252
Perylene[c]	Per	252
Dibenzo[a,j]anthracene[c]	DBajA	278
Indeno[1,2,3-cd]pyrene[a]	IP	276
Dibenzo[ac]anthracene[c]	DBacA	278
Dibenzo[ah]anthracene[a]	DBahA	278
Benzo[b]chrysene[c]	BbC	278
Picene[c]	Pic	278
Benzo[ghi]perylene-d$_{12}$[b]	BghiP-d$_{12}$	288
Benzo[ghi]perylene[a]	BghiP	278
Anthanthrene[c]	Anthan	276
Coronene[c]	Cor	300

[a]16 EPA PAHs; [b]deuterated PAHs; [c]nonpriority PAHs.

for 24 h) quartz fiber filter (Whatman International Ltd.). Before folding the filter, it was spiked with the recovery standards.

Soxhlet extraction was performed by using Büchi extraction system B-811 (BÜCHI, Switzerland), an automated system that can be used to perform extraction according to the original Soxhlet principle. The samples were extracted with hexane using the Soxhlet Warm mode. This mode increases the solubility of the analytes, allowing an optimal extraction in 3 hours [24].

After the extraction process, the extracts of 5 mL were concentrated by a stream of dry nitrogen to a volume less than 0.5 mL. Finally, these extracts were diluted to 1.5 mL with methanol and spiked with deuterated PAHs.

Two-microliters of aliquots from each extract was injected into the GC/MS with split mode. Table 2 collects the timed events and the oven program used in the GC/MS during the validation of the Soxhlet method.

Also, in this method, the GC/MS used simultaneously the SCAN and SIM mode for the identification and quantification of PAHs.

2.4. Sample Collection. Airborne particulate matter (PM$_{10}$) samples were collected on preheated (at 500°C for 24 h) quartz fiber filters (150 mm diameter, Whatman International Ltd., United Kingdom) using a high-volume sampler (Digitel DHA-80, Digitel Elektronik AG, Switzerland) with a flow rate of 30 m$^3\cdot$h^{-1}. DHA-80 stores 15 filters stretched in filter holders that are changed automatically at the preset time. DHA-80 has integrated temperature control in the filter storage section; in this way, the used filters can be stored at low temperatures (in this study at 4°C) after sampling.

Collected filters were put into individual Petri dishes, wrapped in aluminum foil, and kept in a 4°C freezer until analysis (<15 days) according to ISO 12884:2000 [25].

3. Results and Discussion

3.1. Optimization of Thermal Desorption Method. The thermal desorption process can be divided into two main stages: tube and trap desorption. In the first stage, target compounds are thermally desorbed from the sampling tube and transferred to the cold trap, where they are concentrated. After completing the primary desorption, the trapped compounds are released by quick heating of the trap and swept through the heated transfer line to the GC column.

To obtain the best analytical conditions in terms of sensitivity and reproducibility, different parameters in each desorption stage were tested.

For these tests, 1 μL of the 16 EPA PAHs solutions of 20 ng$\cdot\mu$L^{-1} were spiked in sampling tubes packed with portions of quartz fiber.

3.1.1. Primary Desorption (Tube Desorption). The conditions in the tube oven during this stage are key to guarantee an efficient desorption; thus, parameters such as the temperature, time, and flow in the tube oven were studied to optimize this process.

Different values of desorption temperatures, times, and flows were tested, considering factors such as the packing/sample matrix stability, the lability of the components of interest, and the temperature limitations of the system. Figure 1 shows the area of chromatographic peak for each of the 16 EPA PAHs (in %) obtained for each test.

The results demonstrated that an increase in the temperature of the oven tube enhances the desorption of particulate PAHs (Figure 1(a)). This improvement was remarkable for high molecular weight PAHs (IP, DBahA, and BghiP). A value of 320°C was selected as temperature in the first desorption stage. Regarding the time, the lowest value in the test (10 min) clearly showed significantly higher areas for most compounds (Figure 1(b)), indicating a more efficient desorption. This value was selected as desorption time in the tube. Finally, higher desorption flows enable better desorption of PAHs (Figure 1(c)). Flows higher than

TABLE 2: Timed events and oven program used in direct injector mode.

| | Timed event | | | Oven program | | |
Event	Flow (mL·min^{-1})	Time (min)	Ramp	Rate (°C·min^{-1})	Temperature (°C)	Hold (min)
Split	0	−0.51	Initial	0	45	1
Split	50	1	1	20	200	0
Split	20	5	2	4	320	5

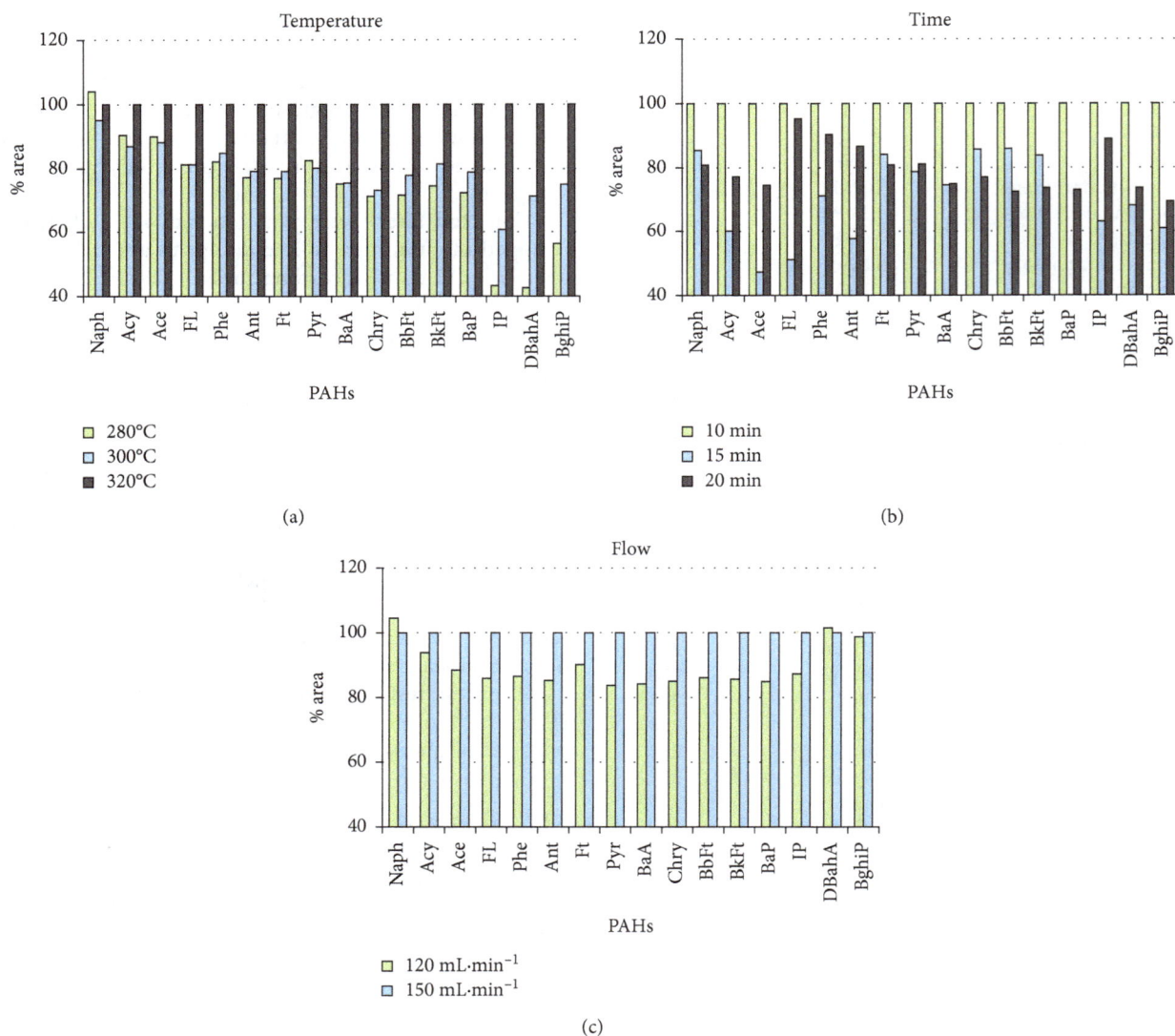

(a)

(b)

(c)

FIGURE 1: Area (in %) for each of the 16 EPA PAH obtained in the study of primary desorption conditions (desorption temperature, time, and flow) for sampling tubes ($n = 5$) packed with filter. % Areas at 280°C and 300°C are compared to % areas at 320°C chosen as 100% (a). % Areas at 15 min and 20 min are compared to % areas at 10 min chosen as 100% (b). % Areas at 120 mL·min^{-1} is compared to % areas at 150 mL·min^{-1} chosen as 100% (c).

150 mL·min^{-1} are not recommended, as they can generate problems in maintaining low temperatures in the trap zone during the first desorption stage [26]. Therefore, a flow of 150 mL·min^{-1} was selected as the optimal value.

3.1.2. Secondary Desorption (Trap Desorption). To enhance PAH desorption from the trap (a quartz tube packed with glass

wool), its high temperature has to be as high as possible. This temperature depends on the trap packing and equipment stability as well as on the target compounds. In this study, a value of 320°C (the value recommended by the manufacturer) was set, while its low temperature (values of −15, −10, and −5°C) and desorption time (values of 4, 6, and 12 min) were tested.

The area of chromatographic peak (in %) for low (2-3 rings: Naph, Ace, Acy, FL, Phe, and Ant), middle (4 rings: Ft,

Optimization and Validation of Thermal Desorption Gas Chromatography-Mass Spectrometry...

67

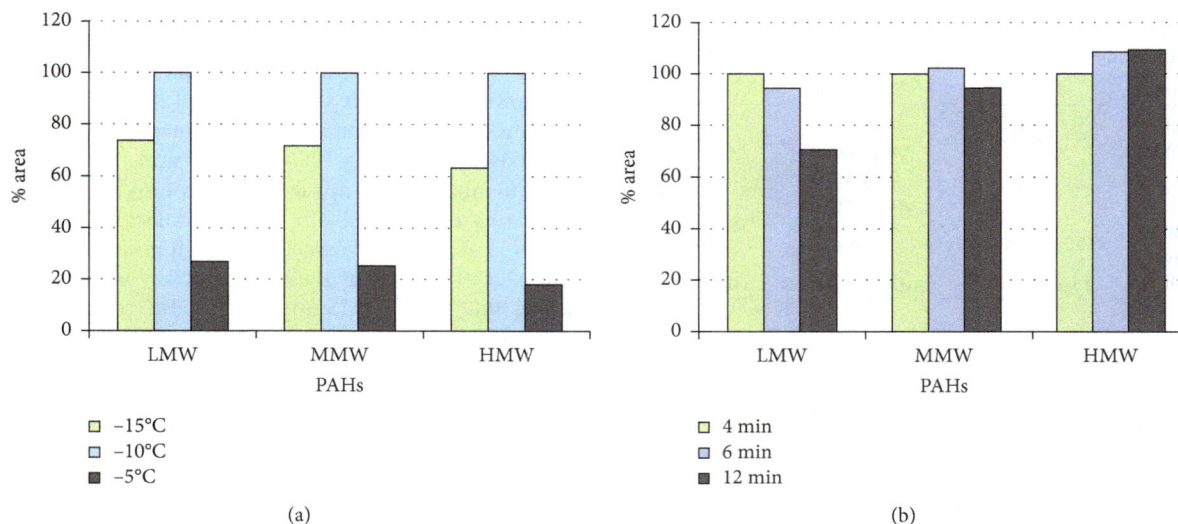

FIGURE 2: Area (in %) for LMW (low molecular weight), MMW (middle molecular weight), and HMW (high molecular weight) PAHs obtained in the study of the low trap temperature (a) and of the trap time (b) for sampling tubes packed with filter ($n = 5$). % Areas at $-15°C$ and $-5°C$ are compared to % areas at $-10°C$ chosen as 100% (a). % Areas at 6 min and 12 min are compared to % areas at 10 min chosen as 100% (b).

Pyr, BaA, and Chry), and high (5-6 rings: BbFt, BkFt, BaP, IP, DBahA, and BghiP) molecular weight PAHs obtained for each test are shown in Figure 2.

The temperature in the Peltier trap is a critical parameter in secondary desorption (Figure 2(a)), showing significant changes in the sensitivity for different values. The temperature of $-10°C$ revealed the best results.

In the study of the trap desorption time (Figure 2(b)), the results demonstrated that longer values do not implicate a higher efficiency and consequently a better detection, 6 min showed a better response than 12. This is especially significant with the lightest PAHs (LMW) which could be affected by the exposure to high temperatures, generating losses. By contrast, the heavier PAHs (MMW and HMW) showed higher concentrations after longer trap desorption times because they could need more time to be completely desorbed. Due to this, a trap duration of 6 min was selected as this value presented good desorption for 16 target PAHs.

3.1.3. Inlet and Outlet Split Flows.

In order to enhance the process of two-stage thermal desorption, a double split mode was used. Therefore, the inlet (split flow as the tube is desorbed) and outlet (split flow as the trap is desorbed) split flows were also tested.

The inlet split flow plays an important role during primary desorption. This maintains a relatively high carrier gas flow through the sample tube, while at the same time establishes a reasonably low flow through the cold trap, aiding the complete removal from the sample tube and analyte retention. The deactivation of the inlet split ($0\,mL \cdot min^{-1}$) generated a significant improvement in PAH desorption because the complete sample, without purge, arrived at the cold trap. With the increase of inlet split, the sensitivity decreased. Although in this study, an inlet split flow of $0\,mL \cdot min^{-1}$ showed the best

results; it is recommended to activate this split in order to avoid the presence of unwanted compounds (permanent gases and water) in the trap. These could reduce the trap lifetime and interfere in the analysis. In order to find a compromise between these rules, an intermediate flow ($23\,mL \cdot min^{-1}$) was considered as the optimal value.

The outlet split flow also plays an important function in the trap desorption: (1) adapting the effluent flow to a capillary column flow, it avoids the system saturation and (2) facilitating the release of the analytes, it guarantees a high enough flow through the trap during desorption. According to the manufacturer, at least $10\,mL \cdot min^{-1}$ of outlet split is necessary to minimize the air/water background on a mass spectrometer when atmospheric samples are analysed [26]. In this study, the results obtained for outlet split flows demonstrated that the increase of this parameter reduces the sensitivity of the technique, with losses becoming significant between 10 and $20\,mL \cdot min^{-1}$. Therefore, the manufacturer's recommended flow ($10\,mL \cdot min^{-1}$) was selected as the optimal value.

Finally, Table 3 summaries the optimized values for thermal desorption.

3.2. Desorption Efficiency.

Once optimized, the efficiency of two-stage thermal desorption was studied. The efficiency was calculated by the following expression:

$$E\,(\%) = \left(\frac{A}{A + A^*}\right) \times 100, \tag{1}$$

where E is the efficiency in %, A is the peak area of the analyte obtained from desorption of the sampling tube (previously loaded with PAHs), and A^* is the peak area of the analyte obtained when the same sampling tube or the trap was desorbed the second time.

TABLE 3: Optimized conditions for thermal desorption system.

Primary desorption		Secondary desorption	
Parameter		Parameter	
Tube temperature	320°C	High trap temperature	320°C
Time	10 min	Low trap temperature	−10°C
Desorption flow	150 mL·min^{-1}	Time	6 min
Inlet split flow	23 mL·min^{-1}	Outlet split flow	10 mL·min^{-1}

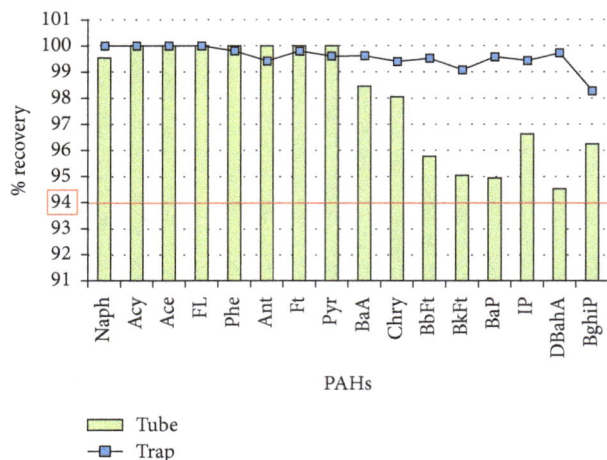

FIGURE 3: Recovery (in %) of the 16 EPA PAHs (particulate phase) in each stage of the thermal desorption.

Figure 3 shows the tube and trap efficiencies obtained for each particulate PAH. The technique demonstrated a good efficiency with recoveries of the PAHs in the tube and trap higher than 94%.

3.3. TD-GC/MS Validation and Comparison with Soxhlet-GC/MS.
In order to determine the performance of the method when applied to atmospheric PM samples, this was tested using the Standard Reference Material (SRM) 1649b urban dust.

The accuracy, repeatability, and recovery of the method were calculated by adding known amounts (approximately 10 mg) of the SRM 1649b to a one-eighth section of blank filters ($n = 10$). Before the analysis, filters were spiked with 1 μL of the deuterated PAH internal standard solution (25 ng·μL^{-1}).

Table 4 shows the results obtained for each PAH by using the TD-GC/MS method, comparing the calculated concentration with the certified values.

Although the column used in this study demonstrated a good resolution for the 16 EPA PAHs, the presence of other PAHs in the urban dust can generate coelution problems with the target compounds [27]. Some PAH pairs such as BbFt and DBahA coeluted with the benzo[j]fluoranthene (BjFt) and dibenzo[a,c]anthracene (DBacA), respectively.

The TD-GC/MS method showed good precision with mean RSD values of 12.2. The accuracy of the TD-GC/MS

method ranged from −22.8% to 25.1%, while the average recovery efficiency was 96.7. These performance parameters of the TD-GC/MS method accomplish the quality objectives for ambient air PAHs stated by ISO 12884:2000 [25], which establishes a precision of ±25%, an accuracy of ±20%, and a recovery efficiency between 75 and 125%. These requirements are accomplished for most PAH; however, there are some exceptions. The lowest molecular weight PAHs (Naph and Acy), with excessively high recoveries, confirmed the overestimation of these compounds when analyzed by using the TD-GC/MS method. These compounds could suffer losses during the sample preparation due to their high volatility. Besides, the presence of interfering compounds in the SRM and the low concentration of Acy could explain these overestimations. Therefore, this method was not applicable to the Naph and Acy analysis.

In order to demonstrate the efficiency of the TD method as compared with other analytical methods, the conventional method (Soxhlet extraction-GC/MS) was also tested (Table 5). Between 300 and 500 mg of the NIST SRM 1649b urban dust was placed on one-eighth of a 150 mm prebaked quartz fiber filter, which was spiked with 1 μL of a solution (0.5 ng·μL^{-1}) of the recovery standards. After the extraction process, the obtained extracts were spiked with 25 μL of a deuterated PAH solution (20 ng·μL^{-1}).

The results of the Soxhlet extraction-GC/MS method showed a good recovery for 4-, 5-, and 6-ring PAHs with values between 72.8 and 131%; whereas the lightest PAHs (2- and 3-ring PAHs), except Phe, showed low recovery (<70%). The loss of these analytes during the extraction process in the Soxhlet Warm mode could be the main reason for these low recoveries. In the case of DBahA, although its coelution with DBacA was considered, its recovery continued to be high (>200%). This indicates an overestimation in the determination of this compound by the Soxhlet process. Regarding precision and accuracy, the Soxhlet extraction-GC/MS showed worse results, with an average precision of 34.9 and values of accuracy out of the limits ±20% for some PAHs.

Comparing both methods, the TD-CG/MS method demonstrated a better performance (good recovery, precision, and accuracy) for the determination of PAHs (except for Naph and Acy). By contrast, the manipulation of the samples in the Soxhlet process meant losses of the light PAHs (2- and 3-ring) and an overestimation of some PAHs, especially of the DBahA.

Regarding the detection limits, the average instrument detection limit (IDL) of the TD-GC/MS method was 0.04 ng and the average method detection limit (MDL), assuming a total sample volume 240 m^3 (30 m^3·h^{-1} for 8 h), was 2.89×10^{-3} ng·m^{-3} [28].

3.4. Extension of the Scope to Other PAHs.
Although most environmental studies are focused on the analysis of 16 PAH listed by US-EPA, it could be interesting to determine other PAHs in order to have a better characterization of these compounds in terms of toxicity and sources. For this reason, besides the 16 EPA PAHs, other 15 PAHs were determinated by using TD-GC/MS. Table 1 shows, according to their

TABLE 4: TD-GC/MS method validation parameters for the 16 EPA PAHs in NIST SRM 1649b urban dust ($n = 10$).

PAH	Experimental mean (ng)[a]	NIST-certified value (ng)[a]	RSD (%)	Recovery (%)	Accuracy[b] (%)
Naph	3694 ± 1082	26.8 ± 3.0	46.3	13809	13709
Acy	6.97 ± 0.61	1.99 ± 0.24	13.9	351	251
Ace	1.57 ± 0.19	2.03 ± 0.41	19.5	77.3	−22.8
FL	2.06 ± 0.21	2.29 ± 0.67	16.4	89.9	−10.1
Phe	42.7 ± 3.3	45.3 ± 0.2	12.0	94.3	−5.70
Ant	12.6 ± 0.8	10.1 ± 0.2	9.82	125	25.1
Ft	59.6 ± 3.2	67.9 ± 0.4	8.56	87.7	−12.3
Pyr	51.8 ± 2.8	51.2 ± 1.4	8.56	101	1.01
BaA	19.3 ± 1.1	21.7 ± 0.5	8.70	88.7	−11.3
Chry	26.2 ± 1.3	31.3 ± 0.3	7.95	83.5	−16.5
BbFt + BjFt	75.0 ± 6.2	81.3 ± 2.3	13.0	92.2	−7.79
BkFt	16.1 ± 1.4	17.5 ± 0.5	13.4	91.9	−8.15
BaP	20.4 ± 1.5	25.4 ± 1.2	11.5	80.2	−19.8
IP	35.5 ± 1.9	29.7 ± 1.7	8.63	119	19.3
DBahA + DBacA	6.35 ± 0.93	6.02 ± 0.11	23.1	105	5.42
BghiP	37.2 ± 2.2	40.8 ± 0.4	9.31	91.1	−8.90
Average[c]	—	—	12.2	96.7	|12.4|

[a]Expanded uncertainty about the mean, with coverage factor, $k = 2$; [b]accuracy = (experimental value − certified value) × 100/certified value; [c]except Naph and Acy.

TABLE 5: Soxhlet extraction-GC/MS method validation parameters for the 16 EPA PAHs in SRM 1649b ($n = 7$).

PAH	Experimental mean (ng)[a]	NIST-certified value (ng)[a]	RSD (%)	Recovery (%)	Accuracy[b] (%)
Naph	67.3 ± 18.6	391 ± 35	33.8	17.2	−82.8
Acy	20.2 ± 5.9	79.9 ± 9.5	35.9	25.2	−74.8
Ace	25.0 ± 5.4	81.6 ± 16.5	26.5	30.7	−69.4
FL	32.6 ± 9.2	92.3 ± 14.4	34.5	35.3	−64.7
Phe	1215 ± 331	1668 ± 24	33.4	72.8	−27.2
Ant	104 ± 25	169 ± 1	32.5	61.8	−38.2
Ft	2392 ± 559	2573 ± 32	31.0	93.0	−7.05
Pyr	1914 ± 398	2054 ± 57	27.6	93.2	−6.79
BaA	808 ± 148	870 ± 20	24.3	92.9	−7.13
Chry	1632 ± 464	1256 ± 11	37.6	129	29.9
BbFt + BjFt	3144 ± 800	3260 ± 91	33.7	96.4	−3.58
BkFt	921 ± 319	702 ± 20	45.9	131	31.3
BaP	1019 ± 271	1018 ± 98	35.2	100	0.07
IP	1109 ± 296	1192 ± 65	35.4	93.1	−6.89
DBahA + DBacA	507 ± 176	241 ± 4	45.9	363	263
BghiP	2143 ± 580	1777 ± 32	35.8	120	20.6
Average[c]	—	—	34.9	120	|36.8|

[a]Expanded uncertainty about the mean, with coverage factor, $k = 2$; [b]accuracy = (experimental value − certified value) × 100/certified value; [c]except Naph, Acy, Ace, and FL.

elution order, the 16 EPA PAHs, the deuterated PAHs, and the 15 nonpriority PAHs.

Because SRM 1649b contains other compounds besides the 16 EPA PAHs, it was used to identify 15 nonpriority PAHs and to quantify them by relative response factors (RRFs). Supplementry Figures S4–S7 show the SIM chromatograms of the target PAHs (16 EPA PAHs + 15 PAHs) obtained in the analysis of SRM 1649b. For quantification, the RRFs for each nonpriority PAH were calculated by the following equation:

$$\text{RRF}_{\text{PAH}} = \frac{A_{\text{PAH}} \cdot C_{\text{ref-PAH}}}{A_{\text{ref-PAH}} \cdot C_{\text{PAH}}}, \qquad (2)$$

where A_{PAH} is the peak area of nonpriority PAH, $A_{\text{ref-PAH}}$ is the peak area of reference PAH compound, C_{PAH} is the

nonpriority PAH concentration in the NIST dust, and $C_{\text{ref-PAH}}$ is the reference PAH concentration in the NIST dust.

Reference PAHs were selected according to the following criterion: the nearest of the 16 EPA PAHs to each new one, which provides the least variation in the RRF. Table 6 collects the reference PAHs, RRFs, and the relative standard deviations (RSDs) for each nonpriority PAH.

The nonpriority PAHs showed a range of RRFs between 0.31 and 4.74, with RSD of less than 15% for most compounds. In the case of Ret, the low chemical similarity between this compound and its reference PAH (BaA) could explain the poor precision in the determination of its RRF (>20%).

3.5. Application to Real Samples. After validation, the method described in this study was applied to extract and analyze samples

TABLE 6: Reference PAH, RRFs, and the relative standard deviations (RSDs) for each nonpriority PAH.

Nonpriority PAH	Reference PAH	RRF	RSD (%)
BghiFt	BaA	1.29	6.68
BcP	BaA	0.76	13.2
CPP	BaA	0.31	15.1
Triph	BaA	0.69	16.1
Ret	BaA	1.13	25.6
BaFt	BkFt	0.99	11.3
BeP	BaP	1.31	4.10
Per	BaP	0.94	4.05
DBajA	IP	4.74	4.49
BbC	IP	1.08	6.16
Pic	IP	0.64	10.2
Anthan	BghiP	0.40	8.54
Cor	BghiP	0.54	15.8

TABLE 7: Descriptive statistics of the individual particle-bound PAH concentrations measured in the city of Bilbao.

PAH	N	Average ($ng \cdot m^{-3}$)	SD ($ng \cdot m^{-3}$)	Min. ($ng \cdot m^{-3}$)	Max. ($ng \cdot m^{-3}$)	5th percentile ($ng \cdot m^{-3}$)	95th percentile ($ng \cdot m^{-3}$)
Ace*	180	0.22	0.27	4.00×10^{-3}	1.69	0.02	0.85
FL*	182	0.08	0.07	0.01	0.61	0.02	0.22
Phe*	182	0.17	0.12	0.03	0.80	0.05	0.41
Ant*	182	0.04	0.05	4.00×10^{-3}	0.48	0.01	0.14
Ft*	182	0.26	0.22	0.03	1.38	0.06	0.73
Pyr*	182	0.27	0.23	0.02	1.48	0.05	0.73
BghiFt	118	0.20	0.19	0.01	0.87	0.03	0.60
BcP	181	0.06	0.06	4.00×10^{-3}	0.40	8.00×10^{-3}	0.20
CPP	73	0.07	0.16	3.00×10^{-3}	0.95	5.00×10^{-3}	0.48
BaA*	182	0.16	0.22	0.01	1.45	0.02	0.62
Triph	164	0.14	0.13	0.02	0.74	0.03	0.43
Chry*	182	0.22	0.24	0.03	1.35	0.04	0.81
BbFt* + BjFt	175	0.50	0.76	0.03	5.98	0.06	2.08
BkFt*	174	0.18	0.23	0.01	1.41	0.03	0.64
BaFt	141	0.05	0.07	3.00×10^{-3}	0.39	4.00×10^{-3}	0.21
BeP	169	0.26	0.32	0.01	1.83	0.03	0.93
BaP*	170	0.16	0.20	0.01	1.16	0.02	0.70
Per	159	0.03	0.04	2.00×10^{-3}	0.19	4.00×10^{-3}	0.14
DBajA	111	0.01	0.01	2.00×10^{-4}	0.06	4.00×10^{-4}	0.03
IP*	161	0.17	0.24	1.00×10^{-3}	1.52	0.01	0.70
DBahA* + DBacA	143	0.05	0.06	3.00×10^{-4}	0.38	3.00×10^{-3}	0.19
BbC	62	0.01	0.01	2.00×10^{-3}	0.05	3.00×10^{-3}	0.04
Pic	63	0.03	0.04	3.00×10^{-3}	0.24	4.00×10^{-3}	0.13
BghiP*	174	0.20	0.19	0.01	1.03	0.04	0.66
Anthan	48	0.03	0.05	3.00×10^{-3}	0.26	3.00×10^{-3}	0.15
Cor	53	0.10	0.12	0.01	0.44	0.02	0.36

*PAH listed as priority pollutant by US-EPA.

collected in the city of Bilbao, Spain (longitude 2°56'56.24''W, latitude 3°15'44.86''N). Bilbao city is the most populated area in the Basque Country and the tenth largest in Spain (approximately 350,000 in the city and 1 million inhabitants in the metropolitan area). In this urban area, local traffic and stationary emissions from the surrounding industries are considered as the major sources of atmospheric pollutants [28].

During seven consecutive days per month, eight-hour PM_{10} samples were collected at a flow rate of $30 \, m^3 \cdot h^{-1}$.

A total of 182 PM_{10} samples were collected over 9 months (between July 2013 and June 2014). Each sample was randomly cut into 8 portions of $1 \, cm^2$ and introduced into the sampling tube and analyzed using the optimized method.

This was performed in the same way as other studies [29, 30], which demonstrated good homogeneity results when using small sections of the filters.

Table 7 shows the descriptive statistics (number of valid data, mean, standard deviation, minimum, maximum, 5th, and 95th percentiles) for individual PAH concentrations measured by using the TD-GC/MS method in the city of Bilbao (urban area). PAHs which showed overestimation in the SRM analysis (Naph and Acy) or poor precision in the RRF determination (Ret) were not measured in the real samples.

The average concentration of individual EPA PAHs in Bilbao ranged from 0.04 ± 0.05 to $0.50 \pm 0.76 \, ng \cdot m^{-3}$, whereas the nonpriority PAHs were between 0.01 ± 0.01 and 0.26 ± 0.32 ng·m^{-3}. The EPA PAHs reported minimum values between 4.00×10^{-3} and 0.03 ng·m^{-3} for most of the compounds, which are between 1.1 and 14.6 times the MDL, showing the suitability of the proposed method to determine particle-bound PAHs in real PM$_{10}$ samples. Although the minimums of IP and DBahA were below their MDL, these values meant only the 5% or less of the measured samples.

Among compounds, BbFt was the major contributor to total PAHs (average concentration of 0.5 ± 0.76 ng·m^{-3}), followed by Pyr (0.27 ± 0.23 ng·m^{-3}), Ft (0.26 ± 0.22 ng·m^{-3}), BeP (0.26 ± 0.32 ng·m^{-3}), and Chry (0.22 ± 0.24 ng·m^{-3}). The high presence of these compounds in PM$_{10}$ fraction has been reported by previous studies [31, 32] in urban areas with traffic loads.

4. Conclusions

The method developed in this study, based on thermal desorption, showed a good efficiency for the determination of particle-bound PAHs. The use of a solvent-free extraction technique has showed numerous advantages (less sample manipulation and analysis time, reduced exposure risk, and higher sensitivity and reliability) that enable a better performance (good recovery, precision, and accuracy) for the determination of particle-bound PAHs; however, the lowest molecular weight PAHs (Naph and Acy) could be overestimated by this methodology.

Parameters such as tube and trap temperature, time, desorption, and split flows (inlet and outlet) were critical in the thermal desorption of PAHs. The optimized TD-GC/MS method showed an efficient desorption of PAHs with recoveries higher than 94%.

The results obtained in the validation of TD-GC/MS by standard reference material (urban dust) demonstrated that this is a reliable method to determine particulate PAHs in aerosol samples (good precision and accuracy), with average recovery efficiency of 96.67 and a mean RSD value of 12.18. Comparing with the conventional method Soxhlet-GC/MS, the TD-CG/MS method demonstrated a better performance for the determination of PAHs. Besides 16 EPA PAHs, the TD-GC/MS method demonstrated its ability to quantify other PAHs in aerosol samples.

Finally, the method was successfully applied for the quantification of PAHs in real PM$_{10}$ samples collected with a time resolution of 8 h.

Conflicts of Interest

The authors declare that there are no conflicts of interest regarding the publication of this paper.

Acknowledgments

The authors gratefully thank the University of the Basque Country UPV/EHU (Ref.: GIU 13/25, GIU 16/03, and UFI 11/47) and the Spanish Ministry of Science and Innovation (MICINN) for financing the project PROMESHAP (Ref.: CTM 2010-20607). Iñaki Elorduy wants to thank the MICINN for his doctoral grant Ministerio de Ciencia e Innovació.

Supplementary Materials

The representative SCAN chromatograms of the 16 EPAs and deuterated PAHs and also the SIM chromatograms of the target PAHs (16 EPA PAH + 15 PAH) obtained in the SRM 1649b analysis are shown in the supplementary material. Figure S1: chromatogram of 16 EPA and deuterated PAHs in SCAN mode, from 0 to 15 min: (1) Naph-d8, (2) Naph, (3) Bph-d10, (4) Acy, (5) Ace, and (6) FL. Figure S2: chromatogram of 16 EPA and deuterated PAHs in SCAN mode, from 15 to 20.5 min: (7) Phe-d10, (8) Phe, (9) Ant, (10) Ft, (11) Pyr-d10, and (12) Pyr. Figure S3: chromatogram of 16 EPA and deuterated PAHs in SCAN mode, from 22.5 to 35.5 min: (13) BaA-d12, (14) BaA, (15) Chry, (16) BbFt, (17) BkFt, (18) BaP-d10, (19) BaP, (20) IP, (21) DBahA, (22) BghiP-d12, and (23) BghiP. Figure S4: PAHs and deuterated PAHs in SIM windows (m/z 226, 240, 228, and 234) in the analysis of NIST SRM 1649b dust. Figure S5: PAHs and deuterated PAHs in SIM windows (m/z 252 and 264) in the analysis of NIST SRM 1649b dust. Figure S6: PAHs and deuterated PAHs in SIM windows (m/z 276, 288, and 278) in the analysis of NIST SRM 1649b dust. Figure S7: PAHs and deuterated PAHs in m/z 300 SIM window in the analysis of NIST SRM 1649b dust. (*Supplementary Materials*)

References

[1] R. H. Peng, A. S. Xiong, Y. Xue et al., "Microbial biodegradation of polyaromatic hydrocarbons," *FEMS Microbiology Reviews*, vol. 32, no. 6, pp. 927–955, 2008.

[2] R. Fernández-González, I. Yebra-Pimentel, E. Martínez-Carballo, J. Simal-Gándara, and X. Pontevedra-Pombal, "Atmospheric pollutants in fog and rain events at the northwestern mountains of the Iberian Peninsula," *Science of the Total Environment*, vol. 497-498, pp. 188–199, 2014.

[3] B. J. Finlayson-Pitts and J. N. Pitts Jr., *Chemistry of the Upper and Lower Atmosphere-Theory, Experiments, and Applications*, Academic Press, San Diego, CA, USA, 2000.

[4] K. Ravindra, R. Sokhi, and R. V. Vangrieken, "Atmospheric polycyclic aromatic hydrocarbons: source attribution, emission factors and regulation," *Atmospheric Environment*, vol. 42, no. 13, pp. 2895–2921, 2008.

[5] M. Howsam and K. C. Jones, "Sources of PAHs in the environment," in *Anthropogenic Compounds: PAHs and Related Compounds*, A. H. Neilson, Ed., pp. 137–174, Springer, Berlin, Germany, 1998.

[6] A. A. Jamhari, M. Sahani, M. T. Latif et al., "Concentration and source identification of polycyclic aromatic hydrocarbons (PAHs) in PM$_{10}$ of urban, industrial and semi-urban areas in Malaysia," *Atmospheric Environment*, vol. 86, pp. 16–27, 2014.

[7] C. E. Boström, P. Gerde, A. Hanberg et al., "Cancer risk assessment, indicators, and guidelines for polycyclic aromatic hydrocarbons in the ambient air," *Environmental Health Perspectives*, vol. 110, no. s3, pp. 451–489, 2002.

[8] United Nations Economic Commission for Europe (UNECE), *Convention on Longrange Trans-Boundary Air Pollution*, United Nations Economic Commission for Europe (UNECE), Aarhus, Denmark, December 2017, http://www.unece.org/env/lrtap/pops_h1.html.

[9] Council Decision 2004/259/EC of 19 February 2004, *Concerning the Conclusion, on Behalf of the European Community,*

of the 1988 Protocol to the 1979 Convention on Long Range Transboundary Air Pollution on Persistent Organic Pollutants [OJ L 81 of 19.03.2004], December 2017, http://eur-lex.europa.eu/legal-content/EN/TXT/?uri=CELEX:32004D0259.

[10] US Environmental Protection Agency, US-EPA, "The U.S. Environmental Protection Agency, list of priority pollutants, Office of water, water quality standards database, Office of the Federal Registration (OFR) appendix A: priority pollutants," *Federal Register*, vol. 47, p. 52309, 1982.

[11] H. L. Sheu, W. J. Lee, S. J. Lin, G. C. Fang, H. C. Chang, and W. C. You, "Particle-bound PAH content in ambient air," *Environmental Pollution*, vol. 96, pp. 369–382, 1997.

[12] N.-D. Dat and M. B. Chang, "Review on characteristics of PAHs in atmosphere, anthropogenic sources and control technologies," *Science of the Total Environment*, vol. 609, pp. 682–693, 2017.

[13] H. Sharma, V. K. Jain, and Z. H. Khan, "Characterization and source identification of polycyclic aromatic hydrocarbons (PAHs) in the urban environment of Delhi," *Chemosphere*, vol. 66, no. 2, pp. 302–310, 2007.

[14] K. Srogi, "Monitoring of environmental exposure to polycyclic aromatic hydrocarbons: a review," *Environmental Chemistry Letters*, vol. 5, no. 4, pp. 169–195, 2007.

[15] M. Tsapakis and E. G. Stephanou, "Occurrence of gaseous and particulate polycyclic aromatic hydrocarbons in the urban atmosphere: Study of sources and ambient temperature effect on the gas/particle concentration and distribution," *Environmental Pollution*, vol. 133, no. 1, pp. 147–156, 2005.

[16] A. Lottmann, E. Cadé, M. L. Geagea et al., "Separation of molecular tracers sorbed onto atmospheric particulate matter by flash chromatography," *Analytical Bioanalytical Chemistry*, vol. 387, no. 5, pp. 1855–1861, 2007.

[17] S. Chantara and W. Sangchan, "Sensitive analytical method for particle-bound polycyclic aromatic hydrocarbons: a case study in Chiang Mai, Thailand," *ScienceAsia*, vol. 35, pp. 42–48, 2009.

[18] D. L. Poster, M. M. Schantz, L. C. Sander, and S. A. Wise, "Analysis of polycyclic aromatic hydrocarbons (PAHs) in environmental samples: a critical review of gas chromatographic (GC) methods," *Analytical and Bioanalytical Chemistry*, vol. 386, no. 4, pp. 859–881, 2006.

[19] L. B. Liu, Y. Liu, J. M. Lin, N. Tang, K. Hayakawa, and T. Maeda, "Development of analytical methods for polycyclic aromatic hydrocarbons (PAHs) in airborne particulates: a review," *Journal of Environmental Sciences*, vol. 19, no. 1, pp. 1–11, 2007.

[20] M. S. Alam, I. J. Keyte, J. Yin, C. Stark, A. M. Jones, and R. M. Harrison, "Diurnal variability of polycyclic aromatic compound (PAC) concentrations: relationship with meteorological conditions and inferred sources," *Atmospheric Environment*, vol. 122, pp. 427–438, 2015.

[21] J. Ringuet, A. Albinet, E. Leoz-Garziandia, H. Budzinski, and E. Villenave, "Diurnal/nocturnal concentrations and sources of particulate-bound PAHs, OPAHs and NPAHs at traffic and suburban sites in the region of Paris (France)," *Science of the Total Environment*, vol. 437, pp. 297–305, 2012.

[22] S. K. Pandey, K. H. Kim, and R. J. C. Brown, "A review of techniques for the determination of polycyclic aromatic hydrocarbons in air," *TrAC Trends in Analytical Chemistry*, vol. 30, no. 11, pp. 1716–1739, 2011.

[23] J. E. Szulejko, K.-H. Kim, R. J. Brown, and M.-S. Bae, "Review of progress in solvent-extraction techniques for the determination of polyaromatic hydrocarbons as airborne pollutants," *TrAC Trends in Analytical Chemistry*, vol. 61, pp. 40–48, 2014.

[24] S. Elcoroaristizabal, A. de Juan, J. A. García, I. Elorduy, N. Durana, and L. Alonso, "Chemometric determination of

PAHs in aerosol samples by fluorescence spectroscopy and second-order data analysis algorithms," *Journal of Chemometrics*, vol. 28, no. 4, pp. 260–271, 2014.

[25] ISO, *International Standard ISO 12884-Ambient air-Determination of Total (gas and particle-phase) PAHs: Collection on Sorbent-Backed Filters with GC/MS Analyses*, ISO, Geneva, Switzerland, 2000.

[26] PerkinElmer, *TurboMatrix Series Thermal Desorbers: User's Guide*, PerkinElmer, Inc., UK, 2007.

[27] L. R. Bordajandi, M. Dabrio, F. Ulberth, and H. Emons, "Optimisation of the GC-MS conditions for the determination of the 15 EU foodstuff priority polycyclic aromatic hydrocarbons," *Journal of Separation Science*, vol. 31, no. 10, pp. 1769–1778, 2008.

[28] I. Elorduy, S. Elcoroaristizabal, N. Durana, J. A. García, and L. Alonso, "Diurnal variation of particle-bound PAHs in an urban area of Spain using TD-GC/MS: influence of meteorological parameters and emission sources," *Atmospheric Environment*, vol. 138, pp. 87–98, 2016.

[29] J. Ringuet, A. Albinet, E. Leoz-Garziandia, H. Budzinski, and E. Villenave, "Reactivity of polycyclic aromatic compounds (PAHs, NPAHs and OPAHs) adsorbed on natural aerosol particles exposed to atmospheric oxidants," *Atmospheric Environment*, vol. 61, pp. 15–22, 2012.

[30] E. Grandesso, P. Pérez Ballesta, and K. Kowalewski, "Thermal desorption GC-MS as a tool to provide PAH certified standard reference material on particulate matter quartz filters," *Talanta*, vol. 105, pp. 101–108, 2013.

[31] J. Aldabe, C. Santamaría, D. Elustondo et al., "Polycyclic aromatic hydrocarbons (PAHs) sampled in aerosol at different sites of the western Pyrenees in Navarra (Spain)," *Environmental Engineering Management Journal*, vol. 11, pp. 1049–1058, 2012.

[32] M. S. Callén, J. M. López, A. Iturmendi, and A. M. Mastral, "Nature and sources of particle associated polycyclic aromatic hydrocarbons (PAH) in the atmospheric environment of an urban area," *Environmental Pollution*, vol. 183, pp. 166–174, 2013.

8

Simultaneous Quantitative Determination of Synthetic Cathinone Enantiomers in Urine and Plasma using GC-NCI-MS

Rashed Alremeithi,[1,2] **Mohammed A. Meetani** ⓘ**,**[1] **Anas A. Alaidaros,**[1] **Adnan Lanjawi,**[2] **and Khalid Alsumaiti**[2]

[1]*Chemistry Department, United Arab Emirates University, P.O. Box 15551, Al-Ain, UAE*
[2]*General Department of Forensic Science and Criminology, Dubai Police, Dubai, UAE*

Correspondence should be addressed to Mohammed A. Meetani; mmeetani@uaeu.ac.ae

Academic Editor: Erwin Rosenberg

Development and validation of sensitive and selective method for enantioseparation and quantitation of synthetic cathinones is reported using GC-MS triple quadrupole mass spectrometry with negative chemical ionization (NCI) mode. Indirect chiral separation of thirty-six synthetic cathinone compounds has been achieved by using an optically pure chiral derivatizing agent (CDA) called (S)-(−)-N-(trifluoroacetyl)pyrrolidine-2-carbonyl chloride (L-TPC), which converts cathinone enantiomers into diastereoisomers that can be separated on achiral columns. As a result of using Ultra Inert 60 m column and performing slow heating rate (2°C/min) on the GC oven, an observed enhancement in enantiomer peak resolution has been achieved. An internal standard, (+)-cathinone, was used for quantitation of synthetic cathinones. Method validation in terms of linearities and sensitivity in terms of limits of detection (LODs), limits of quantitation (LOQs), recoveries, and reproducibilities has been obtained for fourteen selected compounds that examined simultaneously as a mixture after being spiked in urine and plasma. It was found that the LOD of the fourteen synthetic cathinones in urine was in the range of 0.26–0.76 μg/L, and in plasma, it was in the range of 0.26–0.34 μg/L. While the LOQ of the mixture in urine was in the range of 0.86–2.34 μg/L, and in plasma, it was in the range of 0.89–1.12 μg/L. Unlike the electron impact (EI) ion source, NCI showed better sensitivity by two orders of magnitude by comparing the obtained results with the recently published reports for quantitative analysis and enantioseparation of synthetic cathinones.

1. Introduction

From the beginning of the new century till now, governments and forensic science specialists are suffering from a nightmare called new designer substances (NDS), which comprise a risk in society that is growing up day by day. Presently, the latest version of NDS is called "bath salts," and they overrun the drug of abuse market. Bath salts are a group of central nervous system stimulants that consists mainly of synthetic cathinone derivatives [1]. In nature, cathinone (β-keto amphetamine) exists in the leaves of the *Catha edulis* plant, which can be found easily in the region of northeast Africa and the Arabian Peninsula [2]. However, scientists have synthesized cathinones in laboratory when the Germans and the French chemists synthesized methcathinone for the first time in the late 1920s [3]. During the 1930s and 1940s, methcathinone was available in pharmacological markets as an appetite suppressant and antidepressant medicine [4]. Methcathinone abuse spread to the USA at 1991, and as a result of that, it was included in the UN Convention on Psychotropic Substances [5]. In the meantime, drug dealers were looking for new strategies to sell their products and they found it by the "novel psychoactive substances," drugs which contain at least one chemical substance that has similar biological effects as of illegal drugs. For instance, "Explosion" is the trade name of the synthetic cathinone methylone, which emerged for sale in Japan and Netherlands via the Internet in 2004 [6]. In 2007, 4-methyl methcathinone (mephedrone) became one of the most commonly abused drugs in Europe [6]. Thus, concerns about the abuse of novel psychoactive substances especially cathinone-related derivatives grew up in Europe which gave rise to ban of cathinone derivatives in April 2010 by the UK government and by the European Monitoring Centre for Drugs and Drug Addiction (EMCDDA) [7]. Despite all the actions taken by legal authorities, an intense attention by drug dealers has been put on the synthesis of new generations of synthetic cathinone derivatives.

In order to obviate the abuse risks of these psychoactive stimulants, focused studies should be carried out on the neuropharmacological properties of the active compounds. This can be accomplished by separating the enantiomers using a selective and sensitive method. However, the current separation and detection methods are not completely effective; therefore, a new separation and detection method is reported in this work. In nature, cathinone exists as a racemic mixture that contains one chiral center which means that it has two enantiomers and commonly one of them will have greater psychological effect in human biological system than the other enantiomer [8]. For example, it has been found that the stimulating effect of (S)-methcathinone is higher than (R)-methcathinone [8]. However, the literature limitation of the pharmacological data for the new cathinone derivatives racemates lets researchers assume that the case for most phenylalkylamine compounds will be similar to methcathinone. As a result of that, enantioseparation of chiral synthetic cathinones became an attractive and promised field of research where the use of major separation techniques took place such as gas chromatography [8, 9], high-performance liquid chromatography (HPLC) [10–15], and capillary electrophoresis (CE) [15–22].

Generally, the principle of chiral separation can be summarized by two different techniques: direct and indirect chiral separation. The previously mentioned separation techniques can be satisfied by applying chiral separation principles. The use of direct separation technique for the enantiomers implies the use of chiral selector which can be either immobilized on the stationary phase of the column or dissolved in the mobile phase of the separation system as in the case for some HPLC and CE chiral methods [23]. However, indirect chiral separation can be achieved by converting enantiomers to diastereoisomers via derivatization reaction of the targeted compounds with optically pure chiral derivatizing agents (CDAs) [9]. Moreover, the resulted diastereoisomers could be separated on achiral stationary phase column in GC or HPLC system. (S)-(−)-N-(trifluoroacetyl)pyrrolidine-2-carbonyl chloride (L-TPC) is one of the well-known CDAs that is readily available in chemical market and shows impressive results in chiral separation of the phenylalkylamines mainly on GCMS after derivatization reaction [8, 9, 24].

In the literature, only few papers have discussed the chiral separation of L-TPC cathinone derivatives by using GC-EI-MS [8, 9, 25]. Electron impact (EI) is the most preferable ionization source in GC-MS, which provides characteristic and reproducible mass spectrum for each compound. EI is considered as a hard ionization technique which provides mass spectra that are rich with low mass fragments and usually the molecular ion peak is absent [26]. Recently, a short communication on the analysis of twenty-nine synthetic cathinones in GC-MS/MS with positive chemical ionization (PCI) mode has been reported [27]. However, no quantitative assessment was given for these compounds in biological fluids. Unlike EI, determination of molecular weight and structure elucidation can be carried out through the use of chemical ionization source coupled with tandem mass spectrometry [27]. Furthermore, when the investigated compounds are electronegative moieties, the use of NCI mode can dramatically improve the sensitivity of the targeted compounds [28]. In NCI, negative ions are mainly formed by capturing thermal electrons (low-energy electrons with nearly 0–2 eV), and this ionization process is called resonance electron capture. The electrons are produced from the filament and lose their energy by collision and ionization of the reagent gas molecules. If electrons have enough energy (2–15 eV) to break up molecules, fragmentation occurs and this ionization process is called dissociative electron capture. NCI is highly sensitive and selective for compounds with a positive electron affinity. It is a soft ionization method, like PCI, so a NCI spectrum is relatively simple [29].

There are no reports in the literature that discuss the use of GC-MS in negative chemical ionization mode for quantitative analysis of synthetic cathinones. The electrons emitted from a filament lose their energy to become thermal electrons by collision with reagent gas and ionization of reagent gas molecules. Nearly, 0 eV electrons are captured by molecules so that molecular ions are produced (resonance electron capture). If electrons have enough energy to break up molecules, fragmentation occurs (dissociative electron capture) [29].

In this work, a sensitive and selective GC-NCI-MS method has been developed to analyze thirty-six synthetic cathinone compounds after their conversion into diastereoisomers through the derivatization reaction with L-TPC. Quantitative analysis of spiked urine and plasma samples was conducted for fourteen of these synthetic cathinones (Scheme 1), which were analyzed in one mixture simultaneously. The method validation was performed on spiked biological samples and found to produce complete separation of the synthetic cathinone enantiomers on achiral capillary GC column in addition to sensitive detection of low concentrations in the μg/L range better than the previous reported methods that use EI and positive CI ionization mass spectrometry.

2. Experimental

2.1. Chromatographic Conditions. Chromatographic separation was performed on an Agilent 7890A GC coupled to an Agilent 7000 Triple Quad mass selective detector. A commercially available 60 m HP-5MS Ultra Inert capillary column, with 0.25 mm inner diameter and a 0.25 μm film thickness was used as the stationary phase. Chemical ionization (CI) with methane gas (40%, 2.0 mL/min) was employed in the negative ion mode at a voltage of 70 eV. Helium was used as the carrier gas at a constant flow rate of

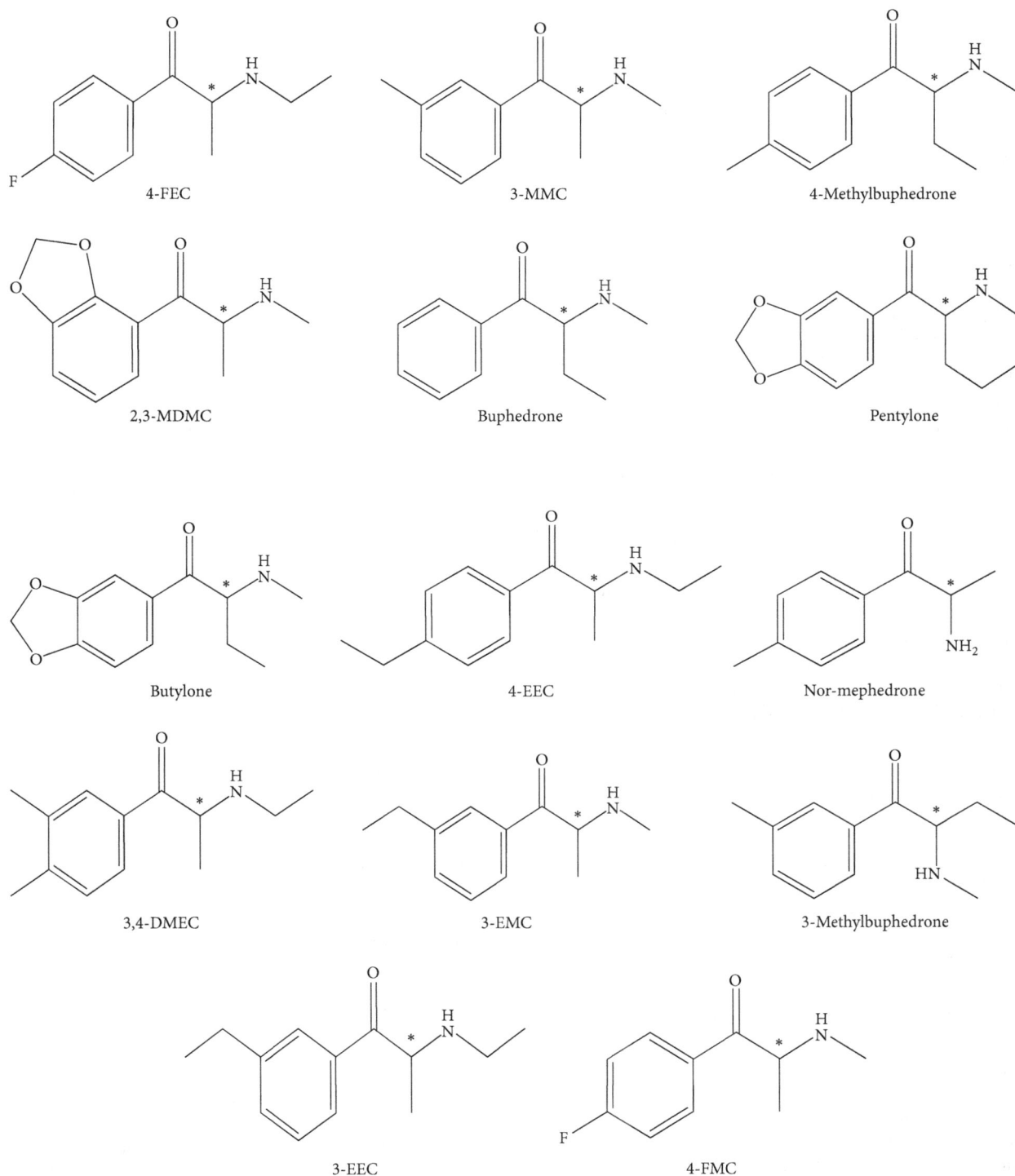

4-FEC

3-MMC

4-Methylbuphedrone

2,3-MDMC

Buphedrone

Pentylone

Butylone

4-EEC

Nor-mephedrone

3,4-DMEC

3-EMC

3-Methylbuphedrone

3-EEC

4-FMC

SCHEME 1

0.8 mL/min. Injection of 3 μl of sample solution was performed automatically in splitless mode. The injector and GC-MS interface temperatures were set at 250 and 280°C, respectively. Data collection was performed in selected ion monitoring (SIM) mode with the selected fragment ions as shown in Table 1, starting at 30 min after injection. The column temperature program was as follows: starting at 160°C and then holding for 5 min, followed by subsequent heating to 260°C at a heating rate of 2°C/min. The final temperature was held at 260°C for 10 min.

2.2. Chemicals and Reagents. All chemicals were of analytical grade. Ethyl acetate, acetic acid, methanol, 2-propanol,

TABLE 1: Time segments table with selected ions used in SIM mode for the analysis of cathinone mixture.

Compound name	Abbreviation	Time	Mass
(+)-Cathinone	—	39.00–41.00	189*, 209, 342
4-Fluoromethcathinone	4-FMC	41.00–42.50	153, 223*, 374
4-Fluoroethcathinone	4-FEC	42.50–44.70	167*, 237, 388
Nor-mephedrone	—	44.70–45.06	189, 209*, 356
Buphedrone	—	45.06–47.13	153*, 223, 370
3-Methylmethcathinone	3-MMC		
Nor-mephedrone	—	47.13–48.00	189*, 209, 356
3-Methylbuphedrone	—	48.00–50.15	153*, 223, 384
4-Methylbuphedrone	—		
3-Ethylmethcathinone	3-EMC		
3-Ethylethcathinone	3-EEC		
4-Ethylethcathinone	4-EEC	50.15–54.00	167*, 237, 398
3,4-Dimethylethcathinone	3,4-DMEC		
2,3-Methylenedioxymethcathinone	2,3-MDMC	54.00–59.00	153*, 223, 400
Butylone	—	59.00–62.00	153, 223*, 414
Pentylone	—	62.00–65.00	156, 223*, 428

*The quantifier mass.

ammonium hydroxide, dichloromethane, 0.1 M solution of (S)-(−)-N-(trifluoroacetyl)pyrrolidine-2-carbonyl chloride (L-TPC) with an enantiomer excess (ee) of 97% (according to the supplier's specification) in methylene chloride, anhydrous sodium sulfate, and sodium phosphate were obtained from Sigma-Aldrich Chemicals (St. Louis, MO, USA). Potassium carbonate was obtained from VWR (Darmstadt, Germany). Doubly deionized water was obtained from Ultra-Pure Millipore system (MS, USA). All chemicals shown in Table 2 were purchased from Cayman Chemicals (Michigan, USA) and were provided as racemic mixtures for individual cathinones (99% purity).

2.3. Sample Preparation

2.3.1. Samples. This investigation conforms to the UAE community guidelines for the use of humans in experiments. The Human Ethics Committee at the Dubai Police approved this study. Blood and urine samples were collected by Dubai Police with the consent of the subjects.

2.3.2. Solid-Phase Extraction (SPE) of Spiked Urine and Plasma Samples. SPE was carried out using "Zymark rapid trace" SPE workstation (Artisan Technology Group, Champaign, IL, USA), and the column was 200MG clean screen CSDAU203 from FluoroChem (Hadfield, UK). Urine samples were diluted in 1 : 2 ratio with doubly deionized water. Diluted urine (3 mL) was spiked with certain concentration of synthetic cathinones and 20 μg/L of IS ((+)-cathinone) in addition to 1 mL of 0.1 M phosphate buffer (pH 6). For the spiking of plasma samples, 1 mL of plasma was spiked with certain concentration of synthetic cathinones and 50 ppm of IS ((+)-cathinone) was added in addition to 3 mL of 0.1 M phosphate buffer (pH 6). Sample was shaken thoroughly for 30 s. The SPE cartridge was conditioned by adding 3 mL of methanol, and the same volume of deionized water was used with 1 mL of 0.1 M phosphate buffer. After that, the spiked urine or plasma sample was loaded to the cartridge and later the cartridge

was washed by 3 mL of methanol followed by 3 mL of deionized water, and finally, 1 mL of 0.1 M acetic acid was added. The column was left for drying for 5 min. Finally, 3 mL of the eluate was collected and evaporated to dryness under nitrogen gas. Solid-phase extraction procedure is summarized in Scheme 2.

2.3.3. Derivatization Step. For the analysis of pure and spiked samples, evaporation step is necessary before derivatization reaction can take place. After the evaporation is done, 100 μl of deionized water was transferred into a glass test tube containing the pure sample together with 125 μl of a saturated aqueous solution of potassium carbonate, 1.5 mL of ethyl acetate, and 12.5 μl of L-TPC. For the analysis of spiked urine and plasma, 50 μl of L-TPC was used. The mixture was covered and stirred for 10 min at room temperature. Afterwards, the upper layer was transferred to a new test tube and dried over anhydrous sodium sulfate. The dried solution was evaporated to completion under a gentle nitrogen stream. The remaining L-TPC derivative was reconstituted in certain amount of ethyl acetate—depending on concentration—prior to injection in GC-MS instrument. Scheme 3 summarizes the L-TPC derivatization process of the synthetic cathinones.

2.4. Method Validation. The combination of SPE with L-TC derivatization proved to be useful for the determination of synthetic cathinones in urine and plasma samples, as no interferences from endogenous and exogenous compounds were observed. During the method validation, various parameters of the method such as linearity, sensitivity, accuracy, recovery, and reproducibility were evaluated according to international criteria.

3. Results

The indirect chiral separation method that has been developed is based on the conversion of synthetic cathinones to L-TPC derivatives. A normal (or achiral) stationary phase

TABLE 2: List of the 36 cathinone-related compounds and their synonyms, in addition to the retention times of the separated two diastereoisomers for each compound analyzed by GC-MS using SIM mode.

	Name	Synonyms	Time (min)		FWHM		Resolution	Selectivity factor (α)
			t_{R1}	t_{R2}	P1	P2		
1	2-Methoxymethcathinone	2-MeOMC	48.98	49.98	0.061	0.062	9.59	1.03
2	3-Fluoroethcathinone	3-FEC	43	43.3	0.066	0.068	2.64	1.01
3	4-Fluoroethcathinone	4-FEC	42.7	43.2	0.084	0.086	3.47	1.01
4	2,3-Methylenedioxymethcathinone	2,3-MDMC	55.1	56.4	0.078	0.075	10.03	1.03
5	2-Methylmethcathinone	2-MMC	45.1	46.2	0.065	0.07	9.61	1.03
6	Nor-mephedrone	—	44.9	47.2	0.078	0.081	17.07	1.06
7	4-Ethylethcathinone	4-EEC	51.6	52.5	0.075	0.076	7.03	1.02
8	3,4-Dimethylethcathinone	3,4-DMEC	52.8	53.6	0.072	0.072	6.56	1.02
9	2-Ethylmethcathinone	2-EMC	47.5	48.6	0.064	0.07	9.69	1.03
10	3-Methoxymethcathinone	3-MeOMC	51.4	51.7	0.059	0.06	2.97	1.01
11	2-Fluoromethcathinone	2-FMC	41.9	43	0.073	0.076	8.71	1.03
12	4-Ethylmethcathinone	4-EMC	50.5	51.7	0.061	0.061	11.61	1.03
13	3-Ethylethcathinone	3-EEC	50.4	51	0.078	0.072	4.72	1.01
14	4-Methylbuphedrone	—	48.96	49.4	0.074	0.078	3.42	1.01
15	2,3-Dimethylmethcathinone	2,3-DMMC	49.7	51.1	0.062	0.069	12.61	1.03
16	3-Ethylmethcathinone	3-EMC	49.8	50	0.078	0.074	1.55	1.00
17	3-Fluoromethcathinone	3-FMC	41.7	41.98	0.07	0.065	2.45	1.01
18	4-Fluoromethcathinone	4-FMC	41.5	41.96	0.092	0.084	3.08	1.01
19	2-Methylethcathinone	2-MEC	46.3	47.6	0.073	0.068	10.88	1.03
20	Buphedrone	—	45.2	45.4	0.079	0.08	1.48	1.01
21	4-Methyl-α-ethylaminobutiophenone	—	49.7	50.3	0.061	0.061	5.80	1.01
22	Pentedrone	—	47.6	47.7	0.057	0.053	1.07	1.00
23	Butylone	—	59.7	60.6	0.091	0.091	5.84	1.02
24	Pentylone	—	62.6	63.2	0.105	0.103	3.40	1.01
25	4-Methylethcathinone	4-MEC	48	49.2	0.058	0.063	11.70	1.03
26	Ethcathinone	—	44.2	44.9	0.067	0.067	6.16	1.02
27	3-Methylmethcathinone	3-MMC	46.3	47	0.078	0.077	5.33	1.02
28	4-Bromomethcathinone	4-BMC	53.96	54.3	0.069	0.068	2.93	1.01
29	3-Bromomethcathinone	3-BMC	42.9	43.6	0.090	0.085	4.72	1.02
30	2,4-Dimethylmethcathinone	2,4-DMMC	48.5	49.96	0.063	0.063	13.67	1.04
31	2,4-Dimethylethcathinone	2,4-DMEC	49.8	51.2	0.065	0.071	12.15	1.03
32	3,4-Methylenedioxy-N-ethylcathinone	Ethylone	58.6	59.9	0.075	0.079	9.96	1.03
33	3-Methylbuphedrone	—	48.4	48.5	0.074	0.074	0.80	1.00
34	N-ethylbuphedrone	NEB	45.9	46.1	0.061	0.061	1.93	1.01
35	2,3-Pentylone isomer	—	59.1	59.9	0.058	0.07	7.37	1.02
36	3-Methylethcathinone	3-MEC	47.4	48.1	0.063	0.061	6.66	1.02

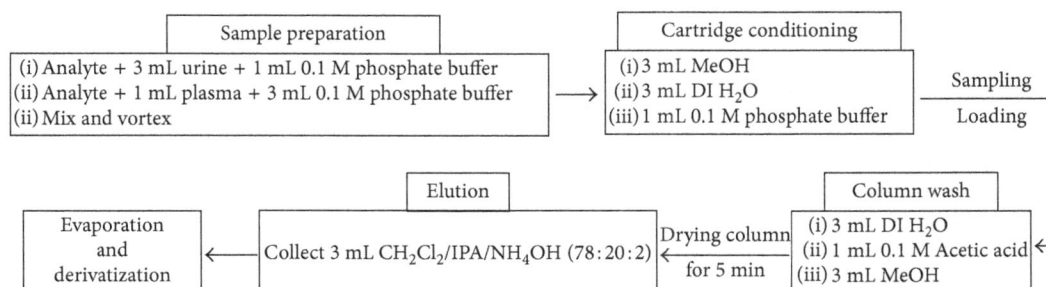

```
┌─────────────────────────┐              ┌─────────────────────────┐
│   Sample preparation    │              │  Cartridge conditioning │
├─────────────────────────┤              ├─────────────────────────┤       ┌──────────┐
│(i) Analyte + 3 mL urine │              │(i) 3 mL MeOH            │       │ Sampling │
│    + 1 mL 0.1 M          │──────────▶   │(ii) 3 mL DI H₂O         │       ├──────────┤
│    phosphate buffer     │              │(iii) 1 mL 0.1 M         │       │ Loading  │
│(ii) Analyte + 1 mL      │              │     phosphate buffer   │       └──────────┘
│    plasma + 3 mL 0.1 M  │              └─────────────────────────┘
│    phosphate buffer     │
│(ii) Mix and vortex      │
└─────────────────────────┘

              ┌─────────────┐                       ┌─────────────────────┐
              │   Elution   │                       │    Column wash      │
┌───────────┐ ├─────────────┤                       ├─────────────────────┤
│Evaporation│ │Collect 3 mL │  Drying column        │(i) 3 mL DI H₂O      │
│   and     │◀─│CH₂Cl₂/IPA/ │◀─  for 5 min          │(ii) 1 mL 0.1 M      │◀─
│derivatiza-│ │NH₄OH        │                       │    Acetic acid      │
│   tion    │ │(78:20:2)    │                       │(iii) 3 mL MeOH      │
└───────────┘ └─────────────┘                       └─────────────────────┘
```

SCHEME 2

capillary column has been used for the separation of the resulting diastereomers due to their different chemical and physical properties. The primary and secondary amine cathinones react with the derivatization reagent L-TPC in the presence of sodium carbonate, and the amidation reaction occurs between the acid chloride in L-TPC and the amine group of the target analytes. The gas chromatogram in Figure 1 shows the separation of the (R) and (S) enantiomers of nor-mephedrone drug after derivatization with L-TPC. Table 2 shows the retention times, resolution, and selectivity factors of the separated enantiomers of all the studied synthetic cathinones. All compounds in Table 2 were analyzed individually on GC-MS using SIM mode, after going through the derivatization step.

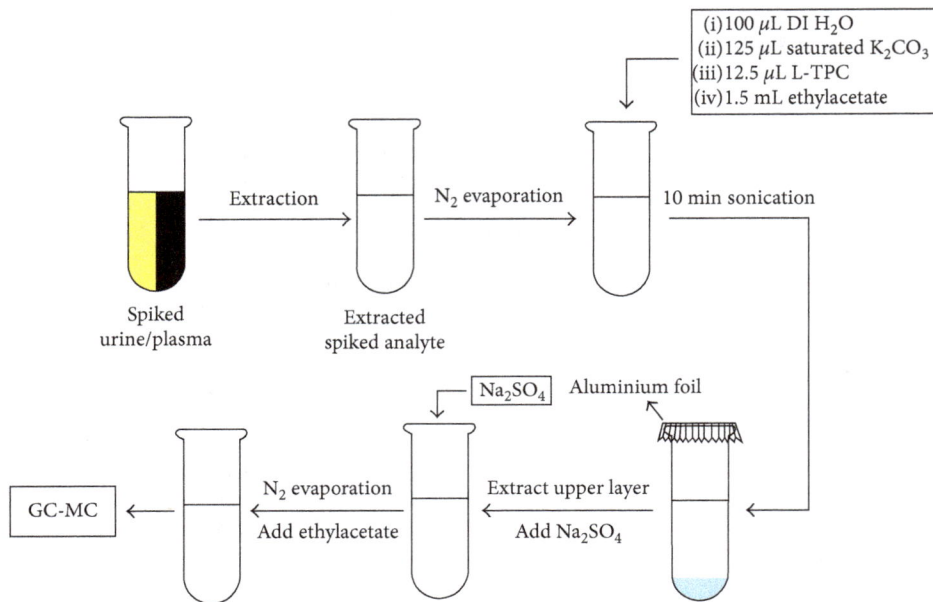

SCHEME 3

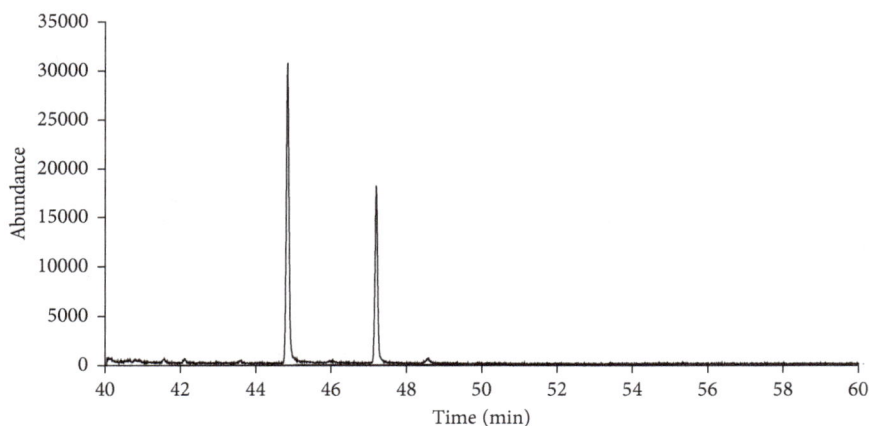

FIGURE 1: Gas chromatogram for separation of the R and S enantiomers of Nor-mephedrone drug in methanol after derivatization with L-TPC.

Figures 2 and 3 show the total ion current chromatogram of the fourteen synthetic cathinones spiked in urine and plasma, respectively. The resulted enantiomer peaks were well separated with good peak resolution. To our knowledge, this is the first example in the literature that demonstrates the separation of fourteen pairs of L-TPC cathinone derivatives in one run analysis of these compounds in complex matrices of urine and plasma.

Validation of the developed method was performed on spiked mixtures successfully. Linearity of the calibration curves, method sensitivity in terms of LOD and LOQ, and recoveries in addition to interday and intraday reproducibilities were collected and summarized in Tables 3–5.

The calibration curves for the fourteen synthetic cathinones derivatives were found to be linear within the tested range of 1 to $100 \, \mu g/L$ in urine and in plasma with mean regression coefficients (R^2; $n = 3$) higher than 0.99. The

regression coefficients and the LOD and LOQ values for the two enantiomers of the synthetic cathinone compounds in the mixture that spiked in urine and plasma are reported in Table 3. Three different concentration levels were tested for each enantiomer of these compounds (20, 60 and $100 \, \mu g/L$) in order to ensure the reproducibility and to provide the recovery study of the new method. The interday and intraday reproducibilities of the cathinones mixture of urine and plasma matrices are shown in Table 4. Moreover, percent error evaluation has been done for the spiked mixture to obtain the recovery studies which are summarized in Table 5.

4. Discussion

L-TPC is considered as a chiral derivatizing agent which can react with the primary and secondary amine enantiomers of synthetic

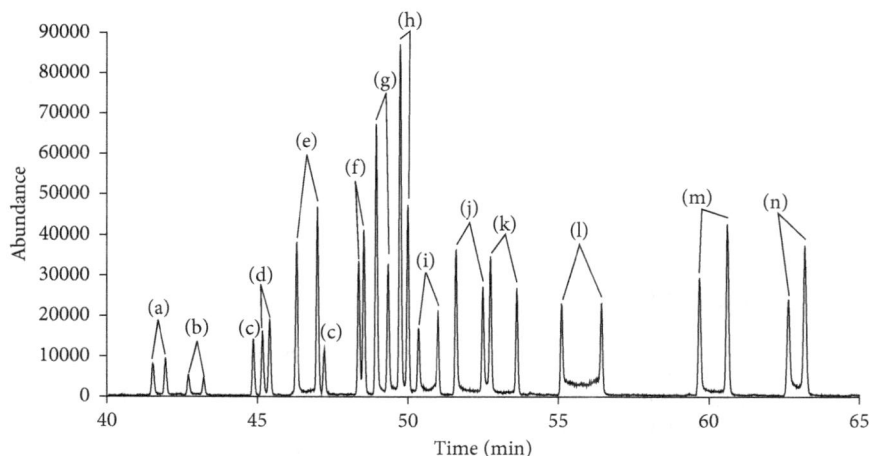

FIGURE 2: Total ion current chromatogram (TIC) of the simultaneous chiral separation of 14 synthetic cathinone compounds spiked in urine and separated as the following L-TPC derivatives: (a) 4-FMC, (b) 4-FEC, (c) nor-mephedrone, (d) buphedrone, (e) 3-MMC, (f) 3-methylbuphedrone, (g) 4-methylbuphedrone, (h) 3-EMC, (i) 3-EEC, (j) 4-EEC, (k) 3,4-DMEC, (l) 2,3-MDMC, (m) butylone, and (n) pentylone.

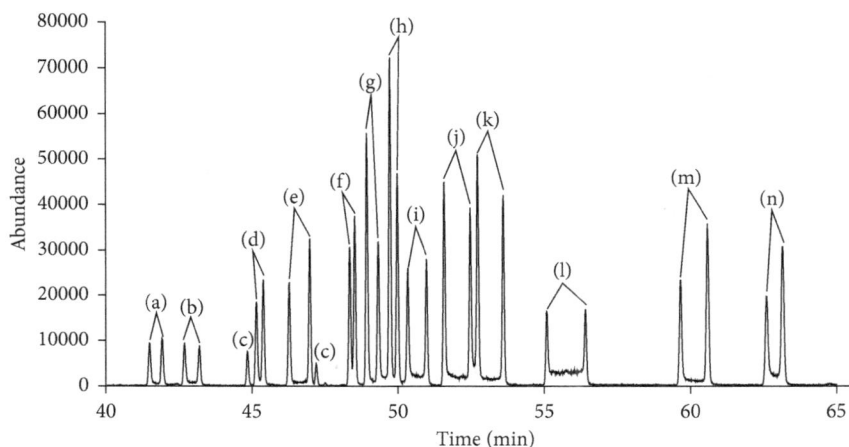

FIGURE 3: Total ion current chromatogram (TIC) of the simultaneous chiral separation of 14 synthetic cathinone compounds spiked in plasma and separated as the following L-TPC derivatives: (a) 4-FMC, (b) 4-FEC, (c) nor-mephedrone, (d) buphendrone, (e) 3-MMC, (f) 3-methylbuphedrone, (g) 4-methylbuphedrone, (h) 3-EMC, (i) 3-EEC, (j) 4-EEC, (k) 3,4-DMEC, (l) 2,3-MDMC, (m) butylone, and (n) pentylone.

cathinones producing two corresponding diastereomers. As a result of the differences in stereochemistry and stability of the formed diastereoisomers, the enantioseparation can occur on achiral stationary phase with different resolutions of product compounds [24, 30]. In this study, chiral separation of 36 racemic mixtures of synthetic cathinone derivatives was carried out: fourteen of them were selected in the spiked mixtures, and each enantiomer was quantitated in urine and plasma as shown in the example of nor-mephedrone in Figure 1. However, the enantioseparations that were obtained showed that there are differences in peak areas for the most resulted diastereoisomers. Mohr et al. assumed that the reason of inequality in the formed peaks is due to (i) racemization of L-TPC during the derivatization reaction, (ii) kinetic resolution of the two enantiomers, and (iii) the difference in diastereoisomers' yields which were explained in terms of keto-enol tautomerization of the analytes. Moreover,

the main reason for enantiomer peak inequality is related to the tested compounds themselves [8].

Interestingly, the fourteen spiked synthetic cathinone derivatives were separated simultaneously in one chromatogram since they have different retention times in the new developed method as shown in Figures 2 and 3, respectively. Moreover, enhancement of the resolution of enantiomers' peaks has been accomplished by using a slow heating rate of 2°C/min in the chromatographic method. Also, the use of Ultra Inert column helped in minimizing the overlap of the two adjacent peaks of the enantiomers. Chemical ionization conditions have allowed better detection of molecular ion peaks (M-H⁻) and minimized the extensive fragmentation of the targeted analytes.

Construction of calibration curves was done for the diastereoisomers based on the peak areas of the following concentration levels: 1, 5, 10, 20, 40, 60, 80, and 100 μg/L. Regression values of the correlation coefficient confirm the

TABLE 3: Results for fourteen cathinone-related compounds spiked in plasma and urine including linearity coefficient, R^2 values, limits of detection, and limits of quantitation for the two enantiomers of each compound.

		| Plasma							| Urine					
		R^2		LOQ (µg/L)		LOD (µg/L)		R^2		LOQ (µg/L)		LOD (µg/L)	
		E1	E2	E1	E2	E1	E2	E1	E2	E1	E2	E1	E2
1	4-FMC	0.9905	0.9962	0.957 ± 0.14	0.924 ± 0.12	0.29 ± 0.08	0.28 ± 0.08	0.9903	0.9900	1.09 ± 0.11	1.06 ± 0.11	0.33 ± 0.08	0.32 ± 0.08
2	4-FEC	0.9908	0.9902	1.023 ± 0.15	1.122 ± 0.13	0.31 ± 0.08	0.34 ± 0.08	0.9959	0.9914	1.82 ± 0.16	2.34 ± 0.16	0.55 ± 0.1	0.71 ± 0.1
3	Nor-mephedrone	0.9902	0.9931	0.891 ± 0.09	1.056 ± 0.14	0.27 ± 0.08	0.32 ± 0.08	0.9949	0.9926	2.15 ± 0.15	2.51 ± 0.16	0.65 ± 0.1	0.76 ± 0.1
4	Buphedrone	0.9933	0.9907	0.891 ± 0.09	0.858 ± 0.08	0.27 ± 0.08	0.26 ± 0.08	0.9931	0.9922	0.96 ± 0.14	0.92 ± 0.10	0.29 ± 0.07	0.28 ± 0.07
5	3-MMC	0.9904	0.9903	0.891 ± 0.09	0.858 ± 0.08	0.27 ± 0.08	0.26 ± 0.08	0.9944	0.9957	1.02 ± 0.15	0.92 ± 0.09	0.31 ± 0.08	0.28 ± 0.07
6	3-Methylbuphedrone	0.9902	0.9914	0.99 ± 0.11	1.023 ± 0.15	0.3 ± 0.08	0.31 ± 0.08	0.9941	0.9931	1.02 ± 0.15	0.99 ± 0.11	0.31 ± 0.07	0.3 ± 0.07
7	4-Methylbuphedrone	0.9910	0.9919	1.056 ± 0.10	1.056 ± 0.10	0.32 ± 0.08	0.32 ± 0.08	0.9902	0.9919	1.12 ± 0.13	1.09 ± 0.11	0.34 ± 0.09	0.33 ± 0.08
8	3-EMC	0.9945	0.9915	1.155 ± 0.11	1.089 ± 0.10	0.35 ± 0.08	0.33 ± 0.08	0.9901	0.9906	1.12 ± 0.11	1.06 ± 0.11	0.34 ± 0.08	0.32 ± 0.08
9	3-EEC	0.9952	0.9905	1.023 ± 0.11	0.957 ± 0.14	0.31 ± 0.08	0.29 ± 0.08	0.9980	0.9921	1.02 ± 0.15	1.09 ± 0.10	0.31 ± 0.07	0.33 ± 0.08
10	4-EEC	0.9942	0.9907	0.957 ± 0.14	0.858 ± 0.10	0.29 ± 0.08	0.26 ± 0.08	0.9907	0.9901	1.09 ± 0.10	1.12 ± 0.11	0.33 ± 0.08	0.34 ± 0.07
11	3,4-DMEC	0.9904	0.9909	0.99 ± 0.10	1.056 ± 0.10	0.3 ± 0.08	0.32 ± 0.08	0.9924	0.9912	1.12 ± 0.11	1.39 ± 0.11	0.34 ± 0.08	0.42 ± 0.09
12	2,3-MDMC	0.9900	0.9918	1.122 ± 0.10	1.089 ± 0.10	0.34 ± 0.08	0.33 ± 0.08	0.9927	0.9916	1.35 ± 0.13	1.35 ± 0.11	0.41 ± 0.09	0.41 ± 0.09
13	Butylone	0.9950	0.9920	0.924 ± 0.09	0.957 ± 0.14	0.28 ± 0.08	0.29 ± 0.08	0.9920	0.9908	0.86 ± 0.11	0.89 ± 0.09	0.26 ± 0.08	0.27 ± 0.07
14	Pentylone	0.9940	0.9969	0.858 ± 0.09	0.957 ± 0.14	0.26 ± 0.08	0.29 ± 0.08	0.9923	0.9901	0.96 ± 0.10	0.86 ± 0.11	0.29 ± 0.07	0.26 ± 0.08

good linearity of the four calibration lines. In order to correct for the loss of analyte during sample inlet or sample preparation, (+)-cathinone has been used as IS as it has a similar structure to synthetic cathinones and shows a good stability. The correlation coefficient (R^2) values were calculated for the mixture components, and they were found to be higher than 0.99 in all cases as shown in Table 3. Additionally, the LODs and LOQs were calculated according to the IUPAC method and are reported in Table 3. The reported values of LODs and LOQs for the synthetic cathinones in this study were in the µg/L range due to the high sensitivity of the analytical technique (GC-NCI-MS). The high mobility electrons that have low mass and energy produced during the NCI process are responsible for enhanced sensitivity when used for a suitably electrophilic compound compared to PCI and EI [29]. The LOD in urine was in the range of 0.26–0.76 µg/L, and in plasma, it was in the range of 0.26–0.34 µg/L. While the LOQ in urine was in the range of 0.89–2.34 µg/L, and in plasma, it was in the range of 0.89–1.12 µg/L (as shown in Table 3).

Three different concentration levels were chosen to test the interday and intraday reproducibility measurements of the synthetic cathinone compounds mixture of urine and plasma as shown in Table 3. In fact, good reproducibility and repeatability were established using the new developed method since most of the coefficients of variance values were below 15% in both urine and plasma matrices for measurements done on the same day or on two different days. In comparison, between spiked urine and spiked plasma samples, urine samples were more reproducible than spiked plasma samples because of the competition between analyte and blood interferences unlike spiked urine samples where the urine was diluted with deionized water prior to the spiking step. Moreover, the presence of proteins and other interferences in plasma can cause difficulty in solid-phase extraction processes and can also create a competition between the targeted analyte and unneeded interferences which will lead to variation in spiked plasma results [31].

SPE efficiency was studied by percent error calculations for the spiked mixture at the following concentration levels: 20, 60, and 100 µg/L. The calculated values in recovery studies were within the acceptable range.

By comparing the results of the GC-EI-MS method recently reported for some of these synthetic cathinones [25] and the current study results using GC-NCI-MS, the latter has shown an enhancement of sensitivity by a magnitude of two orders. The high sensitivity of NCI is due to the low mass and high mobility of the secondary or thermal electrons (low-energy electrons) produced under the CI high pressure conditions in the presence of methane reagent gas, which is responsible for the enhancement factor by nearly 100 times in the sensitivity of NCI compared to that of positive EI or CI for a suitably electrophilic compound [29].

TABLE 4: Interday and intraday reproducibility results in terms of coefficient of variance for fourteen cathinone-related compounds spiked in urine and plasma at three different concentration levels for the two enantiomers of each compound.

| | | | CV% intraday | | | | | | CV% interday | | | | | |
| | | | 20 µg/L | | 60 µg/L | | 100 µg/L | | 20 µg/L | | 60 µg/L | | 100 µg/L | |
			E1	E2	E1	E2	E1	E2	E1	E2	E1	E2	E1	E2
1	4-FMC	U	3.61	3.11	1.04	1.79	0.41	1.33	10.26	11.85	10.28	12.02	8.14	7.6
		P	6	4.79	2.59	5.36	2.26	2.59	11.01	10.6	10.99	13.3	10.95	12.35
2	4-FEC	U	2.22	5.32	1.75	1.53	2.03	1.69	5.67	10.9	14.69	15.74	15.8	16.13
		P	10.57	11.65	5.31	3.89	1.69	2.21	7.44	8.43	7.59	11.25	6.65	7.76
3	Nor-mephedrone	U	0.7	1.01	2.13	1.64	2	2.17	3.07	4.57	3.5	4.43	1.75	4.91
		P	7.59	7.95	3.07	3.51	2.79	6.97	6.01	7.1	14.13	6.48	14.3	14.21
4	Buphedrone	U	1.42	1.6	1.89	2.28	2.12	1.17	5.6	14.08	15.31	16.91	16.62	15.29
		P	11.13	10.88	3.37	6.27	1.32	3.14	8.04	8.49	13.39	14.26	5	8.32
5	3-MMC	U	3.68	2.03	0.93	1.66	0.39	1.34	12.27	9.02	1.72	2.41	0.82	1.29
		P	9.74	12.68	4.55	6.19	2.51	3.69	8.42	9.85	4.03	4.78	16.3	16.97
6	3-Methylbuphedrone	U	1.36	3.07	2.06	4.48	0.93	1.09	5.32	14.73	6.73	11.96	10.04	9.44
		P	10.58	11.42	5.04	5.44	1.19	3.58	7	7.9	10.55	15.34	13.67	14.31
7	4-Methylbuphedrone	U	1.25	2.16	0.35	0.75	0.68	1.14	3.73	11.28	9.25	2.04	10.22	9.93
		P	9.77	11.67	5.41	4.49	2.42	2.82	9.25	11.85	19.55	6.83	18.03	19.59
8	3-EMC	U	2.38	2.57	0.63	2.1	0.74	1.69	14.53	14.81	13.18	3.51	7.62	10.98
		P	11.36	8.27	5.41	4.55	3.01	3.16	11.35	9.79	19.65	7.13	19.04	17.92
9	3-EEC	U	3.22	6.13	1.13	3.6	1.71	2.13	13.29	15.44	8.42	10.94	3.27	4.1
		P	8.81	12.64	4.3	5.28	3.15	3.71	6.62	9.27	10.19	12.67	14.15	14.82
10	4-EEC	U	3.45	1.85	0.93	2.17	0.84	0.82	14.48	11.31	9.31	12.68	7.19	1.8
		P	11.48	10.53	6.12	4.48	4.24	3.59	8.31	9.46	19.81	19.9	10.97	15.42
11	3,4-DMEC	U	3.7	3.5	1.97	1.02	3.44	2.16	16.54	14.77	10.92	14.49	3.07	2.62
		P	12.11	13.14	4.5	6.31	3.26	3.06	9.44	11.05	18.98	20.31	9.66	15.7
12	2,3-MDMC	U	2.07	3.24	3.16	1.43	4.11	5.56	5.5	4.43	13.57	14.03	5.74	6.6
		P	10.94	7.12	2.74	4.02	7.24	7.13	9.2	5.11	3.41	4.46	19.77	19.35
13	Butylone	U	5.14	6.26	1.36	0.5	0.85	0.88	12.57	14.11	14.56	14.9	1.25	1.56
		P	10.78	12.89	4.97	6.67	2.63	3.25	11.42	13.6	3.66	7.62	9.06	9.14
14	Pentylone	U	7.1	1.34	2.59	1.84	1.83	0.19	12.98	5.99	12.77	14.03	2.29	3.86
		P	10.2	18.64	3.74	5.93	2	2.06	10.29	19.91	9.77	6.67	5.01	6.59

U: urine; P: plasma; CV: coefficient of variance; E1: enantiomer 1; E2: enantiomer 2.

TABLE 5: Recovery measurements expressed in percent errors for three different concentrations of the cathinone-related compounds spiked in urine matrix.

| | | Plasma (error %) | | | | | | Urine (error %) | | | | | |
| | | 20 µg/L | | 60 µg/L | | 100 µg/L | | 20 µg/L | | 60 µg/L | | 100 µg/L | |
		E1	E2	E1	E2	E1	E2	E1	E2	E1	E2	E1	E2
1	4-FMC	2.63	5.32	2.48	0.62	2.45	0.26	0.8	10.57	9.49	8.47	1.24	0.68
2	4-FEC	0.17	4.89	5.8	3.1	2.19	1.6	9.28	5.31	2.38	9.21	0.1	0.61
3	Nor-mephedrone	5.64	1.94	3.25	7.64	1.69	2.64	8.67	7.48	1.82	4.48	1.71	0.69
4	Buphedrone	3.18	5.94	0.79	3.78	0.06	1.71	9.12	5.58	4.84	4.88	1.9	0.72
5	3-MMC	1.03	1.47	0.87	0.55	0.74	1.4	3.91	3.02	4.91	3.54	0.25	0.91
6	3-Methylbuphedrone	1.26	7.1	6.41	9.61	4.03	1.71	3.71	9.13	4.09	0.15	1.54	1.77
7	4-Methylbuphedrone	0.69	5.64	1.77	0.73	2.56	1.9	1.15	0.42	0.97	2.54	2.3	4.01
8	3-EMC	0.65	5.73	2.19	4.7	0.56	0.52	3.61	1.27	4.86	8.04	1.55	3.32
9	3-EEC	8.7	7.62	2.22	1.06	0.42	0.34	5.69	4.88	2.1	4.18	2.44	2.22
10	4-EEC	4.38	5.72	7.91	5.68	2.01	2.32	2.37	12.25	0.66	6.64	0.9	1.52
11	3,4-DMEC	4.36	3.85	2.2	2.46	1.89	2.93	9.61	7.83	7.35	8.05	2.07	1.64
12	2,3-MDMC	9.98	8.22	6.17	8.34	1.97	3.51	7.25	10.31	9.45	6.68	0.94	0.54
13	Butylone	1.41	10.98	2.05	5.97	1.62	1.98	2.93	3.78	8.95	8.38	0.83	0.24
14	Pentylone	5.8	10.43	1.07	1.47	1.68	0.22	9.54	0.05	0.14	3.25	2.52	3.14

5. Conclusion

Indirect chiral separation of synthetic cathinones after derivatization with trifluoroacetyl-l-prolyl chloride (L-TPC) was achieved using a new developed method of GC-NCI-MS in SIM mode, which provided high sensitivity and selectivity for the separation and quantitation of the targeted compounds. The use of 60 m HP-5MS Ultra Inert capillary column helps to separate more than thirty-six compounds of synthetic cathinones to their diastereomers. NCI has shown to be an effective ionization method for these cathinones and resulted in lower detection limits when compared to previous reports. A mixture of fourteen cathinone derivatives that were spiked in urine and plasma was separated in one chromatogram simultaneously. For each enantiomer peak in the cathinone mixture chromatogram, calibration curve was constructed using the following concentration levels: 1, 5, 10, 20, 40, 60, 80, and 100μg/L. The developed method was validated in terms of linearities, LOD, LOQ, reproducibilities, and recoveries for all the tested mixtures.

Conflicts of Interest

The authors declare that there are no conflicts of interest.

Acknowledgments

The authors would like to thank the General Department of Forensic Science and Criminology, Dubai Police, for providing the synthetic cathinone standards and the United Arab Emirates University for providing the financial support (31R114, 31S087, and 31S117).

References

[1] D. P. Katz, D. Bhattacharya, S. Bhattacharya et al., "Synthetic cathinones: "a khat and mouse game"," *Toxicology Letters*, vol. 229, no. 2, pp. 349–356, 2014.

[2] M. L. Banks, T. J. Worst, D. E. Rusyniak, and J. E. Sprague, "Synthetic cathinones ("bath salts")," *Journal of Emergency Medicine*, vol. 46, no. 5, pp. 632–642, 2014.

[3] K. Sikk and P. Taba, "Chapter twelve–methcathinone "kitchen chemistry" and permanent neurological damage," in *International Review of Neurobiology*, A. L. Pille Taba and S. Katrin, Eds., vol. 120, pp. 257–271, Academic Press, Cambridge, MA, USA, 2015.

[4] R. A. Glennon, B. R. Martin, T. A. Dal Cason, and R. Young, "Methcathinone ("CAT"): an enantiomeric potency comparison," *Pharmacology Biochemistry and Behavior*, vol. 50, no. 4, pp. 601–606, 1995.

[5] J. DeRuiter, L. Hayes, A. Valaer, C. R. Clark, and F. Noggle, "Methcathinone and designer analogues: synthesis, stereochemical analysis, and analytical properties," *Journal of Chromatographic Science*, vol. 32, no. 12, pp. 552–564, 1994.

[6] M. Bossong, J. Van Dijk, and R. Niesink, "Methylone and mCPP, two new drugs of abuse?," *Addiction Biology*, vol. 10, no. 4, pp. 321–323, 2005.

[7] P. I. Dargan, R. Sedefov, A. Gallegos, and D. M. Wood, "The pharmacology and toxicology of the synthetic cathinone mephedrone (4-methylmethcathinone)," *Drug Testing and Analysis*, vol. 3, no. 7-8, pp. 454–463, 2011.

[8] S. Mohr, J. A. Weiß, J. Spreitz, and M. G. Schmid, "Chiral separation of new cathinone- and amphetamine-related designer drugs by gas chromatography–mass spectrometry using trifluoroacetyl-l-prolyl chloride as chiral derivatization

reagent," *Journal of Chromatography A*, vol. 1269, pp. 352–359, 2012.

[9] J. A. Weiss, S. Mohr, and M. G. Schmid, "Indirect chiral separation of new recreational drugs by gas chromatography-mass spectrometry using trifluoroacetyl-L-prolyl chloride as chiral derivatization reagent," *Chirality*, vol. 27, no. 3, pp. 211–215, 2015.

[10] J. A. Weiss, M. Taschwer, O. Kunert, and M. G. Schmid, "Analysis of a new drug of abuse: cathinone derivative 1-(3,4-dimethoxyphenyl)-2-(ethylamino)pentan-1-one," *Journal of Separation Science*, vol. 38, no. 5, pp. 825–828, 2015.

[11] M. Taschwer, Y. Seidl, S. Mohr, and M. G. Schmid, "Chiral separation of cathinone and amphetamine derivatives by HPLC/UV using sulfated ß-cyclodextrin as chiral mobile phase additive," *Chirality*, vol. 26, no. 8, pp. 411–418, 2014.

[12] D. Wolrab, P. Frühauf, A. Moulisová et al., "Chiral separation of new designer drugs (Cathinones) on chiral ion-exchange type stationary phases," *Journal of Pharmaceutical and Biomedical Analysis*, vol. 120, pp. 306–315, 2016.

[13] S. Mohr, M. Taschwer, and M. G. Schmid, "Chiral separation of cathinone derivatives used as recreational drugs by HPLC-UV using a CHIRALPAK(R) AS-H column as stationary phase," *Chirality*, vol. 24, no. 6, pp. 486–492, 2012.

[14] N. L. Padivitage, E. Dodbiba, Z. S. Breitbach, and D. W. Armstrong, "Enantiomeric separations of illicit drugs and controlled substances using cyclofructan-based (LARIHC) and cyclobond I 2000 RSP HPLC chiral stationary phases," *Drug Testing And Analysis*, vol. 6, no. 6, pp. 542–551, 2014.

[15] L. Li and I. S. Lurie, "Regioisomeric and enantiomeric analyses of 24 designer cathinones and phenethylamines using ultra high performance liquid chromatography and capillary electrophoresis with added cyclodextrins," *Forensic Science International*, vol. 254, pp. 148–157, 2015.

[16] M. Taschwer, J. A. Weiß, O. Kunert, and M. G. Schmid, "Analysis and characterization of the novel psychoactive drug 4-chloromethcathinone (clephedrone)," *Forensic Science International*, vol. 244, pp. e56–e59, 2014.

[17] G. Merola, H. Fu, F. Tagliaro, T. Macchia, and B. R. McCord, "Chiral separation of 12 cathinone analogs by cyclodextrin-assisted capillary electrophoresis with UV and mass spectrometry detection," *Electrophoresis*, vol. 35, no. 21-22, pp. 3231–3241, 2014.

[18] S. Mohr, S. Pilaj, and M. G. Schmid, "Chiral separation of cathinone derivatives used as recreational drugs by cyclodextrin-modified capillary electrophoresis," *Electrophoresis*, vol. 33, no. 11, pp. 1624–1630, 2012.

[19] M. Moini and C. M. Rollman, "Compatibility of highly sulfated cyclodextrin with electrospray ionization at low nanoliter/minute flow rates and its application to capillary electrophoresis/electrospray ionization mass spectrometric analysis of cathinone derivatives and their optical isomers," *Rapid Commun Mass Spectrom*, vol. 29, no. 3, pp. 304–310, 2015.

[20] Z. Aturki, M. G. Schmid, B. Chankvetadze, and S. Fanali, "Enantiomeric separation of new cathinone derivatives designer drugs by capillary electrochromatography using a chiral stationary phase, based on amylose tris(5-chloro-2-methyl-phenylcarbamate)," *Electrophoresis*, vol. 35, no. 21-22, pp. 3242–3249, 2014.

[21] S. Fanali, "Enantioselective determination by capillary electrophoresis with cyclodextrins as chiral selectors," *Journal of Chromatography A*, vol. 875, no. 1-2, pp. 89–122, 2000.

[22] M. Taschwer, M. G. Hofer, and M. G. Schmid, "Enantioseparation of benzofurys and other novel psychoactive compounds by CE and sulfobutylether β-cyclodextrin as chiral

selector added to the BGE," *Electrophoresis*, vol. 35, no. 19, pp. 2793–2799, 2014.

[23] B. Li and D. T. Haynie, "Chiral drug separation," *Encyclopedia of Chemical Processing*, vol. 1, pp. 449–458, 2006.

[24] J. M. Płotka, M. Biziuk, and C. Morrison, "Common methods for the chiral determination of amphetamine and related compounds I. Gas, liquid and thin-layer chromatography," *TrAC Trends in Analytical Chemistry*, vol. 30, no. 7, pp. 1139–1158, 2011.

[25] R. H. Alrumaithi, M. A. Meetani, and S. A. Khalil, "A validated gas chromatography mass spectrometry method for simultaneous determination of cathinone related drugs enantiomers of in urine and plasma," *RSC Advances*, vol. 6, no. 84, pp. 80576–80584, 2016.

[26] D.-X. Li, L. Gan, A. Bronja, and O. J. Schmitz, "Gas chromatography coupled to atmospheric pressure ionization mass spectrometry (GC-API-MS): review," *Analytica Chimica Acta*, vol. 891, pp. 43–61, 2015.

[27] B. Waters, N. Ikematsu, K. Hara et al., "GC-PCI-MS/MS and LC-ESI-MS/MS databases for the detection of 104 psychotropic compounds (synthetic cannabinoids, synthetic cathinones, phenethylamine derivatives)," *Legal Medicine*, vol. 20, pp. 1–7, 2016.

[28] Y.-H. Wu, K.-l. Lin, S.-C. Chen, and Y.-Z. Chang, "Integration of GC/EI-MS and GC/NCI-MS for simultaneous quantitative determination of opiates, amphetamines, MDMA, ketamine, and metabolites in human hair," *Journal of Chromatography B*, vol. 870, no. 2, pp. 192–202, 2008.

[29] J. T. Watson and O. D. Sparkman, *Introduction to Mass Spectrometry: Instrumentation, Applications, and Strategies for Data Interpretation*, John Wiley & Sons, Hoboken, NJ, USA, 2007.

[30] T. Toyo'oka, "Resolution of chiral drugs by liquid chromatography based upon diastereomer formation with chiral derivatization reagents," *Journal of Biochemical and Biophysical Methods*, vol. 54, no. 1-3, pp. 25–56, 2002.

[31] S. L. Prabu and T. N. K. Suriyaprakash, *Extraction of Drug from the Biological Matrix: A Review*, INTECH Open Access Publisher, Rijeka, Croatia, 2012.

9

A Standardized Approach to Quantitative Analysis of Nicotine in e-Liquids based on Peak Purity Criteria using High-Performance Liquid Chromatography

Vinit V. Gholap ⓘ,[1] Leon Kosmider,[1,2] and Matthew S. Halquist ⓘ[1]

[1]Department of Pharmaceutics, School of Pharmacy, Virginia Commonwealth University, Richmond, VA 23298, USA
[2]Center for the Study of Tobacco Products, Virginia Commonwealth University, Richmond, VA 23298, USA

Correspondence should be addressed to Matthew S. Halquist; halquistms@vcu.edu

Academic Editor: Josep Esteve-Romero

The use of electronic cigarettes (e-cigarettes) is a growing trend in population. E-cigarettes are evolving at a rapid rate with variety of battery powered devices and combustible nicotine refills such as e-liquids. In contrast to conventional cigarettes which are studied well for their toxicity and health effects, long-term clinical data on e-cigarettes are not available yet. Therefore, safety of e-cigarettes is still a major concern. Although the Food and Drug Administration (FDA) has recently started regulating e-cigarette products, no limits on nicotine and other ingredients in such products have been proposed. Considering the regulatory requirements, it is critical that reliable and standardized analytical methods for analyzing nicotine and other ingredients in e-cigarette products such as e-liquids are available. Here, we are reporting a fully validated high-performance liquid chromatography (HPLC) method based on nicotine peak purity for accurately quantifying nicotine in various e-liquids. The method has been validated as per ICH Q2(R1) and USP <1225> guidelines. The method is specific, precise, accurate, and linear to analyze nicotine in e-liquids with 1 to >50 mg/mL of nicotine. Additionally, the method has been proven robust and flexible for parameters such as change in flow rate, column oven temperature, and organic phase composition, which proves applicability of the method over wide variety of e-liquids in market.

1. Introduction

The growing popularity of e-cigarettes and vast number of available choices of e-liquids necessitate stringent analytical measurements and controls of these products for quality and regulations. As evolution of e-cigarettes is underway, a large number of e-liquids are being introduced in market [1]. A study carried out on the online market of e-cigarettes by Zhu et al. in [1] showed that there are more than 7700 e-liquid flavors available to customers. Such market is mainly Internet market where there is least control on quality and sale of e-cigarette products [2]. In past few years, several public health organizations and policy makers have expressed concerns over the safety and health impact of e-cigarettes [3, 4]. As a result, effective from August 8, 2016, all products meeting the statutory definition of tobacco products as per Tobacco Control Act are subject to regulations by Food and Drug Administration [5]. Such products also include e-cigarettes and e-liquids [5]. Although e-liquids are now subject to FDA regulations, guidelines and permissible limits of nicotine and other ingredients of e-liquids have not been finalized yet due to several reasons such as lack of definitive and long-term clinical data, assessment of safety claims about e-cigarettes [6–8], and standardized analytical methods [9]. The nicotine in e-liquid products is addictive and can be toxic in high doses [10]. Previous studies have shown that nicotine content in many e-liquids is highly variable than what is mentioned on label claim [11, 12].

Currently, there are several published methods to measure nicotine in e-liquids using gas chromatography with flame ionization detector (GC-FID), gas chromatography-mass spectrometry (GC-MS), and liquid chromatography-mass spectrometry (LC-MS) [13–16]. In contrast, there are few HPLC methods published for measuring nicotine in e-liquids. In Table 1, a summary of current available HPLC methods for measuring nicotine in e-liquids with associated validation parameters and shortcomings is assessed.

GC and mass analyzers are not available across all laboratories and may not be suitable for routine e-liquid analysis. HPLC methods published have several shortcomings as described in Table 1. To address these concerns, we report a standardized and alternative analytical HPLC method for quantification of nicotine in e-liquids of various flavors. As stated above, e-liquids contain a variety of ingredients other than nicotine. Therefore, it become necessary to accurately quantify nicotine without interference from other flavoring ingredients. As mentioned earlier, few HPLC methods have been published for quantification of nicotine in e-liquids. However, none of the method describes the peak purity of nicotine and also do not provide data for full validation of methods. Lack of reliable validation parameters questions the accuracy of data obtained and conclusions derived by such methods [9].

The current paper focuses on quantitative analysis of nicotine based on the peak purity criteria of nicotine using photo diode array (PDA) detector. Peak purity is an algorithm in the chromatographic software. It is analysis of absorbance spectra across a peak to determine similarity or differences between them. Differences in the spectra across a peak indicate that two more compounds are eluting at the same retention time. A peak is said to be pure if its purity angle is less than purity threshold [19]. Although peak purity may not serve as a full proof for chemical purity especially in active pharmaceutical ingredients (API), it serves as an important passing criterion for analytes in formulations. Besides, peak purity criterion is recommended by FDA and commonly used in pharmaceutical industries for analytes in formulations separated by chromatographic methods such as HPLC [19, 20].

The objective of this research was to develop and validate a HPLC analytical method for nicotine analysis in e-liquids. Additionally, taking into consideration the variety of flavoring ingredients in e-liquids, we also aimed to perform robustness study of the method with change in organic phase composition to achieve nicotine peak purity and optimum resolution (≥1.5) [21, 22] between nicotine and other ingredients.

2. Materials and Methods

2.1. Instrumentation. Method development and validation activities were carried out using a Waters Alliance 2695 quaternary pump HPLC equipped with Waters 996 PDA detector, Hypersil Gold Phenyl column (150 mm × 4.6 mm, 3 μm, Thermo Scientific™, USA) and a Security Guard Cartridge Phenyl (4 mm × 2.0 mm, Phenomenex, USA). Waters Empower 2 software was used for processing data.

2.2. Chemicals and Reagents. Nicotine hydrogen tartrate standard (purity 93.14%) was purchased from Glentham Life Sciences, United Kingdom. Nicotine liquid standard

(purity ≥99%) was purchased from Sigma-Aldrich, USA. HPLC grade acetonitrile, methanol, and water were purchased from BDH Chemicals, VWR, USA. Orthophosphoric acid (85%) was purchased from Merck, USA. Triethyl amine and hydrochloric acid (37%) were purchased from Sigma-Aldrich, USA. Sodium hydroxide (10 N) and hydrogen peroxide (30%) were purchased from BDH Chemicals, VWR, USA. Propylene glycol was purchased from Amresco LLC, VWR, USA. USP grade vegetable glycerin was purchased from JT Baker, USA. e-Liquids without nicotine (placebo) with variety of flavors of major categories were used. One e-liquid from each category of tobacco, vanilla, and two e-liquids from fruit flavors were purchased from Avail Vapor, USA. Eight other e-liquids were purchased from Direct Vapor online vape shop, USA, such as two e-liquids from each category of menthol, sweet, tobacco, and one from each category of fruit and coffee flavors. All e-liquids were coded with letters for the analysis.

2.3. Chromatographic Conditions. A gradient method was developed using an HPLC equipped with a quaternary pump system. Mobile phase A was 0.1% (v/v) triethyl amine in water with pH adjusted to 7.6 ± 0.05 by orthophosphoric acid (85%) and sodium hydroxide solution (1 N). Mobile phases B and C were 0.1% (v/v) triethyl amine in methanol and acetonitrile, respectively. Mobile phase D and diluent were 80% (v/v) methanol in water. The chromatographic conditions were run as shown in Tables 2 and 3. Peak purity analysis of nicotine peak was performed using threshold as "noise + solvent angle" calculated from blank and standard response using PDA detector (230 nm–350 nm).

2.4. Preparation of Reagents. Standard stock solution of nicotine hydrogen tartrate (1 mg/mL) in diluent was used for analysis. e-Liquids (8 mg/mL) were prepared by dissolving liquid nicotine standard in each flavored matrix. Similarly, quality control (QC) samples were prepared by dissolving liquid nicotine standard in unflavored matrix of propylene glycol and vegetable glycerin (1 : 1 v/v). Assay samples (80 μg/mL) were prepared by 100-fold dilution of e-liquids and QC in diluent.

2.5. Method Validation. The method for nicotine quantification from e-liquids was validated as per ICH Q2(R1) and USP <1225> guidelines for specificity, linearity, accuracy, precision, LOD, LOQ, and robustness [20, 23].

2.5.1. Specificity. Specificity of the method for nicotine quantification was established by performing forced degradation studies on various e-liquid assay samples, placebos, and blanks. Samples were subjected to acid hydrolysis, base hydrolysis, oxidation, and thermal degradation as mentioned in Table 4. Stressed samples were analyzed for % degradation, nicotine peak purity, and any degradant peak of nicotine.

TABLE 1: HPLC methods published for analysis of nicotine in e-liquids.

Type of detection platform used	Validation parameters evaluated	e-Liquids (brand and number of samples)	Reference	Comments
HPLC, PDA, C 18 (150 × 4.6 mm, 5 μm)	LOD, LOQ, linearity, accuracy, and precision	Smoking everywhere (15), Njoy (5), and CIXI (10)	Trehy et al. [12]	1. Specificity data are not mentioned 2. Although the method uses PDA detector, no information about peak purity is mentioned 3. Chromatographic integration in the chromatograms published by Trehy et al. is improper, which raises concerns over the specificity of the method
HPLC, PDA, C 18 (200 × 4.6 mm, 5 μm)	LOD, LOQ, linearity, accuracy, and precision	Refill fluids (75), do it yourself (1)	Davis et al. And cross-reference of Trehy et al. For the HPLC method [11]	
UPLC	Not applicable	e-Liquids (20). Details not mentioned	Etter et al. [17]	Method not validated for analysis of e-liquids
UPLC, PDA and MS, C 18	Linearity	e-Liquids (6). Details not mentioned	Meruva et al. And cross-reference of Trehy et al. for the HPLC method [18]	1. Additional validation details not provided 2. No data about specificity of the method for nicotine in presence of flavoring chemicals

TABLE 2: HPLC chromatographic conditions of the method.

Chromatographic conditions	
Flow rate	0.8 mL/min
Wavelength	260 nm
Stationary phase	Hypersil Gold Phenyl (150 mm × 4.6 mm, 3 μm)
Column oven temperature	25°C
Injection volume	10 μL
Sample cooler temperature	5°C
Run time	12 min

TABLE 3: HPLC gradient program.

	Pump program			
Time (min)	A (%)	B (%)	C (%)	D (%)
0	60	26	14	0
4	60	26	14	0
4.1	0	0	0	100
7	0	0	0	100
7.1	60	26	14	0
12	60	26	14	0

2.5.2. Linearity. Linearity of the method was established over the range of 0.4 μg/mL to 500 μg/mL of nicotine. Two sets of nicotine standard levels were prepared at 0.4, 10, 50, 100, 200, and 500 μg/mL concentrations for generation of the calibration curve. The residual percent of nicotine was calculated by using the equation of the best fit line. Linearity was evaluated by linear equation, coefficient of variation (r^2), and Y-intercept.

2.5.3. Accuracy. Accuracy was established by analyzing standard and QC samples three times at 50%, 100%, and 150%

of the assay level (80 μg/mL). The averages of the results were calculated against the respective averages of standards prepared at approximately the same concentrations. Equation (1) was used to calculate % nicotine recovery:

$$\%\text{nicotine recovery} = \left(\frac{Ru}{Rs}\right) \times \left(\frac{Cs}{Cu}\right) \times 100, \quad (1)$$

where Ru = peak area of 50%, 100%, or 150% assay level, Rs = average peak area of standard preparations of respective assay level, Cs = concentration of standard preparation, and Cu = concentration of assay level.

2.5.4. Precision. Precision was expressed as the standard deviation or degree of reproducibility or repeatability of the analytical method under normal operating conditions. The % relative standard deviation (% RSD) of nicotine at 50%, 100%, and 150% accuracy samples was calculated to determine repeatability. Intermediate precision was performed by doing repeatability test by a different analyst on a different day. The % RSD of combined results obtained by both analysts was calculated to determine intermediate precision.

2.5.5. Robustness. Robustness of the method performed by analyzing QC sample at assay level ($n = 3$) by making minor changes to the method is as mentioned below:

(i) (±) 10% flow rate adjustment

(ii) (±) 2°C column temperature adjustment

(iii) Change in organic mobile phase ratio to modify polarity (Table 5).

Robustness was evaluated by calculating % RSD of replicate injections at each modified parameter.

TABLE 4: Stressed conditions for e-liquid assay samples, placebos, and blanks.

Sample stress type	Time	Assay sample (mL)	Water (mL)	0.1 N·HCl (mL)	1 N·NaOH (mL)	H₂O₂ (6%) (mL)
Control	N/A	4.5	0.5	0	0	0
Acid hydrolysis	30 min	4.5	0	0.5	0	0
Base hydrolysis	30 min	4.5	0	0	0.5	0
Oxidation	30 min	4.5	0	0	0	0.5
Thermal	2 hrs	4.5	0.5	0	0	0

TABLE 5: Change in organic mobile phase ratio.

Change parameter	% mobile phase A	% mobile phase B	% mobile phase C
Increase in organic polarity	60	24	16
Decrease in organic polarity	60	28	12

2.5.6. LOD and LOQ. Limit of detection (LOD) was calculated based on the standard deviation of the response and the slope. The detection limit (DL) is expressed as mentioned in the following equation:

$$DL = 3.3\left(\frac{\sigma}{S}\right), \tag{2}$$

where σ = standard deviation of the response and S = slope of calibration curve. σ was calculated as standard deviation of analytical background response at the retention time of nicotine and obtained from three placebo samples.

Limit of quantitation (LOQ) was calculated based on visual evaluation where minimum known concentration of nicotine can be analyzed quantitatively with acceptable accuracy and precision.

3. Results and Discussion

3.1. Method Development and Optimization. Based on published literature, earlier HPLC analytical methods have been developed for analysis of nicotine from e-liquids by reversed phase chromatography using a C18 column [11, 12, 18]. Some references for LC-MS method also describe use of a C18 column for separation of nicotine from other ingredients in e-liquids [13, 14]. Therefore, a C18 column was initially used for nicotine analysis. Mobile phase A was water with 0.1% TEA, pH adjusted to 7.6 ± 0.05, and mobile phase B was acetonitrile with 0.1% TEA in isocratic ratio of 70 : 30 (% v/v). Various e-liquids were analyzed for nicotine content. Although nicotine eluted as a single peak in a chromatogram, spectral scans of the nicotine peak were found to be impure in many of the e-liquids such as tobacco and fruit flavors. Several chromatographic runs were performed by varying parameters such as mobile phase composition, pH, and organic phase. However, feasibility of C18 column to separate nicotine peak from multiple flavors of e-liquids was found to be limited in terms of achieving peak purity for nicotine peak. Therefore, the objective of HPLC method development and optimization was based on achieving acceptable peak purity for nicotine from various e-liquid samples using a robust and flexible method.

Since e-liquids contain a variety of flavoring agents (>7700) [1] covering a wide range of chemicals such as unsaturated, aromatic, polycyclic, and so on; phenyl column chemistry was chosen for separation of these compounds from nicotine. Varying compositions of mobile phases, pH, column temperature, and flow rate were carried out using a phenyl column (250 mm × 4.6, 5 μm).

Optimization of the chromatographic parameters was performed using Hypersil Gold Phenyl (150 mm × 4.6 mm, 3 μm) to decrease the run time. Since e-liquids are complex mixtures of compounds, high organic solvent composition was used for elution of late eluting peaks. The proposed method has a run time of 12 min with the nicotine retention time at approximately 5.5 min with a postelution wash. Twelve different flavors of e-liquids from six major categories as mentioned in Section 2.2 were tested using our proposed method. Both methanol and acetonitrile (mobile phase B and mobile phase C) were used to achieve optimum polarity (and resolution) for separation of flavoring agents from nicotine. In all the twelve e-liquids, the nicotine peak was found to be pure.

Since flavoring agents are composed of a variety of chemicals, a fixed composition of mobile phase may not work for separation of nicotine from all flavoring agents. To address this hypothesis, we performed a study on the effects of change in organic phase composition on separation of flavoring agents from nicotine. Based on observations as described in Section 3.6, we propose that a variation in the ratio of mobile phase B (methanol) and C (acetonitrile) provides a window (±10%) for changing the organic polarity to achieve optimum resolution between interfering flavoring agents, if any, and nicotine. The run time can be extended, if required, for optimum postelution phase after each run. The proposed method has been validated for this flexibility and robustness as described in Section 3.6.

3.2. Specificity- and Stability-Indicating Study. Specificity in HPLC analysis is the ability to assess an analyte in the presence of other components in the sample matrix such as impurities, degradation product, and excipients. Forced degradation studies were carried out to test the specificity of the method for the nicotine peak. Ten e-liquid flavors were subjected to various stress conditions such as acid and base hydrolysis, oxidation, and thermal degradation. All samples were checked for peak purity of nicotine and any degradant peak of nicotine. Results of the forced degradation study are as mentioned in Table 6.

Nicotine content from each control sample of each flavor was found to be within the range of 90–110% of labeled claim. Samples were subjected to various stress conditions

TABLE 6: Results of forced degradation of various e-liquid flavors.

Name of sample	Category	% control assay	Stressed condition		
			% acid degradation	% base degradation	% oxidation degradation
Standard	NA	NA	7.75	8.29	7.62
QC	NA	99.39	ND	ND	3.5
e-Liquid flavor P	Sweet	95.98	ND	ND	12.2
e-Liquid flavor Q	Menthol	100.76	1.53	ND	4.14
e-Liquid flavor R	Tobacco	103.09	4.69	5.68	7.53
e-Liquid flavor S	Menthol	100.14	2.04	3.38	10.28
e-Liquid flavor T	Fruit	96.18	1.03	ND	2.06
e-Liquid flavor U	Coffee	97.68	ND	1.59	4.33
e-Liquid flavor A	Tobacco	96.19	1.94	1.00	4.75
e-Liquid flavor C	Vanilla	97.31	1.38	0.69	4.44
e-Liquid flavor E	Fruit	98.13	ND	1.44	14.13
e-Liquid flavor G	Fruit	96.94	6.38	1.00	2.5

Note. % degradation of ±1% is considered as no degradation (ND).

and compared against their respective control assays. The e-liquid flavors have different compositions; therefore, the respective % degradation was found to be different.

Acid and base hydrolysis with 0.1 N hydrochloric acid and 1 N sodium hydroxide, respectively, for 30 min was found to cause more than 5% degradation in one fruit flavored and one tobacco flavored e-liquid, respectively. The nicotine peak was found to pass the peak purity criteria in all samples with base to base separation of the nicotine peak.

Oxidation of e-liquid flavors was carried out with 6% peroxide for 30 min. The percent degradation was found to be more than 5% in each of sweet, tobacco, menthol, and fruit flavored e-liquids. In all oxidation degradations, nicotine peak was found to pass peak purity criteria with base to base separation of nicotine peak.

A standard nicotine solution was found to give more than 5% degradation in all three conditions.

Thermal degradation was carried out at 60°C for 2 hrs. The percent assay values obtained for the thermal degradation was found to be higher than control assays. Thermal degradation was found to concentrate samples possibly due to evaporation of solvent. Therefore, thermal degradation was not considered for specificity.

In conclusion, more than 5% degradation was observed in each of menthol, tobacco, fruit, and sweet flavored e-liquids. In all the degradation patterns, the nicotine peak was found to pass peak purity. No nicotine degradant peaks were observed at specified nicotine absorbance wavelength (260 nm). ICH and USP guidelines for validation and stability testing do not specify the limits of degradation in forced degradation studies. A degradation of 5–20% of analyte in at least one of the stressed conditions is generally the accepted range of forced degradation [24, 25]. Based on the degradation pattern observed, the HPLC method was found to be specific and stable indicating for nicotine. Representative chromatograms are shown in Figure 1.

3.3. Linearity and Range. Linearity is a measure of accuracy over the range of the method. e-Liquids are available in market with nicotine concentration ranging from 1 to >50 mg/mL.

The proposed method is based on "dilute (100-fold) and inject" sample preparation approach. Therefore, linearity of the current method was established in the range of 0.4–500 μg/mL which would cover the wide range of nicotine concentration in e-liquids available in market. Linearity was measured by calibration curve of the nicotine standard. The method was found to be linear over the specified range with linear equation $Y = 13000 X - 266$. The coefficient of variation was found to be $R^2 = 1.000$ with Y-intercept less than 3.0% of the peak area of the assay sample.

3.4. Accuracy. The accuracy of an analytical method is the closeness of test results obtained by that method to the theoretical or labeled value. Accuracy is expressed as percent recovery of known, added amounts of analyte. The results of accuracy parameter of the method are shown in Table 7. The % recovery of nicotine at 50%, 100%, and 150% of assay level was found to be within 98 to 102% with % RSD of triplicate preparations NMT 2.0.

3.5. Precision. The precision of an analytical method is the degree of agreement among individual test results when the procedure is applied repeatedly to multiple samplings of a homogenous sample. The precision of the method is expressed as repeatability under normal operating conditions.

The repeatability results were calculated from the accuracy samples as shown in Table 7. The % RSD of triplicate preparation at each level was found to be NMT 2.0.

Intermediate precision has been determined by repeatability of six preparations of assay samples across different days and different analysts under normal operating conditions. The combined % RSD of response of all twelve preparations was found to be NMT 2.0.

3.6. Robustness. Robustness of the method was evaluated as mentioned in Section 2.5.5. An assay sample was injected three times for each parameter of robustness. The % RSD of triplicate injections at each parameter was found to be no

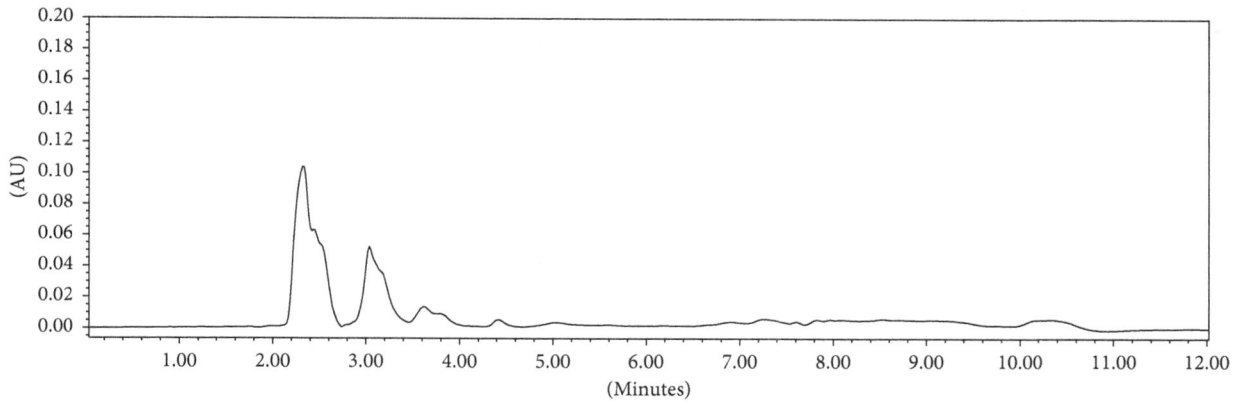

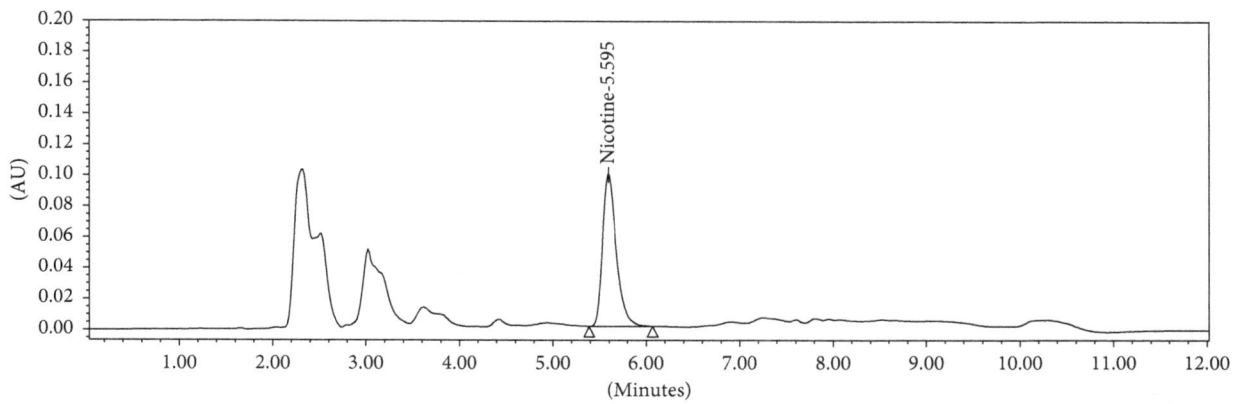

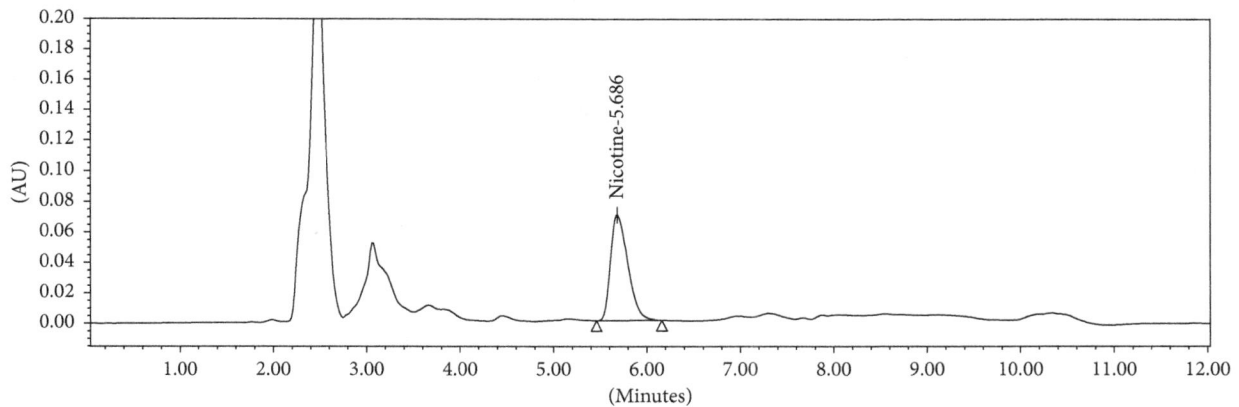

Peak name	RT	Area	USP tailing	USP plate count	Purity angle	Purity threshold
Nicotine	5.595	1019959	1.37	7056	0.544	1.149

(b)

Peak name	RT	Area	USP tailing	USP plate count	Purity angle	Purity threshold
Nicotine	5.686	891825	1.42	4381	0.329	1.153

(c)

FIGURE 1: (a) Representative chromatogram of e-liquid flavor P, a placebo_control. (b) Representative chromatogram of e-liquid flavor P, a sample control. (c) Representative chromatogram of e-liquid flavor P, a sample_oxidation (peroxide degradation).

TABLE 7: Accuracy results of the method.

% spiked level of assay sample	Replicate	% recovery	% mean recovery	% RSD
50 (40 µg/mL)	1	99.95	99.44 (39.78 µg/mL)	0.77
	2	99.81		
	3	98.56		
100 (80 µg/mL)	1	99.41	100.01 (80.01 µg/mL)	0.95
	2	99.52		
	3	101.11		
150 (120 µg/mL)	1	100.27	100.61 (120.73 µg/mL)	0.39
	2	101.04		
	3	100.51		

TABLE 8: Robustness.

Organic phase composition		Sample	Peak area	USP resolution	Purity angle	Purity threshold
% mobile phase B	% mobile phase C					
26	14	e-Liquid flavor V	1031106	1.46	0.729	1.159
			1031686	1.51	0.390	1.106
		e-Liquid flavor W	1023000	1.48	0.984	1.138
			1021702	1.48	0.915	1.144
28	12	e-Liquid flavor V	1030040	2.31	0.192	1.063
			1026349	2.36	0.471	1.104
		e-Liquid flavor W	1023658	2.30	0.134	1.093
			1027008	2.28	0.295	1.128

more than 2.0. Nicotine peak passed the peak purity criteria in all parameters.

In addition to the robustness evaluations, a separate study of the effects of change in organic phase composition on separation of flavoring agents from nicotine was carried out using two different e-liquid flavors V and W of two different categories tobacco and sweet, respectively, which were not tested for specificity study in the validation. The current method was found to be able to separate the nicotine peak from flavoring agents of these two e-liquids, however, with resolution <USP 1.5. ICH and USP recommendation for peak resolution is >2.0, and for accurate quantification, the resolution between peaks should be at least 1.5 [21, 22]. Therefore, to improve peak resolution, same samples were run using a robustness parameter of change in organic mobile phase ratio to modify organic polarity. With decrease in organic polarity, resolution between nicotine and adjacent peak was significantly improved. The results are mentioned in Table 8. Based on these results, it can be concluded that the proposed method can be optimized considering robustness parameter of change in organic phase composition to achieve optimum resolution and peak purity of nicotine peak in those e-liquid flavors which might show interference of flavoring agents at nicotine peak. Representative chromatograms are shown in Figures 2 and 3.

3.7. Limit of Detection and Limit of Quantification (LOD and LOQ). Limit of Detection (LOD) was calculated based on the standard deviation of the response and the slope, as described in Section 2.5.6. The detection limit (DL) is expressed as shown in (2).

Based on the placebo response and linear equation of calibration curve (Section 3.3), LOD was found to be 0.07 µg/mL of nicotine.

Limit of quantitation (LOQ) was calculated based on visual evaluation. The minimum known concentration of nicotine which can be analyzed quantitatively with acceptable accuracy and precision using current method was found to be 0.45 µg/mL of nicotine. The accuracy and precision were performed at LOQ level ($n = 6$). % recovery was found to be within 90 to 110% and % RSD of all six preparations was NMT 10.0. Results are mentioned in Table 9.

3.8. System Suitability. System suitability test was performed by running six injections of standard at assay level. % RSD of the response of six injections was found to be NMT 1.0. USP limits for theoretical plate count (NLT 2000) and peak tailing (NMT 2.0) were applied. The system suitability test of the current method passed all USP criteria (Table 10).

4. Conclusion

An alternative standardized HPLC method for analysis of nicotine in e-liquids has been developed. The method has been fully validated as per ICH and USP guidelines. The method is found to be specific for nicotine as determined by peak purity criteria. The method has been tested for stability, accuracy, and precision for quantification of nicotine in e-liquids. Linearity of the method has been achieved over a wide range of 0.4 to 500 µg/ml of nicotine concentration.

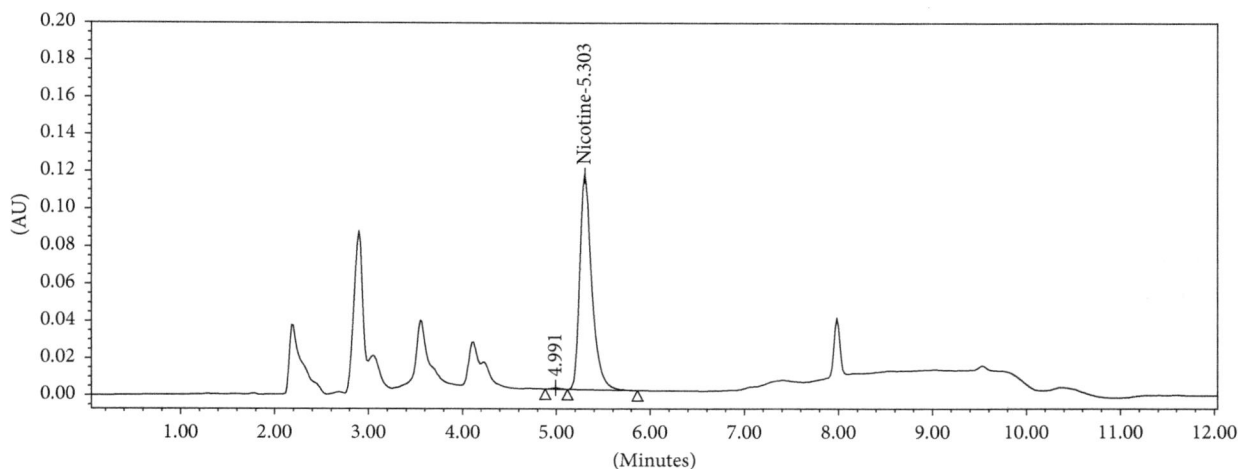

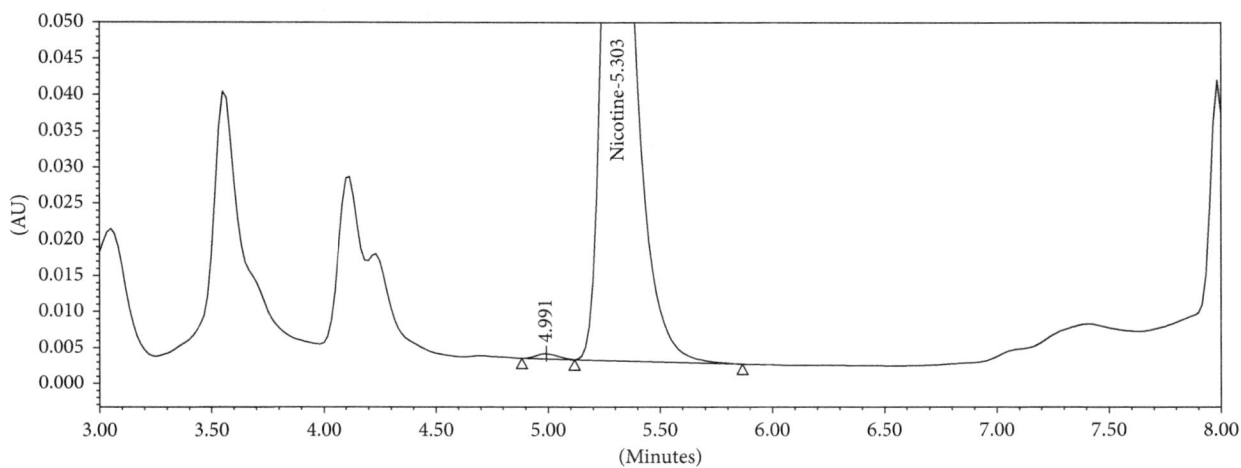

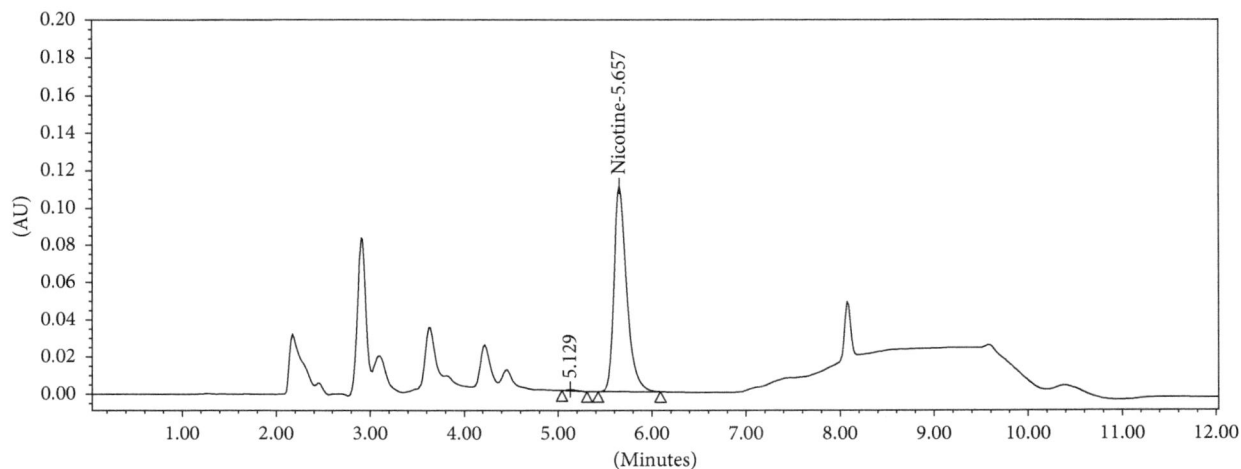

FIGURE 2: Continued.

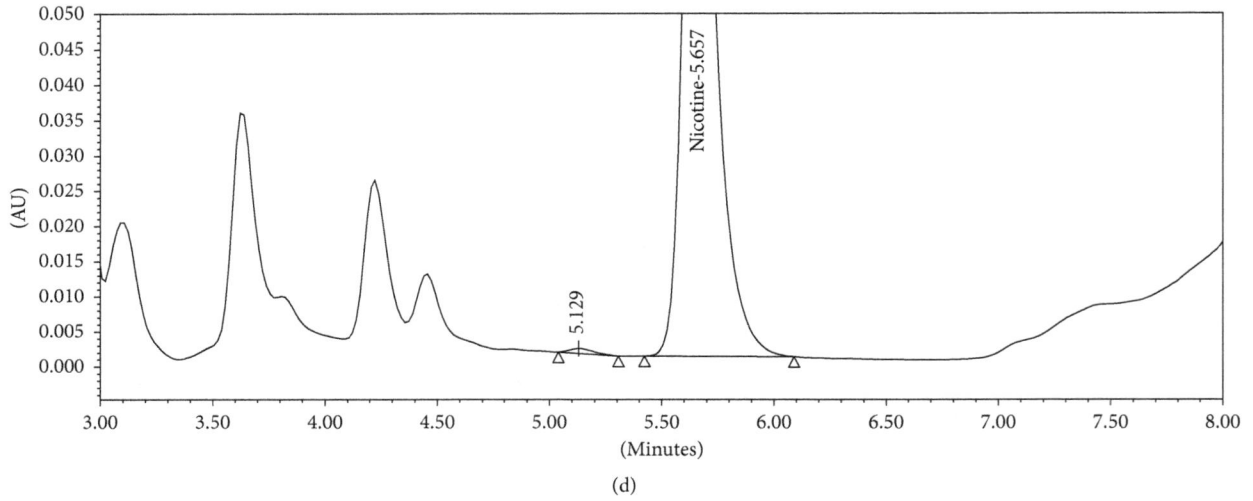

(d)

Figure 2: (a) Full scale representative chromatogram of e-liquid flavor V sample_mobile phase composition B : C, 26 : 14 v/v. (b) Zoomed representative chromatogram of e-liquid flavor V sample_mobile phase composition B : C, 26 : 14 v/v. (c) Full scale representative chromatogram of e-liquid flavor V sample_mobile phase composition B : C, 28 : 12 v/v. (d) Zoomed representative chromatogram of e-liquid flavor V sample_mobile phase composition B : C, 28 : 12 v/v.

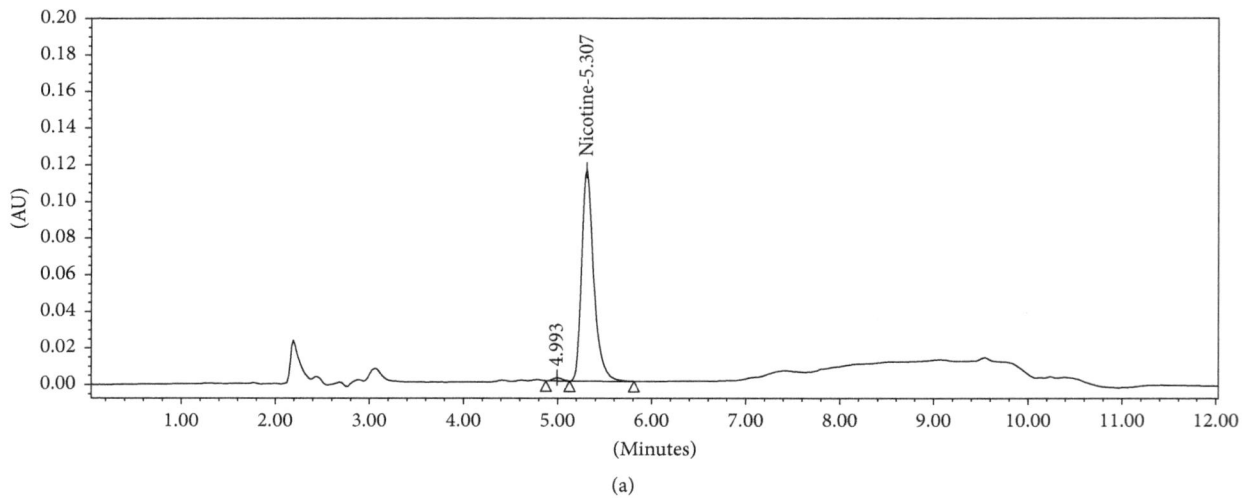

(a)

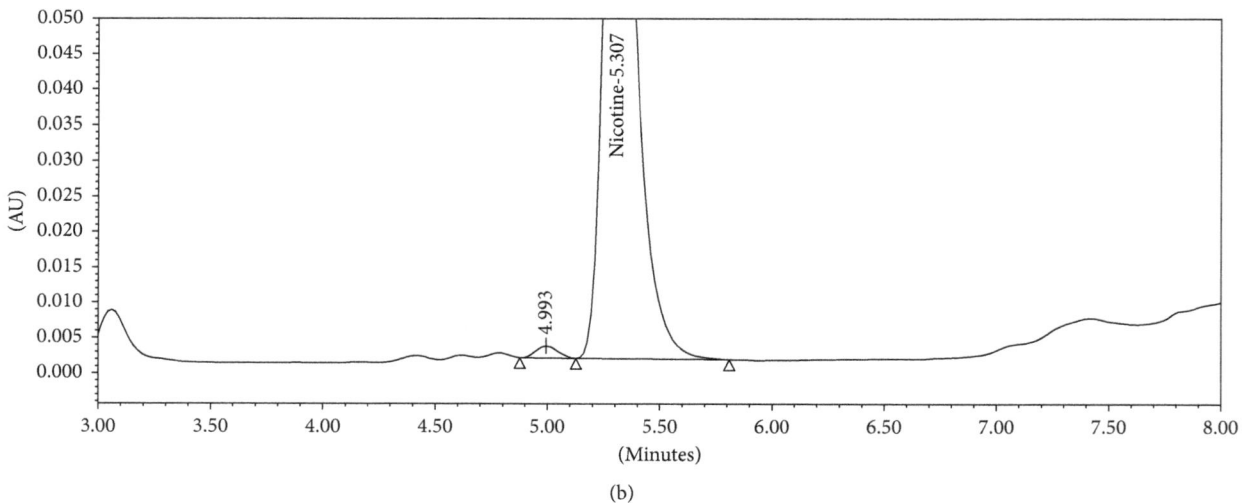

(b)

Figure 3: Continued.

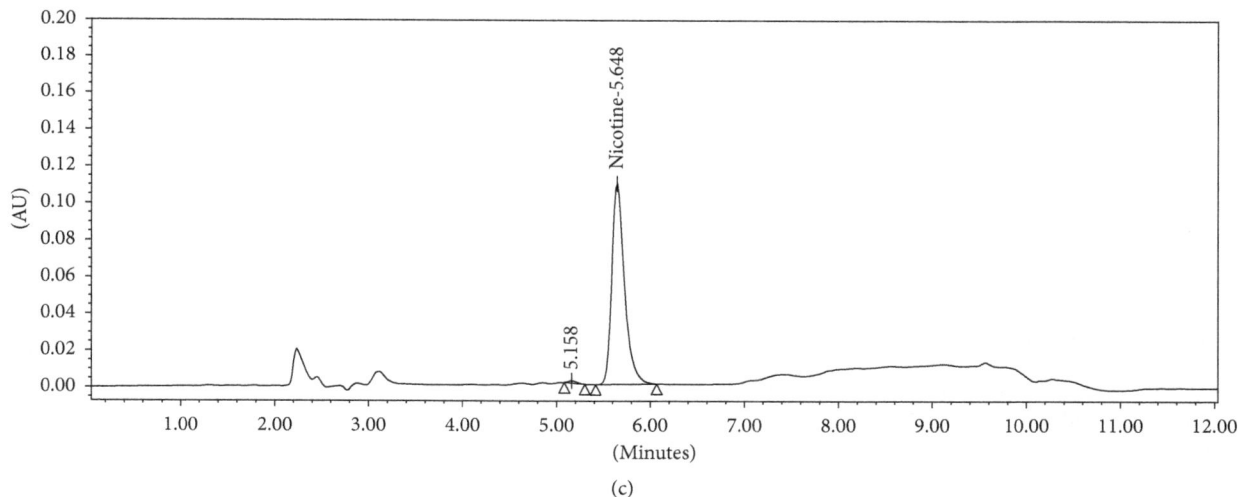

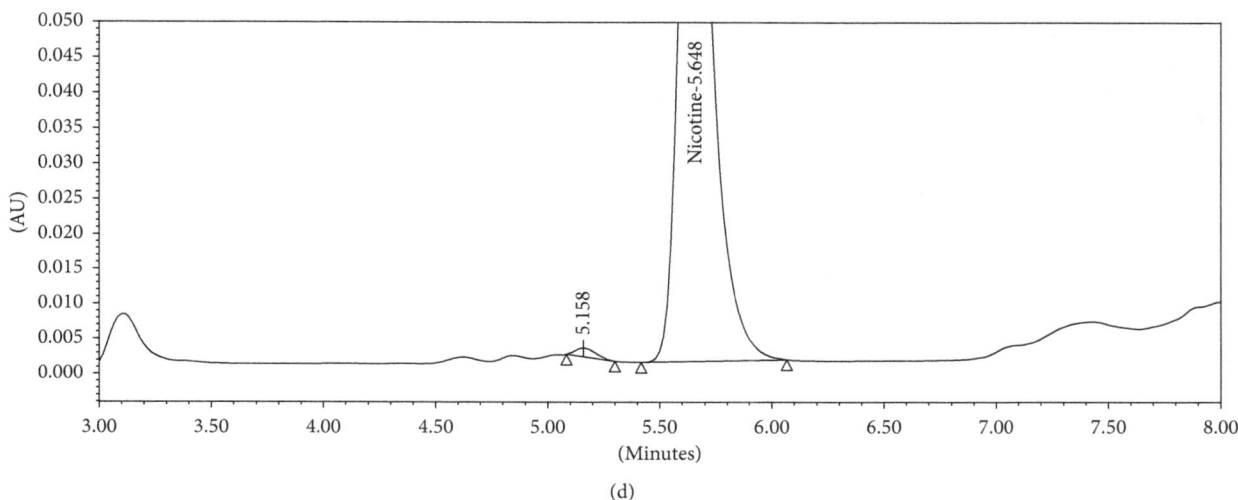

FIGURE 3: (a) Full scale representative chromatograph of e-liquid flavor W sample_mobile phase composition B : C, 26 : 14 v/v. (b) Zoomed representative chromatograph of e-liquid flavor W sample_mobile phase composition B : C, 26 : 14 v/v. (c) Full scale representative chromatograph of e-liquid flavor W sample_mobile phase composition B : C, 28 : 12 v/v. (d) Zoomed representative chromatograph of e-liquid flavor W sample_mobile phase composition B : C, 28 : 12 v/v.

TABLE 9: Accuracy and precision at LOQ level.

% spiked level of assay sample	Replicate	% recovery	% mean recovery	% RSD
LOQ	1	90.89	97.08	6.78
	2	90.63		
	3	99.22		
	4	103.92		
	5	105.17		
	6	92.64		

TABLE 10: System suitability test.

Replicate injection	Area	USP tailing	USP plate count
A1	1299677	1.41	8979
A2	1295191	1.40	9211
A3	1297765	1.40	9132
A4	1294551	1.40	9225
A5	1291896	1.40	9194
A6	1289339	1.39	9035
Mean	1294737	1.40	9129
Std. dev.	3768.89	0.01	101.50
%RSD	0.29	0.45	1.11

Note. No adjacent peaks were observed at the retention time of nicotine. Hence USP resolution criteria is not applicable.

Thus, the method can be used to measure e-liquids with concentration ranging from 1 to >50 mg/mL. Since e-liquids are available in the market with a variety of flavoring combinations, a fixed composition of mobile phase may not work for separation of nicotine from all flavoring agents. Therefore, we tested the method for robustness parameters of change in flow rate, column oven temperature, and organic phase composition. After achieving the nicotine peak purity by successfully passing the robustness parameters, we are concluding that the proposed HPLC method for analysis of nicotine in e-liquids is flexible with accuracy over a wide variety of e-liquids in the market.

Abbreviations

NMT: Not more than
NLT: Not less than
ND: Not detected
RSD: Relative standard deviation
LOD: Limit of detection
LOQ: Limit of quantitation.

Conflicts of Interest

Leon Kosmider works as an expert for the Polish National Committee for Standardization and for the European Committee for standardization of requirements and test methods for e-liquids and emissions. Leon Kosmider was also an employee of the Institute of Occupational Medicine and Environmental Health in Poland. One of the institute's objectives is outsourcing for the industrial sector, including manufacturers of e-cigarettes. However, this has no influence on the study design. Other authors declare no conflicts of interest.

Acknowledgments

Leon Kosmider is supported by the National Institute on Drug Abuse of the National Institutes of Health under Award no. P50DA036105 and the Center for Tobacco Products of the U.S. Food and Drug Administration.

References

[1] S. H. Zhu, J. Y Sun, E. Bonnevie et al., "Four hundred and sixty brands of e-cigarettes and counting: implications for product regulation," *Tobacco Control*, vol. 23, no. 3, pp. iii3–iii9, 2014.

[2] J. K. Noel, V. W. Rees, and G. N. Connolly, "Electronic cigarettes: a new "tobacco" industry?," *Tobacco Control*, vol. 20, no. 1, p. 81, 2011.

[3] A. Bhatnagar, L. P. Whitsel, K. M. Ribisl et al., "Electronic cigarettes: a policy statement from the American Heart Association," *Circulation*, vol. 130, no. 16, pp. 1418–1436, 2014.

[4] M. Blanding, *The E-Cig Quandary|Harvard Public Health Magazine|Harvard T.H. Chan School of Public Health*, Harvard Public Health, Boston, MA, USA, 2016, https://www.hsph.harvard.edu/magazine/magazine_article/the-e-cig-quandary/.

[5] Food And Drug Administration, *Guidance for Industry Listing of Ingredients in Tobacco Products*, Food and Drug Administration, Silver Spring, MD, USA, 2017.

[6] P. Callahan-Lyon, "Electronic cigarettes: human health effects," *Tobacco Control*, vol. 23, no. 2, pp. ii36–ii40, 2014.

[7] L. Manzoli, C. La Vecchia, M. E. Flacco et al., "Multicentric cohort study on the long-term efficacy and safety of electronic cigarettes: study design and methodology," *BMC Public Health*, vol. 13, no. 1, p. 883, 2013.

[8] M. B. Drummond and D. Upson, "Electronic cigarettes: potential harms and benefits," *Annals of the American Thoracic Society*, vol. 11, no. 2, pp. 236–242, 2014.

[9] M. A. Kaisar, S. Prasad, T. Liles, and L. Cucullo, "A decade of e-cigarettes: limited research & unresolved safety concerns," *Toxicology*, vol. 365, pp. 67–75, 2016.

[10] U.S. Department of Health and Human Services, *How Tobacco Smoke Causes Disease: The Biology and Behavioral Basis for Smoking-Attributable Disease*, Centers for Disease Control and Prevention, Atlanta, GA, USA, 2010.

[11] B. Davis, M. Dang, J. Kim, and P. Talbot, "Nicotine concentrations in electronic cigarette refill and do-it-yourself fluids," *Nicotine & Tobacco Research*, vol. 17, no. 2, pp. 134–141, 2015.

[12] M. L. Trehy, W. Ye, M. E. Hadwiger et al., "Analysis of electronic cigarette cartridges, refill solutions, and smoke for nicotine and nicotine related impurities," *Journal of Liquid Chromatography & Related Technologies*, vol. 34, no. 14, pp. 1442–1458, 2011.

[13] J. W. Flora, C. T. Wilkinson, K. M. Sink, D. L. McKinney, and J. H. Miller, "Nicotine-related impurities in e-cigarette cartridges and refill e-liquids," *Journal of Liquid Chromatography & Related Technologies*, vol. 39, no. 17-18, pp. 821–829, 2016.

[14] X. Liu, P. Joza, and B. Rickert, "Analysis of nicotine and nicotine-related compounds in electronic cigarette liquids and aerosols by liquid chromatography-tandem mass spectrometry," *Beiträge zur Tabakforschung International/Contributions to Tobacco Research*, vol. 27, no. 7, pp. 154–167, 2017.

[15] J. Aszyk, P. Kubica, A. Kot-Wasik, J. Namieśnik, and A. Wasik, "Comprehensive determination of flavouring additives and nicotine in e-cigarette refill solutions. Part I: liquid chromatography-tandem mass spectrometry analysis," *Journal of Chromatography A*, vol. 1519, pp. 45–54, 2017.

[16] B. J. S. Herrington, C. Myers, and A. Rigdon, *Analysis of Nicotine and Impurities in Electronic Cigarette Solutions and Vapor*, Restek, State College, PA, USA, 2015.

[17] J. F. Etter, E. Zäther, and S. Svensson, "Analysis of refill liquids for electronic cigarettes," *Addiction*, vol. 108, no. 9, pp. 1671–1679, 2013.

[18] N. K. Meruva, M. E. Benvenuti, G. E. Cleland, and J. A. Burgess, *Simultaneous Determination of Nicotine and Related Impurities in E-Liquids and E-Cigarettes Using UPLC-UV-MS*, Waters, UK, 2016.

[19] Waters Corporation, *Waters 996 Photodiode Detector: Peak Purity I What is Peak Purity Analysis? Peak Purity Analysis*, Waters Corporation, Milford, MA, USA, 1998.

[20] International Conference on Harmonization, *ICH Topic Q2 (R1) Validation of Analytical Procedures: Text and Methodology*, Waters, UK, 2005.

[21] A. Bose, "HPLC calibration process parameters in terms of system suitability test," *Austin Chromatography*, vol. 1, no. 2, pp. 1–4, 2014.

[22] The United States Pharmacopeia, "General chapter, <621> chromatography," in *USP40-NF35*, pp. 424–434, USP, Rockville, MD, USA, 2015.

[23] The United States Pharmacopeia, "1225 validation of compendial procedures," in *USP34-NF29*, pp. 778–782, USP, Rockville, MD, USA, 2011.

[24] K. M. Alsante, A. Ando, R. Brown et al., "The role of degradant profiling in active pharmaceutical ingredients and drug products," *Advanced Drug Delivery Reviews*, vol. 59, no. 1, pp. 29–37, 2007.

[25] M. K. Sharma and M. Murugesan, "Forced degradation study an essential approach to develop stability indicating method," *Journal of Chromatography and Separation Techniques*, vol. 8, no. 1, pp. 8–10, 2017.

10

Detection of Organophosphorus Pesticides in Wheat by Ionic Liquid-Based Dispersive Liquid-Liquid Microextraction Combined with HPLC

Wei Liu ⓘ,[1] Ji Quan,[2] and Zeshu Hu[1]

[1]School of Resources and Environmental Engineering, Wuhan University of Technology, No. 122 Luoshi Road, Wuhan 430070, China
[2]School of Management, Wuhan University of Technology, No. 122 Luoshi Road, Wuhan 430070, China

Correspondence should be addressed to Wei Liu; liuwei86@whut.edu.cn

Academic Editor: Luca Campone

Food safety issues closely related to human health have always received widespread attention from the world society. As a basic food source, wheat is the fundamental support of human survival; therefore, the detection of pesticide residues in wheat is very necessary. In this work, the ultrasonic-assisted ionic liquid-dispersive liquid-liquid microextraction (DLLME) method was firstly proposed, and the extraction and analysis of three organophosphorus pesticides were carried out by combining high-performance liquid chromatography (HPLC). The extraction efficiencies of three ionic liquids with bis(trifluoromethylsulfonyl)imide (Tf_2N) anion were compared by extracting organophosphorus in wheat samples. It was found that the use of 1-octyl-3-methylimidazolium bis(trifluoromethylsulfonyl)imide ([OMIM][Tf_2N]) had both high enrichment efficiency and appropriate extraction recovery. Finally, the method was used for the determination of three wheat samples, and the recoveries of them were 74.8–112.5%, 71.8–104.5%, and 83.8–115.5%, respectively. The results show that the method proposed is simple, fast, and efficient, which can be applied to the extraction of organic matters in wheat samples.

1. Introduction

Food cultivation is the main part of agricultural production. In the process of agricultural cultivation, pesticide spraying is the dominant approach to protect the healthy growth of plants. Organophosphorus pesticides (OPPs), as an inexpensive, stable, and efficient pesticide, are usually and widely used in agricultural production in the world [1]. OPPs can inhibit the activity of acetylcholinesterase, and acetylcholine in body is thereby accumulated, which can have a serious effect on central nervous system, can cause symptoms of poisoning, and can even lead one to death. Because of their toxicity and abuse, the pollution of water and land by OPPs has also become a serious environmental problem, which at all times threatens people's lives. As the main and basic food crop for human beings, wheat is the source of daily food that people often come into contact with. The pesticide residue in wheat must be controlled and monitored.

Therefore, in order to ensure food safety and human health, the detection of OPPs in wheat is very necessary [2].

In the pesticide residue analysis, the commonly used detection methods are mainly gas chromatography (GC) [3], gas chromatography-mass spectrometry (GC-MS) [4], high-performance liquid chromatography (HPLC) [5], and liquid chromatography tandem mass spectrometry (LC-MS/MS) [6]. Although LC-MS/MS and GC-MS show excellent detection capability, the high cost still inhibits their widespread use. Compared to them, HPLC with its convenience, efficiency, and durability, is the most extensive means of pesticide detection.

The traditional methods of extraction of pesticides in wheat are mainly liquid-phase extraction (LLE) [7], liquid-phase microextraction (LPME) [8], supercritical fluid extraction (SFE) [9], and so on. In recent years, a rapid, simple, and convenient dispersive liquid-liquid microextraction (DLLME) [10] method has been proposed and then is rapidly applied to various drug extraction studies, especially in the field of pesticide extraction. In the process of DLLME, the extractant, dispersant, and water form a three-phase system, and the analytes and the extractant are deposited at the bottom of the centrifuge tube by centrifugation, followed by quantitative analysis by means of analytical instruments; the whole process is very simple and efficient, thus, it is a promising method for trace extraction. In the traditional DLLME, the commonly used extractants are chlorobenzene [11], dichloromethane [12], dibromoethane [13], and so on. They are toxic, hazardous, flammable, and environmentally damaging organic solvents and difficult to reuse [14]. The use of green, low-toxic, and highly-efficient extractants is an inevitable trend for DLLME.

As a class of green solvent, ionic liquids (ILs) are gaining huge attention since their unique properties especially negligible vapour pressure, wide range of solubility, miscibility, and stability at high temperatures; they are good replacements for conventional volatile and toxic organic solvents in chemical processes [15]. One of the advantages arising from the chemical structures of ILs is that alteration of the cation or anion can cause changes in properties such as viscosity, melting point, water miscibility, and density [14]. Due to the ILs often showing great capability of dissolving both organic and nonorganic compounds, it is easy to separate an IL from the reaction system [16], the application of ILs in DLLME is getting a growing interest in drugs and pesticides extraction in analytical chemistry [17–19].

ILs with 1-alkyl-3-methylimidazolium cation and anions of bromide (Br^-), hexafluorophosphate (PF_6^-) and bis(trifluoromethylsulfonyl)imide (Tf_2N^-) are most commonly used as extraction solvents in many literatures. For example, 1-butyl-3-methylimidazolium hexafluorophosphate ([BMIM] [PF_6]) was used to extract benzodiazepines [20], 1-hexyl-3-methylimidazolium bis(trifluoromethylsulfonyl)imide ([HMIM] [Tf_2N]) was used to extract bisphenol A [21], 1-hexyl-3-methylimidazolium hexafluorophosphate ([HMIM][PF_6]) was used to extract hexachlorophene [22], and 1-butyl-3-methylimidazolium bromide ([BMIM] Br) was the extraction solvent of brazilin and protosappanin B [23]. The ILs showed good extraction efficiency in their extraction systems.

There are two factors influencing the solubility and miscibility of a given ionic liquid. One is the type of its anion and the other is the length of the alkyl chain of the cation. As literature reported [15], increase of the alkyl chain causes increase of the capacity and hydrophobicity of the ILs, and the bigger the anion size, the stronger the hydrophobicity. The size of Tf_2N anion is bigger than PF_6^- and BF_4^-, thus, the Tf_2N anion-based ILs exhibit high hydrophobicity and capacity, which with the same effect will save the use of IL volume. In addition, the strong delocalization and diffuse nature of the negative charge in the S–N–S core of [Tf_2N]$^-$ leads to a reduction in cation-anion interactions [24]. Due to these properties, Tf_2N anion-based ILs have attracted broad attention and have been widely applied in chemical processes such as recovery of metal [25], organics extraction [26], and CO_2 capture [27], and they are promising extraction agents for pesticides extraction.

Although there are many literatures employing Tf_2N anion-based ILs and DLLME for metal ion extraction [19],

the application of them for OPPs extraction is rare. Furthermore, IL-DLLME for pesticide extraction is often applied in water or liquid samples. As far as we know, there is no report about using Tf_2N anion-based ILs as extraction solvent in the DLLME process to extract OPPs from wheat yet.

Understanding of the structure and features of Tf_2N-based ILs is of great interest due to their exclusive physicochemical properties. This work compared the OPPs extraction capacity of ILs with Tf_2N anion and three different cations ([HHIM], [OMIM], and [BeOIM]) in wheat samples for the first time, and applied the ILs in the DLLME method followed by HPLC analysis. Different factors influencing the extraction efficiency including the extractant type and volume, dispersant type and volume, and temperature were investigated. The proposed method in this work fills the gap in the extraction method of pesticide residues in wheat samples by Tf_2N-based ILs.

2. Experimental

2.1. Reagents, Standards, and Materials. Organophosphorus pesticides (OPPs) of fenitrothion, fenthion, and phoxim were purchased from Beijing Agricultural Environmental Protection Center, the structure of them are shown in Figure 1. Acetonitrile and methanol (HPLC grade) were obtained from Tianjin Siyou Fine Chemicals Co., Ltd. (Tianjin, China). The 1-octyl-3-methylimidazole bis(trifluoromethylsulfonyl)imide ([OMIM] [Tf_2N]), 1,3-dihexylimidazole bis(trifluoromethylsulfonyl)imide ([HHIM][Tf_2N]), and 1-benzyl-3-octylimidazole bis(trifluoromethylsulfonyl)imide ([BeOIM][Tf_2N]) were laboratory made. Standard solutions with the concentration of 1 mg/ml were prepared by dissolving each OPP standard (0.0100 g) into 10.0 mL acetonitrile and stored at 4°C. Wheat samples were purchased from Henan Agricultural Sciences Institute (Zhengzhou, Henan).

2.2. Apparatus. HPLC analysis was carried out by the Shimadzu HPLC system which was equipped with an LC-10AT pump (Shimadzu, Japan), an SPD-10A UV-VIS detector (Shimadzu, Japan), a Shimadzu VP-ODS column (150 mm × 4.6 mm i.d., 5 μm), and a Rheodyne 7725i six-way valve injector with 20 μL sample loop (Rheodyne, Rohnert Park, CA, USA). The mobile phase was the mixture of methanol and water (70 : 30, v/v) with the flow rate of 1 mL/min. The wavelength was set at 254 nm. A JP010/S ultrasonic cleaner (Shenzhen Jie UNITA Cleaning Equipment Co., Ltd., Shenzhen, China) was used for extracting OPPs from wheat sample into the acetonitrile phase. An 80-1 centrifuge (Huafeng Instrument Co. Ltd., Jintan, China) was used for centrifuging.

2.3. Extraction Procedure

2.3.1. Wheat Sample Extraction (Step 1). Wheat samples were grinded into fine powder, and 1 g powder was placed in a 10 mL centrifuge tube, then a solution of 5 mL methanol containing 110 μL [OMIM][Tf_2N] was added into the centrifuge tube and followed by ultrasonic treatment for 8 min.

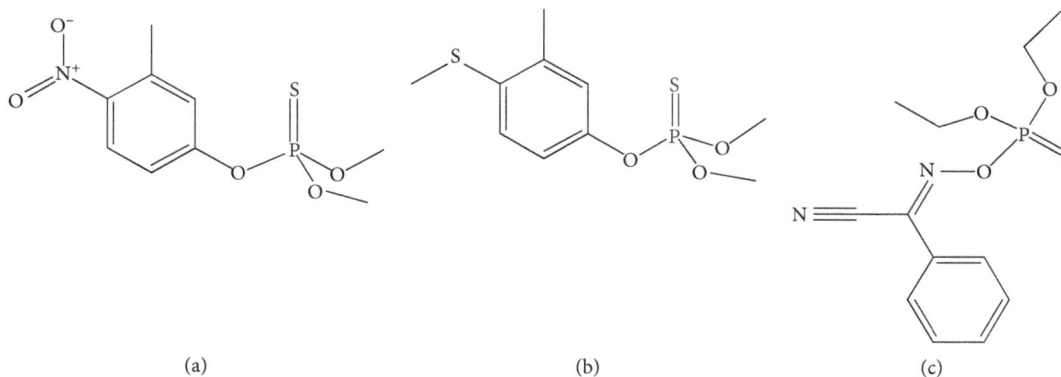

(a)   (b)   (c)

FIGURE 1: Structures of three OPPs. (a) Fenitrothion. (b) Fenthion. (c) Phoxim.

2.3.2. DLLME Procedure (Step 2). After centrifugation for 5 min, the obtained methanol solution containing OPPs and ionic liquid was taken out into a centrifuge tube including 5 mL distilled water, and a cloudy solution was formed in the tube. Then, it was centrifuged for 5 min. Finally, 5 μL of [OMIM][Tf$_2$N] sedimentary facies formed in the bottom of the centrifuge tube was injected to the HPLC system for analysis.

2.4. Enrichment Factor and Extraction Recovery. Enrichment factor (EF) and extraction recovery (ER) were the two evaluating indicators of this developed method. They were calculated by the equation as follows:

$$EF = \frac{C_{sed}}{C_0},$$

$$ER = \frac{C_{sed} \times V_{sed}}{C_0 \times V_{aq}} \times 100\% = EF \times \frac{V_{sed}}{V_{aq}} \times 100\%. \quad (1)$$

C_0 and C_{sed} express the concentration of the OPPs in the DLLME procedure and the concentration in sedimentary facies, respectively. V_{aq} and V_{sed} stand for the volume of aqueous solution and sedimentary facies.

Experimental data were the average of three repetitions in each case.

3. Results and Discussion

3.1. Method Optimization

3.1.1. Selection of Solvent in Ultrasonic Extraction. The selection of the solvent in ultrasonic extraction is very important for the efficient extraction of the target from the wheat sample. In fact, the solvent in ultrasonic extraction plays a dual role in the whole extraction process, and it acts as an extractant for wheat samples and simultaneously as a dispersant in the process of DLLME.

Therefore, the solvent in the ultrasonic extraction needs to satisfy the following conditions:

(1) Effective extraction for the targets in wheat

(2) Excellent dispersibility for extractants in the process of DLLME

In order to select the appropriate solvent, methanol, ethanol, and acetonitrile were investigated experimentally. The results are shown in Figure 2. It can be seen from the figure that when methanol is used as the solvent, the EFs of the three OPPs are the highest, and the ER values are slightly lower than when acetonitrile is used. Although, the ERs of the targets are the highest when acetonitrile is used as the solvent, the EFs are the lowest. When using ethanol, both the EFs and ERs are low. In order to ensure both EFs and ERs are higher, ultimately, methanol was chosen as the solvent in the ultrasonic extraction process, which also acts as a dispersant for the DLLME process.

3.1.2. Selection of Methanol Volume. The choice of methanol volume in the ultrasonic extraction process requires two aspects: on one hand, there is a need for sufficient methanol to facilitate the extraction of the OPPs from the wheat sample as much as possible; on the other hand, a suitable volume of methanol is required to better disperse the ionic liquid in the DLLME to obtain a suitable deposition phase volume for subsequent injection analysis. Therefore, it is necessary to examine the volume of methanol.

The effect of extraction on three OPPs was investigated when the volume of methanol was changed from 0.6 to 1.4 mL (containing 22 μL [OMIM][Tf$_2$N]). The results are shown in Figure 3.

As can be seen from the figure, with the increase in methanol volume, EFs show an increasing trend, on the contrary, ERs show a downward trend. This can be explained by the fact that the increase in the volume of methanol is better to extract the targets from the wheat sample and that increases the solubility of the extractant (ionic liquid) in the DLLME, thus, decreases the volume of the deposition phase from 13 μL to 7.5 μL, so EFs increase, while ERs become smaller.

Considering the EF and ER values, 1.0 mL of methanol was selected to use.

3.1.3. Selection of Ionic Liquid Species. In this paper, ionic liquid was employed for the extractant in the DLLME

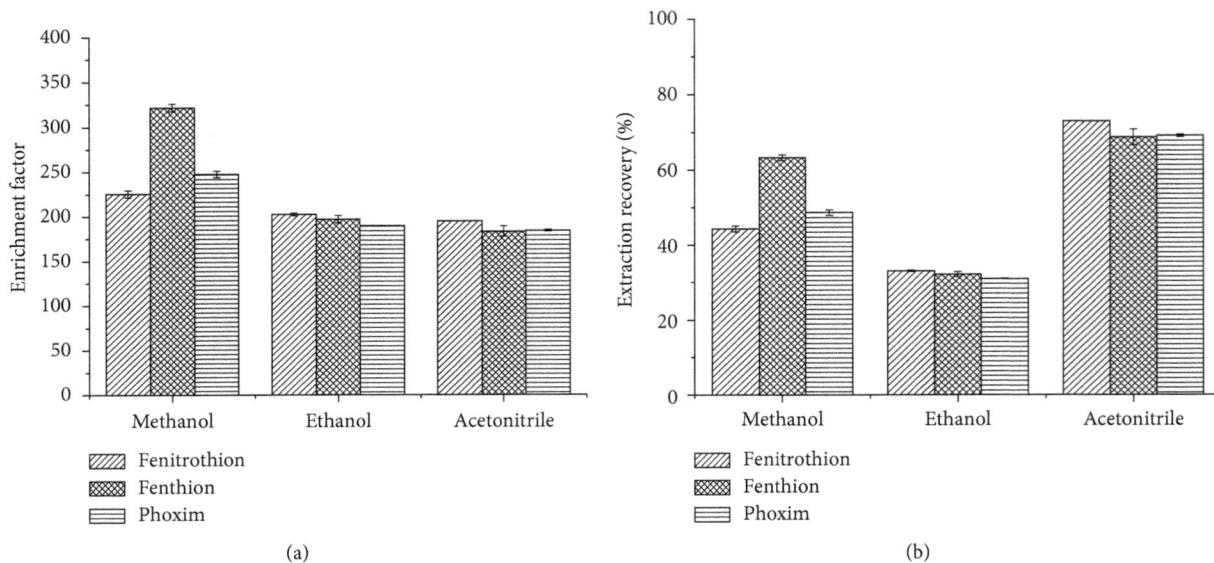

FIGURE 2: Effects of different reagents on EF (a) and ER (b). Extraction conditions: solvents, methanol, ethanol, and acetonitrile, respectively; solvent volume, 1 mL; extractant, [OMIM][Tf$_2$N]; extractant volume, 22 μL; extraction temperature, room temperature; ultrasonic time, 8 min; centrifugal time, 5 min.

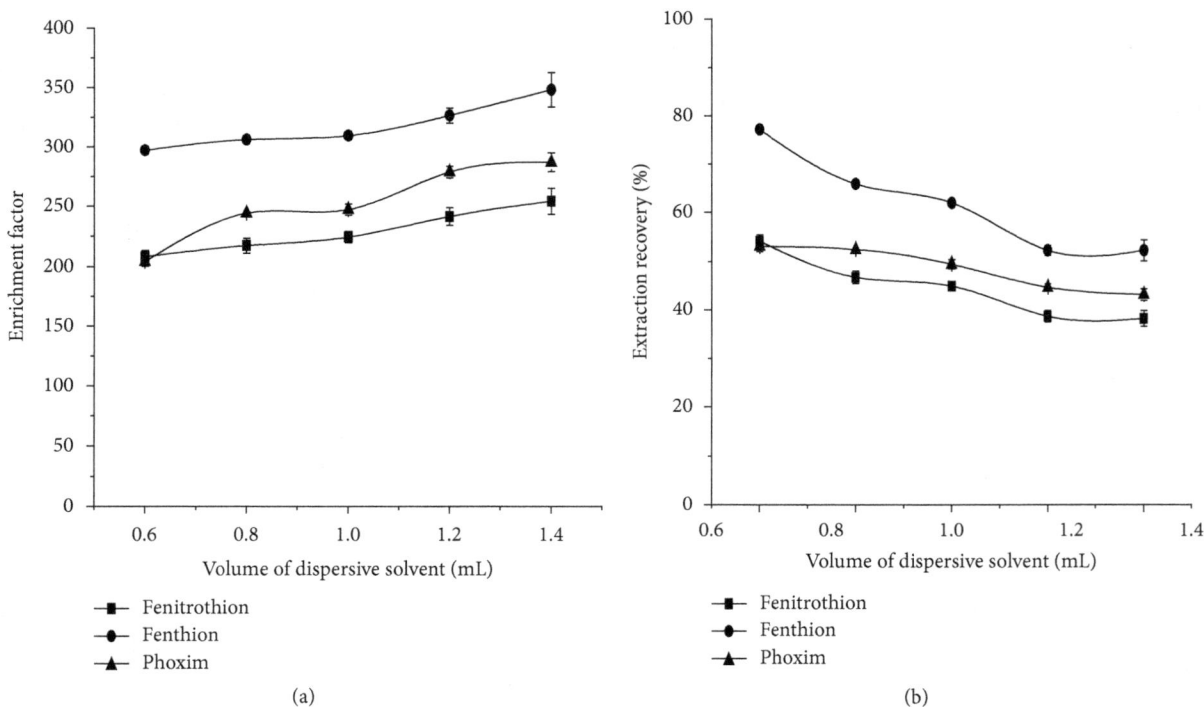

FIGURE 3: Effect of methanol volume on EFs (a) and ERs (b). Extraction conditions: solvent, methanol; solvent volume, 0.6 mL, 0.8 mL, 1 mL, 1.2 mL, and 1.4 mL, respectively; extractant, [OMIM][Tf$_2$N]; extractant volume, 22 μL; extraction temperature, room temperature; ultrasonic time, 8 min; centrifugal time, 5 min.

process. The choice of ionic liquid type is very important. Because different ionic liquids have different solubilities and extraction abilities, the dispersion effects in aqueous solution are different, which can affect the volume of sedimentary facies and finally affect the extraction results.

In order to select the appropriate IL type, the results of the extraction of three ILs ([OMIM][Tf$_2$N], [HHIM][Tf$_2$N], and [BeOIM][Tf$_2$N], resp.) with the same anions ([Tf$_2$N$^-$]) were compared using the same volume (22 μL). The results are shown in Figure 4. The figure shows that using [OMIM]

(a)

(b)

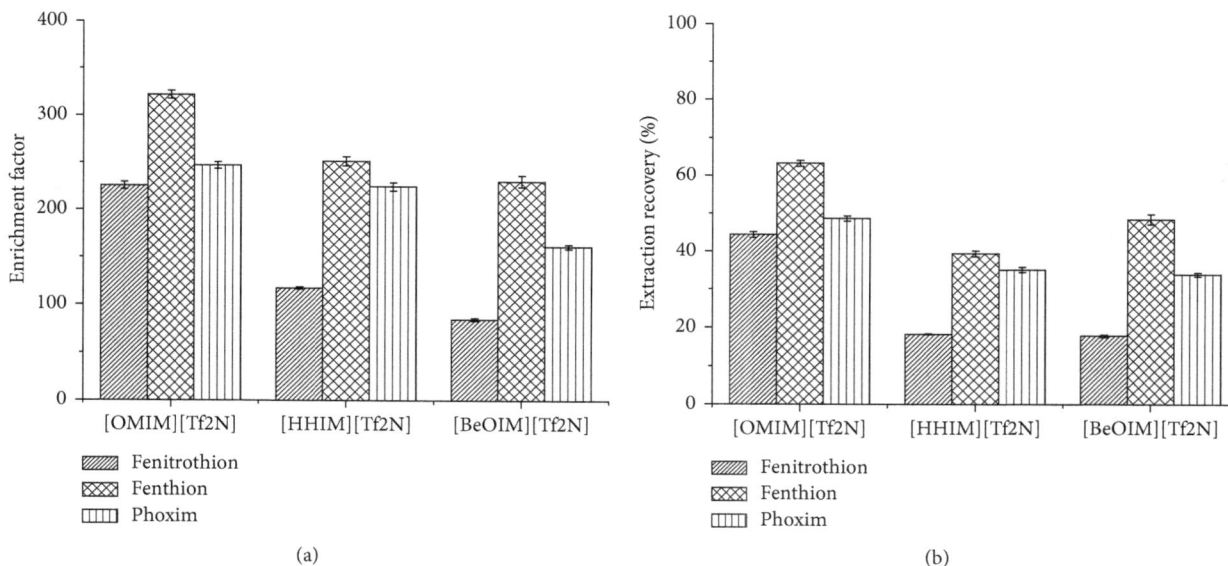

FIGURE 4: Effects of IL species on EF (a) and ER (b). Extraction conditions: solvent, methanol; solvent volume, 1 mL; extractant, [OMIM] [Tf$_2$N], [HHIM][Tf$_2$N], and [BeOIM][Tf$_2$N], respectively; extractant volume, 22 μL; extraction temperature, room temperature; ultrasonic time, 8 min; centrifugal time, 5 min.

[TF$_2$N] as the extractant has the highest EF and ER values. So [OMIM][TF$_2$N] was selected as the DLLME extractant.

3.1.4. Selection of [OMIM][TF$_2$N] Volume.

In the DLLME process, the volume of the extractant will affect the volume of the deposited phase and also affect the formation of the dispersant-extractant-water three-phase suspension system, thereby affecting the extraction effect. In order to obtain the optimized amount of IL, the effect of [OMIM][Tf$_2$N] volume from 18–26 μL on the extraction results was investigated during the DLLME process. It can be seen from Figure 5 that the EFs of the three OPPs decrease as the volume of [OMIM] [TF$_2$N] increases while the ERs increase. This is because the larger the [OMIM][TF$_2$N] volume was used, the poorer the dispersion effect in the water formed and the larger the volume of the deposited phase, from 7.16 μL to 15.75 μL, was obtained, resulting in a decrease in the EFs and an increase in the ERs. Considering the EFs and ERs, the volume of [OMIM][TF$_2$N] was chosen to be 22 μL.

3.1.5. Selection of Ultrasonic Time.

Ultrasonic time will affect the dissolution of OPPs from wheat samples to methanol solution, which will affect the extraction of DLLME and ultimately affect EFs and ERs of the OPPs. The effects of ultrasonic extraction time from 2 min to 14 min on the extraction efficiency were investigated. Results are shown in Figure 6. It can be seen from the figure, when the time of ultrasound is set to 8 min, both EFs and ERs are high, so 8 min was selected as the ultrasonication time.

3.1.6. Selection of Temperature in DLLME.

The temperature of DLLME may affect the mass transfer efficiency of OPPs in wheat, thus affecting the extraction results. Therefore, it is necessary to examine the effect of temperature on the extraction efficiency of the targets. The effect of temperature on the extraction efficiency was investigated by changing the temperature of the aqueous solution (10°C–50°C). The results are shown in Figure 7. The figure indicates that when the temperature of the aqueous solution is 20°C, the EFs and ERs of the OPPs are higher; when the temperature is 50°C, the solubility of [OMIM][TF$_2$N] increased and the volume of sedimentary facies decreased slightly from 10.5 μL to 9.75 μL, which reduced ERs slightly. Because the DLLME process is very short, extraction can be done in an instant, the overall effect of aqueous solution temperature is not significant for simple experimental operation, and DLLME was carried out at room temperature.

3.1.7. Selection of Centrifugal Time.

In the process of DLLME, in order to separate the ionic liquid-phase from the aqueous phase, it is necessary to separate the extractant-dispersant-water three-phase system. The length of the centrifugal time will affect the volume of the deposited phase, resulting in changes in EFs and ERs. In this work, consistent with literature [28], 5 min is selected for centrifugation because 5 min of centrifugation is enough to ensure the deposition of [OMIM][Tf$_2$N] ionic liquid owing to its high hydrophobicity.

3.2. Evaluation of Method Performance.

In order to evaluate the proposed method for extracting three OPPs from the wheat sample, parameters including linearity, repeatability, and limits of detection were obtained and investigated through a series of experiments under the optimized conditions. Table 1 shows the results. Under optimized conditions, the EFs range from 203.8 to 332.4. The method has

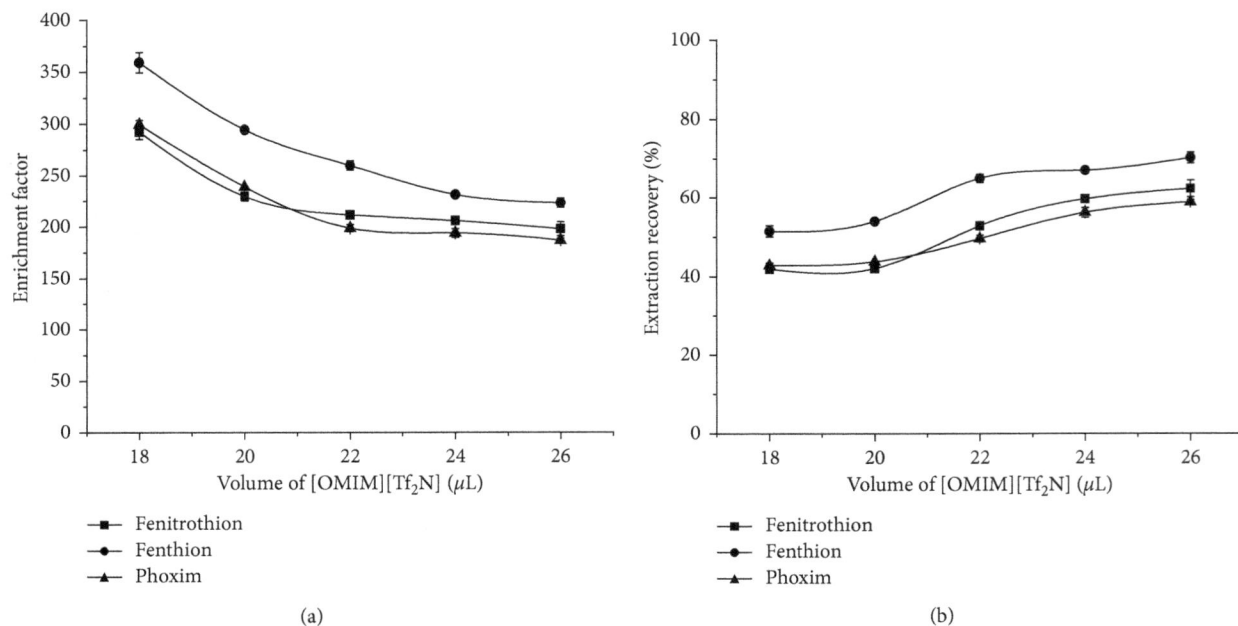

(a)   (b)

FIGURE 5: Effects of [OMIM][TF$_2$N] volume on EFs (a) and ERs (b). Extraction conditions: solvent, methanol; solvent volume, 1 mL; extractant, [OMIM][Tf$_2$N]; extractant volume, 18 μL, 20 μL, 22 μL, 24 μL, and 26 μL, respectively; extraction temperature, room temperature; ultrasonic time, 8 min; centrifugal time, 5 min.

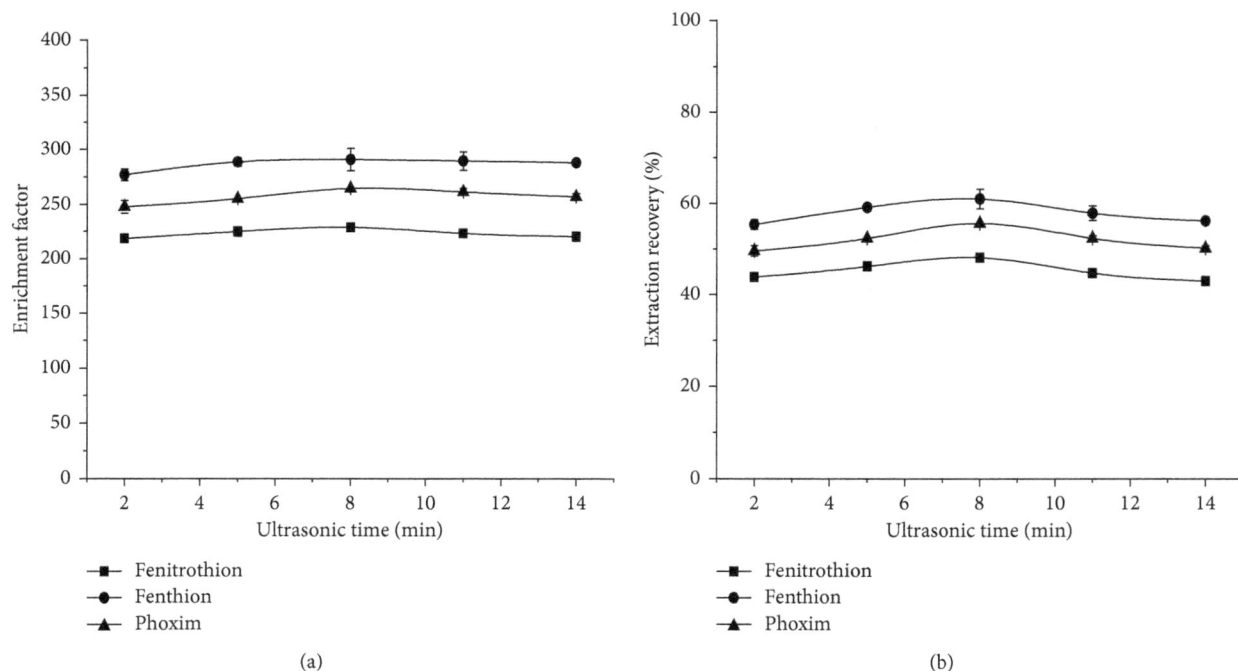

(a)   (b)

FIGURE 6: Effects of ultrasonic time on EFs (a) and ERs (b). Extraction conditions: solvent, methanol; solvent volume, 1 mL; extractant, [OMIM][Tf$_2$N]; extractant volume, 22 μL; extraction temperature, room temperature; ultrasonic time, 2 min, 5 min, 8 min, 11 min, and 14 min, respectively; centrifugal time, 5 min.

a good linear relationship between 0.1 and 100.0 μg/g, and the range of correlation coefficient (r^2) is 0.9973–0.9998. The relative standard deviations (RSDs) were between 0.6% and 6.3% ($n = 5$). The limits of detection (LODs), based on

a signal-to-noise ratio (S/N) of 3, were 0.1 μg/kg for all analytes. The results show that this method has high sensitivity, good reproducibility, and wide linear range when used in wheat samples determination.

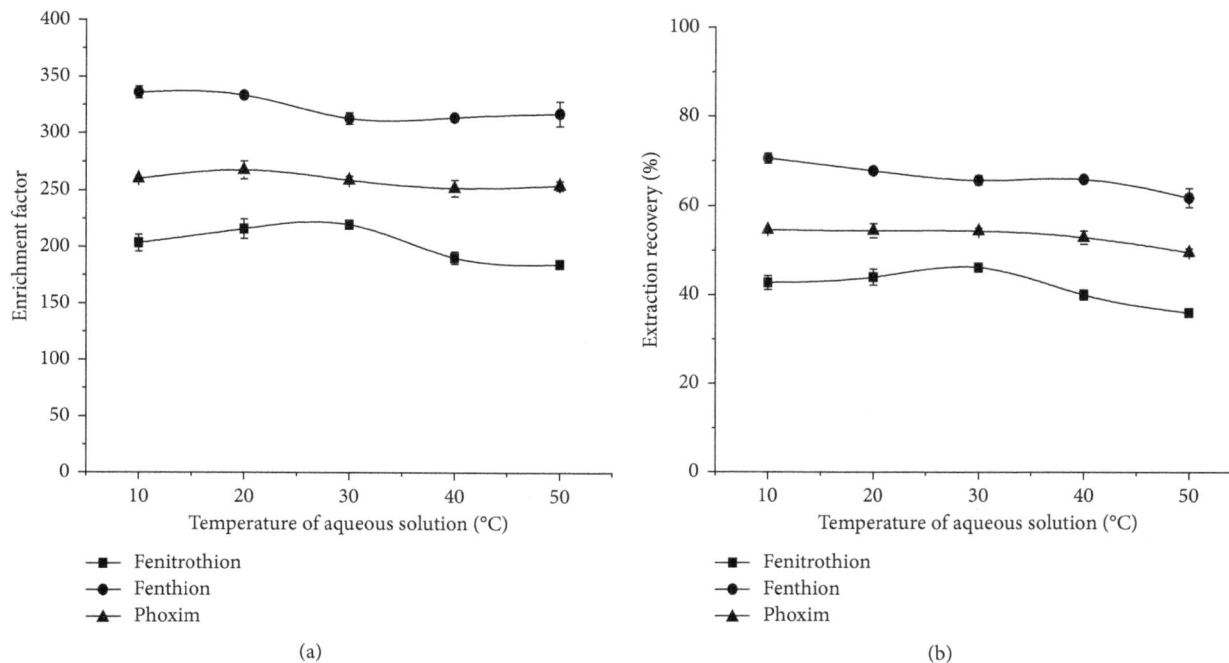

(a)   (b)

FIGURE 7: Effect of aqueous solution temperature on EFs (a) and ERs (b). Extraction conditions: solvent, methanol; solvent volume, 1 mL; extractant, [OMIM][Tf$_2$N]; extractant volume, 22 μL; extraction temperature, 10°C, 20°C, 30°C, 40°C, and 50°C, respectively; ultrasonic time, 2 min, 5 min, 8 min, 11 min, and 14 min, respectively; centrifugal time, 5 min.

TABLE 1: Evaluation of method performance.

Compounds	Linearity range (μg/g)	1 μg/g spiked		5 μg/g spiked		50 μg/g spiked		LOD (μg/kg)
		EF	RSD (%) ($n = 5$)	EF	RSD (%) ($n = 5$)	EF	RSD (%) ($n = 5$)	
Fenitrothion	0.01–100	332.4	1.8	250.4	2.5	203.8	4.6	0.1
Fenthion	0.01–100	251.4	3.8	297.8	5.5	214.3	6.3	0.1
Phoxim	0.01–100	328.4	3.4	265.0	5.9	264.0	0.6	0.1

TABLE 2: Analysis of wheat samples in three different years.

Compounds	Spike level (μg/g)	Wheat in 2002		Wheat in 2006		Wheat in 2010	
		RR (%)	RSD (%) ($n = 3$)	RR (%)	RSD (%) ($n = 3$)	RR (%)	RSD (%) ($n = 3$)
Fenitrothion	1	94.6	4.8	106.8	0.8	99.5	4.2
	5	106.9	3.4	112.5	3.4	100.0	3.7
	50	74.8	1.5	72.9	0.7	83.3	2.8
Fenthion	1	71.8	1.1	90.2	4.2	98.3	3.4
	5	76.9	3.7	74.7	1.8	104.5	3.1
	50	82.7	4.2	74.9	2.8	87.5	3.0
Phoxim	1	115.5	2.3	106.7	3.9	106.9	1.0
	5	83.8	3.2	101.6	5.2	102.3	0.8
	50	108.3	1.8	85.4	3.4	84.1	3.5

3.3. Analysis of Real Samples. The method was applied to the wheat samples in three different years. The recoveries of the methods were determined by adding 3 different concentrations of OPPs (1, 5, and 50 μg/g). The results are shown in Table 2. The experimental results show that the recoveries of the three wheat species are between 74.8 and 112.5%, 71.8 and 104.5%, and 83.8 and 115.5%, respectively, which indicates that this method can be used accurately and reliably for the extraction and determination of the actual wheat

samples. The chromatograms of the wheat samples adding 5 μg/g of three OPPs before and after extractions are shown in Figure 8. After extracting wheat with methanol, and directly injecting the methanol, the analytes cannot be detected, while after the [OMIM][Tf$_2$N]-DLLME process, OPPs can be effectively detected.

3.4. Comparison with Other Analytical Methods. To characterize the extraction performance of the proposed method

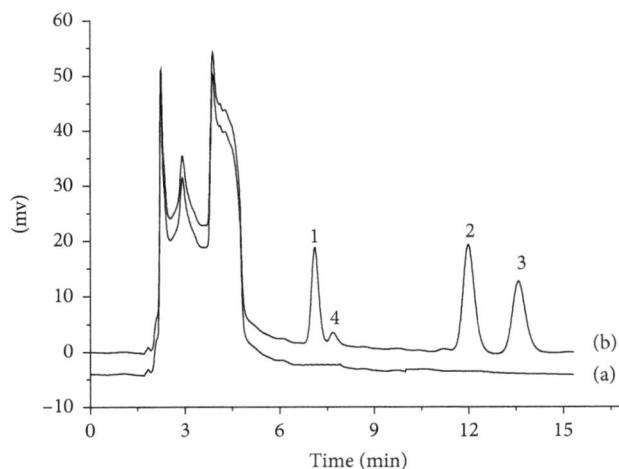

FIGURE 8: Wheat samples with 5 μg/g concentration of three OPPs were analyzed before (a) and after (b) [OMIM][Tf$_2$N]-DLLME. Peaks 1, 2, 3, and 4 represent fenitrothion, fenthion, phoxim, and impurity, respectively.

TABLE 3: Comparison of this method with previous works.

Method	IL type	IL volume	Process time	Linear range (μg/L) (μg/Kg)	LOD (μg/L) (μg/Kg)	RSD (%)	Matrix	Reference
SBME	[OMIM][PF6]	Not mentioned	>60 min	1–200	<0.026	<3	Water	[29]
LPME	[HMIM][PF6]	50 μL	>50 min	1–100	<0.29	<3	Water	[30]
DLLME	[ODMIM]/[HMIM] [Nf2T] mix IL	54 μL	~13 min	2.5–500	<0.69	<0.69	Fruit juices	[18]
DLLME	[OMIM][PF6]	50 μL	~6 min	2-100 μg/Kg	<0.73	<5.7	Apple and pear	[28]
DLLME	[OMIM][Tf2N]	22 μL	~13 min	0.01–100	0.1	<5.9	Wheat	This work

in this work, a comparison of the [OMIM][Tf$_2$N] IL-DLLME with other methods is summarized in Table 3. From it, we can see that the advantages of the present method can be described as follows: (i) the amount of the IL used is the least; (ii) the operation time is shorter than most methods; and (iii) the matrix is wheat, and the method shows good linear range and RSD. The results indicate that this method is simple, time-saving, and with satisfactory extraction effect for wheat samples.

4. Conclusions

This work firstly proposed a method that used Tf$_2$N anion-based ionic liquids (ILs) and dispersive liquid-liquid mircoextraction (DLLME) for extracting OPPs in wheat samples by ultrasonic-assisted and combined with HPLC analysis. By investigating the influencing factors in the extraction process, the optimum conditions were determined. [OMIM] [Tf$_2$N] ionic liquid exhibited best extraction performance for OPPs among three Tf$_2$N-based ILs in the extraction system due to its unique properties. Compared with other methods, the results of this method show that it is a time-saving, simple but efficient method with high sensitivity and reliability. Furthermore, it has good recoveries, wider LRs, and lower LODs, and RSDs indicate that the method is

satisfactory for OPPs extraction by using Tf$_2$N-based IL as an extractant in wheat samples. It is a promising method to be used for the trace determination of various organic compounds in complex matrices in the future.

Conflicts of Interest

The authors declare that they have no conflicts of interest.

Acknowledgments

This work was supported by the National Natural Science Foundation of China (Grant no. 71501149) and the Fundamental Research Funds for the Central Universities, China (Grant no. WUT: 185208007).

References

[1] U. Uygun, B. Senoz, and H. Koksel, "Dissipation of organophosphorus pesticides in wheat during pasta processing," *Food Chemistry*, vol. 109, no. 2, pp. 355–360, 2008.

[2] I. A. T. Khan, Riazuddin, Z. Parveen, and M. Ahmed, "Multiresidue determination of synthetic pyrethroids and organophosphorus pesticides in whole wheat flour using gas chromatography," *Bulletin of Environmental Contamination and Toxicology*, vol. 79, no. 4, pp. 454–458, 2007.

[3] S. M. Yousefi, F. Shemirani, and S. A. Ghorbanian, "Deep eutectic solvent magnetic bucky gels in developing dispersive solid phase extraction: application for ultra trace analysis of organochlorine pesticides by GC-micro ECD using a large-volume injection technique," *Talanta*, vol. 168, pp. 73–81, 2017.

[4] D. S. Chormey, C. Buyukpinar, F. Turak, O. T. Komesli, and S. Bakırdere, "Simultaneous determination of selected hormones, endocrine disruptor compounds, and pesticides in water medium at trace levels by GC-MS after dispersive liquid-liquid microextraction," *Environmental Monitoring and Assessment*, vol. 189, no. 6, p. 277, 2017.

[5] D. Harshit, K. Charmy, and P. Nrupesh, "Organophosphorus pesticides determination by novel HPLC and spectrophotometric method," *Food Chemistry*, vol. 230, pp. 448–453, 2017.

[6] I. Timofeeva, A. Shishov, D. Kanashina, D. Dzema, and A. Bulatov, "On-line in-syringe sugaring-out liquid-liquid extraction coupled with HPLC-MS/MS for the determination of pesticides in fruit and berry juices," *Talanta*, vol. 167, pp. 761–767, 2017.

[7] Riazuddin, M. F. Khan, S. Iqbal, and M. Abbas, "Determination of multi-residue insecticides of organochlorine, organophosphorus, and pyrethroids in wheat," *Bulletin of Environmental Contamination and Toxicology*, vol. 87, no. 3, pp. 303–306, 2011.

[8] M. A. Gonzalez-Curbelo, J. Hernandez-Borges, T. M. Borges-Miquel, and M. Á. Rodríguez-Delgado, "Determination of organophosphorus pesticides and metabolites in cereal-based baby foods and wheat flour by means of ultrasound-assisted extraction and hollow-fiber liquid-phase microextraction prior to gas chromatography with nitrogen phosphorus detection," *Journal of Chromatography A*, vol. 1313, pp. 166–174, 2013.

[9] S. H. W. P. Kevin and N. T. Norman, "Supercritical fluid extraction and quantitative determination of organophosphorus pesticide residues in wheat and maize using gas chromatography with flame photometric and mass spectrometric detection," *Journal of Chromatography A*, vol. 907, no. 1-2, pp. 247–255, 2001.

[10] M. Rezaee, Y. Assadi, M. R. M. Hosseinia, E. Aghaee, F. Ahmadi, and S. Berijani, "Determination of organic compounds in water using dispersive liquid-liquid microextraction," *Journal of Chromatography A*, vol. 1116, no. 1-2, pp. 1–9, 2006.

[11] R. Sousa, V. Homem, J. L. Moreira, L. M. Madeira, and A. Alves, "Optimisation and application of dispersive liquid–liquid microextraction for simultaneous determination of carbamates and organophosphorus pesticides in waters," *Analytical Methods*, vol. 5, no. 11, p. 2736, 2013.

[12] G. Cinelli, P. Avino, I. Notardonato, and M. V. Russo, "Ultrasound-vortex-assisted dispersive liquid–liquid microextraction coupled with gas chromatography with a nitrogen–phosphorus detector for simultaneous and rapid determination of organophosphorus pesticides and triazines in wine," *Analytical Methods*, vol. 6, no. 3, pp. 782–790, 2014.

[13] M. A. Farajzadeh, M. R. Afshar Mogaddam, S. Rezaee Aghdam, N. Nouri, and M. Bamorrowat, "Application of elevated temperature-dispersive liquid-liquid microextraction for determination of organophosphorus pesticides residues in aqueous samples followed by gas chromatography-flame ionization detection," *Food Chemistry*, vol. 212, pp. 198–204, 2016.

[14] S. Pandey, "Analytical applications of room-temperature ionic liquids: a review of recent efforts," *Analytica Chimica Acta*, vol. 556, no. 1, pp. 38–45, 2006.

[15] A. Marciniak, "Influence of cation and anion structure of the ionic liquid on extraction processes based on activity coefficients at infinite dilution. A review," *Fluid Phase Equilibria*, vol. 294, no. 1-2, pp. 213–233, 2010.

[16] B. Kudlak, K. Owczarek, and J. Namiesnik, "Selected issues related to the toxicity of ionic liquids and deep eutectic solvents–a review," *Environmental Science and Pollution Research*, vol. 22, no. 16, pp. 11975–11992, 2015.

[17] C. Zhang, C. Cagliero, S. A. Pierson, and J. L. Anderson, "Rapid and sensitive analysis of polychlorinated biphenyls and acrylamide in food samples using ionic liquid-based in situ dispersive liquid-liquid microextraction coupled to headspace gas chromatography," *Journal of Chromatography A*, vol. 1481, pp. 1–11, 2017.

[18] H. Zeng, X. Yang, M. Yang et al., "Ultrasound-assisted, hybrid ionic liquid, dispersive liquid-liquid microextraction for the determination of insecticides in fruit juices based on partition coefficients," *Journal of Separation Science*, vol. 40, no. 17, pp. 3513–3521, 2017.

[19] M. Tuzen, O. D. Uluozlu, D. Mendil et al., "A simple, rapid and green ultrasound assisted and ionic liquid dispersive microextraction procedure for the determination of tin in foods employing ETAAS," *Food Chemistry*, vol. 245, pp. 380–384, 2018.

[20] M. De Boeck, S. Missotten, W. Dehaen, J. Tytgat, and E. Cuypers, "Development and validation of a fast ionic liquid-based dispersive liquid-liquid microextraction procedure combined with LC-MS/MS analysis for the quantification of benzodiazepines and benzodiazepine-like hypnotics in whole blood," *Forensic Science International*, vol. 274, pp. 44–54, 2017.

[21] M. Faraji, M. Noorani, and B. N. Sahneh, "Quick, easy, cheap, effective, rugged, and safe method followed by ionic liquid-dispersive liquid-liquid microextraction for the determination of trace amount of bisphenol A in canned foods," *Food Analytical Methods*, vol. 10, no. 3, pp. 764–772, 2017.

[22] R. Q. Liu, Y. Liu, C. S. Cheng, and Y. Yang, "Magnetic solid-phase extraction and ionic liquid dispersive liquid-liquid microextraction coupled with high-performance liquid chromatography for the determination of hexachlorophene in cosmetics," *Chromatographia*, vol. 80, no. 5, pp. 783–791, 2017.

[23] Z. Xia, D. Li, Q. Li, Y. Zhang, and W. Kang, "Simultaneous determination of brazilin and protosappanin B in *Caesalpinia sappan* by ionic-liquid dispersive liquid-phase microextraction method combined with HPLC," *Chemistry Central Journal*, vol. 11, no. 1, p. 114, 2017.

[24] M. H. Kowsari and M. Fakhraee, "Influence of butyl side chain elimination, tail amine functional addition, and C2 methylation on the dynamics and transport properties of imidazolium-based [Tf$_2$N$^-$] ionic liquids from molecular dynamics simulations," *Journal of Chemical and Engineering Data*, vol. 60, no. 3, pp. 551–560, 2015.

[25] C. Deferm, J. Luyten, H. Oosterhof, J. Fransaer, and K. Binnemans, "Purification of crude in(OH)3 using the functionalized ionic liquid betainium bis(trifluoromethylsulfonyl) imide," *Green Chemistry*, vol. 20, no. 2, pp. 412–424, 2018.

[26] P. F. Requejo, N. Calvar, Á. Domínguez, and E. Gómez, "Application of the ionic liquid tributylmethylammonium bis (trifluoromethylsulfonyl)imide as solvent for the extraction of benzene from octane and decane at T = 298.15 k and atmospheric pressure," *Fluid Phase Equilibria*, vol. 417, pp. 137–143, 2016.

[27] A. Tagiuri, K. Z. Sumon, and A. Henni, "Solubility of carbon dioxide in three [Tf$_2$N] ionic liquids," *Fluid Phase Equilibria*, vol. 380, pp. 39–47, 2014.

[28] L. Zhang, F. Chen, S. Liu et al., "Ionic liquid-based vortex-assisted dispersive liquid-liquid microextraction of organophosphorus pesticides in apple and pear," *Journal of Separation Science*, vol. 35, no. 18, pp. 2514–2519, 2012.

[29] Y. Zhang, R. Wang, P. Su, and Y. Yang, "Ionic liquid-based solvent bar microextraction for determination of organophosphorus pesticides in water samples," *Analytical Methods*, vol. 5, no. 19, p. 5074, 2013.

[30] Q. Zhou, H. Bai, G. Xie, and J. Xiao, "Trace determination of organophosphorus pesticides in environmental samples by temperature-controlled ionic liquid dispersive liquid-phase microextraction," *Journal of Chromatography A*, vol. 1188, no. 2, pp. 148–153, 2008.

11

Method for Analyzing the Molecular and Carbon Isotope Composition of Volatile Hydrocarbons (C_1–C_9) in Natural Gas

Chunhui Cao ⓘ, Zhongping Li, Liwu Li, and Li Du

Key Laboratory of Petroleum Resources, Gansu Province/Key Laboratory of Petroleum Resources Research, Institute of Geology and Geophysics, Chinese Academy of Sciences, Lanzhou 730000, China

Correspondence should be addressed to Chunhui Cao; caochunhui@lzb.ac.cn

Academic Editor: Guido Crisponi

Solid-phase microextraction (SPME) coupled with gas chromatography-isotope ratio mass spectrometry (GC-IRMS) has already been applied to collect and identify volatile light hydrocarbons in oil and source rocks. However, this technology has not yet been used to analyze volatile light hydrocarbons in dry gas (natural gas with C_1/C_{2+} > 95%). In this study, we developed a method to measure the molecular and carbon isotope composition of natural gas using divinylbenzene/carboxen/polydimethylsiloxane (DVB/CAR/PDMS) fiber. This fiber proved to be suitable for extracting C_1–C_9 hydrocarbons from natural gas without inducing carbon isotopic fractionation. Notably, the extraction coefficents of the analytes were not the same but rather increased with the increasing carbon number of the hydrocarbons. Nevertheless, we successfully identified 24 hydrocarbons from the in-lab standard natural gas, while also obtaining the carbon isotope composition of C_1 to C_9 hydrocarbons with satisfying repeatability. The relative standard deviation (RSD) of the molecular composition data was in the range of 0.06–0.74%, with the RSDs of the carbon isotope composition data not exceeding 1‰. Finally, seven natural gas samples, collected from different sedimentary basins, were successfully analyzed and the stable carbon isotope compositions of C_1–C_9 hydrocarbons present in these were determined through this method. Overall, the new approach provides a simple but useful technique to obtain more geochemical information about the source and evolution of natural gas.

1. Introduction

Volatile light hydrocarbons (C_1–C_9) are important components of crude oil because they possess a large scope of geochemical information that is of great significance to oil and natural gas exploration. Furthermore, these components can be applied to classify source rocks and oil types [1, 2], identify source rock evolution [3–6], study oil-oil or oil-source correlations [1, 7–11], and estimate the thermal maturity of source rocks and crude oil [12–14]. However, restrained by analytical methods, previous studies focused on the measurement of light hydrocarbons only in source rocks and crude oil while not being able to acquire their carbon isotope compositions. Natural gas (particularly dry gas, e.g., shale gas), as opposed to petroleum and source rocks, mainly contains methane and ethane and has extremely low content of C_{4+} hydrocarbons. Nevertheless, despite the scarcity, volatile light hydrocarbons (C_{4+}) in

natural gas (derived from oil and/or source rocks) and their carbon isotope compositions carry abundant and significant geochemical information. Employing this information assists in the determination of the maturity of natural gas, recognition of gas accumulation suffering from washing or biodegradation, tracing the source of natural gas, and classification of the origin types of natural gas [14–19]. Because natural gas is generated from the oil and/or source rocks, its molecular and carbon isotope composition retains the information about the oil and source rocks [20, 21]. Considering that natural gas, oil, and source rocks contain light hydrocarbons (C_1–C_9), we expect that, instead of using an indirect deduction via the molecular and isotope fractionation theory [22–24], a direct study of these components would allow the detection of the correlation between gas, oil, and source rocks [15]. However, implementing this new method to measure the molecular and carbon isotope compositions of light hydrocarbons in natural gas is difficult, due to the limit of detection (LOD) values of analytical instruments.

Solid-phase microextraction (SPME) is an innovative, solvent-free sample preparation approach that is fast, economical, and versatile and requires only a small amount of sample. A general SPME device (Figure 1) resembles a modified syringe, consisting of an SPME holder and SPME head with a built-in fiber inside a needle [25]. The SPME holder, which consists of a plunger, stainless steel barrel, and adjustable depth gauge, is usually designed to be used with reusable and replaceable fiber assemblies [26]. The SPME head includes a spring, sealing septum, and piercing needle. The fiber inside the needle is coated with a special polymeric stationary phase, which could concentrate the organic analytes from the sample matrix [26]. Notably, this device has been widely applied in sample pretreatment technology [27], as well as to analyze the hydrogen isotope composition of light hydrocarbons in natural gas [28] and crude oil [29].

In this study, we successfully applied SPME technology to enriched light hydrocarbons (C_1–C_9) in natural gas. In addition, we combined SPME with a gas chromatography (GC) or gas chromatography-isotope ratio mass spectrometry (GC-IRMS) system in an effort to measure the molecular and carbon isotope compositions of a series of light hydrocarbons (C_1–C_9) in natural gas. Considering that thermogenic natural gas is derived from the cracking process of organic materials (e.g., source rocks and kerogen), the geochemical data of light hydrocarbons in natural gas could definitely provide clues on their source and evolution, which would, in turn, be very relevant to the research of oil/natural gas geochemical scientists.

2. Experimental

2.1. Materials. For this study, we selected a type of fiber coating with divinylbenzene/carboxen/polydimethylsiloxane (DVB/CAR/PDMS) to enrich the light hydrocarbons in natural gas. An in-lab standard natural gas was employed to test the characteristics of this fiber, and the enrichment conditions were optimized. The in-lab standard natural gas was collected from Ordos Basin, China, but it is not a standard for other laboratories. We analyzed its molecular composition and compared the result with that of three other laboratories to confirm that it mainly contains 24 types of hydrocarbons: methane, ethane, propane, isobutane, *n*-butane, neopentane, isopentane, *n*-pentane, methyl cyclopentane, cyclohexane, 2-methylpentane, 3-methylpentane, *n*-hexane, 2,2/3,3-dimethyl pentane, methyl cyclohexane, 2/3-methyl hexane, 2,3-dimethyl pentane, *n*-heptane, benzene, *n*-octane, methyl benzene, *n*-nonane, ethyl benzene, *p*-xylene, and *o*-xylene. Several separate parallel samples were prepared from the in-lab standard natural gas for the experimental condition optimization. Each sample was collected in a fixed-volume (600 mL) glass bottle that was subsequently sealed with a rubber stopper.

2.2. Conditions for the Analysis Instruments. The molecular composition of the gas samples was determined using a gas

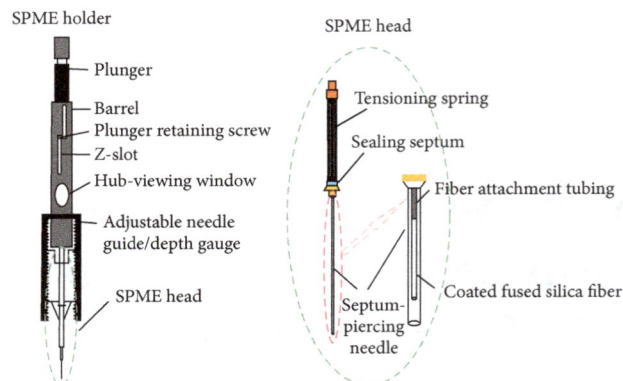

FIGURE 1: Schematic diagram of the SPME holder and fiber (Supelco Data Sheet No. T713019 A, 1998).

chromatograph (6890A, Agilent Technologies, USA) equipped with a flame ionization detector. The individual hydrocarbon gas components (C_1–C_9) were separated using an AT-Al_2O_3 capillary column (50 m × 0.53 mm × 20 μm, Agilent Technologies, USA). The GC oven temperature was adjusted according to the following procedure: 35°C for 3 min, increased to 100°C at a rate of 7°C/min, kept at that temperature for 5 min, 7°C/min ramped to 160°C, kept at that temperature for 10 min, increased to 200°C at a rate of 15°C/min, maintained at that temperature for 30 min, 25°C/min ramped to 220°C, maintained at that temperature for 85 min. The GC injection port temperature was set to 350°C, and the split ratio was 1:1. The carrier gas (He) was in a constant-flow mode, with a flow velocity of 2 mL/min.

The stable carbon isotope ratios were measured on a gas chromatography-isotope ratio mass spectrometry system (GC-IRMS, isotope ratio mass spectrometer interfaced with an Agilent 6890A gas chromatograph). The individual hydrocarbon gas components (C_1–C_9) were separated on a 6890A gas chromatograph using the instrumental conditions mentioned above. Then, the separated compounds were injected into a combustion furnace for oxidizing at 950°C. The produced H_2O was removed using a water trap, and CO_2 was injected into the Delta plus XP mass spectrometer (Thermo-Fisher, Bremen, Germany) for isotopic analysis. An electron impact (EI) ion source was used for the mass spectrometer, with a filament emission current of 1.3 mA and electron energy of 100 eV. CO_2 (purity ≥99.99%) with a carbon isotopic value of $\delta^{13}C_{CO2} = -20.9‰$ (±0.5‰) was used as a reference gas. The analytical error in the $\delta^{13}C$ values was <0.4‰ ($n = 6$) for standard natural gas. The stable carbon isotopic values were reported as δ-notation in per mil (‰) relative to the Vienna Pee Dee Belemnite (VPDB) standard with a measurement precision for $\delta^{13}C$ of ±0.5‰.

2.3. SPME Fiber Selection. The SPME fiber is coated with relatively thin films of several polymeric stationary phases, which are conventionally used as coating materials in chromatography. This film acts like a sponge, concentrating the organic analytes from the sample matrix [26]. The coating on the fiber can consist of a variety of materials, including carbowax template resin, polydimethylsiloxane, polydimethylsiloxane divinylbenzene, polyacrylate, carboxen polydimethylsiloxane, and carbowax divinylbenzene [30, 31]. The carboxen/polydimethylsiloxane (CAR/PDMS) fiber was first used in 2014 to measure the carbon isotope composition of volatile light hydrocarbons in natural gas by combining SPME and GC-IRMS [32]. This research work opened the way to a new application of SPME technology into the natural gas study area. Recently, a type of SPME fiber coating, i.e., divinylbenzene/carboxen/polydimethylsiloxane (DVB/CAR/PDMS), which can be used to extract C_3–C_{20} volatiles, was found suitable for enriching trace light hydrocarbons in natural gas samples [28, 29]. It has already been applied to the analysis of the hydrogen isotopic composition ($\delta^{13}D$) of volatile light hydrocarbons in natural gas [28] and crude oil [29]. In this study, DVB/CAR/PDMS fiber was employed to extract trace light hydrocarbon compounds in natural gas samples and analyze their molecular and carbon isotope compositions ($\delta^{13}C$).

2.4. Procedures for the Extraction of Trace Hydrocarbons.

When extracting light hydrocarbons, the septum-piercing needle of the SPME device (Figures 1 and 2) was directly inserted into the glass bottle (600 ml) containing the sample through the rubber stopper. Then, the plunger was pushed to make the coated fiber stretch out of the needle, and the fiber was immersed directly into the natural gas sample to expose the coating to the hydrocarbons to be sampled (Figure 2). This would initiate the absorption of the analyte molecules onto the coating. The transport of analytes from the matrix into the coating begins as soon as the coated fiber has been placed in contact with the sample. After trace hydrocarbons are trapped onto the coating by an equilibrium mechanism, the fiber was retracted into the needle and taken out of the bottle [33]. Then, the fiber can be inserted into the injection port of a GC (or GC-IRMS) (Figure 2) and quickly desorbed by the heat of the port, resulting in a rapid transfer of all absorbed components into GC (or GC-IRMS) for molecular composition (or isotope composition) analysis.

The extraction time was found to be a critical parameter in the SPME sampling process. The parallel samples of the in-lab standard natural gas were extracted the light hydrocarbons by SPME while applying different extraction times (2, 5, 10, 30, 60, and 120 min). First, DVB/CAR/PDMS fiber was applicated to extract trace light hydrocarbon compounds in one of the in-lab standard natural gas samples at room temperature for 2 min. After finishing trace hydrocarbons extraction by using the SPME device, we have acquired the C_1-C_9 hydrocarbons on the DVB/CAR/PDMS fiber. In order to desorb the analytes that were extracted on the fiber and analyze their molecular composition with GC, we should take the SPME device out of the sample bottle and introduce it into the GC injector port where the adsorbed analytes are thermally desorbed at 280°C (most commonly used temperature in GC analysis [34]) and consequently be routed into the GC column for composition analysis [33]. And then, the other five in-lab standard natural gas samples were analyzed using the same procedures, but the extraction

time was selected as 5, 10, 30, 60, and 120 min, respectively. As a result, we found that the DVB/CAR/PDMS fiber had an unstable adsorption of hydrocarbons in the first 10 min (Figure 3). The extracted amount of trace hydrocarbons onto the fiber reached a peak value at 2 min, after which it decreased. Between 10 and 30 min, the extracted amount rapidly increased from 144.21 to 181.39 × 10^5 mV, respectively. After 30 min, the content of extracted analytes did not change much (Figure 3). Therefore, we select to extract the light hydrocarbons at room temperature for 30 min.

2.5. Desorption Temperature.

The injection port temperature of GC was found to be a key parameter as it affected the volume of each analyte entering the instruments used for molecular or carbon isotope composition analyses. Different temperatures (150, 200, 250, 300, 350, 400, 450, and 500°C) were selected for the injection port of the GC in order to find the optimal desorption conditions. Eight in-lab standard natural gas samples contained in glass bottles were prepared for the optimal temperature condition test. One of the eight in-lab standard natural gas samples was adsorbed by DVB/CAR/PDMS fiber, thermally desorbed in the injector port of GC at 150°C, and then introduced into the GC column by the carrier gas for molecular composition analysis, respectively. And then, the other seven in-lab standard natural gas samples were analyzed using the same procedures, but the injection port temperature was selected 200, 250, 300, 350, 400, 450, and 500°C, respectively. Notably, most of the compounds displayed a degassing peak at 300 and 350°C (Figure 4). The desorption ratio at 350°C (defined as $\sum_{\text{desorption amount before 400°C}} / \sum_{\text{desorption amount at each temperature}}$) of all hydrocarbons exceeded 60%, except for n-hexane, which exhibited a ratio of 42.47%. In addition, the amount of each analyte desorbed at 350°C was enough for molecular and carbon isotope composition analyses to be performed. Thus, in order to desorb as much amount of each analyte as possible and ensure that the DVB/CAR/PDMS fiber does not age quickly, 350°C was chosen as the most suitable desorption temperature.

2.6. Extraction Performance.

By combining SPME and GC, we analyzed the molecular composition of the in-lab standard natural gas using the experimental conditions and operation procedures discussed above. The obtained results were then compared with the ones from the sample without extraction by SPME. Figure 5(a) shows the chromatogram of the origin in-lab standard natural gas analyzed by GC, while Figure 5(b) displays the chromatogram after the sample was extracted by SPME with the DVB/CAR/PDMS fiber. As can be seen from Figure 5, only C_1–C_5 compounds were detected in the natural gas sample without the SPME extraction, whereas, after the extraction by SPME, 24 types of hydrocarbon compounds (C_1–C_9) were enriched and measured. The standard deviation (SD) of all compounds, except methylbenzene (0.66%), did not exceed 0.5%. According to the standard spectrum diagram (at the same experimental conditions) obtained with the AT-Al_2O_3 capillary column, 24 types of hydrocarbon compounds were identified

Headspace microextraction   Sample desorption

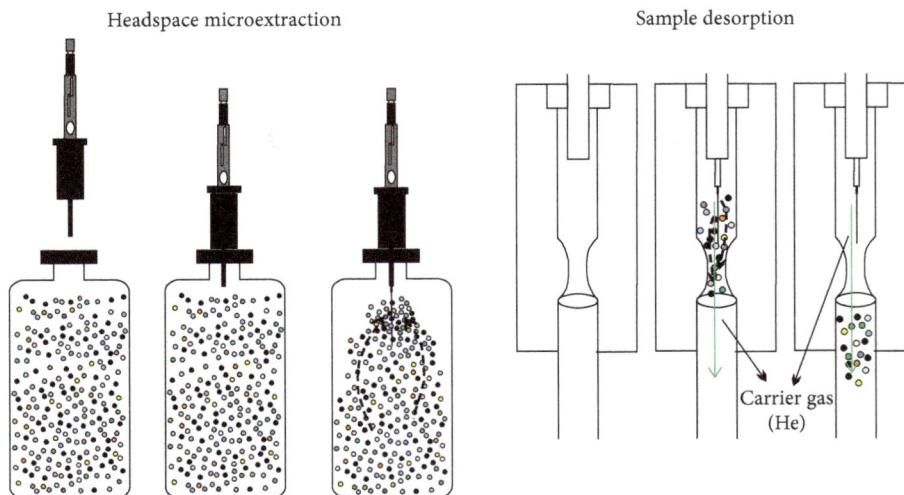

FIGURE 2: Extraction and desorption processes using the SPME fiber. The dashed red arrows represent the movement of the hydrocarbons in sample, while solid green arrows represent the direction of the carrier gas.

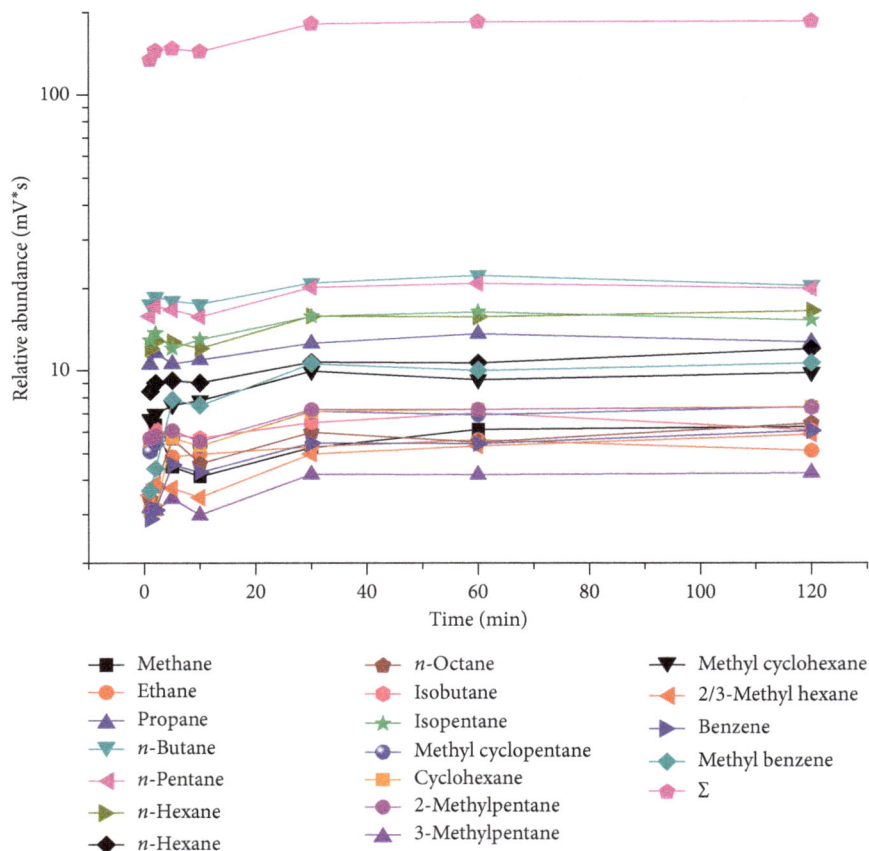

FIGURE 3: Relationship between the adsorption capacity and adsorption time (desorption time: 2 min).

(Figure 5(b)). After comparing Figures 5(a) and 5(b), we found that the concentrations of methane, ethane, and propane markedly decreased after the extraction, while those of n-butane, n-pentane, and C_{6+} compounds increased dramatically. These results imply that the DVB/CAR/PDMS fiber exhibited various adsorption abilities on the different hydrocarbon compounds, thus being suitable for extracting

C_{4+} hydrocarbons, which are trace compounds in natural gas [15].

Although the DVB/CAR/PDMS fiber has a diverse adsorption ability for different compounds, it has a particularly strong extraction ability for C_5–C_9 compounds, especially on n-alkanes (Figures 5 and 6). As can be seen from the analytical results, the concentration of n-hexane, n-heptane,

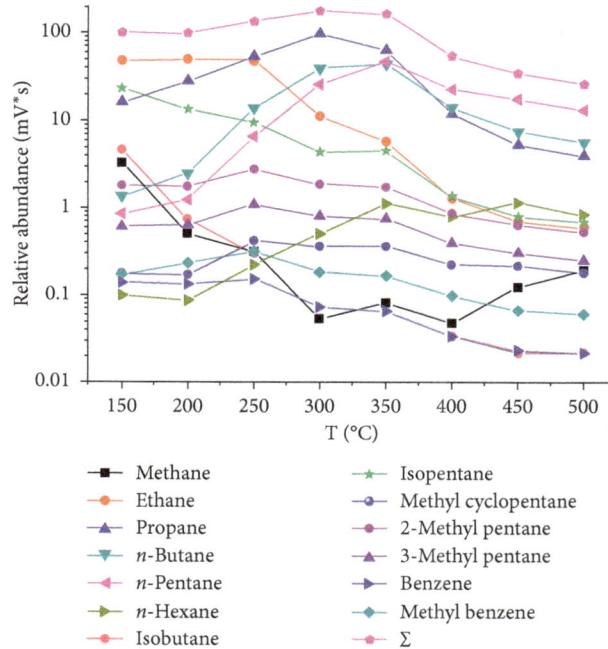

FIGURE 4: Relationship between the desorption amount and desorption temperature (desorption time: 2 min).

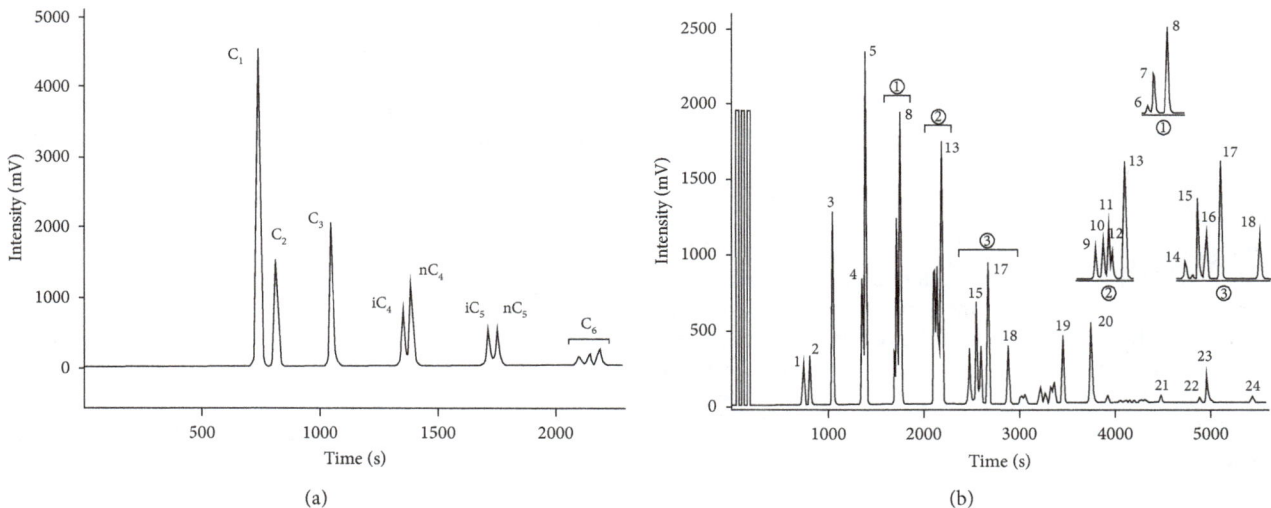

(a)

(b)

FIGURE 5: Chromatograms of the light hydrocarbons present in the (a) in-lab standard natural gas and (b) analytes extracted using SPME. 1, Methane; 2, ethane; 3, propane; 4, isobutane; 5, n-butane; 6, neopentane; 7, isopentane; 8, n-pentane; 9, methyl cyclopentane; 10, cyclohexane; 11, 2-methylpentane; 12, 3-methylpentane; 13, n-hexane; 14, 2,2/3,3-dimethyl pentane; 15, methyl cyclohexane; 16, 2/3-methylhexane +2,3-dimethyl pentane; 17, n-heptane; 18, benzene; 19, n-octane; 20, methylbenzene; 21, n-nonane; 22, ethylbenzene; 23, p-xylene; 24, o-xylene.

n-octane, and n-nonane increased dramatically. The concentrations of CH_4, C_2H_6, C_3H_8, and n-C_4H_{10} in the extracted analytes were lower than those in the original natural gas sample, while the concentration of C_{5+} in natural gas was enriched after the extraction with DVB/CAR/PDMS fiber, to reach the LOD of analytical instruments. Reportedly [15], the concentration of C_1-C_3 is higher in natural gas, while that of C_{5+} is usually quite low. Considering that the concentration of C_1-C_3 in the extracted analytes was much lower than that of the natural gas sample and the

concentration of C_{5+} is raised apparently, we concluded that the fiber had a balancing effect on the hydrocarbons in natural gas. Therefore, the fiber coated with DVB/CAR/PDMS could efficiently overcome the challenge of measuring trace C_{5+} compounds in natural gas.

2.7. Extraction Coefficient (f). The extraction coefficient, which is defined as the ratio of each component's relative concentration (vol/%) before and after extraction,

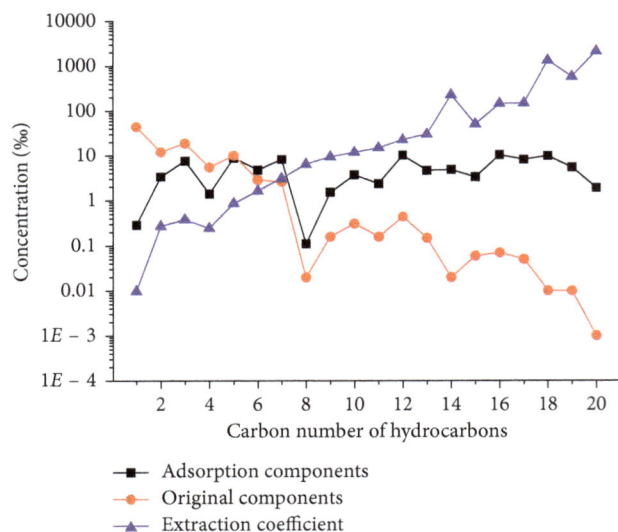

FIGURE 6: Comparison of the content of each analyte before and after the adsorption by SPME. 1, Methane; 2, ethane; 3, propane; 4, isobutane; 5, *n*-butane; 6, isopentane; 7, *n*-pentane; 8, methyl cyclopentane; 9, cyclohexane; 10, 2-methylpentane; 11, 3-methylpentane; 12, *n*-hexane; 13, methyl cyclohexane; 14, 2/3-methylhexane; 15, *n*-heptane; 16, benzene; 17, *n*-octane; 18, methylbenzene; 19, *n*-nonane; 20, *p*-xylene. Adsorption components: the relative concentration of extracted analytes. Original components: the relative concentration of original analytes. Extraction coefficient: the ratio of each component's relative concentration (vol/%) before and after the extraction.

respectively, can be used to describe the fiber's enrichment ability for each hydrocarbon. We used the in-lab standard natural gas as a sample to test the enrichment ability of the DVB/CAR/PDMS fiber. The relative concentration of extracted analytes ("Adsorption components" in Figure 6) was measured by SPME+GC, and the relative concentration of original analytes ("Original components" in Figure 6) was analyzed by standard GC techniques. The extraction coefficient (*f*) of each analyte can be described using the following equation:

$$f_x = \frac{C_{x(\text{after})}}{C_{x(\text{before})}}, \tag{1}$$

where f_x is the extraction coefficient of component x, and $C_{x(\text{before})}$ and $C_{x(\text{after})}$ are the relative concentrations of component x before and after, respectively, extraction by DVB/CAR/PDMS fiber. As can be seen from Figure 6, the extraction coefficient increased significantly as the carbon number of the analytes became larger. The extraction coefficients of C_1–C_4 hydrocarbons were less than 1, which implies that the fiber (coating with DVB/CAR/PDMS) could not enrich these hydrocarbons. In particular, the original concentration of CH_4 in the in-lab standard natural gas was about 50%. After the extraction, its concentration in the analytes absorbed on the fiber dropped to around 0.3%, thereby affording a CH_4 extraction coefficient of about 0.006. Nevertheless, the extracted contents of C_1–C_4 hydrocarbons by the fiber were enough for these compounds to be detected

and for the carbon isotope composition analysis to be successful. Furthermore, after the extraction, the relative concentration (%) of C_{5+} hydrocarbons (trace amount components in the in-lab standard natural gas sample) increased significantly, with their extraction coefficients ranging from 10 to 2000 (Figure 6). Thus, after the extraction, almost every hydrocarbon reached a concentration between 1% and 10%, which indicated that the fiber (coated with DVB/CAR/PDMS) can balance the content of C_1–C_9 hydrocarbons in natural gas. Although this characteristic of the DVB/CAR/PDMS fiber allowed the detection of trace components in natural gas, it also caused difficulties in the calculation of the real content of each analyte. In this respect, more tests are needed to develop a mathematical model for the calculation of the real content of trace hydrocarbons in natural gas, especially C_6–C_8 components, which are useful for gas source studies. Nevertheless, the content balance effect makes it easier to simultaneously measure the carbon isotope compositions of C_1–C_9 hydrocarbons in natural gas.

3. Application of Carbon Isotopic Analysis

3.1. Stability of Carbon Isotopic Analysis. GC-IRMS paired with SPME was used to analyze the carbon isotope composition of light hydrocarbons in natural gas. Since the analytes underwent physical adsorption and desorption processes before the carbon isotope analysis, several experiments had to be conducted to detect whether carbon isotopic fractionation has occurred during the SPME extraction processes. In this respect, Li et al. [29] have found that the extraction time, extraction temperature, and desorption temperature have no significant effect on $\delta^{13}D$ when using the DVB/CAR/PDMS fiber to analyze hydrocarbons in oil. Therefore, in this work, we selected the following extraction conditions: room temperature for 30 min and desorption at 350°C. In contrast, different desorption durations were selected (1, 2, 5, 10, 30, 60, and 120 min) to detect the effect of the desorption time on $\delta^{13}C$ of each analyte in the in-lab standard natural gas sample. First, one of the in-lab standard natural gas samples (collected in the glass bottle, Section 2.1) was extracted by SPME device using the procedures illustrated in Section 2.4. Then, the SPME needle was introduced into the GC-IRMS injector port where the adsorbed analytes are thermally desorbed for 1 min and routed into the GC column for composition separating. Eventually, individual hydrocarbon gas components (C_1–C_9) were oxidized into CO_2 and H_2O, and the CO_2 was let into IRMS for carbon isotope composition analysis. Then, the other six in-lab standard natural gas samples were analyzed using the same procedures, but the duration of desorption was selected 2, 5, 10, 30, 60, and 120 min, respectively. The variation range of the obtained carbon isotopic values of the analytes in the in-lab standard natural gas samples did not exceed 1‰, with a standard deviation not exceeding 0.7‰. As can be seen from Figure 7, the carbon isotope data obtained for each analyte desorbed in different durations overlapped almost perfectly, thus indicating that the desorption time would not cause an apparent isotope fractionation.

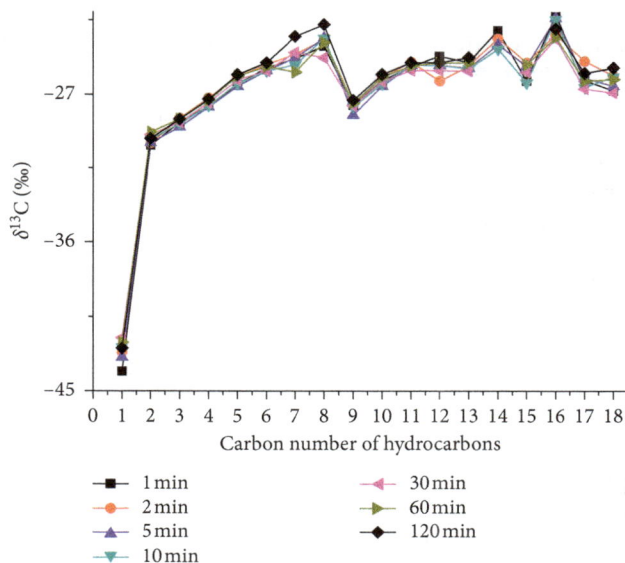

FIGURE 7: Relationship between the carbon isotope ratios and desorption times of each hydrocarbon. 1, Methane; 2, ethane; 3, propane; 4, isobutane; 5, n-butane; 6, isopentane; 7, n-pentane; 8, methyl cyclopentane; 9, 2-methylpentane; 10, n-hexane; 11, methyl cyclohexane; 12, 2/3-methylhexane; 13, n-heptane; 14, benzene; 15, n-octane; 16, methylbenzene; 17, n-nonane; 18, p-xylene.

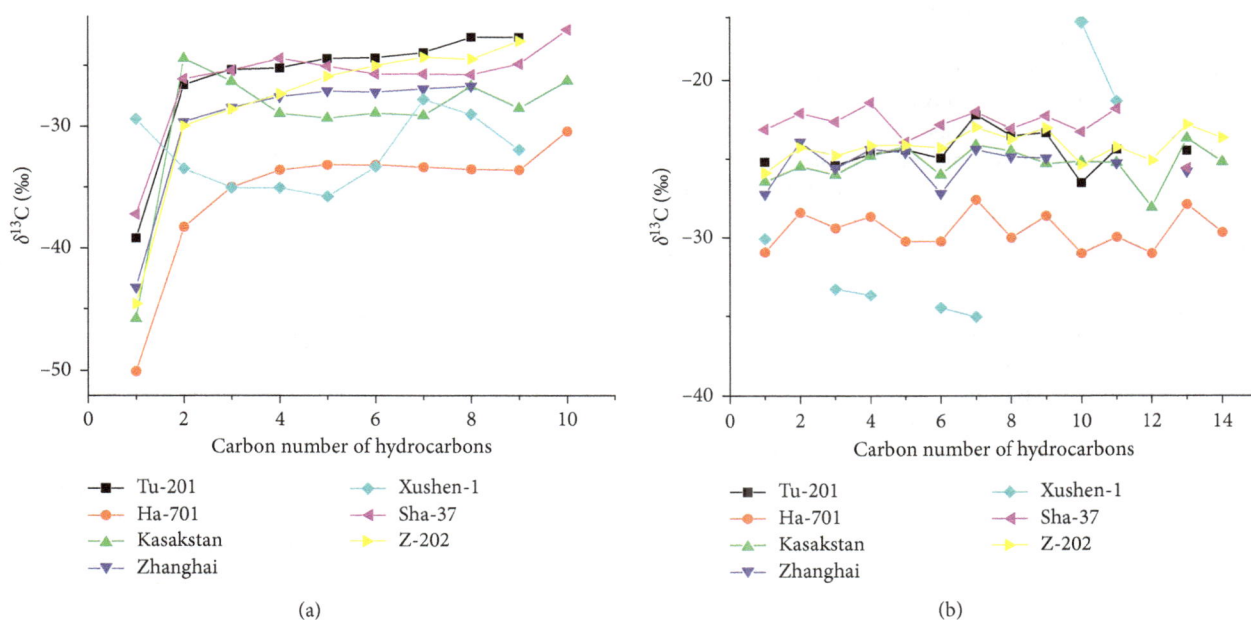

(a)

(b)

FIGURE 8: Carbon isotope composition of hydrocarbons in natural gas. (a) 1, Methane; 2, ethane; 3, propane; 4, n-butane; 5, n-pentane; 6, n-hexane; 7, n-heptane; 8, n-octane; 9, n-nonane; 10, benzene. (b) 1, Isobutane; 2, neopentane; 3, isopentane; 4, methyl cyclopentane; 5, cyclohexane; 6, 2-methylpentane; 7, 3-methylpentane; 8, 2,2/3,3-dimethyl pentane; 9, methyl cyclohexane; 10, 2/3-methylhexane +2,3-dimethyl pentane; 11, methylbenzene; 12, ethylbenzene; 13, p-xylene; 14, o-xylene.

3.2. Geological Sample Analysis. Seven natural gas samples collected from different sedimentary basins (Tarim Basin, Qaidam Basin, Songliao Basin, Huanghua Depression, Tuha Basin, and Bohai Bay Basin) were analyzed using the abovementioned method in an effort to detect the stable carbon isotope compositions of hydrocarbons (C_1–C_9) contained therein. The carbon isotope distribution pattern of n-alkanes, branched alkanes, and cycloalkanes is shown in Figure 8. More specifically, Figure 8(a) shows the carbon isotope distribution pattern of n-alkanes and benzene, while Figure 8(b) displays the patterns for branched alkanes and cycloalkanes. The natural gas form Xushen-1 in Songliao Basin exhibited a heavier $\delta^{13}C_1$ value (−29.40‰) than other natural gases and was characterized by a reversed carbon isotopic distribution pattern ($\delta^{13}C_1 > \delta^{13}C_2 > \delta^{13}C_3 > \delta^{13}nC_4 > \delta^{13}nC_5$), which suggested that it is an abiogenic gas [35].

The other natural gases ($\delta^{13}C_1$ ranging from $-50‰$ to $-37‰$) were recognized as thermal genetic gases or mixtures of thermal genetic gases and abiogenic gas and/or secondary cracking gas.

The thermal genetic methane ($\delta^{13}C_1 = -45.78‰$) in the natural gas from Kazakhstan and the partially reversal carbon isotopic distribution pattern ($\delta^{13}C_1 < \delta^{13}C_2 > \delta^{13}C_3 > \delta^{13}nC_4 > \delta^{13}nC_5$), which was due to secondary alterations under high temperature such as oil, gas cracking, and diffusion, indicate the complicate organic thermal evolution process.

4. Conclusions

The SPME technique has brilliant application prospects as its specific extraction characteristics provide a simple and useful method to analyze the carbon isotope compositions of trace hydrocarbons (C_1–C_9) in natural gas. In this respect, an SPME fiber coating with DVB/CAR/PDMS is suitable for the extraction of C_1–C_9 hydrocarbons from natural gas, thereby allowing the extraction of 24 types of hydrocarbons, with an extraction coefficient increasing significantly upon increasing the carbon number. However, the various enrichment abilities of each hydrocarbon make it difficult to calculate the real content of hydrocarbons, while the content balance effect makes it easier to simultaneously measure the carbon isotope composition of C_1–C_9 hydrocarbons in natural gas.

The herein presented analysis method has an excellent repeatability and high precision. Carbon isotope stability experiments of trace hydrocarbons proved that the desorption time did not induce any isotopic fractionation. Furthermore, this method was applied to seven natural gas samples in an effort to determine the carbon isotope compositions of C_1–C_9 hydrocarbons, which could help to further understand the origins and evolution of natural gas. Therefore, this experimental method can be successfully applied to analyze geological samples, thereby providing important information on the geochemical characteristics of natural gas, which would be very useful for geochemists.

Conflicts of Interest

The authors declare that there are no conflicts of interest regarding the publication of this paper.

Acknowledgments

The research was financially supported by the National Science Foundation of China (Grant No. 41502143) and the Project of Gansu Key Lab (1309RTSA041).

References

[1] W. Odden, R. L. Patience, and G. W. Van Graas, "Application of light hydrocarbons (C_4-C_{13}) to oil/source rock correlations:," *Organic Geochemistry*, vol. 28, pp. 823–847, 1998.

[2] G. Hu, X. Luo, Z. Li et al., "Geochemical characteristics and origin of light hydrocarbons in biogenic gas," *Science China Earth Sciences*, vol. 53, no. 6, pp. 832–843, 2010.

[3] D. H. Welte, H. Kratochvil, J. Rullkötter, H. Ladwein, and R. G. Schaefer, "Organic geochemistry of crude oils from the Vienna Basin and an assessment of their origin," *Chemical Geology*, vol. 35, no. 1-2, pp. 33–68, 1982.

[4] B. Horsfield, H. J. Schenk, N. Mills, and D. H. Welte, "An investigation of the in-reservoir conversion of oil to gas: compositional and kinetic findings from closed-system programmed-temperature pyrolysis," *Organic Geochemistry*, vol. 19, no. 1-3, pp. 191–204, 1992.

[5] A. Akinlua, T. R. Ajayi, and B. B. Adeleke, "Niger Delta oil geochemistry: insight from light hydrocarbons," *Journal of Petroleum Science and Engineering*, vol. 50, no. 3-4, pp. 308–314, 2006.

[6] D. M. Jones, I. M. Head, N. D. Gray et al., "Crude-oil biodegradation via methanogenesis in subsurface petroleum reservoirs," *Nature*, vol. 451, no. 7175, pp. 176–180, 2007.

[7] F. D. Mango, "An invariance in the isoheptanes of petroleum," *Science*, vol. 237, no. 4814, pp. 514–517, 1987.

[8] F. D. Mango, "The origin of light hydrocarbons in petroleum: a kinetic test of the steady-state catalytic hypothesis," *Geochimica et Cosmochimica Acta*, vol. 54, no. 5, pp. 1315–1323, 1990.

[9] F. D. Mango, "The origin of light hydrocarbons in petroleum: ring preference in the closure of carbocyclic rings," *Geochimica et Cosmochimica Acta*, vol. 58, no. 2, pp. 895–901, 1994.

[10] F. D. Mango, "The origin of light hydrocarbons," *Geochimica et Cosmochimica Acta*, vol. 64, no. 7, pp. 1265–1277, 2000.

[11] M. J. Whiticar and L. R. Snowdon, "Geochemical characterization of selected Western Canada oils by C_5-C_8 compound specific isotope correlation (CSIC)," *Organic Geochemistry*, vol. 30, no. 9, pp. 1127–1161, 1999.

[12] R. G. Schaefer and J. Höltkemeier, "Direkte analyse von Dimethylnaphthalinen in erdölen mittels zweidimensionaler Kapillar-gas-chromatographie," *Analytica Chimica Acta*, vol. 260, no. 1, pp. 107–112, 1992.

[13] R. G. Schaefer and R. Littke, "Maturity-related compositional changes in the low-molecular-weight hydrocarbon fraction of Toarcian shales," *Organic Geochemistry*, vol. 13, no. 4–6, pp. 887–892, 1988.

[14] Q. Meng, X. Wang, X. Wang et al., "Biodegradation of light hydrocarbon (C_5-C_8) in shale gases from the Triassic Yanchang Formation, Ordos basin, China," *Journal of Natural Gas Science and Engineering*, vol. 51, pp. 183–194, 2018.

[15] J. X. Dai and P. J. Li, "Carbon isotope study of light hydrocarbon (C_5-C_8) in natural gas-bearing basins in China," *Chinese Science Bulletin*, vol. 39, no. 22, p. 2071, 1994, in Chinese.

[16] D. Gong, R. Ma, G. Chen et al., "Geochemical characteristics of biodegraded natural gas and its associated low molecular weight hydrocarbons," *Journal of Natural Gas Science and Engineering*, vol. 46, pp. 338–349, 2017.

[17] G. Hu, W. Peng, and C. Yu, "Insight into the C_8 light hydrocarbon compositional differences between coal-derived and oil-associated gases," *Journal of Natural Gas Geoscience*, vol. 2, no. 3, pp. 157–163, 2017.

[18] G. Hu, J. Li, J. Li et al., "Preliminary study on the origin identification of natural gas by the parameters of light hydrocarbon," *Science in China Series D: Earth Sciences*, vol. 51, no. S1, pp. 131–139, 2008.

[19] S. Huang, X. Fang, D. Liu, C. Fang, and T. Huang, "Natural gas genesis and sources in the Zizhou gas field, Ordos Basin, China," *International Journal of Coal Geology*, vol. 152, pp. 132–143, 2015.

[20] Z. W. Zhi, W. Z. Yun, Z. S. Chang et al., "Successive generation of natural gas from organic materials and its significance in future exploration," *Petroleum Exploration and Development*, vol. 32, no. 2, pp. 1–7, 2005.

[21] W. Zhao, Z. Wang, H. Wang, Y. Li, G. Hu, and C. Zhao, "Further discussion on the connotation and significance of the natural gas relaying generation model from organic matter," *Petroleum Exploration and Development*, vol. 38, no. 2, pp. 129–135, 2011.

[22] C. Clayton, "Carbon isotope fractionation during natural gas generation from kerogen," *Marine and Petroleum Geology*, vol. 8, no. 2, pp. 232–240, 1991.

[23] Y. Ni, F. Liao, J. Dai et al., "Studies on gas origin and gas source correlation using stable carbon isotopes—a case study of the giant gas fields in the Sichuan Basin, China," *Energy Exploration and Exploitation*, vol. 32, no. 1, pp. 41–74, 2014.

[24] Y. Tang, J. K. Perry, P. D. Jenden, and M. Schoell, "Mathematical modeling of stable carbon isotope ratios in natural gases," *Geochimica et Cosmochimica Acta*, vol. 64, no. 15, pp. 2673–2687, 2000.

[25] O. Holm and W. Rotard, "Effect of radial directional dependences and rainwater influence on CVOC concentrations in tree core and Birch Sap samples taken for phytoscreening using HS-SPME-GC/MS," *Environmental Science and Technology*, vol. 45, no. 22, pp. 9604–9610, 2011.

[26] G. F. Ouyang, *SPME Devices Integrating Sampling with Sample Preparation for On-Site Analysis, Handbook of Sample Preparation*, John Wiley & Sons, Inc, Hoboken, NJ, USA, 2011.

[27] J. Pawliszyn, "Solid-Phase Microextraction in Perspective," in *Handbook of Solid Phase Microextraction*, pp. 1–12, Elsevier, Atlanta, GA, USA, 2012.

[28] X. Wang, Z. Li, L. Xing et al., "Development of a new method for hydrogen isotope analysis of trace hydrocarbons in natural gas samples," *Journal of Natural Gas Geoscience*, vol. 1, no. 6, pp. 481–487, 2016.

[29] Z. Li, L. Li, L. Xing et al., "Development of new method for D/H ratio measurements for volatile hydrocarbons of crude oils using solid phase micro-extraction (SPME) coupled to gas chromatography isotope ratio mass spectrometry (GC-IRMS)," *Marine and Petroleum Geology*, vol. 89, pp. 232–241, 2018.

[30] Z. Zhang, M. J. Yang, and J. Pawliszyn, "Solid-phase microextraction. a solvent-free alternative for sample preparation," *Analytical Chemistry*, vol. 66, no. 17, pp. 844A–853A, 2008.

[31] J. Pawliszyn, Z. Y. Zhang, and T. Gorecki, *Theory and Practice of Solid Phase Microextraction, United States*, 1997.

[32] Z. Li, X. Wang, L. Li et al., "Development of new method of $\delta^{13}C$ measurement for trace hydrocarbons in natural gas using solid phase micro-extraction coupled to gas chromatography isotope ratio mass spectrometry," *Journal of Chromatography A*, vol. 1372, pp. 228–235, 2014.

[33] M. Ábalos, J. M. Bayona, and F. Ventura, "Development of a solid-phase microextraction GC-NPD procedure for the determination of free volatile amines in wastewater and sewage-polluted waters," *Analytical Chemistry*, vol. 71, no. 16, pp. 3531–3537, 1999.

[34] R. Kubinec, V. G. Berezkin, R. Górová, G. Addová, H. Mračnová, and L. Soják, "Needle concentrator for gas chromatographic determination of BTEX in aqueous samples," *Journal of Chromatography B*, vol. 800, no. 1-2, pp. 295–301, 2004.

[35] J. X. Dai, C. N. Zou, S. C. Zhang et al., "Identification of inorganic genesis and organic origin alkane gas," *Science in China, Series D: Earth Sciences*, vol. 38, no. 11, pp. 1329–1341, 2008, in Chinese.

12

HPLC Method for Quantification of Caffeine and its Three Major Metabolites in Human Plasma using Fetal Bovine Serum Matrix to Evaluate Prenatal Drug Exposure

Rosa del Carmen Lopez-Sanchez, Victor Javier Lara-Diaz, Alejandro Aranda-Gutierrez ⓘ, Jorge A. Martinez-Cardona ⓘ, and Jose A. Hernandez ⓘ

Tecnologico de Monterrey, Escuela de Medicina y Ciencias de la Salud, Ave. Morones Prieto 3000, 64710 Monterrey, NL, Mexico

Correspondence should be addressed to Jose A. Hernandez; j.a.hernandez@itesm.mx

Academic Editor: Valdemar Esteves

Caffeine is recognized as the first-line therapeutic agent for apnea of prematurity. The dosage regimen is 10 mg/kg loading dose and 2.5 mg/kg maintenance dose. However, the plasma concentration achieved, not always, is therapeutically useful. It makes necessary to increase the doses to reach plasma concentration up to 30 or 35 µg/mL or even higher to attain therapeutic effect. To study why neonates have these differences, and whether these effects are linked to prenatal caffeine exposure, we had to develop an analytical method for an accurate measurement of caffeine and metabolites concentration. The analysis was carried out using fetal bovine serum (FBS) as biological matrix in a high-performance liquid chromatography with an ultraviolet detector method. This method allows acceptable chromatographic resolution between analytes in 15 minutes. It was validated and proved to be linear in the 0.1–40 µg/mL range for caffeine, paraxanthine, theobromine, and theophylline in the same chromatographic analysis. Accuracy for quality control samples for intra- and interday assays was ranged from 96.5 to 105.2% and 97.1 to 106.2%. Precision had CV no more than 10% in all concentration levels for all analytes. No differences were observed between quantification in human and FBS. This method was applied to quantify plasma drug concentration in mothers and their newborns in a Mexican northeast population. In our study, we confirmed self-reported caffeine maternal intake in 85.2% ($n = 23$); meanwhile, in their newborn's plasma, it was detected only in 78% ($n = 21$). Caffeine plasma concentrations in mother and newborn had a linear relationship, and no differences were observed between groups (mothers versus children). These results suggest that our analytical method and substitution of biological matrix was linear, precise, and accurate for caffeine quantification and could be used for measuring prenatal exposure and let us to study, in the future, concentration differences observed during apnea clinical treatment.

1. Introduction

Caffeine (1,3,7-trimethylpurine-2,6-dione) (Figure 1) is among the most consumed legal psychostimulants nowadays and is present in many and diverse kinds of foods and beverages [1, 2].

This drug has psychoactive properties. It is a slightly dissociative and stimulant drug because of its nonselective antagonist action against adenosine receptors [3, 4].

Clinically, caffeine is recognized as the first-line therapeutic agent for apnea of prematurity [5], because of its safer clinical profile compared to the older drug

theophylline and its oral form aminophylline [6]. For this indication, there is an internationally proposed dosage regimen for caffeine base (10 mg/kg load dose and 2.5 mg/kg maintenance dose) [7, 8]. Effective concentrations can be found among 5 to 20 µg/mL [7, 8]. In some cases, it is needed to increase plasma caffeine levels up to 30 or 35 µg/mL or even higher to be able to attain therapeutic effect [7, 9]. It has been stated that there is no need to monitor caffeine concentrations due to the high therapeutic margin that it has [9], although others suggest that serum concentrations should be monitored periodically because caffeine clearance and half-life rapidly change in

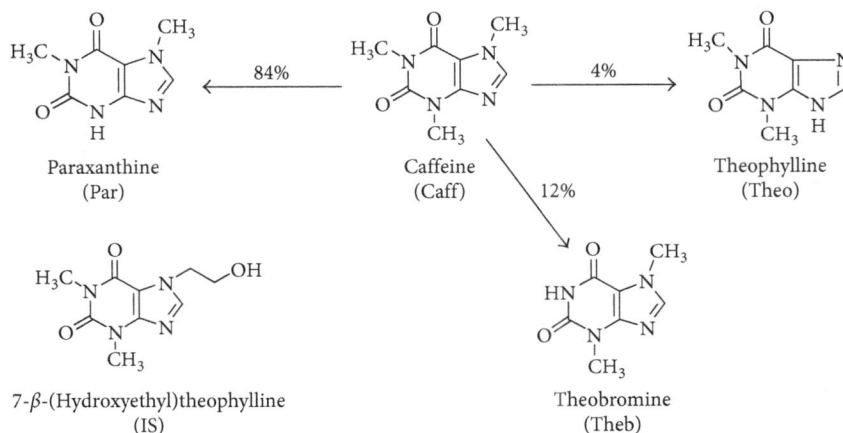

FIGURE 1: Caffeine and its major metabolites and internal standard structures.

the postnatal period [10, 11], and under those circumstances, a method such as the one proposed here can be very useful.

The presence of multiple cytochrome P-450 isoforms (CYP450) may explain the species, strain, age, tissue, and sex differences as well as the effect of inducers, nutritional status, and human drug metabolism. In humans, caffeine is metabolized through the liver microsomal drug-metabolizing enzyme system CYP450. In neonates, the expression of this enzymatic system is dependent on their prenatal and postnatal age [10, 12]. It has been suggested that the therapeutic effects of caffeine are affected by the expression and type of CYP1A2 isoform present (main metabolic pathway), which affects not only plasma concentration of caffeine (Caf) and its metabolites (theophylline (Theo), theobromine (Theb), and paraxanthine (Par)) (Figure 1) but also its therapeutic effects [13–15].

The analytical methods used for the quantification of caffeine and its main metabolites are immunoassay with monoclonal antibodies, which presents as main disadvantage the selectivity and sensitivity of the method [16], and liquid chromatography coupled to mass spectrometry detector (LC-MS/MS), which although is a very powerful tool, but its high cost and availability for clinical application has limited its use [17, 18]. The most common analytical method used is liquid chromatography with ultraviolet detection (LC-UV) that provides adequate sensitivity and selectivity [19, 20].

Another analytical problem is to obtain caffeine-free human plasma from volunteers. Some authors have tried to use substitutes, human plasma treated with activated charcoal to remove caffeine, or the manufacture of synthetic plasma devoid of caffeine [19]. We evaluate the possibility of using bovine whole serum because it is more representative than using human plasma or serum treated with activated charcoal to deplete substances or synthetic plasma. Likewise, we consider that bovine serum is more accessible and easily freer of caffeine than that derived from a human donor [19].

Recent studies have reported that intake of caffeine during pregnancy (plasma concentration estimated by self-reported consumption) is associated with the development of sickness such as obesity and hypertension, besides others [21, 22].

Our group is interested in quantifying plasma levels of caffeine and its metabolites to determine what is the prenatal exposure to caffeine in our population and if this exposure has an impact on therapeutic response (apnea) or if it increases risk or predisposition to other diseases [21, 23].

For this reason, initially we had to develop an analytical method to quantify caffeine and its metabolites in plasma at our laboratory.

In this study, we present the results on the use of fetal bovine serum (FBS) for the development and validation of an analytical method by HPLC for the quantification of caffeine and three of its main metabolites in samples of neonates and their mothers to perform the previously mentioned studies and results of its clinical application.

2. Materials and Methods

2.1. Chemicals and Reagents. Caffeine (1,3,7-trimethylxanthine) (C0750), paraxanthine (1,7-dimethylxanthine) (D5385), theobromine (3,7-dimethylxanthine) (T4500, >98%), and 7-(β-hydroxyethyl)theophylline (IS) (H9006) were obtained from Sigma (St. Louis, MO, USA). Theophylline (1,3-dimethylxanthine) (J1H052) was USP standard (Rockville, MD, USA). Acetonitrile and methanol were of HPLC grade, and acetic acid was of reagent grade; all were acquired from J.T. Baker (Xalostoc, Mexico). Water used during this study was of HPLC grade (Fermont laboratories, Monterrey, Mexico). Fetal bovine serum (FBS) was used as biological matrix to prepare calibration curves and quality control samples used during method validation and quantification of newborn samples (Gibco, Life Technologies).

2.2. Equipment. HPLC system consisted in a quaternary pump with a degasser, and it was coupled to an autosampler and DAD–UV detector (Agilent 1200 series) (Agilent Technologies Mexico, S. de R.L. de C.V.). Separation was performed on a reverse-phase column Zorbax® SB-Aq narrow bore RR (2.1 × 100 mm, 3.5 μm) (Agilent Technologies). The column oven was maintained at 40°C, while the autosampler was set at

TABLE 1: Final gradient program for HPLC sample analysis.

Time (min)	A (%)	B (%)	Flow rate (ml/min)
0	97	3	0.7
8.5	97	3	0.7
12	92	8	0.7
13	97	3	0.7

Mobile phase A: 10 mM phosphate buffer, pH 6.8. Mobile phase B: acetonitrile 100%.

room temperature. Fifteen microliters of processed sample was injected into the HPLC system.

2.3. Chromatographic Conditions.

For optimization of chromatographic conditions, the effects of various method parameters such as mobile phase, column, flow rate and solvent ratio, and detection system were evaluated, and the chromatographic parameters such as asymmetric factor, resolution, and column efficiency were calculated. The best results were obtained with a mixture of 10 mM phosphate buffer, pH 6.8, and acetonitrile, in a gradient phase mobile composition obtained using a gradient program (Table 1) at 0.7 mL/min. Chromatograms were recorded at 273 nm with a run time of 15 min.

At the end of each work day, the column was washed with acetonitrile : water (90 : 10 v/v) during 30 min. Chromatographic data were processed using Chemstation for LC systems software (Agilent Technologies).

2.4. Preparation of Stock and Working Solutions.

Stock solutions of Par, Theo, and Caf (4 mg/mL) were prepared separately, dissolving an appropriate amount of each drug in diluent (Milli-Q water). Theb solution was prepared at half concentration than the others (2 mg/mL) due to theobromine having the lowest aqueous solubility compared to the rest of caffeine alkaloids. This solution was prepared adding Theb to diluent (Milli-Q water) and heating and mixing the solution prior to obtain its total volume. The solution was cooled to room temperature before it was adjusted to its final volume.

Working solutions for the different points in the calibration curve or quality control samples were prepared simultaneously for all interest drugs in water. Those solutions were obtained mixing the necessary reagent volume to attain different concentrations; 1, 3, 5, 10, 25, 50, 100, 200, 300, and 400 μg/mL.

IS working solution (7-(β-hydroxyethyl)theophylline) was prepared at 20 μg/mL in Milli-Q water. It was stored until use.

2.5. Standard Calibration Curves and Quality Control Samples.

Seven level calibration curves were constructed by spiking water or drug-free FBS with known amounts of Caf, Theo, Par, and Theb to reach concentrations of 0.1, 0.3, 1, 2.5, 10, 20, and 40 μg/mL. Three quality control samples were prepared at low, middle, and high level of the calibration curve (0.5, 5, and 30 μg/mL). In all cases, the biological matrix dilution did not exceed 10%. These solutions were

vortexed for one min, and then 0.5 mL aliquots were transferred into 1.5 mL Eppendorf microcentrifuge tubes and stored at -60°C until use.

2.6. Clinical Design.

The analytical method was developed, validated, and challenged in a clinical trial. The clinical study was focused to quantify prenatal caffeine exposure, and later the analytical method will be applied to study clinical pharmacokinetics of caffeine in preterm neonates. This study reports only the results of the pilot trial of prenatal caffeine. The clinical protocol was reviewed and approved by our Institutional Research and Ethics Board registered according to Mexican law to authorize and oversee the conduction of clinical trials (13CI19039138 and CONBIOETICA-19-CEI-011-20161017). The protocol was conducted in accordance with the Declaration of Helsinki. All participants (fathers, mothers, or legal responsible) agreed to be included in the study, by signing an informed consent form.

Inclusion criteria are as follows: women 18 years old or older, pregnant with single or multiple products, gestational age less than 34 completed weeks (based on dates and confirmed with an ultrasound examination), and without a history of preeclampsia or any neonatal abnormality.

We excluded women with known fetal genetic or major malformations, very critical condition, or fetal demise.

2.7. Plasma Samples Collection.

Blood samples (1 mL) were collected during the 15 minutes immediately after delivery from umbilical cord (neonates) or from venous puncture (mothers). Samples were collected in heparinized Vacutainer® tubes and centrifuged for 10 min at 3500 rpm under a controlled temperature of 10°C. The plasma supernatant was carefully transferred to two polypropylene tubes and frozen at -60°C until they were analyzed.

2.8. Samples Preparation.

One hundred microliters of the IS working solution was added to 150 μL of plasma samples (calibration standards, quality control samples, or mother or neonate samples) in a 1.5 mL Eppendorf tube. To this solution, we added 350 μL of a 10% (v/v) acetic acid solution and 400 μL of water. The tube was vortexed for 20 s and then centrifuged at 12000 rpm for 10 min at 4°C. The supernatant clear layer was collected and preprocessed in a solid-phase extraction (SPE) system.

2.9. SPE Cleaning.

The sample pretreatment procedure was carried out on plasma by means of solid-phase extraction (SPE) on polymeric 96-well plates Strata-X™ (30 mg, 1 mL) (Phenomenex®, Torrance, CA). The plate was placed in a 96-well plate vacuum manifold system (Phenomenex, Torrance, CA). The vacuum pressure was adjusted between -15 and -20 mmHg, and each well was activated by washing with 1 mL methanol followed by 1 mL water.

One milliliter of each sample supernatant (previous step) was placed on its corresponding well and allowed to pass through low vacuum (-5 to -6 mmHg). The wells were then washed with 1 mL 5% methanol at vacuum pressure between

−15 and −20 mmHg. Caffeine and its metabolites were eluted with 1 mL methanol-2% acetic acid (70 : 30) solution. The eluent was dried under N_2 gas at 40°C. The residue was dissolved in 50 μL mobile phase and filtered through 0.22 mm filters, and an aliquot (15 μL) was injected into the HPLC.

2.10. Assay Validation. Validation was carried out following the criteria established in the Mexican regulatory guidelines [24], which are similar to the Guidance for Industry Bioanalytical Method Validation by FDA [25]. Inter- and intraday precision and accuracy were determined at low, medium, and high concentration levels on the calibration curve. Selectivity, stability, limit of quantification, and limit of detection were also evaluated [26]. Absolute extraction recoveries of Caf, metabolites, and IS were determined by comparing their respective response from quality control samples after the extraction process against their non-extracted samples above aqueous solutions. Validation results were obtained using chromatographic results and processed by SPSS software version 22.0.

2.11. Clinical Study Analysis. Results of caffeine and metabolites for prenatal exposure pilot study were expressed as plasma concentration for mothers and newborns. Plasma concentration differences by each mother/child pair (M/C) were calculated and used to construct a Bland–Altman plot for each compound. The differences between each M/C pair plasma concentrations were analyzed for means difference with Student's *t*-test, with a statistical significance level (α) of 0.05 and assuming normal data distribution. All statistical analyses were performed using SPSS software version 22.0.

3. Results and Discussion

3.1. Chromatography. We tested different analytical methods reported for caffeine and metabolite quantification [27] in order to adjust them to our laboratory conditions. However, none proved to be useful for our purposes, quantifying in a single step caffeine and its three major primary metabolites. At the beginning, we tried to develop an analytical method by fluorescence detection in an Acquity™ ultra-performance liquid chromatography (UPLC; Waters Corp., Milford, MA, USA), but it did not have enough sensitivity.

It is well known that fluorescence is more sensitive than UV for analysis of various compounds. However, fluorescence sensitivity to detect caffeine in biological samples remains a challenge.

At the beginning of our experimental process, we found a report that described fluorescence parameters for caffeine analysis. We obtained the same values of fluorescence spectrum by scanning a caffeine aqueous solution at 40 μg/mL, and excitation wavelength (λexc) and emission wavelength (λemis) were 272 nm and 385 nm, respectively (data not shown). Recently, Weldegebreal et al. reported fluorescence caffeine wavelengths similar to those we found [28]. However, in our

experiments, at the highest concentration of caffeine in the calibration curve, the signal recorded was too low. This is similar to what was reported by Weldegebreal at a greater concentration [28].

We believe that another problem could be related with the caffeine fluorescence in biological samples. Some researchers have described that caffeine and its metabolites (theophylline and theobromine) can quench fluorescence signals for proteins, amino acids, and hemoglobin [29]. But, could those analytes (protein, aromatic amino acids, and hemoglobin) quench caffeine fluorescence? Currently, we do not have data regarding this, but we believe it is possible because the static quenching mechanism is based on the noncovalent interaction characteristic of a π-stacked complexes [29].

Recently, a "light traffic" detector has been developed to improve caffeine fluorescence detection. This device uses a compound called "caffeine orange" to conjugate caffeine and increase about 250-fold the fluorescence signal. However, the main setback is that it is not selective and can increase fluorescence for caffeine, theobromine, theophylline, and other methylxanthines. This result suggests that fluorescence intensity signal by caffeine is too low for direct detection and needs derivatization to increase response [30].

Conversely, the HPLC/UV method had enough sensitivity for 0.1 μg/ml for each analyte in solution. We tested C8 and C18 columns with different pH values or mobile phase composition. However, retention time for each analyte was too short (losing resolution) or too long (increasing analytical run time). Our analytical method had an excellent chromatographic behavior. We were able to separate Caf, Theo, Theb, Par, and IS in 15 minutes using a chromatographic column designed to retain hydrophilic compounds using highly aqueous mobile phase without "phase collapse." This method was faster than other HPLC-UV methods that have been reported previously with a running time higher than 20 min or even up to 60 min [31]. Caf, metabolites, and IS had a continuous resolution factor of about 1.2 between contiguous peaks. Retention times were 3.8, 4.8, 5.5, 7.0, and 12.0 min for Theb, Par, Theo, internal standard (IS), and Caf, respectively. Figure 2 shows representative chromatograms of FBS blank (Figure 2 (a)); FBS + IS (Figure 2(b)); FBS spiked with caffeine and metabolites at all calibration curve levels lower (0.1 μg/mL) and upper (40.0 μg/mL) (Figure 2(c)) and zoom to show the lower limit of quantification (Figure 2(d)); plasma obtained from a mother volunteer and her child (Figure 2(e)) without caffeine and metabolites (volunteer 3); and plasma obtained from a mother volunteer and a pair of her twins T1 and T2 (Figure 2(f)) containing caffeine and metabolites (volunteer 5). In all conditions, chromatographic traces were completely clean and none exhibited any type of interference throughout the entire running time.

3.2. Linearity and Lower Limit of Quantification (LLOQ). The calibration curves were linear over the concentration range of 0.1–40 μg/mL, in a first-order model for all analytes.

(a)

(b)

(c)

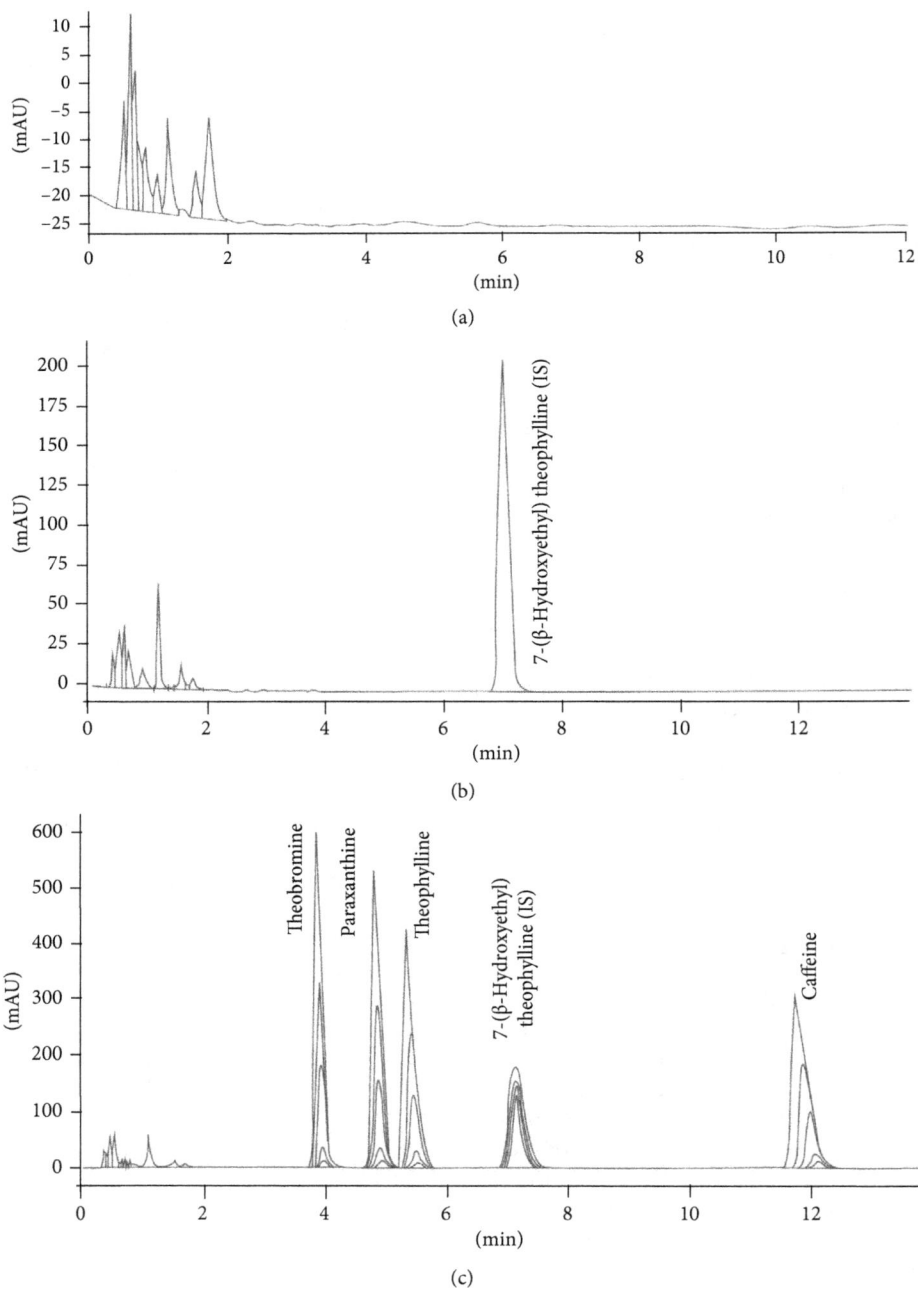

FIGURE 2: Continued.

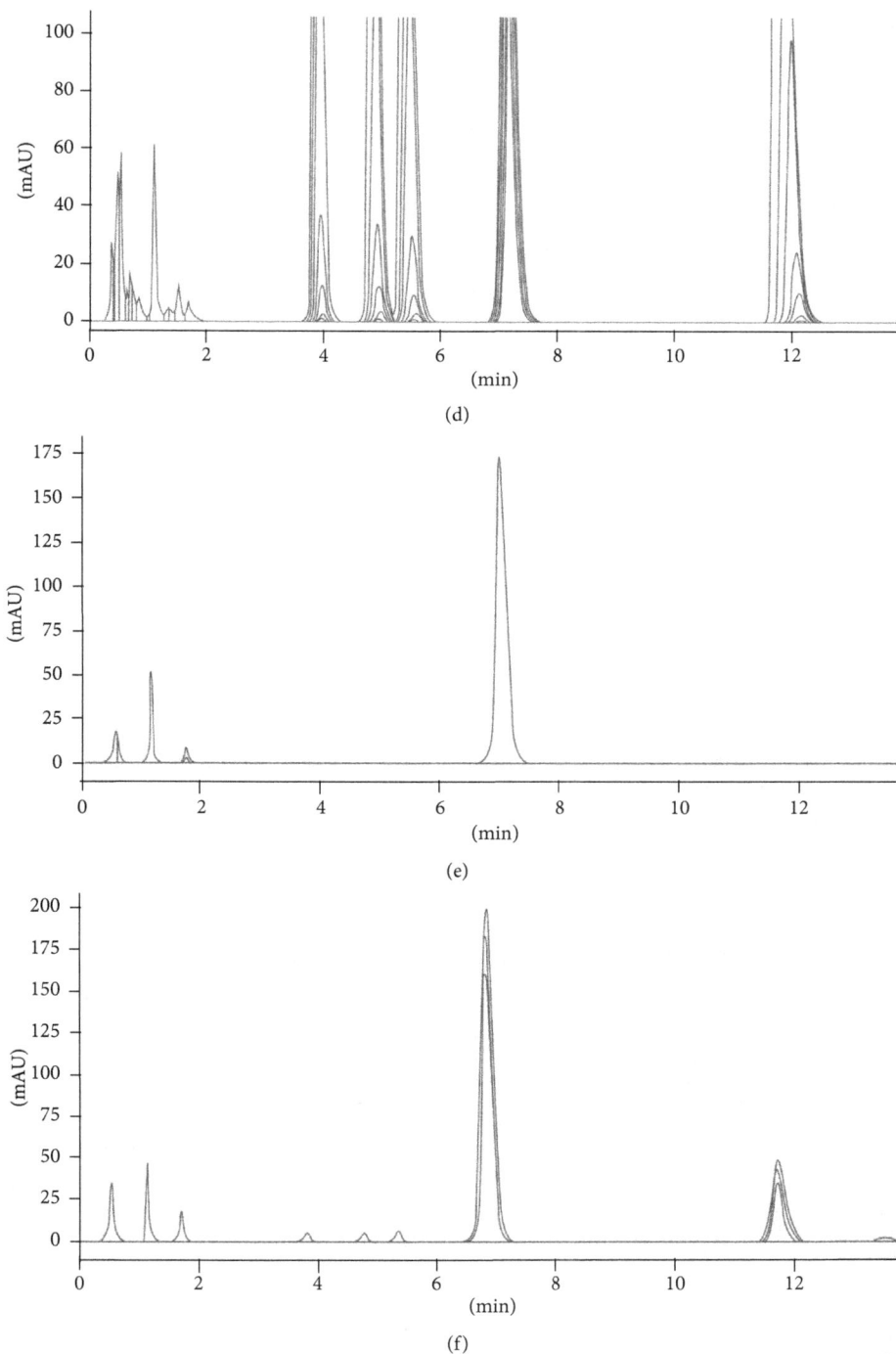

FIGURE 2: Representative chromatograms of FBS blank (a), FBS + IS (b), FBS spiked with caffeine and metabolites at all calibration curve levels lower (0.1 μg/mL) and upper (40.0 μg/mL) (c), and zoomed-in view showing the lower limit of quantification (d); plasma obtained from a mother volunteer and her child (e) without caffeine and metabolites (volunteer 3); and plasma obtained from a mother volunteer and a pair of her twins T1 and T2 (e) containing caffeine and metabolites (volunteer 5). Theb: theobromine (RT 3.8 min), Par: paraxanthine (RT 4.8 min), Theo: theophylline (RT 5.5 min), IS: 7-β-(hydroxyethyl)theophylline (RT 7.0 min), and caffeine (RT 12 min). RT: retention time.

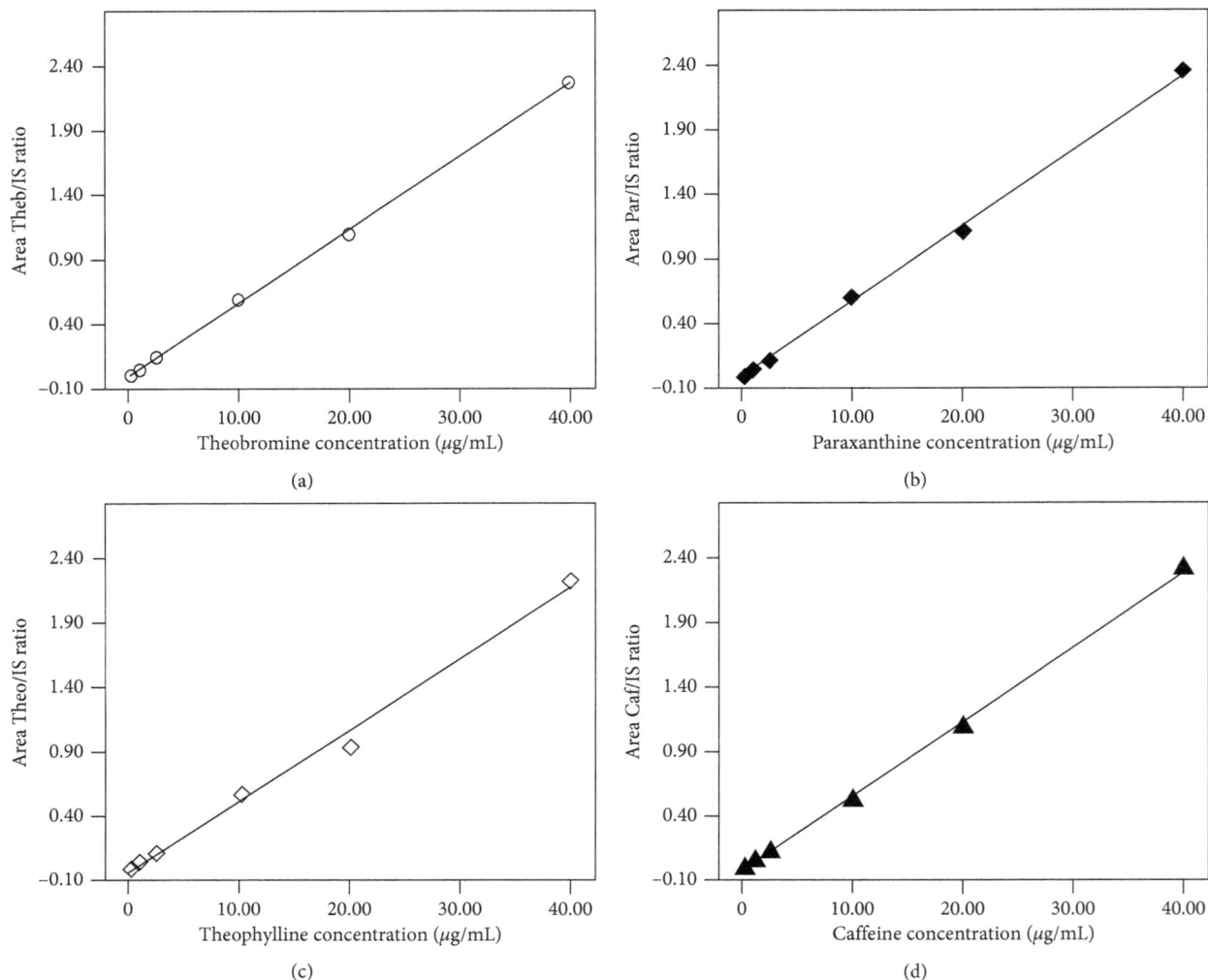

FIGURE 3: Typical calibration curves for quantification of theobromine (a), paraxanthine (b), theophylline (c), and caffeine (d) in human plasma. Theobromine (-O-), paraxanthine (-◆-), theophylline (-◊-), and caffeine (-▲-).

TABLE 2: Results of intra- and interday variability during validation of the HPLC method. Accuracy is expressed as percentage of the nominal value. Precision is expressed in terms of the percentage of the coefficient of variation.

Concentration (μg/mL)	Interday, $n = 9$, accuracy (%)/precision (CV)				Intraday, $n = 5$, accuracy (%)/precision (CV)			
	Theb	Par	Theo	Caf	Theb	Par	Theo	Caf
0.5	103.7 (8.8)	96.6 (4.0)	96.5 (2.3)	100.73 (8.0)	101.2 (8.7)	97.1 (3.5)	97.5 (4.1)	98.4 (6.3)
5	100.9 (5.7)	103.2 (4.9)	99.0 (4.6)	102.2 (6.6)	100.6 (6.8)	102.1 (5.0)	99.9 (5.0)	103.4 (5.6)
30	101.9 (7.5)	102.1 (6.4)	103.2 (5.5)	105.2 (5.7)	102.9 (8.7)	101.0 (5.5)	103.0 (5.4)	106.2 (4.8)

Interday results are the accumulation of determination of 3 quality control samples in three different days for each concentration level.

The linear equation for each assayed analyte was as follows: for theobromine (Theb/IS) = $-0.001 + 0.056$ (concentration), $r^2 = 0.998$ (Figure 3(a)); for paraxanthine (Par/IS) = $0.006 + 0.056$ (concentration), $r^2 = 0.997$ (Figure 3(b)); for theophylline (Theo/IS) = $0.015 + 0.055$ (concentration), $r^2 = 0.983$ (Figure 3(c)); and for caffeine (Caf/IS) = $-0.017 + 0.055$ (concentration), $r^2 = 0.995$ (Figure 3(d)).

For all analytes, LLOQ in BFS was 0.1 μg/mL. The LLOQ had a signal-to-noise ratio higher than $10:1$ (CV < 15%) (Figure 2(d)).

3.3. Recovery, Accuracy, and Precision. The optimal extraction recovery was obtained with SPE cartridges as it was described previously. To calculate absolute recovery of caffeine, metabolites, and IS, 5 sets of samples (0.5, 5, and 30 μg/mL) were prepared in FBS, in human plasma, or in mobile phase, and peak areas were compared. Mean analytical recovery for all concentration levels of Theb, Par, Theo, and Caf was 92, 96, 95, and 102%, respectively, and 95% for the IS. Inter- and intraday accuracy and precision values of the assay are presented in Table 2.

TABLE 3: Differences between concentrations of caffeine and metabolites in human plasma or fetal bovine serum spiked sample.

Concentration (μg/mL)	Human plasma mean (CV)			Fetal bovine serum mean (CV)			p value
	0.5	5.0	30.0	0.5	5.0	30.0	
Theb	0.50 (3.4)	5.21 (3.4)	31.49 (4.1)	0.52 (8.8)	5.04 (5.7)	30.57 (7.5)	NS
Par	0.50 (2.6)	5.01 (4.4)	30.92 (7.4)	0.48 (4.0)	5.16 (4.9)	30.62 (6.3)	NS
Theo	0.49 (4.2)	4.98 (4.6)	31.10 (5.2)	0.48 (2.3)	4.95 (4.6)	30.95 (5.5)	NS
Caf	0.52 (6.7)	5.03 (7.0)	32.28 (4.9)	0.50 (8.0)	5.11 (6.6)	31.57 (5.7)	NS

NS: no differences besides groups for each analyte at the same concentration.

TABLE 4: t-Test analysis for mean comparison between maternal and neonatal caffeine and metabolite concentration. Statistical analysis was done by comparing difference of concentrations mother-child versus zero difference.

Analyte	Mean difference	SE	95% CI difference	Diff (mother-neonate) versus zero	
Theobromine	−0.1462	0.087	(−0.327 to 0.035)	$p = 0.109$	NS
Paraxanthine	−0.0352	0.037	(−0.113 to 0.042)	$p = 0.361$	NS
Theophylline	−0.0044	0.028	(−0.064 to 0.055)	$p = 0.880$	NS
Caffeine	−0.0194	0.097	(−0.221 to 0.182)	$p = 0.844$	NS

Caffeine and its metabolites were measured by our HPLC method; CI: confidence interval; Diff: difference; SE: standard error.

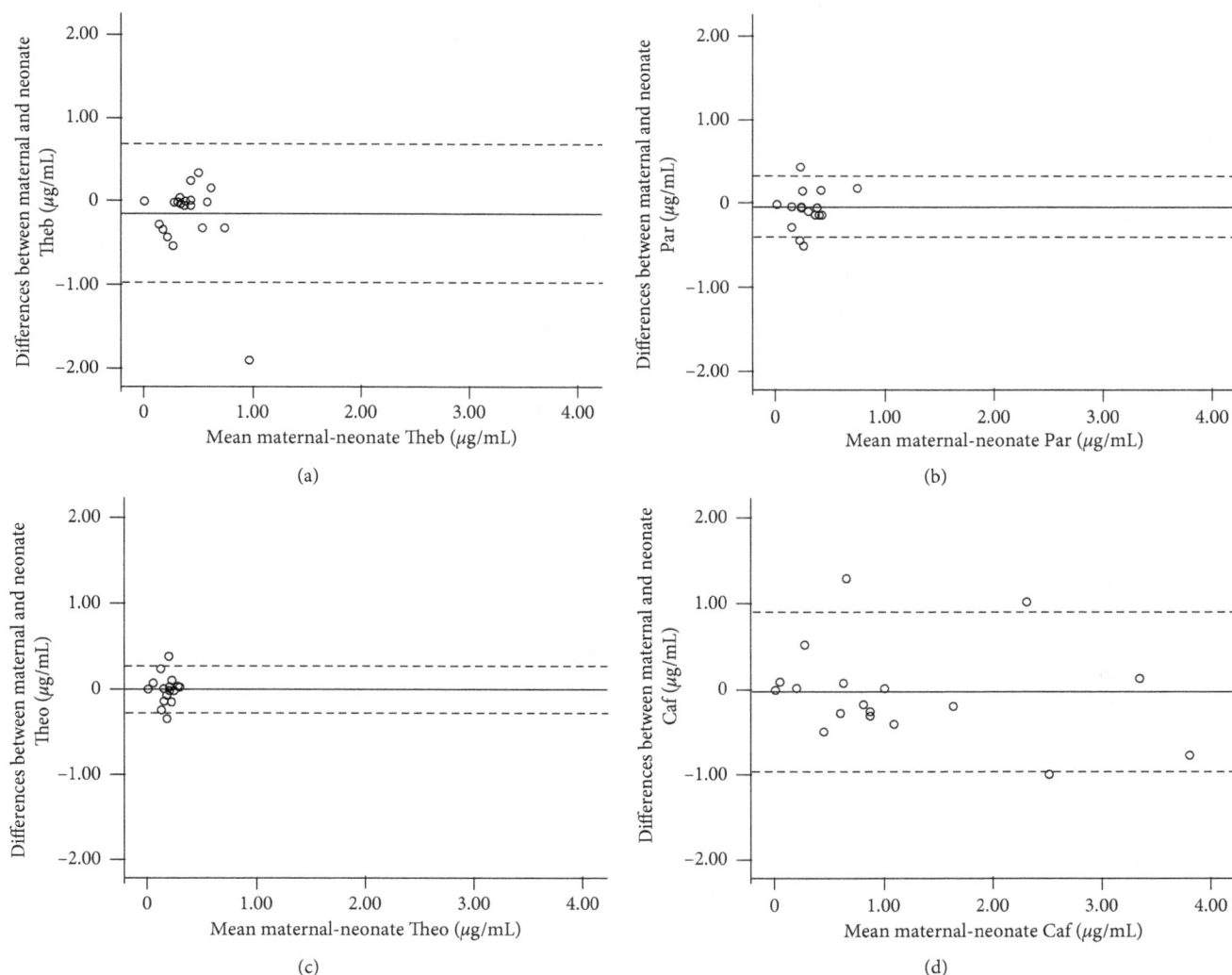

FIGURE 4: Bland-Altman plot for theobromine (a), paraxanthine (b), theophylline (c), and caffeine (d) plasma concentrations for mother versus newborn samples. Differences between mother-child pairs. Continuous line represents mean of difference between maternal concentration and neonatal concentration for each analyte. Dotted lines represent lower and upper limits of 95% confidence interval. Theb: Theobromine, Par: paraxanthine, Theo: theophylline, and Caf: caffeine.

No differences were observed between relative recovery of caffeine and its metabolites from FBS samples compared with human plasma assessed by Bland–Altman analysis and Student's t-test (Table 3). This result indicates that the method in FBS was reliable within the studied concentration range and its application to human samples analysis.

3.4. Stability Assays and Selectivity. Caf, Theo, Theb, Par, and IS proved to be stable in biological samples for at least 6 months at −60°C (final mean recovery of 101.7, CV 5.7%); samples were also stable for at least two freeze and thaw cycles (98%; CV 4.3%) and for at least 8 h on the worktable at room temperature (25°C; 96.5%; CV 4.5%). This was enough time according to our preparation sample method. Analytes were also stable for at least 13 h in the autosampler (101.5%; CV 6.0%), which made it possible to analyze about 50 samples in a row.

There were no endogenous compounds that interfered either with caffeine or its metabolites or with IS, even in hemolyzed or jaundiced samples. Other results of selectivity evaluation demonstrated that there were no interfering peaks from any of the following drugs: acetaminophen, ibuprofen, erythromycin, furosemide, or heparin, which are commonly used during neonatal pharmacotherapy.

3.5. Prenatal Caffeine Exposure. The trial was conducted from July 2013 through December 2013 in three medical centers located in Northeastern Mexico: the Hospital Regional Materno Infantil (Center 1), the Hospital Metropolitano Dr. Bernardo Sepúlveda (Center 2), and the Hospital Zambrano-Hellion (Center 3). The first two are public hospitals that harbor level II-III neonatal intensive care units and belong to the Servicios de Salud Network in Nuevo León, Mexico. The last is a private hospital with a level III neonatal intensive care unit and belongs to the academic and research health branch of Tecnológico de Monterrey in Monterrey, Nuevo León, Mexico. Seventy-three pregnant females that were admitted to the obstetric ward during labor were eligible to participate. Forty-six were excluded for different reasons: fetal death (19), malformation (1), or transfer to another unit (4), and 22 eligible mothers who declined the invitation to participate. Thus, only 27 pregnant volunteers participated in this study.

In our study, we were able to demonstrate the maternal consumption of caffeine in 85.2% of the volunteers ($n = 23$); meanwhile, in newborns' plasma, caffeine was only quantified in 78% ($n = 21$). Means of plasma Caf concentration were 0.87 ± 0.22 and $0.89 \pm 0.24\,\mu$g/mL in mother and newborn groups, respectively. No differences were observed in occurrence and level of caffeine from other populations [32]; however, it is necessary to increase sample size to establish better conclusions. No differences were detected between groups or each pair of mother/child samples evaluated by Bland–Altman plot (Table 4, Figure 4(d)). Likewise, means of plasma concentration for Theb, Par, and Theo were 0.27 ± 0.05 and $0.42 \pm 0.08\,\mu$g/mL, 0.17 ± 0.05 and $0.21 \pm 0.04\,\mu$g/mL, and 0.12 ± 0.03 and $0.12 \pm 0.03\,\mu$g/mL for mother and newborn groups, respectively. No differences were

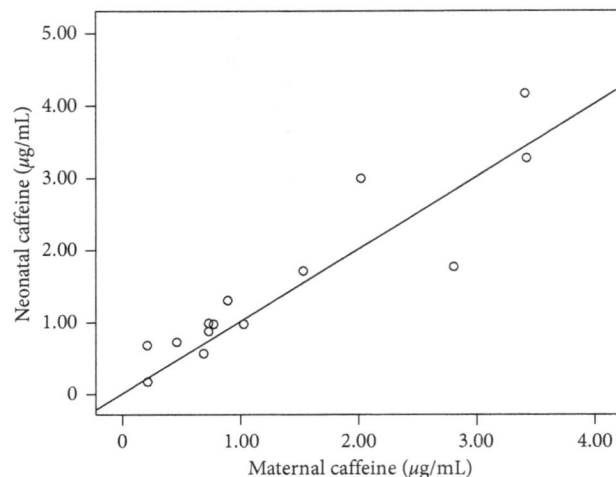

FIGURE 5: Linear regression between plasma caffeine concentrations in mother versus newborn samples.

detected between groups assessed by t-test or Bland–Altman plot (Table 4, Figures 4(a)–4(c)).

For Caf in maternal and newborn plasma concentration, there was a positive linear regression (Figure 5). However, for Theb, Par, and Theo, the same relationship was not found. An explanation could be that those metabolites had a different transplacental transfer rate perhaps related to gestational age [33] or changes in the metabolic pathway rate for caffeine and metabolites, also in relation to gestational age as it has been reported for other drugs [34]. We need to perform a deeper study to explain differences between maternal caffeine and its metabolite concentrations and their corresponding concentrations in newborns.

4. Conclusion

We developed a rapid and sensitive analytical method for the simultaneous determination of caffeine and its three principal metabolites in adult and newborn human plasma using HPLC technique after solid-phase extraction with Strata-X™ plates. This method had an acceptable accuracy (96.5–105.2% for intraday and 97.1–106.2% for interday) and precision with a CV less than 10% in all concentration levels used as quality control samples for all analytes. This method is suitable for use in individuals in which blood extraction volumes are limited (150 μL of plasma) due to their body size, as is the case in newborns, and it can be used for repeated sample procurement such as in pharmacokinetics studies as well as in therapeutic drug monitoring of caffeine and its metabolites in human neonates and perhaps also in small mammals (biological models).

Conflicts of Interest

The authors report no conflicts of interest.

Acknowledgments

This work was supported by Tecnologico de Monterrey Fund, GEE Investigación en Cáncer (Cátedra Hematología y Cáncer) and GEE Genética Humana (Cátedra Crecimiento y Desarrollo del Ser Humano). The authors would like to thank Juan Carlos Carrazco Arroyo for his technical assistance in improving the quality of chromatogram images.

References

[1] C. J. Reissig, E. C. Strain, and R. R. Griffiths, "Caffeinated energy drinks-A growing problem," *Drug and Alcohol Dependence*, vol. 99, no. 1-3, pp. 1-10, 2009.

[2] R. D. Goldman, "Caffeinated energy drinks in children," *Canadian Family Physician*, vol. 59, no. 9, pp. 947-948, 2013.

[3] B. B. Fredholm, J. F. Chen, R. A. Cunha, P. Svenningsson, and J. M. Vaugeois, "Adenosine and brain function," *International Review of Neurobiology*, vol. 63, pp. 191-270, 2005.

[4] J.-F. Chen, H. K. Eltzschig, and B. B. Fredholm, "Adenosine receptors as drug targets—what are the challenges?," *Nature Reviews Drug Discovery*, vol. 12, no. 4, pp. 265-286, 2013.

[5] J. Zhao, F. Gonzalez, and D. Mu, "Apnea of prematurity: from cause to treatment," *European Journal of Pediatrics*, vol. 170, no. 9, pp. 1097-1105, 2011.

[6] G. M. Pacifici, "Clinical pharmacology of theophylline in preterm infants: effects, metabolism and pharmacokinetics," *Current Pediatric Reviews*, vol. 10, no. 4, pp. 297-303, 2014.

[7] A. Aldridge, J. V Aranda, and A. H. Neims, "Caffeine metabolism in the newborn," *Clinical Pharmacology and Therapeutics*, vol. 25, no. 4, pp. 447-453, 1979.

[8] G. Natarajan, M. Lulic-Botica, and J. V. Aranda, "Pharmacology review clinical pharmacology of caffeine in the newborn," *Neoreviews*, vol. 8, no. 5, pp. e214-e221, 2007.

[9] J. E. Scanlon, K. C. Chin, M. E. Morgan, G. M. Durbin, K. A. Hale, and S. S. Brown, "Caffeine or theophylline for neonatal apnoea?," *Archives of Disease in Childhood*, vol. 67, no. 4, pp. 425-428, 1992.

[10] C. Cazeneuve, G. Pons, E. Rey et al., "Biotransformation of caffeine in human liver microsomes from foetuses, neonates, infants and adults," *British Journal of Clinical Pharmacology*, vol. 37, no. 5, pp. 405-412, 1994.

[11] G. Pons, O. Carrier, M. O. Richard et al., "Developmental changes of caffeine elimination in infancy," *Developmental Pharmacology and Therapeutics*, vol. 11, no. 5, pp. 258-264, 1988.

[12] N. Y. Jordan, J. Y. Mimpen, W. J. M. van den Bogaard, F. M. Flesch, M. H. M. van de Meent, and J. S. Torano, "Analysis of caffeine and paraxanthine in human saliva with ultra-high-performance liquid chromatography for CYP1A2 phenotyping," *Journal of Chromatography B*, vol. 995-996, pp. 70-73, 2015.

[13] B. Schmidt, R. Roberts, and P. Davis, "Long-term effects of caffeine therapy for apnea of prematurity," *New England Journal of Medicine*, vol. 357, no. 19, pp. 1893-1902, 2007.

[14] B. G. Charles, S. R. Townsend, P. A. Steer, V. J. Flenady, P. H. Gray, and A. Shearman, "Caffeine citrate treatment for extremely premature infants with apnea: population pharmacokinetics, absolute bioavailability, and implications for therapeutic drug monitoring," *Therapeutic Drug Monitoring*, vol. 30, no. 6, pp. 709-716, 2008.

[15] P. H. Gray, V. J. Flenady, B. G. Charles, and P. A. Steer, "Caffeine citrate for very preterm infants: effects on development, temperament and behaviour," *Journal of Paediatrics and Child Health*, vol. 47, no. 4, pp. 167-172, 2011.

[16] J. J. Carvalho, M. G. Weller, U. Panne, and R. J. Schneider, "A highly sensitive caffeine immunoassay based on a monoclonal antibody," *Analytical and Bioanalytical Chemistry*, vol. 396, no. 7, pp. 2617-2628, 2010.

[17] H. Li, C. Zhang, J. Wang, Y. Jiang, J. P. Fawcett, and J. Gu, "Simultaneous quantitation of paracetamol, caffeine, pseudoephedrine, chlorpheniramine and cloperastine in human plasma by liquid chromatography-tandem mass spectrometry," *Journal of Pharmaceutical and Biomedical Analysis*, vol. 51, no. 3, pp. 716-722, 2010.

[18] A. S. Ptolemy, E. Tzioumis, A. Thomke, S. Rifai, and M. Kellogg, "Quantification of theobromine and caffeine in saliva, plasma and urine via liquid chromatography-tandem mass spectrometry: a single analytical protocol applicable to cocoa intervention studies," *Journal of Chromatography B*, vol. 878, no. 3-4, pp. 409-416, 2010.

[19] S. N. Alvi and M. M. Hammami, "Validated HPLC method for determination of caffeine level in human plasma using synthetic plasma: application to bioavailability studies," *Journal of Chromatographic Science*, vol. 49, no. 4, pp. 292-296, 2011.

[20] V. Perera, A. S. Gross, and A. J. McLachlan, "Caffeine and paraxanthine HPLC assay for CYP1A2 phenotype assessment using saliva and plasma," *Biomedical Chromatography*, vol. 24, no. 10, pp. 1136-1144, 2010.

[21] M. A. Klebanoff and S. A. Keim, "Maternal serum paraxanthine during pregnancy and offspring body mass index at ages 4 and 7 years," *Epidemiology*, vol. 26, no. 2, pp. 185-191, 2015.

[22] D.-K. Li, J. R. Ferber, and R. Odouli, "Maternal caffeine intake during pregnancy and risk of obesity in offspring: a prospective cohort study," *International Journal of Obesity*, vol. 39, no. 4, pp. 658-664, 2015.

[23] F. P. Vitti, N. Adati, H. Bettiol et al., "Association of caffeine consumption with preterm birth and low birthweight in RibeirÃo Preto, São Paulo, Brazil," *Archives of Disease in Childhood*, vol. 99, no. 2, Ribeirão Preto Medical School, University of São Paulo, Ribeirão Preto, Brazil, pp. A223.3-A224, 2014.

[24] COFEPRIS, NOM-177-SSA1-2013, 2013.

[25] FDA, *Guidance for Industry: Bioanalytical Method Validation*, U.S. Department of Health and Human Services, Washington, DC, USA, 2013.

[26] G. Marcelín-Jiménez, J. Hernández, A. P. Ángeles et al., "Bioequivalence evaluation of two brands of ketoconazole tablets (Onofin-K® and Nizoral®) in a healthy female Mexican population," *Biopharmaceutics & Drug Disposition*, vol. 25, no. 5, pp. 203-209, 2004.

[27] P. N. Patil, "Caffeine in various samples and their analysis with HPLC-a review," *International Journal of Pharmaceutical Sciences Review and Research*, vol. 16, no. 2, pp. 76-83, 2012.

[28] B. Weldegebreal, M. Redi-Abshiro, and B. S. Chandravanshi, "Development of new analytical methods for the determination of caffeine content in aqueous solution of green coffee beans," *Chemistry Central Journal*, vol. 11, no. 1, 2017.

[29] M. Makarska-Bialokoz, "Spectroscopic evidence of xanthine compounds fluorescence quenching effect on water-soluble porphyrins," *Journal of Molecular Structure*, vol. 1081, pp. 224-232, 2015.

[30] W. Xu, T.-H. Kim, D. Zhai et al., "Make caffeine visible: a fluorescent caffeine "traffic light" detector," *Scientific Reports*, vol. 3, no. 1, 2013.

[31] H. Kanazawa, R. Atsumi, Y. Matsushima, and J. Kizu, "Determination of theophylline and its metabolites in biological samples by liquid chromatography-mass spectrometry,"

Journal of Chromatography A, vol. 870, no. 1-2, pp. 87–96, 2000.

[32] L. M. Grosso, E. Triche, N. L. Benowitz, and M. B. Bracken, "Prenatal caffeine assessment: fetal and maternal biomarkers or self-reported intake?," *Annals of Epidemiology*, vol. 18, no. 3, pp. 172–178, 2008.

[33] S. K. Griffiths and J. P. Campbell, "Placental structure, function and drug transfer," *Continuing Education in Anaesthesia Critical Care & Pain*, vol. 15, no. 2, pp. 84–89, 2015.

[34] K. C. Worley, S. W. Roberts, and R. E. Bawdon, "The metabolism and transplacental transfer of oseltamivir in the ex vivo human model," *Infectious Diseases in Obstetrics and Gynecology*, vol. 2008, Article ID 927574, 5 pages, 2008.

13

Analytical Method Development and Validation for the Quantification of Acetone and Isopropyl Alcohol in the Tartaric Acid Base Pellets of Dipyridamole Modified Release Capsules by using Headspace Gas Chromatographic Technique

Sriram Valavala ⓘ,[1] Nareshvarma Seelam ⓘ,[1] Subbaiah Tondepu ⓘ,[1] V. Shanmukha Kumar Jagarlapudi ⓘ,[1] and Vivekanandan Sundarmurthy ⓘ[2]

[1]Department of Chemistry, K L University, Green Fields, Vaddeswaram, Guntur 522502, Andhra Pradesh, India
[2]Research and Development, Bluefish Pharmaceuticals Private Limited, Bangalore 560115, Karnataka, India

Correspondence should be addressed to Subbaiah Tondepu; tsubbaiah@yahoo.com

Academic Editor: Guido Crisponi

A simple, sensitive, accurate, robust headspace gas chromatographic method was developed for the quantitative determination of acetone and isopropyl alcohol in tartaric acid-based pellets of dipyridamole modified release capsules. The residual solvents acetone and isopropyl alcohol were used in the manufacturing process of the tartaric acid-based pellets of dipyridamole modified release capsules by considering the solubility of the dipyridamole and excipients in the different manufacturing stages. The method was developed and optimized by using fused silica DB-624 (30 m × 0.32 mm × 1.8 μm) column with the flame ionization detector. The method validation was carried out with regard to the guidelines for validation of analytical procedures Q2 demanded by the International Council for Harmonisation of Technical Requirements for Pharmaceuticals for Human Use (ICH). All the validation characteristics were meeting the acceptance criteria. Hence, the developed and validated method can be applied for the intended routine analysis.

1. Introduction

Dipyridamole (2,6-bis-(diethanolamino)-4,8-dipiperidino-(5,4-d)-pyrimidine) displays antithrombotic and antiaggregatory activity. The dipyridamole (Figure 1) is used in combination with "blood thinners" such as warfarin to avoid clot formation after heart valve replacements. Clots are a serious complication that can cause strokes, heart attacks, or blocked blood vessels in the lungs (pulmonary embolisms). Dipyridamole is an antiplatelet drug. Dipyridamole is an odourless yellow crystalline powder, having a bitter taste.

Dipyridamole exhibits a relatively short biological half-life of less than one hour. Therefore, extended release formulations of dipyridamole, which provide a continual administration of active ingredient over time, are preferred. Dipyridamole is soluble in acidic mediums with a pH below 4 and is practically insoluble in water. Therefore,

dipyridamole is readily absorbed in the more acidic regions of the upper gastrointestinal tract, but remains insoluble in the more basic regions of the intestine. To obtain a constant level of dipyridamole in the blood, it is advantageous to formulate a dipyridamole dosage form that releases dipyridamole at a controlled rate and at a defined pH. Acidic components can be coadministered with dipyridamole to maintain a defined pH level throughout administration. Dipyridamole can also be administered with other active ingredients, such as aspirin. Aspirin (acetylsalicylic acid) is an inhibitory substance which counteracts the aggregation of human blood platelets by inhibiting cyclooxygenase and thereby inhibiting the biosynthesis of the aggregation promoting thromboxane A2 [1, 2]. The residual solvents present in tartaric acid-based pellets of dipyridamole modified release capsules are classified as class 3 solvents as per the ICH Q3C guidelines.

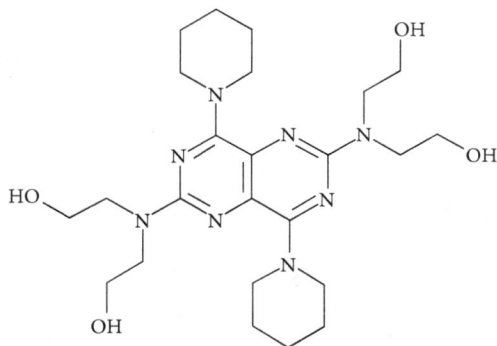

FIGURE 1: Structure of dipyridamole.

The dipyridamole in tartaric acid-based pellets of dipyridamole modified release 150 mg and 200 mg capsules are available in the market. Each capsule contains dipyridamole 200 mg and 150 mg respective dosage strength. The adults including the elders recommended dose is one capsule twice daily, usually one in the morning and one in the evening preferably with meals. The capsules should be swallowed whole without chewing as per eMC [3].

In the literature survey, quite a few GC method have been reported from the determination of the residual solvents in dipyridamole API [4], few liquid chromatographic methods have been reported for determination of dipyridamole in pharmaceutical preparation [4–6], and few methods have been reported for dipyridamole and its degradation product [7, 8]. However, several methods were reported for determination of dipyridamole in combination with other drugs [9–12]. Estimation of dipyridamole and its metabolites in human plasma by liquid chromatographic-mass spectroscopy and HPLC has been performed [13–15].

The aim of this study is to develop the simple and fast analytical method for estimation of residual solvents in the tartaric acid-based pellets of dipyridamole modified release capsules, and the method can be used for the routine analysis. The developed method was subjected for the analytical validation with respect to specificity, linearity, precision, accuracy, limit of detection (LOD), limit of quantification (LOQ), robustness, and ruggedness as per the ICH guidelines [16].

2. Materials and Methods

2.1. Chemicals and Reagents.
The GC grade N, N-dimethylsulfoxide, isopropyl alcohol, acetone, HPLC grade water, nitrogen gas, air, and hydrogen. The dipyridamole drug substance, placebo samples of dipyridamole modified release capsules, and samples of dipyridamole modified release capsules were supplied by Bluefish Pharmaceuticals Pvt. Ltd, Bangalore, India.

2.2. Equipment.
The analytical method was developed by using the Agilent 7890A coupled with G1888 network headspace sampler, analytical balance from Mettler Toledo, micropipette from Eppendorf, headspace crimp vials, and suitable glass apparatus for solution preparations.

2.3. Chromatographic Conditions.
The method was developed and validated by using fused silica DB-624 (30 m × 0.32 mm × 1.8 µm) column with the flame ionization detector (FID) and the chromatographic parameters are given in Table 1. The chromatographic retention time of acetone and isopropyl alcohol is given in Table 2.

2.4. Preparation of Solutions

2.4.1. Diluent Solution.
A mixture of N,N-dimethylsulfoxide and water was used as diluent.

2.4.2. Blank Solution.
Transfer 5 mL of diluent into 20 mL headspace and crimp cap immediately.

2.4.3. Preparation of Standard Solution (Stock).
Weigh and transfer about 500 mg of isopropyl alcohol and 500 mg of acetone into 50 mL volumetric flask containing about 30 mL of diluent mix and make up to the mark with diluent.

2.4.4. Preparation of Standard Solution.
Pipette out 5 mL of the above standard stock solution into 100 mL volumetric flask and make up to the mark with diluent. Transfer 5 mL of above solution into 20 mL headspace vials, and crimp cap immediately.

2.4.5. Preparation of Sample Solution.
Open five capsules and crush the pellets using mortar pestle. Weigh and transfer 500 mg of crushed powder into 20 mL headspace vial, add 5 mL of diluent, and crimp the cap immediately.

2.5. System Suitability Criteria.
The present relative standard deviation of standard peak area for six replicate injections should not be more than 10.

3. Results and Discussion

3.1. Method Development and Optimization.
The analytical method development was initiated by using the Agilent 7890A coupled with G1888 network headspace sampler, fused silica DB-624 (30 m × 0.32 mm × 1.8 µm) column with the flame ionization detector (FID), carrier gas as helium. The front injector conditions (injector temperature 140°C, carrier gas flow 2.0 mL/min, and split ration 5 : 1), front detector conditions (detector temperature 260°C, hydrogen flow 40 mL/minute, and air flow 300 mL/minute), oven conditions (40°C for 10 minutes and increasing the temperature to 250°C at the rate of 30°C/minute and hold for 10 minutes), headspace oven temperature 100°C, loop temperature 105°C, transfer line temperature 110°C. The dimethylsulfoxide used as diluent for solution preparations. Based on the above experiment, we found the very less resolution between acetone and isopropyl alcohol. The further experiments were conducted by altering the chromatographic conditions to achieve satisfactory resolution between acetone and isopropyl alcohol.

TABLE 1: Chromatographic parameters.

Column	Fused silica DB-624 (30 m × 0.32 mm × 1.8 μm)	Makeup gas flow	25 mL/minute
Detector	Flame ionization detector (FID)	Oven temperature	40°C for 10 minutes and increasing the temperature to 210°C at the rate of 35°C/minute and kept for 5 minutes
Mode	Constant flow	*Headspace sampler condition*	
Carrier gas	Nitrogen	Oven temperature	80°C
Front Injector condition		Loop temperature	90°C
Inject temperature	140°C	Transfer line temperature	100°C
Carrier gas flow	1.0 mL/min	GC cycle time	34 minutes
Split ratio	20 : 1	Vial equilibration time	20 minutes
Injector type	Split	Pressurization time	1.0 minutes
Front detector condition		Loop fill time	0.2 minutes
Detector temperature	260°C	Loop equilibration time	0.05 minutes
Hydrogen flow	30 mL/minute	Injection time	1.0 minutes
Air flow	300 mL/minute	Shake	High

TABLE 2: Retention times of acetone and isopropyl alcohol.

Serial Number	Impurity name	RT (about)
1	Acetone	6.8
2	Isopropyl alcohol	7.3

The further experiment was conducted by using the above experiment chromatographic parameters and by changing the front injector condition and headspace sampler condition. Based on the above experiment, we found the very less resolution between acetone and isopropyl alcohol and found the placebo peak interference was observed at the retention time of acetone. The aim is to resolve the issue of placebo peak interference, needs to modify the diluent or headspace oven temperature conditions to finalize the method conditions.

The further experiment was conducted by using the fused silica DB-624 (30 m × 0.32 mm × 1.8 μm) column with the flame ionization detector (FID), carrier gas as nitrogen. The front injector conditions (injector temperature 140°C, carrier gas flow 1.0 mL/min, and split ration 20 : 1), front detector conditions (detector temperature 260°C, hydrogen flow 30 mL/minute, and air flow 300 mL/minute), oven conditions (40°C for 10 minutes and increasing the temperature to 210°C at the rate of 35°C/minute and hold for 5 minutes), headspace oven temperature 80°C, loop temperature 90°C, transfer line temperature 100°C. The misture of N,N-dimethylsulfoxide and water used as diluent for solution preprations. Based on the above experiment, it was found that no placebo interference was observed at the retention time of acetone and isopropyl alcohol, and resolution was found satisfactory.

Based on the optimization of the trials, the above-mentioned chromatographic conditions were finalized for the quantification of the acetone and isopropyl alcohol in tartaric acid-based pellets of dipyridamole modified release capsules. Hence, this method can be validated and introduced for the routine analysis.

3.2. *Method Validation.* The developed analytical method for quantification of the residual solvents in the tartaric acid-based pellets of dipyridamole modified release capsules was validated as per International Council for Harmonisation of Technical Requirements for Pharmaceuticals for Human Use (ICH) [15]. The validation parameters [17, 18] specificity, estimation of limit of detection (LOD), and limit of quantification (LOQ), accuracy, precision, linearity, range, ruggedness, and robustness were examined [17, 18].

3.2.1. *System Suitability.* To check the system suitability criteria, the solutions were prepared and injected as per the test method. All the parameters were found well within the acceptance criteria (Table 3).

3.2.2. *Limit of Detection and Limit of Quantification.* The limit of detection (LOD) and limit of quantification (LOQ) were established by the signal-to-noise ratio method by preparing the known concentrations of acetone and isopropyl alcohol and injected into gas chromatography headspace instrument as per the test method. The limit of detection and limit of quantification for each matrix were determined from the signal-to-noise ratio (S/N) method of 3 : 1 and 10 : 1 by injecting the standard solutions (Table 4). The LOD and LOQ were verified by analysis of spiked standard solutions predefined acceptance criteria. The percent relative standard deviation for area response found for acetone at LOD concentration is 7.1 and LOQ is 3.5 and for isopropyl alcohol at LOD concentration is 5.2 and LOQ is 3.7.

3.2.3. *Specificity.* Specificity was accomplished by injecting the samples as per the test method and as a part of the specificity study. Blank, acetone, isopropyl alcohol solvent, and placebo were prepared and injected as per test method. No peak interference at the retention time of acetone and

TABLE 3: System suitability criteria and results

Parameter	Acceptance criteria	Result
The present relative standard deviation of acetone peak area for six replicate injections.	≤10.0	0.9
The present relative standard deviation of isopropyl alcohol peak area for six replicate injections.	≤10.0	1.3
The present relative standard deviation of acetone retention time for six replicate injections.	≤10.0	0.0
The present relative standard deviation of isopropyl alcohol retention time for six replicate injections.	≤10.0	0.0

TABLE 4: Limit of detection and Limit of quantification.

Name	LOD Concentration (μg/mL)	LOQ Concentration (μg/mL)
Acetone	3	7
Isopropyl alcohol	11	27

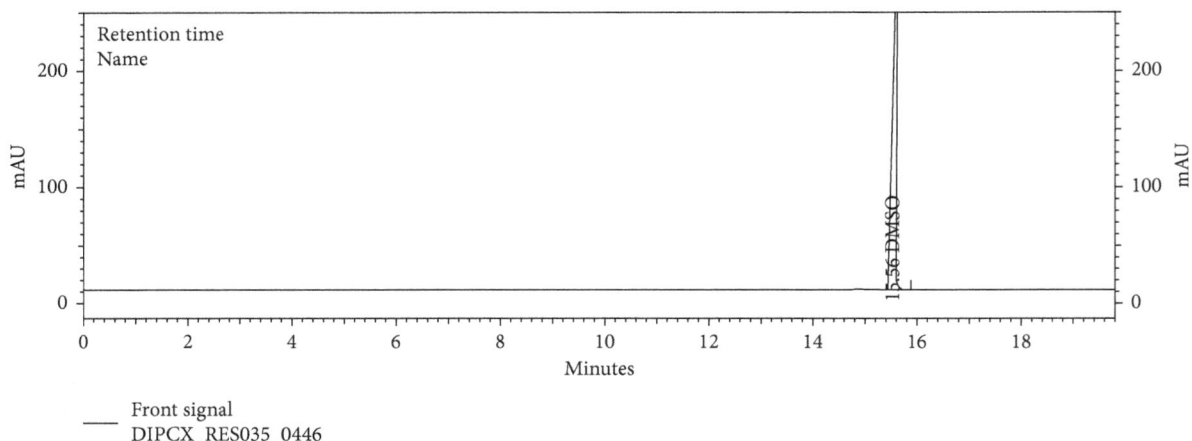

FIGURE 2: Typical chromatogram of blank.

isopropyl alcohol was observed. Therefore, we conclude that this method is selective and suitable for the identification and quantification of the acetone and isopropyl alcohol in the dipyridamole modified release capsules.

The chromatograms are given for the blank (Figure 2), placebo (Figure 3), acetone (Figure 4), isopropyl alcohol (Figure 5), standard (Figure 6), as-such sample (Figure 7), and spiked blend solution (Figure 8).

3.2.4. Method Precision (Repeatability).
The method precision or intraassay precision was performed by preparing the six replicate test preparations ($n = 6$) of dipyridamole 200 mg modified release capsules by spiking acetone and isopropyl alcohol specification level (Table 5 and Figure 9) analyzed as per the test method. The concentration in parts per million was calculated and found to be within the acceptance criteria. The relative standard deviations obtained for acetone were 1.0% and isopropyl alcohol 1.4%. The graphs for acetone and isopropyl alcohol are shown in Figure 9.

3.2.5. Accuracy.
Accuracy of the proposed analytical procedure was evaluated from the assay results of the acetone and isopropyl alcohol as per the test method. A series of sample solutions were prepared in triplicate (six replicate test preparations for LOQ and about 200% levels) by spiking the acetone and isopropyl alcohol in placebo except LOQ level in the range of about 25%, 50%, 100%, and 150% of specification level and injected into HPLC system and analyzed as per the test method. The concentrations of acetone are 7.5 μg/mL, 1300 μg/mL, 2738 μg/mL, 5050 μg/mL, and 7694 μg/mL and of isopropyl alcohol are 28 μg/mL, 1304 μg/mL, 2745 μg/mL, 5063 μg/mL, and 7715 μg/mL. Individual % recovery, mean % recovery, % RSD, and squared correlation coefficient for linearity of the test method were calculated, and the results were found to be within the acceptance criteria (Table 6). The linearity graphs from accuracy results for acetone and isopropyl alcohol are shown in Figures 10 and 11, respectively.

3.2.6. Linearity.
The linearity was studied by analyzing the standard solutions. A series of solutions of acetone and

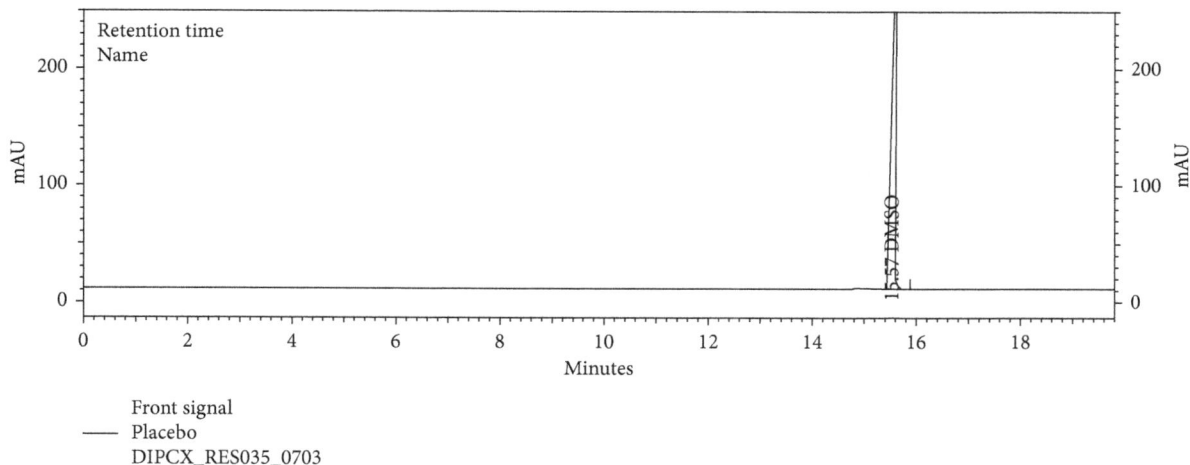

FIGURE 3: Typical chromatogram of placebo.

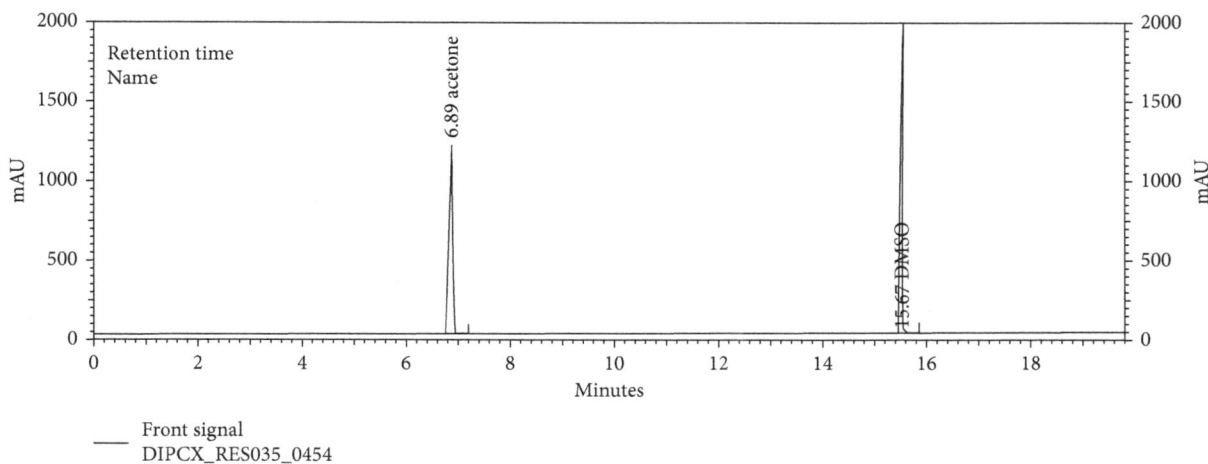

FIGURE 4: Typical chromatogram of acetone.

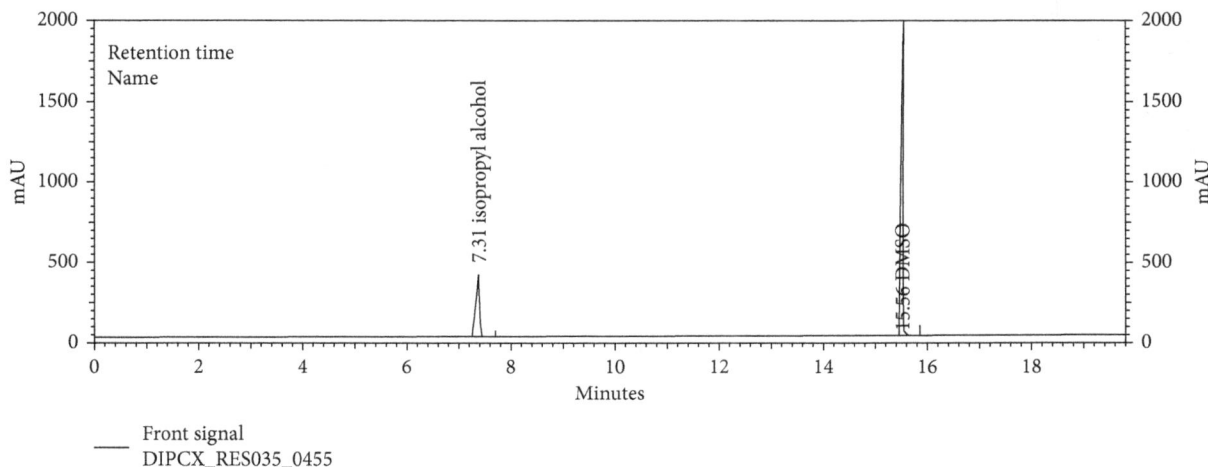

FIGURE 5: Typical chromatogram of isopropyl alcohol.

isopropyl alcohol solutions were prepared in the range of LOQ to about 150% of specification level and injected into the HPLC system. Linearity of detector response was established by plotting a graph between concentration versus response of acetone and isopropyl alcohol peaks. The detector response was found to be linear from about LOQ to

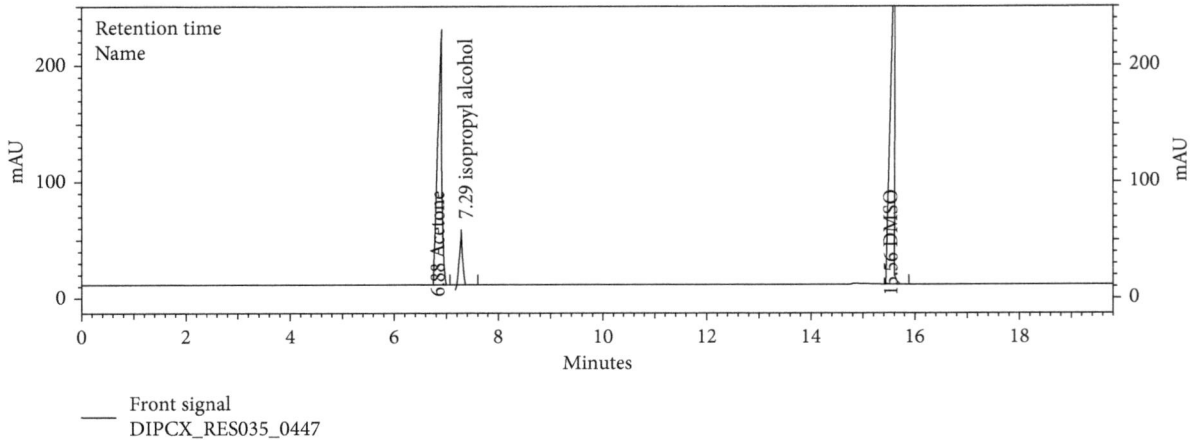

Figure 6: Typical chromatogram of standard solution.

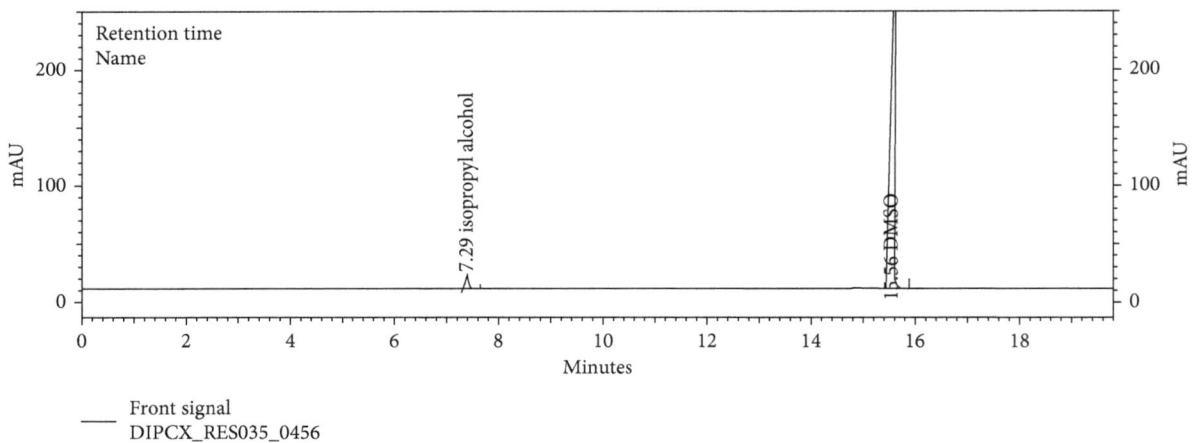

Figure 7: Typical chromatogram of as-such sample.

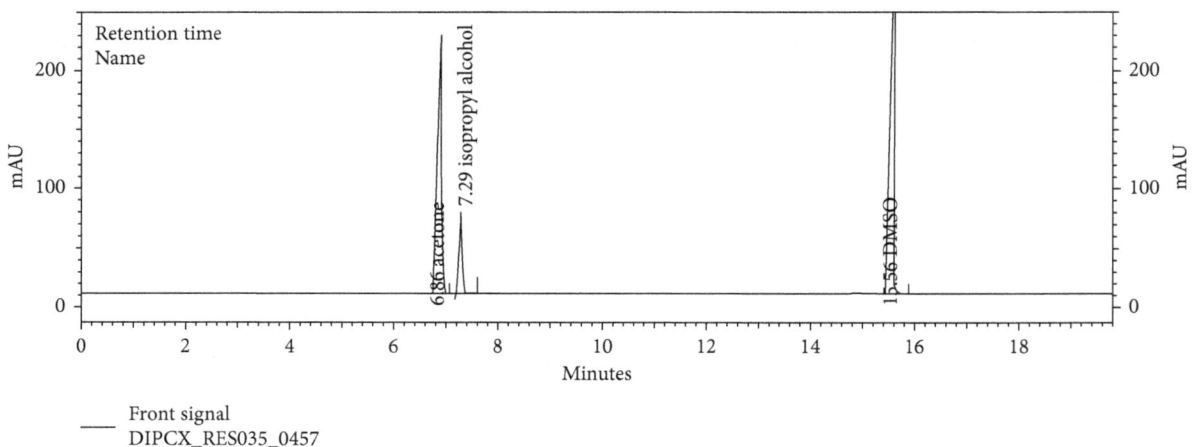

Figure 8: Typical chromatogram of spiked sample.

150% of specification level and injected into HPLC system and analyzed as per the test method. The concentrations of acetone are 7 μg/mL, 1260 μg/mL, 3024 μg/mL, 5040 μg/mL, and 7560 μg/mL and of isopropyl alcohol are 27 μg/mL, 1261 μg/mL, 3027 μg/mL, 5045 μg/mL, and 7567 μg/mL.

The square of correlation coefficient, slope, and % y-intercept at 100% level, intercept, and residual sum of

TABLE 5: Method precision data for spiked sample.

Sample number	Acetone	Isopropyl alcohol
1	5242	6391
2	5346	6538
3	5369	6542
4	5289	6469
5	5240	6315
6	5281	6413
Mean	*5294*	*6445*
% RSD	*1.0*	*1.4*

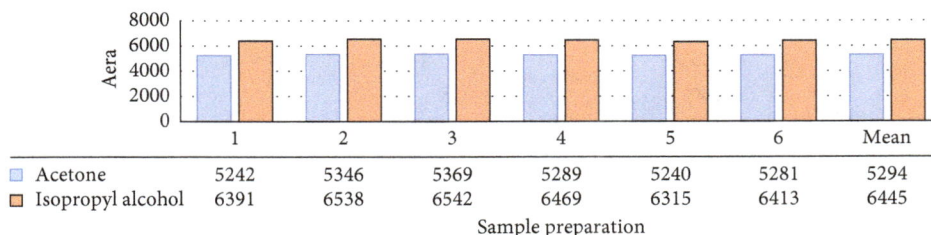

	1	2	3	4	5	6	Mean
Acetone	5242	5346	5369	5289	5240	5281	5294
Isopropyl alcohol	6391	6538	6542	6469	6315	6413	6445

Sample preparation

FIGURE 9: Method precision.

TABLE 6: Accuracy data of acetone and isopropyl alcohol.

Spike level		% recovery and relative standard deviation of acetone and isopropyl alcohol	
		Acetone	Isopropyl alcohol
Level-1 LOQ	Mean	108.2	109.9
	% RSD	2.6	2.0
Level-2 LOQ	Mean	94.2	106.8
	% RSD	0.2	1.6
Level-3 LOQ	Mean	94.7	104.9
	% RSD	1.5	1.7
Level-4 LOQ	Mean	101.2	110.1
	% RSD	0.6	0.1
Level-5 LOQ	Mean	101.3	108.1
	% RSD	0.5	1.0

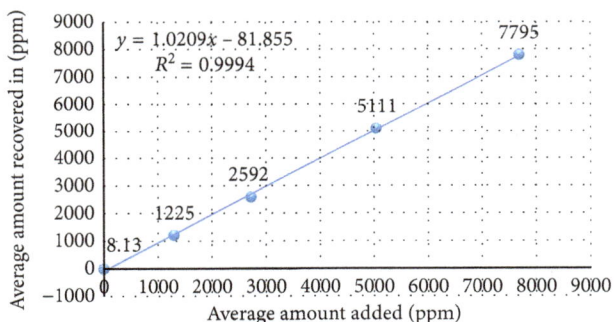

FIGURE 10: Linearity plot for acetone from accuracy results.

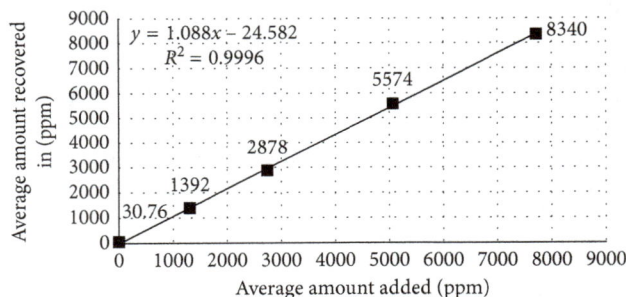

FIGURE 11: Linearity plot for isopropyl alcohol from accuracy results.

squares were calculated, and the results were found to be within the acceptance criteria (Table 7).The linearity graphs from accuracy results for acetone and isopropyl alcohol are shown in Figures 12 and 13, respectively.

3.2.7. *Ruggedness (Intermediate Precision).* Intermediate precision was performed by preparing the six replicate test preparations ($n = 6$) of dipyridamole 200 mg

TABLE 7: Linearity data of acetone and isopropyl alcohol.

Description	Dipyridamole and known impurities	
	Acetone	Isopropyl alcohol
Square of correlation coefficient (R^2)	0.998	0.997
Slope	1385.9	357.7
Y-intercept	100,186.4	39,555.8
% Y-intercept	1.4	2.1

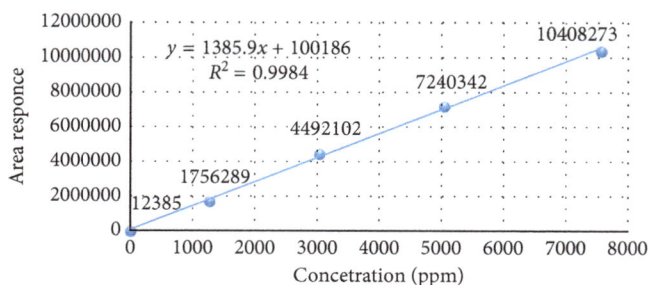

FIGURE 12: Linearity plot for acetone.

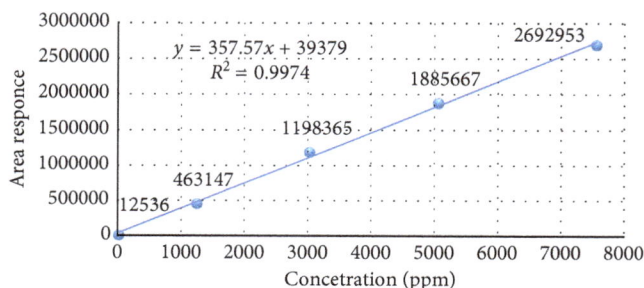

FIGURE 13: Linearity plot for isopropyl alcohol.

modified release capsules by spiking acetone and isopropyl alcohol at specification level and analyzed as per the test method by using different Headspace gas chromatography system, different column of same make by different analyst on different day. The concentration in parts per million was calculated and found to be within the acceptance criteria. The overall concentration in parts per million for replicate preparations ($n = 12$) of method precision and intermediate precision was calculated and found to be within the acceptance criteria (Table 8). The relative standard deviations obtained for acetone was 1.4% and isopropyl alcohol was 3.1%.

3.2.8. Solution Stability. The solution stability of acetone and isopropyl alcohol was determined by keeping sample solution and standard solutions at room temperature for 1 day and 2 day and measured against freshly prepared standard solution. The standard solution and sample solutions were found stable for 2 days at room temperature.

TABLE 8: Ruggedness data.

Impurity	% RSD for six individual preparation	The overall % RSD ($n = 12$)
Acetone	0.6	1.4
Isopropyl alcohol	0.7	3.1

3.2.9. Robustness. Robustness of the proposed method was performed by keeping the chromatographic conditions constant with the following deliberate variations:

(i) Change in carrier gas flow rate

(ii) Change in column oven temperature

(iii) Change in headspace sampler vial equilibration time

(iv) Change in headspace vial oven temperature

The standard solution was injected six times in replicate for each abovementioned change. The system suitability parameters like % relative standard deviation for area response and % relative standard deviation for retention time

TABLE 9: Robustness data.

Parameter variation	The present relative standard deviation of peak area for six replicate injections should not more than 10.0		The present relative standard deviation of retention time for six replicate injections should not more than 10.0	
	Acetone	Isopropyl alcohol	Acetone	Isopropyl alcohol
Flow 0.8 mL/min	0.4	0.6	0.0	0.0
Flow 1.2 mL/min	0.5	0.7	0.1	0.0
Column oven temperature 35°C	0.3	1.0	0.0	0.0
Column oven temperature 45°C	0.4	0.6	0.0	0.0
Headspace vial equilibration time 15 min	1.2	3.1	0.0	0.0
Headspace vial equilibration time 25 min	0.2	0.7	0.0	0.1
Headspace vial oven temperature 75°C	0.5	0.8	0.0	0.1
Headspace vial oven temperature 85°C	0.2	0.4	0.0	0.0

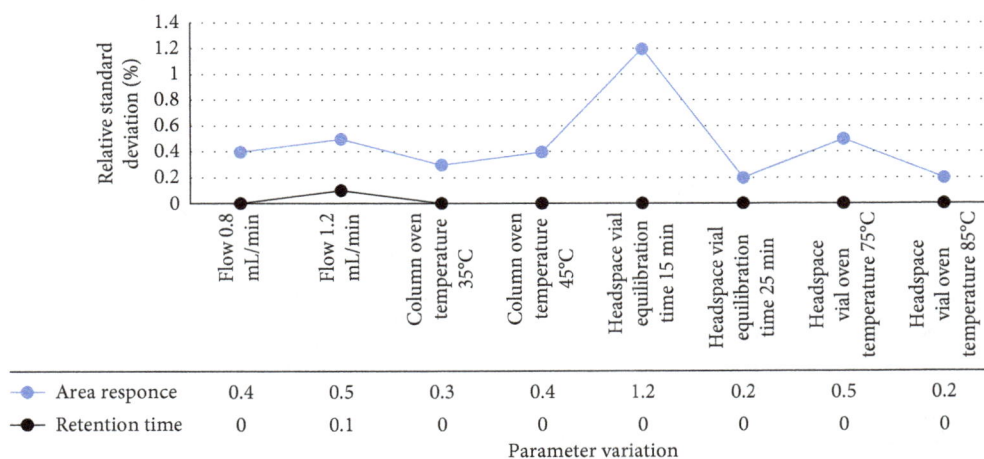

	Flow 0.8 mL/min	Flow 1.2 mL/min	Column oven temperature 35°C	Column oven temperature 45°C	Headspace vial equilibration time 15 min	Headspace vial equilibration time 25 min	Headspace vial oven temperature 75°C	Headspace vial oven temperature 85°C
Area responce	0.4	0.5	0.3	0.4	1.2	0.2	0.5	0.2
Retention time	0	0.1	0	0	0	0	0	0

FIGURE 14: Robustness data for acetone.

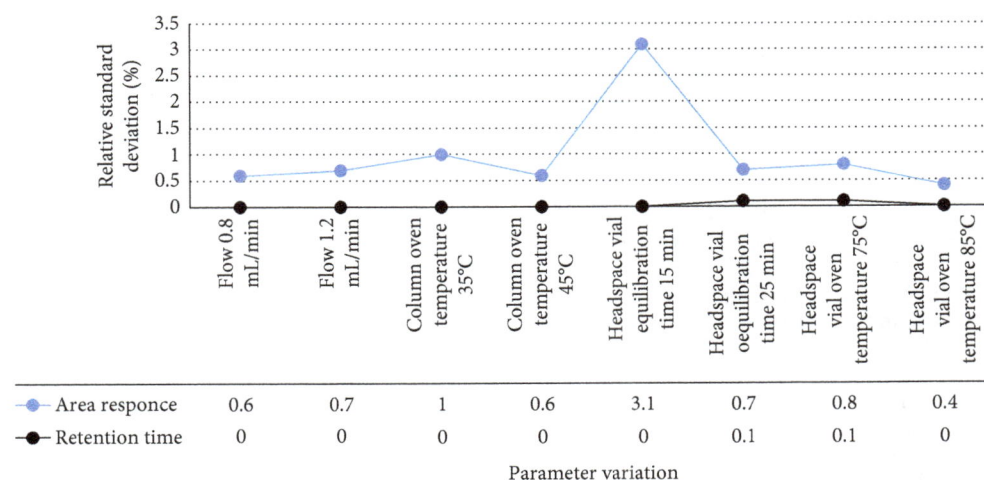

	Flow 0.8 mL/min	Flow 1.2 mL/min	Column oven temperature 35°C	Column oven temperature 45°C	Headspace vial equilibration time 15 min	Headspace vial oequilibration time 25 min	Headspace vial oven temperature 75°C	Headspace vial oven temperature 85°C
Area responce	0.6	0.7	1	0.6	3.1	0.7	0.8	0.4
Retention time	0	0	0	0	0	0.1	0.1	0

FIGURE 15: Robustness data for isopropyl alcohol.

were recorded for acetone and isopropyl alcohol and found well within the acceptance criteria. The results are given in Table 9, and the graphs for acetone and isopropyl alcohol are shown in Figures 14 and 15, respectively.

3.2.10. Application of the Proposed Method. The developed analytical method was applied to the analysis of real samples from the manufacturing unit. All the analytical validation parameters could be confirmed, and the method was proven to be suitable for routine analysis regarding rapid and accurate results.

4. Conclusions

A simple, sensitive, accurate, robust headspace gas chromatographic method was developed for the quantitative determination of residual solvents in tartaric acid-based pellets of dipyridamole modified release capsules. The proposed method was validated and found to be precise, accurate, linear, robust, and rugged, and all the validation parameter results were found satisfactory. The described method is suitable for routine analysis of production samples at laboratories.

Conflicts of Interest

The authors declare that there are no conflicts of interest regarding the publication of this paper.

Acknowledgments

The authors wish to thank the management of Bluefish Pharmaceuticals Private Limited for supporting and encouragement.

References

[1] V. Karuppiah, N. Kannappan, and R. Manavalan, "In-vitro and in-vivo dissolution of dipyridamole extended release capsules," *International Journal of Pharmaceutical Sciences Review and Research*, vol. 13, no. 1, 2012.

[2] S. U. Ahmed, P. R. Katikaneni, and Y. Zu, "Pharmaceutical capsules comprising extended release dipyridamole pellets," WO2009097156A1, 2009.

[3] Persantin Retard 200 mg, Summary of Product Characteristics (SmPC), The electronic Medicines Compendium (emc), https://www.medicines.org.uk/emc/product/897/smpc.

[4] S. S. Raju, U. Vidyamani, P. Jayapal, and P. Naga Raju, "Development and validation of a head space gas chromatographic method for the determination of ethylene oxide content in dipyridamole API," *World Journal of Pharmaceutical Research*, vol. 4, no. 11, pp. 1127–1139.

[5] A. R. Zoest, J. E. Watson, C. T. Hung, and S. A. Wanwimolruk, "Rapid isocratic HPLC assay for dipyridamole using a microbore column technique," *Journal of Liquid Chromatography*, vol. 14, no. 10, pp. 1967–1975, 1991.

[6] A. Sreedhara Rao, M. Krishnaji Rao, A. S. Dadichand, A. M. L. Punna Rao, and B. Balaswami, "Development and validation of RP-HPLC method for assay of dipyridamole in formulations," vol. 8, no. 5, pp. 256–259, 2016.

[7] J. H. Bridle and M. T. Brimble, "A stability indicating method for dipyridamole," *Informa Healthcare Drug Development and Industrial Pharmacy*, vol. 19, no. 3, pp. 371–381, 1993.

[8] J. Zhang, R. B. Miller, and R. Jacobus, "Development and validation of a stability-indicating HPLC method for the determination of degradation products in dipyridamole injection," *Chromatographia*, vol. 44, no. 5-6, pp. 247–252, 1997.

[9] B. K. Vagela, S. Singh Rao, and P. Sunil Reddy, "Development and validation of a stability-indicating RP-LC method for the estimation of process related impurities and degradation products of dipyridamole retard capsules," *International Journal of Pharmacy and Pharmaceutical Sciences*, vol. 4, no. 1, 2012.

[10] K. Prakash, R. Rao Kalakuntla, and J. Reddy Sama, "Rapid and simultaneous determination of aspirin and dipyridamole in pharmaceutical formulations by reversed- phase high performance liquid chromatography (RP-HPLC) method," *African Journal of Pharmacy and Pharmacology*, vol. 5, no. 2, pp. 244–251, 2011.

[11] A. P. Rajput and C. M. Sonanis, "Development and validation of a rapid RP-UPLC method for the determination of aspirin and dipyridamole in combined capsule formulation," *International Journal of Pharmacy and Pharmaceutical Sciences*, vol. 3, no. 2, 2011.

[12] H. H. Hammud, F. A. Yazbib El, M. E. Mahrousc, G. M. Sonjib, and N. M. Sonjib, "Stability-indicating spectro fluorimetric and RP-HPLC methods for the determination of aspirin and dipyridamole in their combination," *Open Spectroscopy Journal*, vol. 2, no. 1, pp. 19–28, 2008.

[13] Z. Kopitar and H. Weisenberger, "Specific binding of dipyridamol on human serum protein. Isolation, identification and characterization as alpha-1-acidic glycoprotein," *Arzneimittel-Forschung*, vol. 21, no. 6, pp. 859–862, 1971.

[14] D. B. Bandarabadi, M. P. Hamedani, M. Amini, and A. Shafiee, "High performance liquid chromatographic method for determination of dipyridamole in human plasma," *DARU Journal of Pharmaceutical Sciences*, vol. 7, no. 2, 1999.

[15] T. Qin, F. Qin, N. Li, S. Lu, W. Liu, and F. Li, "Quantitative determination of dipyridamole in human plasma by high-performance liquid chromatography–tandem mass spectrometry and its application to a pharmacokinetic study," *Biomedical Chromatography*, vol. 24, no. 3, pp. 268–273, 2010.

[16] ICH Q2 (R1), "Validation of analytical procedures: text and methodology", Harmonised Tripartite Guideline, in *Proceedings of the International Conference on Harmonisation of Technical Requirements for Registration of Pharmaceuticals for Human Use*, Chicago, USA, 2005.

[17] P. S. Reddy, V. S. K. Jagarlapudi, and C. B. Sekharan, "Determination of edoxaban in bulk and in tablet dosage form by stability indicating high-performance liquid chromatography," *Pharmaceutical Sciences*, vol. 22, no. 1, pp. 35–41, 2016.

[18] V. K. Nekkala, J. S. Kumar, D. Ramachandran, and G. Ramanaiah, "Development and validation of stability indicating RP-LC method for estimation of calcium dobesilate in pharmaceutical formulations," *Der Pharmacia Lettre*, vol. 8, no. 11, pp. 236–242, 2016.

14

Development of Simultaneous Analysis of Thirteen Bioactive Compounds in So-Cheong-Ryong-Tang using UPLC-DAD

Ji Hyun Jeong⑩, **Seon Yu Lee**⑩, **Bo Na Kim**⑩, **Guk Yeo Lee**⑩, and **Seong Ho Ham**⑩

National Development Institute of Korean Medicine, Udae land gil 288, Jangheung-gun, Jeollanam-do 59338, Republic of Korea

Correspondence should be addressed to Seong Ho Ham; phd_ham@nikom.or.kr

Academic Editor: Pablo Richter

So-Cheong-Ryong-Tang, which is a standardized Korean medicine of the National Health Insurance, is a traditional prescription for the treatment of allergic rhinitis, bronchitis, and bronchial asthma. Simultaneous analysis and development of SCRT is essential for its stability, efficacy, and risk management. In this study, a simple, reliable, and accurate method using ultrahigh-performance liquid chromatography (UPLC) fingerprinting with a diode array detector (DAD) was developed for the simultaneous analysis. The chromatographic separation of the analytes was performed by an ACQUITY UPLC BEH C18 column (1.7 μM, 2.1 × 100 mm, Waters) with a mobile phase of water containing 0.01% (v/v) phosphoric acid and acetonitrile containing 0.01% (v/v) phosphoric acid. The flow rate and detection wavelength were set at 0.4 mL/min and 215, 230, 254, and 280 nm. All calibration curves of the thirteen components showed good linearity ($R^2 > 0.999$). The limit of detection and limit of quantification ranged 0.001–0.360 and 0.004–1.200 μg/mL, respectively. The relative standard deviation (RSD) of intra- and interday was less than 2.60%, and the recoveries were within the range 76.08–103.79% with an RSD value of 0.03–1.50%. The results showed that the developed method was simple, reliable, accurate, sensitive, and precise for the quantification of bioactive components of SCRT.

1. Introduction

Traditional Korean medicines, because of their high effectiveness and low toxicity, have been used for thousands of years for the prevention and treatment of various kinds of human diseases. Various ingredients in these herbs cause the efficacy of traditional Korean medicines, which consist of many herbal combinations. Therefore, the consistency of the composition and proportion of the composition are the key to quality control in safety, efficacy, and risk management. In general, analyzing a single marker compound is simple and convenient, but it does not provide sufficient quantitative information on other components in Korean medicines. Thus, over the past decade, the chromatographic fingerprinting method has been considered one of the most important and acceptable approaches for the identification and quality evaluation of Korean medicines [1, 2]. So-Cheong-Ryong-Tang, standardized as a Korean medicine of the National Health Insurance, is a traditional prescription for the treatment of allergic rhinitis, bronchitis, and bronchial

asthma [3]. Recently, SCRT was reported to show therapeutic effects in *in vivo* experiments on the respiratory system for allergic rhinitis and asthma [3–9]. So-Cheong-Ryong-Tang (SCRT, Xiao-Qing-Long-Tang in Chinese, Shoseiryu-to in Japanese) is composed of eight herbal preparations (*Ephedrae herba, Paeoniae radix, Glycyrrhizae radix, Zingiberis rhizoma, Cinnamomi ramulus, Schisandrae fructus, Pinelliae rhizoma*, and *Asiasari radix*) [4–6, 8].

To optimize the quality control of SCRT, thirteen bioactive compounds from eight herbal preparations were chosen. Among the 13 standard compounds, ephedrine and catechin were found in *Ephedrae herba*, which is known for its efficacy as a sympathomimetic and its antiobesity effects [10]. Albiflorin, paeoniflorin, benzoic acid, PGG, and methyl gallate are the major constituents of *Paeoniae radix*, which has anti-inflammatory, analgesic, antispasmodic, liver protection, and immune regulatory functions [11, 12]. Liquiritin, Liquiritin apioside, and Glycyrrhizin from *Glycyrrhizae Radix* were used, which is an effective detoxifying agent, presenting neuroprotective effect, antiviral activity, and

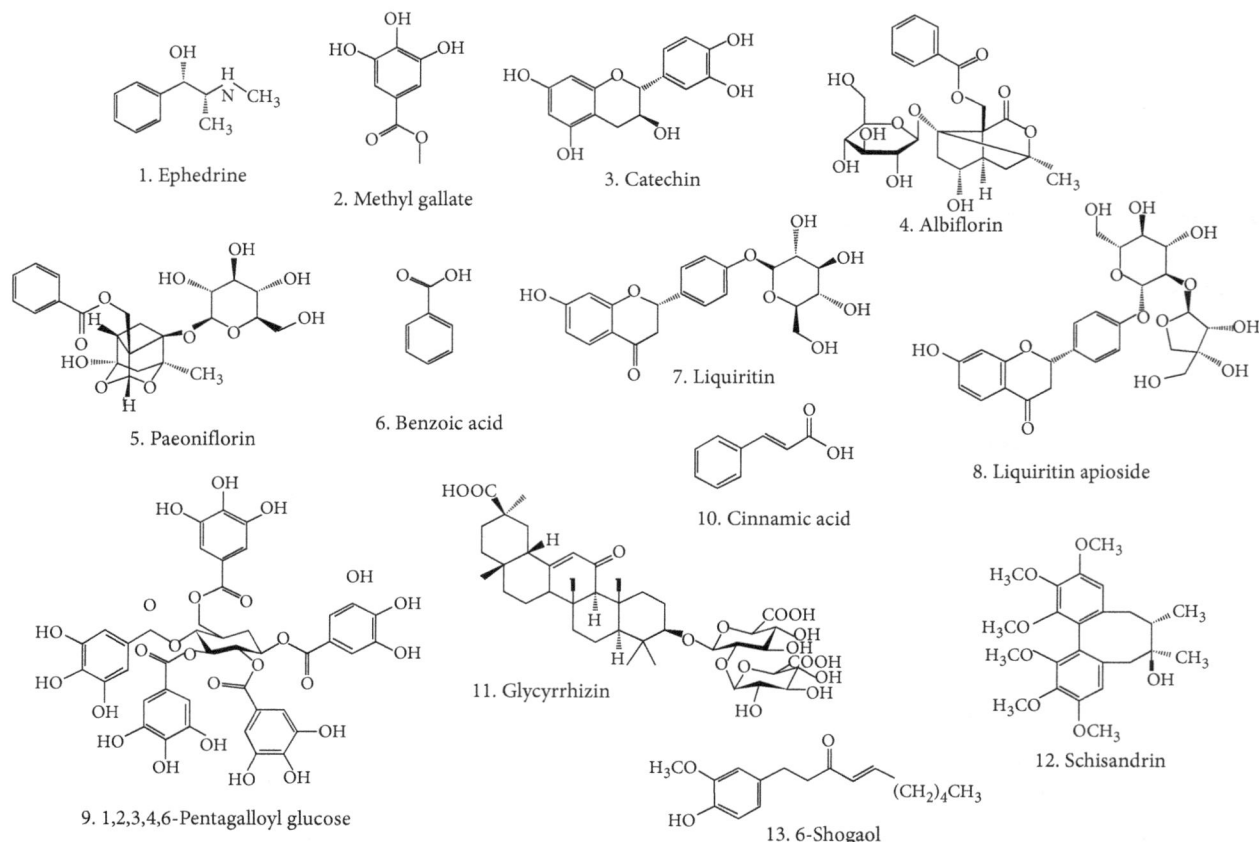

FIGURE 1: Chemical structures of the 13 marker compounds.

anti-inflammatory, antitumor, and antibiosis effects [13]. 6-Shogaol from *Zingiberis Rhizoma* was found to have various pharmacological activities, including antioxidative, anti-tumorigenic, and immunomodulatory effects, and is an effective antimicrobial and antiviral agent [14]. *Cinnamomi ramulus*, which includes cinnamic acid, has been found to be able to effectively attenuate influenza virus, inflammations, human platelet aggregation, and arachidonic acid meta-bolism [15] and is known for its antimicrobial activity against [16]. The major constituent of *Schisandrae fructus*, which includes schisandrin, was found to have liver protective, hypoglycemic, antioxidant, antiaging, immune regulatory, an-titumor, and bactericidal effects and plays a role in regulating the central nervous system [17]. *Pinelliae rhizoma* has antitussive, antiemetic, glandular secretion-inhibiting, and antitumor effects [18]. *Asiasari radix* has been used as an analgesic, antitussive, or antiallergic agent [19].

Several studies of these compounds have been developed for qualitative and quantitative analyses using high-performance liquid chromatography-diode array detector (HPLC-DAD) and mass spectrometry (HPLC-ESI-MS) [4, 5, 9]. However, these methods cannot offer simulta-neous analysis of the multiple bioactive compounds in SCRT. Although an HPLC method for simultaneous de-termination of the four marker constituents of SCRT has been developed, there are limitations to the quantitative and qualitative analyses of many compounds in SCRT. Therefore,

methods for simultaneously detecting these biomarkers in SCRT are essential to ensure efficient quality control and pharmaceutical evaluation. In this study, it was necessary to find more accurate, efficient, and stable solvent extraction conditions prior to simultaneous analysis. We performed simultaneous determination of the thirteen marker compounds.

2. Experimental Materials and Reagents

2.1. Chemicals and Reagents. Ephedrine, methyl gallate, catechin, albiflorin, paeoniflorin, benzoic acid, liquiritin, liquiritin apioside, 1,2,3,4,6-pentagalloyl glucose (PGG), cinnamic acid, glycyrrhizin, schisandrin, and 6-shogaol were purchased from Sigma-Aldrich Co. (St. Louis, MO, USA). The purity of all standards was >97%. Figure 1 shows the chemical structures of the thirteen bioactive compounds. HPLC-grade acetonitrile and methanol were purchased from J. T. Baker Inc. (Phillipsburg, NJ, USA). Deionized water was prepared using an ultrapure water production apparatus (Human Corporation, Seoul, Korea). SCRT medicines were purchased from Kyoungbang Medicinal Herbs (Incheon, Korea).

2.2. Preparation of Standard and Sample Solutions. Standard stock solutions of the thirteen bioactive standards—ephedrine,

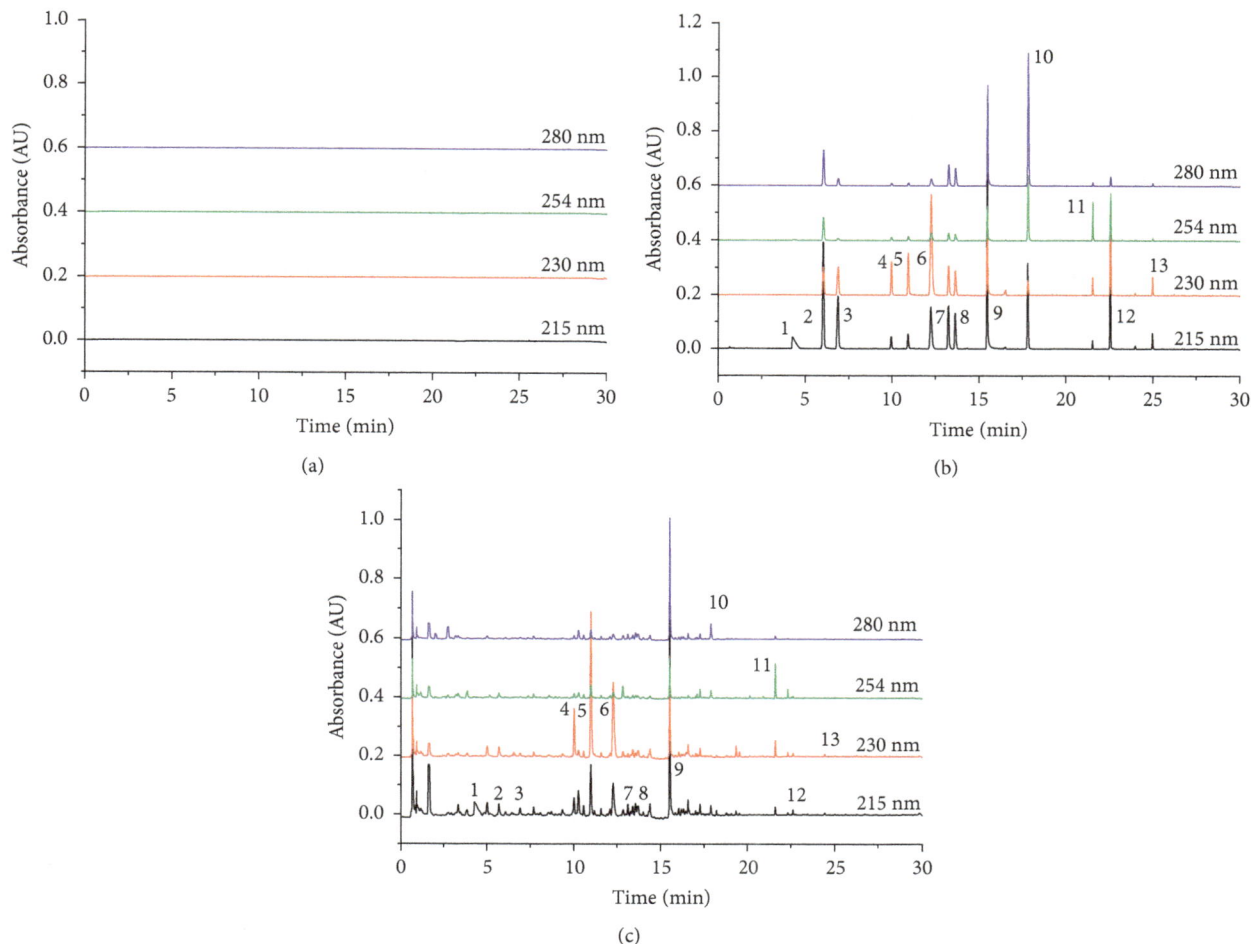

FIGURE 2: UPLC chromatograms of the (a) blank solvent, (b) standard mixture, and (c) SCRT extract.

methyl gallate, catechin, albiflorin, paeoniflorin, benzoic acid, liquiritin, liquiritin apioside, 1,2,3,4,6-pentagalloyl glucose (PGG), cinnamic acid, glycyrrhizin, schisandrin, and 6-shogaol—were prepared by accurately weighing appropriate amounts of reference compounds and dissolving in methanol. The thirteen bioactive standards were mixed in stock solutions and then diluted serially to seven concentrations for the construction of calibration curves. All the solutions were stored at 4°C.

2.3. Extraction Method.

The herbal medicine prepared by water extraction contains a water-soluble component and some lipoid-soluble substances, and most of the high molecular weight polymers are contained in a suspended state. It is necessary to estimate and optimize the extraction conditions for the 13 marker components in the SCRT sample. In this study, the liquid extraction method was selected, and aqueous methanol (20, 50, and 80%) was tried and examined as the extraction solvent to evaluate the optimal extraction solvent. Second, the volume (50 and 100 mL) of the extraction solvent was investigated. Finally, extraction methods using ultrasonic or reflux were investigated [20–22].

2.4. Preparation of Sample Solution.

The SCRT powder (2.4 g, 1 dose) was weighed precisely and extracted with 80% methanol-water (v/v) solution in an ultrasonic water bath for 10 min at room temperature. Then, the samples were refluxed twice at 80°C for 30 min, followed by filtration and making up to volume in a volumetric flask. The samples were centrifuged (4,000 rpm, 10 min, 18°C), and the supernatant was filtered with a 0.2 μm membrane filter, prior to injection. All working solutions and sample solutions were stored at 4°C before use.

2.5. Chromatographic Conditions by UPLC-DAD.

The simultaneous determination of thirteen bioactive compounds in SCRT was performed on the UPLC-DAD system equipped with a pump, an autosampler, and a photodiode array detector, and the amount of data were calculated using Empower software. Chromatographic separation was carried out using an ACQUITY UPLC BEH C18 column (1.7 μm, 2.1 × 100 mm, Waters), and the column temperature was kept at 40°C. The mobile phase consisted of (A) water (0.01% phosphoric acid, v/v) and (B) acetonitrile (0.01% phosphoric acid, v/v). The gradient solvent was optimized and performed as 98% A (0-1 min), 98–84% A (1–14 min),

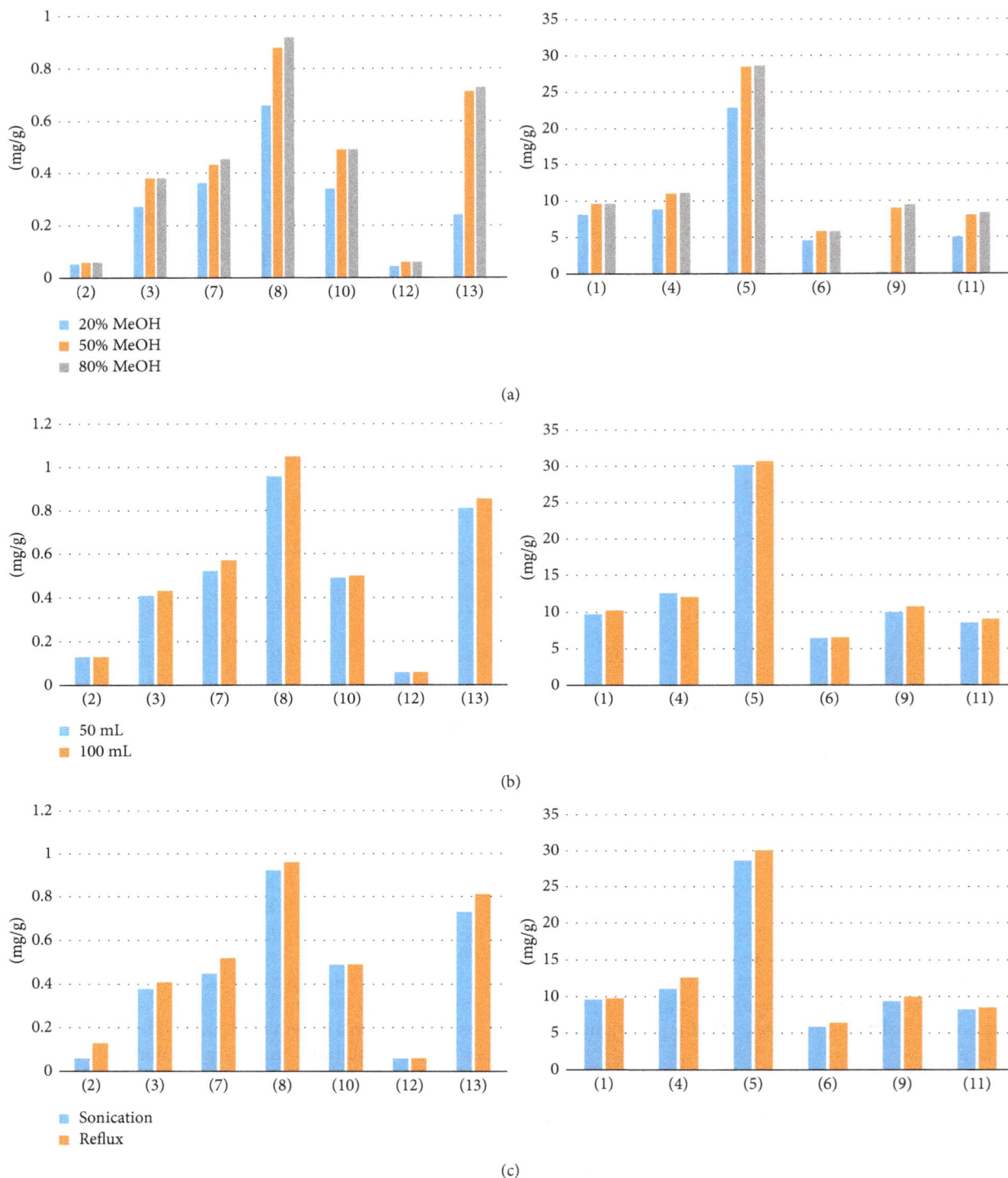

FIGURE 3: Efficiencies of the extraction for the nine compounds in SCRT using different (a) extraction solvents, (b) volumes of the extraction solvent, and (c) extraction methods.

84–60% A (14–21 min), 60–20% A (21–28 min), 20% A (28–29 min), and 20–98% A (29-29.5 min), at a flow rate of 0.4 mL/min. The detection wavelengths for analytes were set at 215, 230, 254, and 280 nm, and the injection volume of each sample was 2 μL.

2.6. Method Validation. The method was validated for linearity, limit of detection (LOD), limit of quantification (LOQ), specificity, precision (interday, intraday, and repeatability), and accuracy (recovery), following the guideline on Bioanalytical Method Validation [23–26].

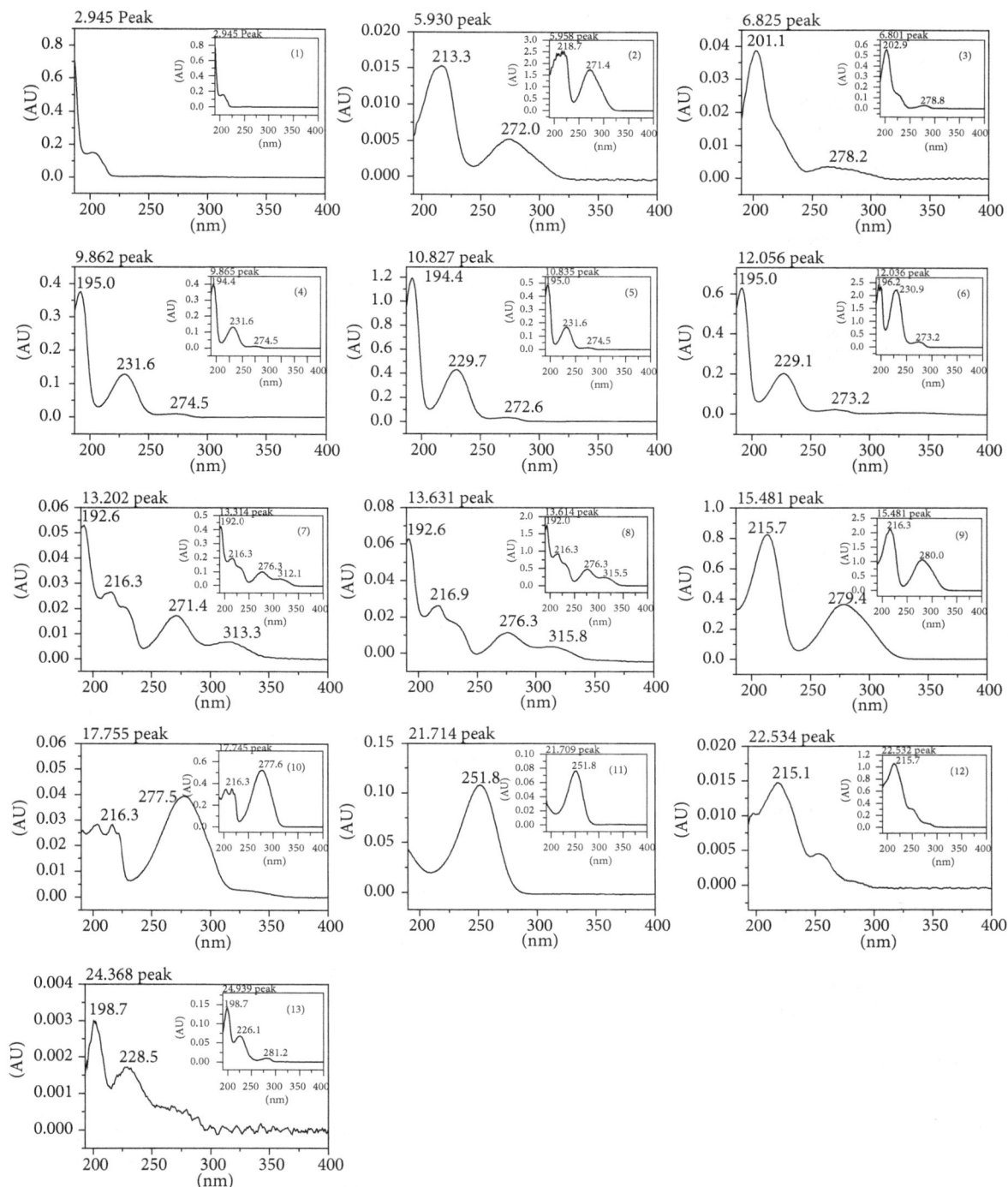

FIGURE 4: UV spectra of the thirteen marker compounds in SCRT corresponding to the standard solution (inset).

2.6.1. Linearity, Limits of Detection (LODs), and Limits of Quantification (LOQs).

The standard solutions of the 13 compounds were prepared by serially diluting the stock solution to appropriate concentrations for plotting the calibration curves [28, 29]. These solutions of the 13 compounds were analyzed in triplicate [28]. The calibration standard curve was plotted with the peak area (*y*-axis) versus concentration (*x*-axis) for each analyte in that range [30]. All calibration curves were required to have a correlation value of at least 0.995. The limit of detection (LOD) and limit of quantification (LOQ) were determined on the basis of signal-to-noise ratio (S/N):

$$LOD = \frac{amount * 3.3}{(S/N)},$$

$$LOQ = \frac{amount * 10}{(S/N)}.$$

(1)

TABLE 1: Regression data, LODs, and LOQs for the marker compounds analyzed by UPLC.

	Compound	Regression equation	R^2	LOQ (μg/mL)	LOD (μg/mL)
1	Ephedrine	$y = 6{,}569.4x - 2{,}539.7$	0.9996	0.070	0.233
2	Methyl gallate	$y = 49{,}631x - 14{,}901$	0.9995	0.007	0.026
3	Catechin	$y = 28{,}138x - 4{,}098.4$	0.9992	0.008	0.029
4	Albiflorin	$y = 6{,}199.8x - 2{,}341.7$	0.9996	0.040	0.134
5	Paeoniflorin	$y = 8{,}506.1x - 1{,}169.5$	0.9994	0.018	0.062
6	Benzoic acid	$y = 30{,}844x - 11{,}922$	0.9995	0.021	0.070
7	Liquiritin	$y = 19{,}772x - 3{,}660$	0.9995	0.017	0.056
8	Liquiritin apioside	$y = 15{,}249x - 1{,}630.9$	0.9996	0.145	0.483
9	1,2,3,4,6-Pentagalloyl glucose	$y = 29{,}863x - 17{,}029$	0.9994	0.009	0.031
10	Cinnamic acid	$y = 25{,}479x - 5{,}716.6$	0.9994	0.008	0.027
11	Glycyrrhizin	$y = 3{,}546.4x - 422.55$	0.9994	0.073	0.244
12	Schisandrin	$y = 86{,}584x - 4{,}328.7$	0.9995	0.001	0.004
13	6-Shogaol	$y = 1{,}025.4x + 1{,}176.6$	0.9988	0.360	1.200

R^2 = correlation coefficient; y = peak area; x = sample concentration (μg/mL).

2.6.2. Precision, Accuracy, and Recovery. The precision of the method was evaluated by both intra- and interday tests. Three different concentrations of each biomarker in five replicates on the same day (intraday) and on three consecutive days (interday) were prepared to verify the precision and accuracy of the analytical method [30]. The precision was expressed as relative standard deviation (RSD, %); a value of RSD within ±15% is generally considered acceptable [28]. The accuracy of the assay is the closeness of the observed concentration to the nominal concentration [31]. The recoveries of analytes were determined by adding different concentrations of the 13 marker components into the SCRT sample solution (2.4 mg/mL). Recovery (%) was calculated with the following equation:

$$\text{Recovery}(\%) = \frac{(\text{amount found} - \text{original amount})}{\text{amount spiked}} \times 100\%.$$

(2)

3. Results and Discussion

3.1. Optimization of Chromatographic Condition. A chromatogram of SCRT was obtained using an UPLC-PDA [4]. Figure 2 shows typical chromatograms corresponding to the mixed standard and SCRT. To obtain accurate, valid, and optimal separation, the UPLC conditions were investigated with regard to column, mobile phase (water-acetonitrile with different modifiers including acetic acid, formic acid, and phosphoric acid), detection wavelength (215, 230, 254, and 280 nm), and mobile-phase flow rate (0.5, 0.4, 0.3, and 0.2 mL/min) [32] The best UPLC conditions were obtained from the ACQUITY UPLC BEH C18 column, which had better resolution than the others. The gradient solvent system consisted of 0.01% phosphoric acid in water (A) and 0.01% phosphoric acid in acetonitrile (B), at a column temperature of 40°C, with a flow rate of 0.4 mL/min. Four detection wavelengths (215, 230, 254, and 280 nm) were finally selected to achieve the goal of high detection sensitivity and small interference because the maximum absorption of the 13 reference compounds was different.

3.2. Optimization of the Sample Extraction Protocol. The extraction conditions, for example, extraction solvent,

method, and volume, can easily influence the efficiency of the extraction. In this paper, aqueous methanol (20, 50, and 80%) was examined as the extraction solvent for SCRT by ultrasonication for 30 min [33]. Figure 3(a) shows the results, which indicate that 80% methanol was the best extraction solvent. Second, the volume of solvent (50 and 100 mL) was investigated, and Figure 3(b) shows the results. The volume of the extraction solvent was 100 mL. The extraction efficiency was better at 100 mL volume in low-content compounds. Finally, Figure 3(c) shows the results of the extraction method using ultrasound and reflux, which indicate that the extraction method using reflux showed high efficiency. In the end, the suitable extraction conditions were as follows: the samples were extracted by reflux with 80% methanol in a volume of 100 mL.

3.3. Method Validation

3.3.1. Specificity. The specificity was determined by comparing the peak purity of the 13 markers with the extracted samples and the standard. Figure 4 shows the UV spectra of individual marker compounds, which confirm that the peaks are pure and there is no interference from the impurities [34].

3.3.2. Linearity, LOD, and LOQ. Table 1 summarizes the calibration curves, LOQ, and LOD of the thirteen analytes. The linearity of the developed method was assessed using seven concentrations: ephedrine, albiflorin, paeoniflorin, benzoic acid, PGG, glycyrrhizin (0.1~100 μg/mL), methyl gallate, catechin, liquiritin, liquiritin apioside, cinnamic acid (0.05~50 μg/mL), and schisandrin (0.02~20 μg/mL) with correlation coefficients $R^2 \geq 0.999$. The values of LOD and LOQ were in the ranges of 0.001–0.360 μg/mL and 0.004–1.200 μg/mL, respectively. The results showed that the calibration curves were within the adequate range and exhibited good sensitivity for the analysis of the thirteen bioactive components [28].

3.3.3. Precision, Accuracy, and Recovery. Table 2 shows the results of the intra- and interday precision tests. The values of RSD (%) for intra- and interday tests were within the

TABLE 2: Precision (intra- and interday) and accuracy of the thirteen analytes.

Analyte concentration (μg/mL)	Intraday ($n = 5$)			Interday ($n = 3$)		
	Measured amount (mean ± SD, Dg/mL)	RSD (%)	Accuracy (%)	Measured amount (mean ± SD, Dg/mL)	RSD (%)	Accuracy (%)
Ephedrine						
25	5.03 ± 0.02	0.43	100.59	4.97 ± 0.04	0.82	99.30
50	25.59 ± 0.01	0.03	102.36	25.33 ± 0.23	0.90	101.31
100	100.46 ± 0.08	0.08	100.46	100.46 ± 0.31	0.31	100.46
Methyl gallate						
2.5	2.48 ± 0.04	1.49	99.36	2.51 ± 0.01	0.35	100.57
12.5	12.29 ± 0.00	0.03	98.34	12.44 ± 0.12	1.00	99.48
50	49.14 ± 0.31	0.63	98.27	50.03 ± 1.00	1.99	100.06
Catechin						
2.5	2.53 ± 0.00	0.12	101.09	2.52 ± 0.02	0.75	100.65
12.5	12.50 ± 0.00	0.02	100.00	12.50 ± 0.00	0.02	100.02
50	49.69 ± 0.21	0.43	99.38	50.14 ± 0.48	0.96	100.28
Albiflorin						
5	5.02 ± 0.03	0.61	100.36	5.12 ± 0.07	1.42	102.42
25	25.05 ± 0.04	0.15	100.20	25.12 ± 0.11	0.43	100.47
100	98.68 ± 0.16	0.16	98.68	99.78 ± 0.80	0.80	99.78
Paeoniflorin						
5	5.05 ± 0.10	1.94	101.10	5.13 ± 0.04	0.73	102.62
25	25.24 ± 0.08	0.30	100.96	25.11 ± 0.27	1.07	100.44
100	101.46 ± 1.96	0.93	101.46	100.46 ± 0.15	0.15	100.46
Benzoic acid						
1	1.01 ± 0.01	1.35	100.92	1.01 ± 0.02	1.61	101.28
5	5.02 ± 0.02	0.39	100.31	4.99 ± 0.04	0.80	99.80
100	96.82 ± 2.43	2.51	96.82	100.59 ± 0.28	0.28	100.59
Liquiritin						
2.5	2.49 ± 0.01	0.57	99.64	2.44 ± 0.04	1.64	97.67
12.5	12.14 ± 0.01	0.05	97.10	12.37 ± 0.21	1.66	98.99
50	50.20 ± 0.29	0.58	100.39	50.17 ± 0.48	0.96	100.35
Liquiritin apioside						
2.5	2.53 ± 0.03	1.08	101.11	2.51 ± 0.00	0.16	100.45
12.5	12.56 ± 0.00	0.02	100.48	12.51 ± 0.04	0.34	100.07
50	49.91 ± 0.36	0.72	99.81	49.95 ± 0.09	0.17	99.90
1,2,3,4,6-Pentagalloyl glucose						
5	4.94 ± 0.07	1.45	98.73	5.05 ± 0.03	0.68	101.00
25	25.26 ± 0.02	0.08	101.04	25.15 ± 0.10	0.38	100.60
100	99.89 ± 0.45	0.45	99.89	100.27 ± 0.37	0.37	100.27
Cinnamic acid						
0.5	0.50 ± 0.00	0.38	100.16	0.50 ± 0.00	0.33	99.45
12.5	12.50 ± 0.00	0.02	99.96	12.51 ± 0.02	0.13	100.06
50	50.03 ± 0.01	0.02	100.06	50.15 ± 0.19	0.38	100.31
Glycyrrhizin						
5	5.03 ± 0.02	0.30	100.62	4.99 ± 0.11	2.27	99.84
25	25.04 ± 0.11	0.42	100.14	25.09 ± 0.01	0.05	100.35
100	100.63 ± 0.30	0.29	100.63	100.65 ± 0.17	0.17	100.65
Schisandrin						
1	1.00 ± 0.00	0.11	100.03	1.00 ± 0.00	0.40	100.40
5	5.09 ± 0.00	0.01	101.83	5.04 ± 0.05	0.90	100.89
20	19.33 ± 0.42	2.17	96.64	20.12 ± 0.18	0.87	100.60
6-Shogaol						
2	2.04 ± 0.04	2.00	101.93	2.05 ± 0.05	2.60	102.50
50	50.36 ± 0.04	0.09	100.71	50.27 ± 0.18	0.35	100.54
200	197.59 ± 0.09	0.05	98.79	199.67 ± 1.72	0.86	99.83

SD = standard deviation; RSD (%) = (SD/mean) × 100; accuracy (%) = (C_{obs}/C_{nom}) × 100.

TABLE 3: Determination of recoveries of the 13 compounds (1–13) in SCRT.

Compounds	Spiked amount	Measured amount	Recovery (%)	RSD (%)
Ephedrine	25	25.20 ± 0.19	100.78	0.73
	50	51.86 ± 0.78	103.72	1.50
	100	100.23 ± 0.59	100.23	0.59
Methyl gallate	5	4.47 ± 0.01	89.33	0.30
	12.5	10.94 ± 0.04	87.53	0.33
	50	45.63 ± 0.03	91.26	0.07
Catechin	5	4.98 ± 0.02	99.53	0.40
	12.5	11.42 ± 0.09	91.39	0.76
	50	45.53 ± 0.30	91.05	0.65
Albiflorin	5	5.15 ± 0.02	103.09	0.36
	25	25.90 ± 0.11	103.59	0.42
	100	98.23 ± 1.06	98.23	1.08
Paeoniflorin	10	10.05 ± 0.02	100.54	0.21
	25	24.64 ± 0.08	98.55	0.34
	100	100.97 ± 0.35	100.97	0.35
Benzoic acid	25	24.43 ± 0.09	97.70	0.39
	50	47.29 ± 0.17	94.59	0.37
	100	97.28 ± 0.10	97.28	0.10
Liquiritin	5	5.05 ± 0.05	100.93	0.91
	12.5	12.54 ± 0.12	100.31	0.97
	50	42.98 ± 0.22	85.96	0.52
Liquiritin apioside	5	4.94 ± 0.04	98.79	0.79
	12.5	12.55 ± 0.03	100.43	0.27
	50	48.99 ± 0.45	97.98	0.93
1,2,3,4,6-Pentagalloyl glucose	10	10.00 ± 0.13	99.98	1.30
	50	46.89 ± 0.25	93.77	0.53
	100	99.17 ± 0.56	99.17	0.56
Cinnamic acid	12.5	12.61 ± 0.06	100.88	0.49
	25	24.70 ± 0.02	98.81	0.09
	50	52.28 ± 0.22	104.57	0.42
Glycyrrhizin	25	24.13 ± 0.16	96.54	0.66
	50	50.36 ± 0.14	100.72	0.27
	100	102.79 ± 0.52	102.79	0.50
Schisandrin	1	0.98 ± 0.01	97.78	0.87
	5	4.65 ± 0.04	92.93	0.77
	20	17.98 ± 0.03	89.90	0.18
6-Shogaol	20	15.22 ± 0.02	76.08	0.14
	100	85.23 ± 0.05	85.23	0.06
	200	177.66 ± 0.05	88.83	0.03

ranges of 0.01–2.51% and 0.02–2.60%, with accuracy from 96.82 to 102.36% and from 97.67 to 102.62%, respectively [2]. The results indicated that the accuracy and precision of the proposed method were accurate and reliable for determination of the thirteen compounds in the sample of SCRT [28]. The average recovery (%) of the 13 marker compounds shows the accuracy of this analytical method. The recovery ranged from 76.08 to 103.79%, and the values of RSD were in the range 0.03–1.50%. Table 3 summarizes these results; they indicate that the proposed method enables highly accurate simultaneous analysis of the thirteen compounds [35].

4. Discussion

Standardization and analysis of the marker compounds in herbal medicines are necessary for safety, efficacy, and risk management. A simple, reliable, and accurate method using ultrahigh-performance liquid chromatography (UPLC) fingerprinting with a diode array detector (DAD) was developed for the simultaneous qualitative and quantitative analyses of the thirteen biomarkers: ephedrine, methyl gallate, catechin, albiflorin, paeoniflorin, benzoic acid, liquiritin, liquiritin apioside, 1,2,3,4,6-pentagalloyl glucose (PGG), cinnamic acid, glycyrrhizin, schisandrin, and 6-shogaol in SCRT.

To optimize the quality control of SCRT, thirteen bioactive compounds from eight herbal preparations (*Ephedrae herba, Paeoniae radix, Glycyrrhizae radix, Zingiberis rhizoma, Cinnamomi ramulus, Schisandrae fructus, Pinelliae rhizoma*, and *Asiasari radix*) were chosen.

However, *Pinelliae rhizoma* and *Asiasari radix* were not detected in the SCRT extract sample. Because the contents of bioactive compounds in herbal medicines can differ as a function of the collection period, region, species, and preparation method, or the medicine contains a concentration lower than the LOD value, further studies should be required to perform quantification and qualification analyses, using a standard addition method or ultrahigh-performance liquid chromatography-tandem mass spectrometry (UPLC-MS/MS).

The results show that all biomarker components were detected (Figure 3), and identified based on the UV absorbance spectra (Figure 4) and retention times, by comparison with standard compounds. The UPLC-DAD method was validated, and the results showed good linearity, LOD, LOQ, precision, and accuracy, with RSD < 2.51%. Furthermore, the method did not interfere with other chemical constituents in SCRT. The results show that our method is accurate and reliable for quantification and qualification of the bioactive components of SCRT.

5. Conclusion

The developed UPLC-DAD fingerprinting method for the simultaneous determination of thirteen biomarkers in SCRT, which include ephedrine, methyl gallate, catechin, albiflorin, paeoniflorin, benzoic acid, liquiritin, liquiritin apioside, PGG, cinnamic acid, glycyrrhizin, schisandrin, and 6-shogaol, proved an efficient tool for quality control and pharmaceutical evaluation. The results demonstrated that the developed UPLC-DAD method is simple, reliable, accurate, sensitive, and precise for quantification and qualification analyses of SCRT.

Conflicts of Interest

The authors declare that there are no conflicts of interest regarding the publication of this paper.

Acknowledgments

This paper was supported by a grant from the Modernization Project of Korean Medicinal Preparations and the National Development Institute of Korean Medicine (NICOM) and by a grant from the Ministry of Health and Welfare, Republic of Korea.

References

[1] D. K. Sharma, S. G. Kim, R. Lamichhane, K. H. Lee, A. Poudel, and H. J. Jung, "Development of UPLC fingerprint with multicomponent quantitative analysis for quality consistency evaluation of herbal medicine "hyangsapyeongwisan"," *Journal of Chromatographic Science*, vol. 54, no. 4, pp. 536–546, 2016.

[2] L. Zheng and D. Dong, "Development and validation of an HPLC method for simultaneous determination of nine active components in "Da-Chai-Hu-Tang"," *Chinese Medicine*, vol. 2, no. 1, pp. 20–28, 2011.

[3] C. Y. Lim, H. W. Kim, B. Y. Kim, and S. I. Cho, "Genome wide expression analysis of the effect of Socheongryong Tang in asthma model of mice," *Journal of Traditional Chinese Medicine*, vol. 35, no. 2, pp. 168–174, 2015.

[4] M. Y. Lee, C. S. Seo, J. Y. Kim, and H. K. Shin, "Evaluation of a water extract of So-Cheong-Ryong-Tang for acute toxicity and genotoxicity using in vitro and in vivo tests," *BMC Complementary and Alternative Medicine*, vol. 15, no. 1, pp. 1–8, 2015.

[5] N. H. Yim, A. Kim, Y. P. Jung, T. Kim, C. J. Ma, and J. Y. Ma, "Fermented So-Cheong-Ryong-Tang (FCY) induces apoptosis via the activation of caspases and the regulation of MAPK signaling pathways in cancer cells," *BMC Complementary and Alternative Medicine*, vol. 15, no. 1, pp. 1–11, 2015.

[6] E. Ko, S. Rho, C. Cho et al., "So-Cheong-Ryong-Tang, tradititional Korean medicine, suppresses Th2 lineage development," *Biological & Pharmaceutical Bulletin*, vol. 27, no. 5, pp. 739–743, 2004.

[7] H. W. Kim, C. Y. Lim, B. Y. Kim, and S. I. Cho, "So-Cheong-Ryong-Tang, a herbal medicine, modulates inflammatory cell infiltration and prevents airway remodeling via regulation of interleukin-17 and GM-CSF in allergic asthma in mice," *Pharmacognosy Magazine*, vol. 10, no. 39, pp. 506–511, 2014.

[8] C. Park, S. H. Hong, G. Y. Kim, and Y. H. Choi, "So-Cheong-Ryong-Tang induces apoptosis through activation of the intrinsic and extrinsic apoptosis pathways, and inhibition of the PI3K/Akt signaling pathway in non-small-cell lung cancer A549 cells," *BMC Complementary and Alternative Medicine*, vol. 15, no. 1, pp. 1–13, 2015.

[9] M. K. Lee, K. Y. Lee, J. Park, and S. H. Sung, "Simultaneous determination of paeoniflorin, trans-cinnamic acid, schisandrin and glycyrrhizin in So-Cheong-Ryong-Tang by HPLC-DAD and HPLC-ESI-MS," *Natural Product Sciences*, vol. 16, no. 1, pp. 26–31, 2010.

[10] H. J. Kim, J. M. Park, J. A. Kim, and B. P. Ko, "Effect of herbal *Ephedra sinica* and *Evodia rutaecarpa* on body composition and resting metabolic rate: a randomized, double-blind clinical trial in Korean premenopausal women," *Journal of Acupuncture and Meridian Studies*, vol. 1, no. 2, pp. 128–138, 2008.

[11] C. Kim, J. H. Lee, W. Kim, and S. K. Kim, "The suppressive effects of Cinnamomi Cortex and its phytocompound coumarin on oxaliplatin-induced neuropathic cold allodynia in rats," *Molecules*, vol. 21, no. 9, pp. 1–12, 2016.

[12] X. Li, W. Wang, Y. Su, Z. Yue, and J. Bao, "Inhibitory effect of an aqueous extract of Radix Paeoniae Alba on calcium oxalate nephrolithiasis in a rat model," *Renal Failure*, vol. 39, no. 1, pp. 120–129, 2017.

[13] D. Wu, J. Chen, H. Zhu, and X. Huang, "UPLC-PDA determination of paeoniflorin in rat plasma following the oral administration of Radix Paeoniae Alba and its effects on rats with collagen-induced arthritis," *Experimental and Therapeutic Medicine*, vol. 7, no. 1, pp. 209–217, 2013.

[14] X. Li, W. Chen, and D. Chen, "Protective effect against hydroxyl-induced DNA damage and antioxidant activity of radix glycyrrhizae (liquorice root)," *Advanced Pharmaceutical Bulletin*, vol. 3, no. 1, pp. 167–173, 2013.

[15] S. Chrubasik, M. H. Pittler, and B. D. Roufogalis, "Zingiberis rhizoma: a comprehensive review on the ginger effect and efficacy profiles," *Phytomedicine*, vol. 12, no. 9, pp. 684–701, 2005.

[16] M. T. Liang, C. H. Yang, S. T. Li et al., "Antibacterial and antioxidant properties of Ramulus Cinnamomi using supercritical CO2 extraction," *European Food Research and Technology*, vol. 227, no. 5, pp. 1387–1396, 2008.

[17] Z. Hou, G. Xu, X. Han, and L. An, "Pharmacological research of *Schisandra chinensis* (Turcz.) Baill," in *Proceedings of ASME International Manufacturing Science and Engineering*, pp. 116–119, Los Angeles, CA, USA, June 2015.

[18] X. Zhang, Y. Cai, L. Wang, H. Liu, and X. Wang, "Optimization of processing technology of Rhizoma Pinelliae Praeparatum and its anti-tumor effect," *African Health Sciences*, vol. 15, no. 1, pp. 101–106, 2015.

[19] J. B. Park, J. E. Lee, S. H. Jin, Y. Ko, and S. H. Jeong, "Evaluation of the effects of a low dose of Asiasari radix on

stem cell morphology and proliferation," *Journal of Korean Medicine*, vol. 37, no. 2, pp. 85–92, 2016.

[20] M. Li, X. F. Hou, J. Zhang, S. C. Wang, Q. Fu, and L. C. He, "Applications of HPLC/MS in the analysis of traditional Chinese medicines," *Journal of Pharmaceutical Analysis*, vol. 1, no. 2, pp. 81–91, 2011.

[21] A. A. Boligon, R. B. Freitas, T. F. Brum, and L. F. Bauermann, "Antiulcerogenic activity of *Scutia buxifolia* on gastric ulcers induced by ethanol in rats," *Acta Pharmaceutica Sinica B*, vol. 4, no. 5, pp. 358–367, 2014.

[22] A. Bahrami, F. Ghamari, Y. Yamini, F. G. Shahna, and A. Moghimbeigi, "Hollow fiber supported liquid membrane extraction combined with HPLC-UV for simultaneous preconcentration and determination of urinary hippuric acid and mandelic acid," *Membranes*, vol. 7, no. 1, pp. 1–13, 2017.

[23] L. Y. Ma, Y. B. Zhang, Q. L. Zhou, Y. F. Yang, and X. W. Yang, "Simultaneous determination of eight ginsenosides in rat plasma by liquid chromatography–electrospray ionization tandem mass spectrometry; application to their pharmaco-kinetics," *Molecules*, vol. 20, no. 12, pp. 21597–21608, 2015.

[24] L. Yi, L. W. Qi, P. Li, Y. H. Ma, Y. J. Luo, and H. Y. Li, "Simultaneous determination of bioactive constituents in Danggui Buxue Tang for quality control by HPLC coupled with a diode array detector, an evaporative light scattering detector and mass spectrometry," *Analytical and Bioanalytical Chemistry*, vol. 389, no. 2, pp. 571–580, 2007.

[25] H. Bae, G. K. Jayaprakasha, J. Jifon, and B. S. Patil, "Extraction efficiency and validation of an HPLC method for flavonoid analysis in peppers," *Food Chemistry*, vol. 130, no. 3, pp. 751–758, 2012.

[26] D. Q. Tang, X. X. Zheng, X. Chen, D. Z. Yang, and Q. Du, "Quantitative and qualitative analysis of common peaks in chemical fingerprint of Yuanhu Zhitong tablet by HPLC-DAD-MS/MS," *Journal of Pharmaceutical Analysis*, vol. 4, no. 2, pp. 96–106, 2014.

[27] J. H. Kim, C. S. Seo, S. S. Kim, and H. K. Shin, "Quality assessment of Ojeok-San, a traditional herbal formula, using high-performance liquid chromatography combined with chemometric analysis," *Journal of Analytical Methods in Chemistry*, vol. 2015, Article ID 607252, 11 pages, 2015.

[28] H. J. Yang, N. H. Yim, K. J. Lee, and J. Y. Ma, "Simultaneous determination of nine bioactive compounds in Yijin-tang via high-performance liquid chromatography and liquid spectrometry," *Integrative Medicine Research*, vol. 5, no. 2, pp. 140–150, 2016.

[29] Y. Wang, Q. Lin, and F. Ikegami, "Development and validation of an HPLC-DAD method for the simultaneous quantification of 8 characteristic components in kakkonoto decotion," *Journal of Traditional Medicine*, vol. 29, no. 4, pp. 195–202, 2012.

[30] L. H. Shaw, L. C. Lin, and T. H. Tsai, "HPLC-MS/MS analysis of a traditional Chinese medical formulation of Bu-Yang-Huan-Wu-Tang and its pharmacokinetics after oral administration to rats," *Pharmacokinetics of Herbal Formulation*, vol. 7, no. 8, pp. 1–13, 2012.

[31] Z. Yu, X. Gao, H. Yuan, and K. Bi, "Simultaneous determination of safflor yellow A, puerarin, daidzein, ginsenosides (Rg1, Rb1, Rd), and notoginsenoside R1 in rat plasma by liquid chromatography–mass spectrometry," *Journal of Pharmaceutical and Biomedical Analysis*, vol. 45, no. 2, pp. 327–336, 2007.

[32] A. Poudel, S. G. Kim, R. Lamichhane, Y. K. Kim, H. K. Jo, and H. J. Jung, "Quantitative assessment of traditional oriental herbal formulation samhwangsasim-tang using UPLC

technique," *Journal of Chromatographic Science*, vol. 52, no. 2, pp. 176–185, 2014.

[33] W. Jia, C. Wang, Y. Wang et al., "Qualitative and quantitative analysis of the major constituents in Chinese medical preparation Lianhua-Qingwen capsule by UPLC-DAD-QTOF-MS," *The Scientific World Journal*, vol. 2015, Article ID 731765, 19 pages, 2015.

[34] S. G. Kim, A. Poudel, Y. K. Kim, H. K. Jo, and H. J. Jung, "Development of simultaneous analysis for marker constituents in Hwangryunhaedok-tang and its application in commercial herbal formulas," *Journal of Natural Medicines*, vol. 67, pp. 390–398, 2013.

[35] D. Tang, D. Yang, A. Tang, and X. Yin, "Simultaneous chemical fingerprint and quantitative analysis of *Ginkgo biloba* extract by HPLC–DAD," *Analytical and Bioanalytical Chemistry*, vol. 396, no. 8, pp. 3087–3095, 2010.

15

Determination of β-Agonist Residues in Animal-Derived Food by a Liquid Chromatography-Tandem Mass Spectrometric Method Combined with Molecularly Imprinted Stir Bar Sorptive Extraction

Jiwang Tang⬩,[1,2] Jianxiu Wang,[1] Shuyun Shi,[1] Shengqiang Hu,[1] and Liejiang Yuan[2]

[1]College of Chemistry and Chemical Engineering, Central South University, Changsha 410083, China
[2]Hunan Testing Institute Product and Commodity Supervison, Changsha 410007, China

Correspondence should be addressed to Jiwang Tang; tjw-mail@163.com

Academic Editor: Valdemar Esteves

A novel clenbuterol molecularly imprinted polymer (MIP)-coated stir bar was prepared and applied to the determination of six β-agonists in animal-derived food. Characterization and various parameters affecting adsorption and desorption behaviours were investigated. The extraction capacities of clenbuterol, salbutamol, ractopamine, mabuterol, brombuterol, and terbutaline for MIP coating were 3.8, 2.9, 3.1, 3.5, 3.2, and 3.3 times higher, respectively, than those of the NIP coating, respectively. The method of MIP-coated SBSE coupled with HPLC-MS/MS was developed. The recoveries in pork and liver samples were 75.8–97.9% with RSD from 2.6 to 5.3%. Limits of detection (LODs) and limits of quantification (LOQs) were 0.05–0.15 μg/kg and 0.10–0.30 μg/kg, respectively. Good linearities were obtained for six β-agonists with correlation coefficients (R^2) higher than 0.994. These results indicated the superiority of the proposed method in the analysis of β-agonists in a complex matrix.

1. Introduction

β-adrenergic agonists are a kind of chemical synthesized from the benzyl ethanol amines. These compounds have been extensively used as growth promoters in livestock production, with advantages of increasing the animal's lean meat, improving feed efficiency, and reducing the number of days to market [1, 2]. Consumption of animal tissues with high β-adrenergic agonist content by humans leads to symptoms such as muscle tremor, muscle pain, nausea, and dizziness and can pose a serious threat to life [3]. Owing to illegal β-agonist use and its potential hazardous effects on human health, some countries have banned its use in farm animals [4]. The wide variety of these compounds and their extensive use in farm animals give rise to enormous challenges of daily supervision and routine monitoring for the government. It is therefore necessary to develop a validated method with characteristics of high throughput and high sensitivity for the detection and quantification of β-adrenergic agonists.

Recently, several analytical methods have been developed for the determination of β-agonist drugs in foodstuffs and animal urine [5–13]. The analytical methods are mainly based on two different immune analysis techniques [6, 7, 9, 11] and on the chromatographic technique [5, 8, 10, 12, 13]; both have advantages and drawbacks. Immune analysis techniques such as enzyme-linked immunosorbent assay (ELISA) exhibit the features of rapid and simple analysis but come with drawbacks of giving rise to false positives. Chromatographic techniques such as HPLC, GC-MS, and LC-MS/MS require expensive instrumentation and cumbersome sample pre-treatments but have advantages of high accuracy and sensitivity. High-performance liquid chromatography-mass spectrometry/mass spectrometry (HPLC-MS/MS) is a feasible method that is commonly used due to its advantages such as sensitivity and accuracy [14]. For all mass spectrometry analytical procedures, the sample matrix effect is a major problem that influences the accuracy and precision of the testing results [15, 16]. Therefore, sample preparation is important and crucial, occupying nearly seventy percent of the total analysis time.

Research into sample preparation techniques has become a frontier of modern analytical chemistry, motivated by the demand for preparation methods with advantages of reduced or eliminated solvent, high selectivity, simplicity, speed, and so on. Traditional techniques such as liquid-liquid extraction (LLE) [17, 18], immune-affinity chromatography [19], and solid-phase extraction (SPE) [20, 21] are time-consuming and labour-intensive. The advanced and nonexhaustive techniques such as SPME (solid-phase microextraction) are solvent-free sample preparation techniques and combine sampling, analyte enrichment, and purification into one step, therefore substantially reducing the cost and total time of analysis [22]. The SPME contains various extraction concepts such as stirrers, vessels, coated fibres, and membranes [23]. Among these techniques, stir bar sorptive extraction (SBSE), first introduced in 1999 by Baltussenetal [24], is an environmentally friendly sample preparation technique using a stir bar coated with polydimethylsiloxane (PDMS). However, SBSE with a PDMS-coated stir bar shows low selectivity and specificity for target analyte and is only suitable for the analysis of nonpolar and weakly polar compounds [25]. Molecularly imprinted polymer (MIP) coatings for SBSE were proposed to solve the problem of polar molecules and selectivity. MIPs are prepared by copolymerization of functional monomers and cross-linkers in the presence of template molecules (target analytes). Extraction of the template molecules leaves the recognition sites in the polymers with specific size, shape, and functional group complementarity to the original print molecule [26]. Thus, MIPs show selective recognition and high affinity to the template molecule and its structurally related compounds [27]. Moreover, the imprinted polymers are robust, stable, and resistant to a wide range of pH, temperature, and solvents [28, 29]. The MIP-SBSE technology that combines MIP with SBSE plays an important role in the aspects of removing or decreasing sample matrix effect. The combination of MIP-SBSE technology with liquid chromatography has also been reported [30–38]. However, the combination of MIP-SBSE technology with liquid chromatography-tandem mass spectrometry has been rarely reported.

The aim of this present work was to prepare an MIP coating on the glass stir bar and apply it to the selective SBSE of clenbuterol and its analogue compounds. The preparation parameters and extraction conditions, which affected the extraction performance of the stir bar, were studied in detail. The extraction capacity and selectivity of the clenbuterol MIP-coated stir bar were also evaluated. Then, a method of determination of structural-related β-agonists by MIP-SBSE followed by HPLC-MS/MS was developed. The proposed method was successfully applied to the analysis of β-agonists in animal-derived food.

2. Experimental Section

2.1. Reagents and Standards. Clenbuterol, salbutamol, ractopamine, mabuterol, brombuterol, and terbutaline were purchased from Dr. Ehrenstorfer GmbH (Augsburg, Germany). The functional monomer methacrylic acid (MAA) and the

free radical initiator azoisobutyronitrile (AIBN) were supplied by Qiangsheng Chemical Co. Ltd. (Jiangsu, China). The cross-linking agent ethylene glycol dimethacrylate (EGDMA) was supplied by Aladdin Industrial Co. (Shanghai, China). The silane coupling agent KH-570 was purchased from Sinopharm Chemical Reagent Co. Ltd. (Shanghai, China). HPLC-grade methanol, toluene, and acetonitrile were purchased from Merck (Darmstadt, Germany). Water for HPLC-MS/MS was purified using a Milli-Q water purification system (Millipore, USA). Glass capillary (2 mm diameter, 100 mm length) was obtained from Guangzhou Fine Packaging Equipment Co., Ltd.

2.2. Apparatus. The coating surface of the stir bar was characterized by a VEGA3 scanning electron microscope (TESCAN, Czech Republic), and the composition of the coating was investigated using a Fourier transform infrared (FT-IR) spectrometer (Nicolet iS5, Thermo, USA). Measurements were taken using a TSQ Quantum Ultra triple quadrupole mass spectrometer (Thermo, USA) equipped with an electrospray ion source and coupled with an Accela liquid chromatography (Thermo, USA), equipped with a Hypersil GOLD (100 mm × 2.1 mm i. d., 3 μm) column.

2.3. HPLC-MS/MS Conditions. The mobile phase consisted of mobile phase A (0.1% formic acid in water) and mobile phase B (acetonitrile) at the flow rate of 0.3 mL·min^{-1} using a gradient elution program. The solvents were degassed by an in-line vacuum degasser. The gradient conditions were as follows: initial time, 10% B; 1 min, 40% B; 3 min, 95% B; 5 min, 100% B, 5.1 min, 10% B; and re-equilibration to 7 min. The column was kept at 35°C. The injection volume was 10 μL.

The compound was ionized in an electrospray ionization (ESI) instrument operated in positive mode. Capillary voltage was set at 3.5 kV. The sheath gas and aux gas were nitrogen at the pressures of 45 and 10 arb, respectively. The collision gas was argon at the vacuum pressure of 1.5 mTorr. Desolvation and source temperatures were set at 300°C and 350°C, respectively. To ensure high specificity and sensitivity, two or three of the most abundant precursor → fragment ion transitions (selected reaction monitoring, SRM) were observed for each analyte. SRM transitions, collision energies, and fragmentor for each analyte are given in Table S1.

2.4. Stir Bar Preparation. The glass capillary (2 mm diameter, 100 mm length) was cut into 20 mm long substrate. Then, a 16 mm iron rod was inserted in the substrate, and both of its ends were sealed by flame to generate a stir bar (20 mm × 2 mm). The pretreatment of the stir bar was carried out by successively placing it into 1 mol/L sodium hydroxide for 10 h and 0.1 mol/L hydrochloric acid for 1 h. Then, the stir bar was dried in an oven at 150°C for 1 h and immersed in a 30% (v/v) KH-570 solution in acetone for 3 h. Then, it was pulled out and cleaned with methanol.

Approximately 139.0 mg clenbuterol and 0.20 mL MAA were dissolved in 5 mL acetonitrile. The solution was mixed thoroughly and incubated overnight at room temperature. Then, 1.60 mL EGDMA and 20.0 mg AIBN were added, and the mixture was degassed by an ultrasonic device for 5 min. Then, a silylated stir bar was inserted into a glass tube, and 0.4 mL of solution was transferred into the tube. The glass tube was purged with nitrogen for 3 min and sealed by flame. The polymerization was accomplished in a water bath at 70°C for 24 h. The coating procedure was repeated two times. Then, the stir bar was pulled out by peeling off the outside glass tube and washed with 10% (v/v) acetic solution in methanol to remove the template molecules. The nonimprinted polymer-coated stir bar was prepared under the same conditions but without the addition of template clenbuterol in the synthesis.

2.5. Extraction Experiment.

A round-bottom flask was used to conduct an extraction experiment in order to reduce the wear of the stir bar. The schematic diagram for the experiment is shown in Figure S1. The stir bar was immersed in solution and stirred at 500 rpm for 60 min at room temperature. Subsequently, the stir bar was removed from the sample solution, inserted in a 200 μL glass vial with a conical insert, and desorbed with 100 μL 10% (v/v) acid in water solution by an ultrasonic bath for 10 min. Then, 10 μL desorption solution was injected into HPLC-MS/MS for analysis.

2.6. Sample Preparation.

Samples of pork and liver were purchased from a local supermarket. A total of 2.0 g of each homogenized sample was spiked with β-agonist mixed standard solution. The spiked sample was extracted with 10 mL 2% (v/v) trichloroacetic acid solution in an ultrasonic bath for 10 min. The extraction procedure was repeated twice. The combined extracted solution was centrifuged at 5000 rpm for 5 min, and the pH of the supernatant was adjusted to 9-10 by 1 mol/L sodium hydroxide. Then, the solution was centrifuged again, and the supernatant was transferred to another Teflon centrifugal tube and extracted by 10 mL isopropyl alcohol/ethyl acetate (60 : 40, v/v) two times. The organic layers were combined and dried with nitrogen steam. Then, 5 mL toluene was added to dissolve the residual for SBSE.

3. Results and Discussion

3.1. Preparation of the Stir Bar.

The pretreatment and silanization of the glass bar were carried out to immobilize the MIP coatings on the bar. Soaking of the glass bar in alkaline solution is performed in order to expose the maximum number of silanol groups on the surface to promote silylation with the silane coupling agent. Thus, double bonds were present in the molecular structure, forming an active site for the MIP coating. After the overnight self-assembled polymerization of the template molecule and functional monomer through intermolecular hydrogen bonding interactions, the cross-linking agent and initiator were added to the prepolymerization solution, and thus, robust coatings were formed on the glass bar through thermal initiation

polymerization. Then, the cavity structure with high selectivity for a target molecule was formed after eluting the template molecule by breaking the hydrogen bonding interaction between the functional monomer and template. The preparation schematic is illustrated in Figure 1.

The experimental parameters of MIP coating preparation such as the polymerization reagents and their proportion, temperature, and time were optimized. The experiment finally confirmed that the optimum mole ratio of template (clenbuterol), functional monomer (MAA), and cross-linker (EGDMA) was 1 : 4 : 16. The coatings of the MIP-coated stir bars prepared under optimum conditions were uniform and compact with a certain thickness (as shown in Figure S2).

3.2. Characterization of MIP Coatings.

The morphological structure of the stir bar coating surface was characterized by scanning electron microscopy. Figure 2 shows the surface structure of NIP coatings and MIP coatings under the magnification of 200x and 5000x. It is obvious that both MIP coatings and NIP coatings show homogeneous surfaces (Figures 2(a) and 2(c)), whereas the MIP coating exhibited a more porous structure than the NIP coating (Figures 2(b) and 2(d)). This is because the porous surface of the MIP coatings was more beneficial for analyte adsorption and desorption.

The molecular type and structure of the stir bar coatings were investigated by infrared spectroscopy. As shown in Figure 3, the absorption peaks of the NIP coating (Figure 3(b)), MIP coating before eluting template (Figure 3(c)), and MIP coating after eluting template (Figure 3(d)) show no obvious differences except for the peak at 1520 cm^{-1}, which was only present in the spectra of template (Figure 3(a)) and MIP coating before eluting template (Figure 3(c)). The absorption peak at 1520 cm^{-1} was derived from a C = C stretching vibration in the benzene ring of the template (clenbuterol). Therefore, it was reasonable to conclude that the template (clenbuterol) did not participate in the polymerization but only interacted with the functional monomer by hydrogen bonding.

3.3. Investigation of the Extraction Capability and Selectivity

3.3.1. Extraction Capability.

The adsorption capacity of the MIP-coated stir bar was studied with various clenbuterol standard solutions in toluene through static adsorption experiments. The NIP-coated stir bar was used for comparison. The extraction solution volume was 5 mL. The extraction time and desorption time were 60 and 10 min, respectively. As shown in Figure 4, the extraction amount of both the MIP-coated stir bar and NIP-coated stir bar increased with increasing clenbuterol concentration in the 1.0–60.0 μg/L range, and the extraction amounts were balanced over the concentration of 35.0 μg/L. The saturated adsorption amounts of the MIP coating and NIP coating were 67.0 and 17.8 ng, respectively. The adsorption amount of the MIP coating was approximately 3.76 times higher than that of the NIP coating. The excellent adsorption capacity of

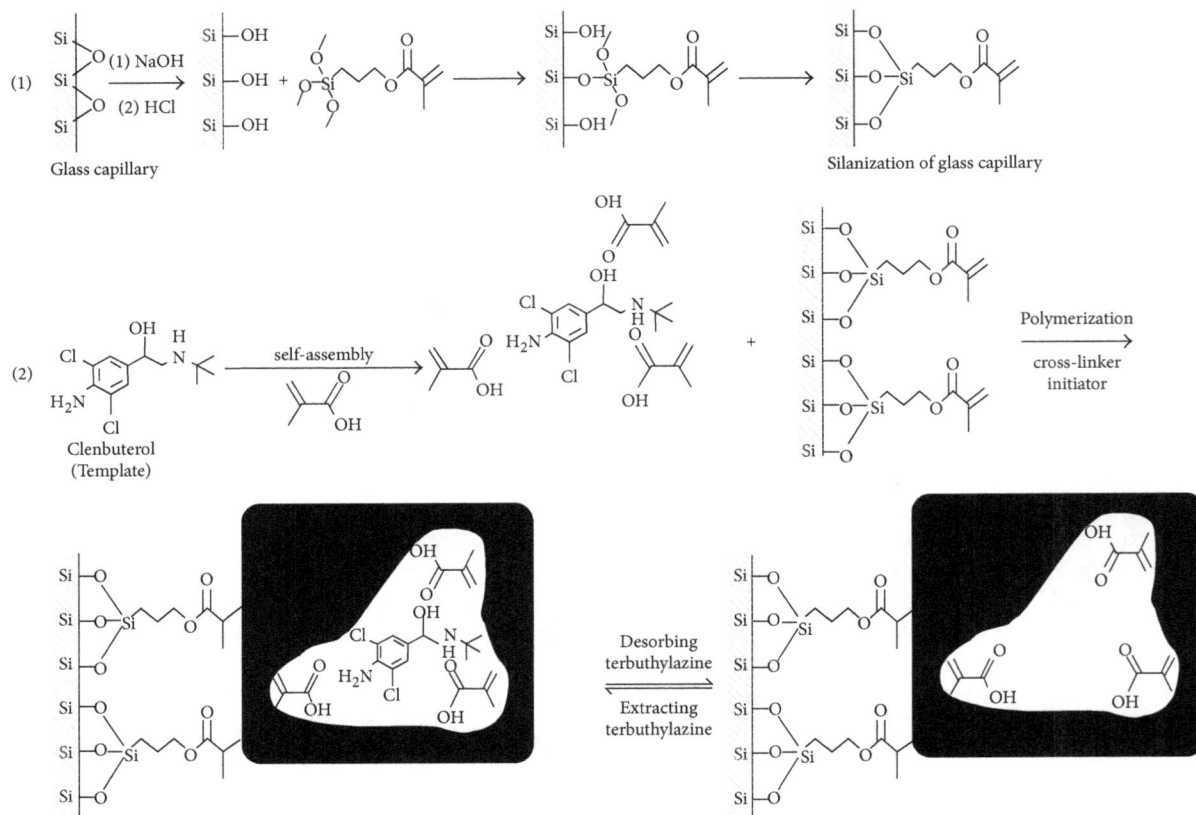

FIGURE 1: Schematic of preparation of molecularly imprinted stir bar.

the MIP coating could be attributed to the molecular imprinting function of the clenbuterol template. The cavities of the MIP coating have a special affinity to the template, while no such cavities were present in the NIP coating that interacts with the template by nonspecific sorption only.

3.3.2. Selectivity Evaluation. Clenbuterol and its structural analogues including ractopamine, salbutamol, mabuterol, terbutaline, and brombuterol were applied to extraction evaluation. Two reference compounds, namely, benzyl-alcohol and acrylamide, were used for comparison. All analytes were prepared individually with toluene to avoid competitive adsorption. The concentrations of all analytes were 10 μg/L, and the extraction time was 60 min. As shown in Figure 5, the extraction amounts of the MIP coatings for clenbuterol, salbutamol, ractopamine, mabuterol, terbutaline, and brombuterol were 3.8, 2.9, 3.1, 3.5, 3.3, and 3.2 times higher, respectively, than those of the NIP coatings. By contrast, there was hardly any sorption of both kinds of coatings for benzyl-alcohol and acrylamide, which show different characteristics and structure. The selectivity of the MIP coating was closely related to the shape and size of the cavity, and the strength of interaction between the target molecules and the binding sites [39]. The steric shape complementary between the analytes and the recognition sites of MIPs plays a key role in the molecular recognition process. As a result, the MIP-coated stir bars exhibited

higher selectivity toward the template of clenbuterol than its structural analogues. The analogues of salbutamol, mabuterol, terbutaline, and brombuterol with molecular shape and size similar to the template showed better imprinted effect than ractopamine, which had a little different molecular structure and functional groups. The reference compounds of benzyl-alcohol and acrylamide with less similarity with the template could hardly be extracted by both coatings.

3.4. Investigation of SBSE Conditions. To optimize the SBSE conditions, factors that affected the extraction results, such as solvent, time, stirring speed, and temperature, were also investigated in this work. The clenbuterol template was dissolved in the acetone, chloroform, acetonitrile, methanol, and toluene solvents to prepare individual solutions with a concentration of 10 μg/L. The effect of extraction solvents on the extraction capacity was investigated. As shown in Figure S3, the maximum extraction amount was attained in the case of toluene solvent. The above results indicate that the primary driving forces behind the rebinding process, namely, the hydrophobic interaction and hydrogen bonding, were strongly related to the polarity of the extraction solvents. The hydrogen bonding was liable to be formed in noncovalently imprinted polymers, and higher adsorptive capacity was obtained in the solvents with weaker polarity. Thus, the matrix effect of water should be removed, which influenced the absorption efficiency of MIP-coated stir bars.

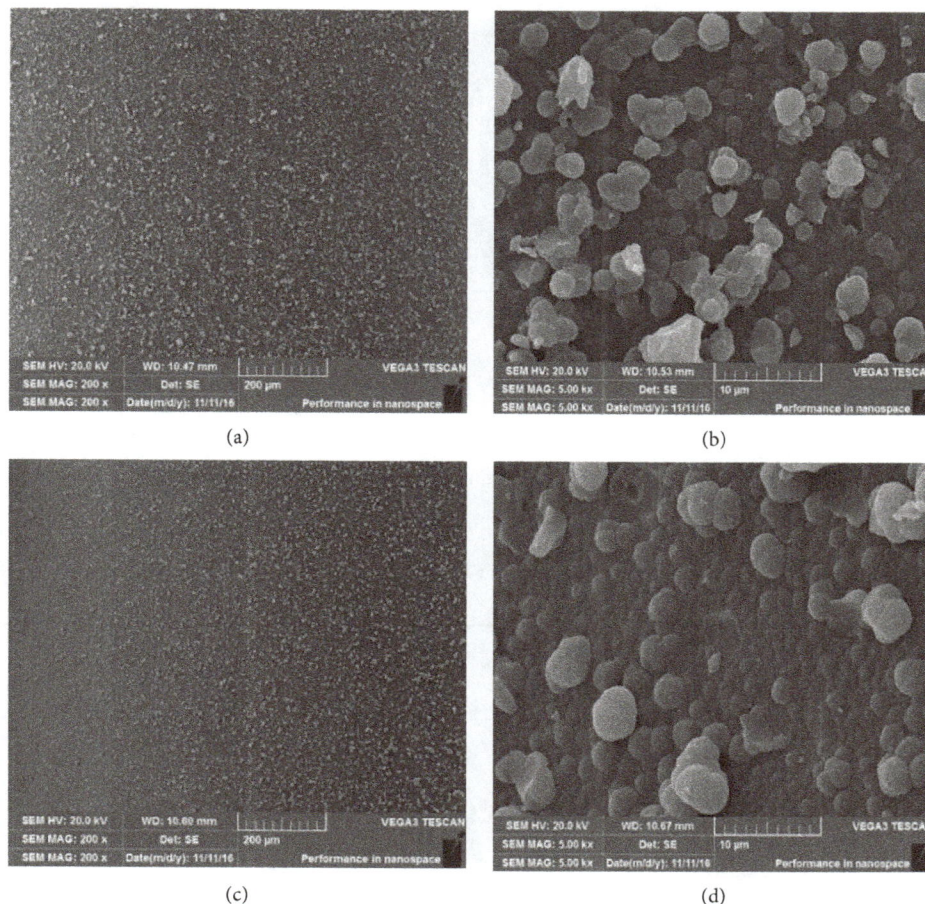

(a)

(b)

(c)

(d)

FIGURE 2: Scanning electron micrographs of (a) MIP-coated stir bar with the magnifications of 200x; (b) MIP-coated stir bar with the magnifications of 5000x; (c) NIP-coated stir bar with the magnifications of 200x; (d) NIP-coated stir bar with the magnifications of 5000x.

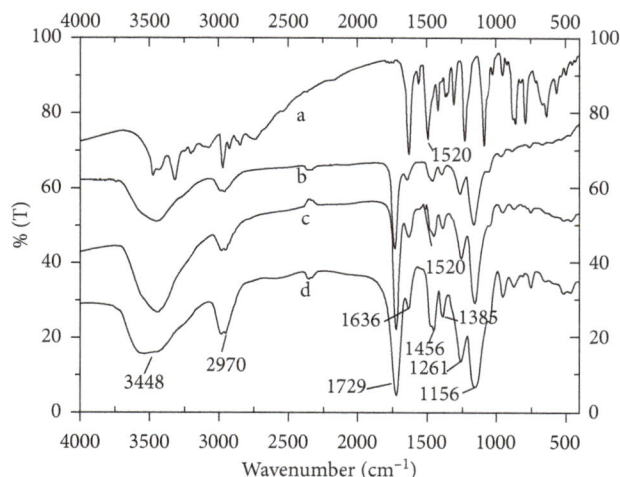

FIGURE 3: FT-IR spectra of clenbuterol moleculerly imprinted polymer coatings. (a) Clenbuterol, (b) NIP coating, and (c) MIP coating before eluting template; (d) MIP coating after eluting template clenbuterol.

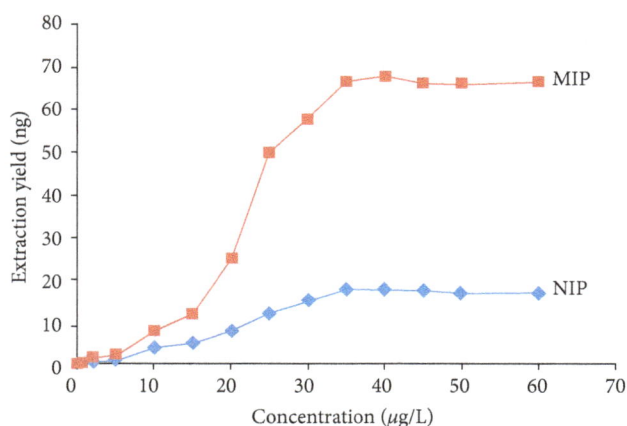

FIGURE 4: Extraction yield curves of MIP- and NIP-coated stir bars to various clenbuterol concentrations.

The water, methanol, methanol-acetic acid (9 : 1, v/v), and water-acetic acid (9 : 1, v/v) solvents were selected for the investigation of the solvent effect on the desorption. The desorbed amounts were 26.5, 25.6, 27.5, and 30.5 ng, respectively. Therefore, water-acetic acid (9 : 1, v/v) was chosen as the best desorption solvent.

The extraction and desorption amounts increased with increasing time, and the times at which each reached equilibrium were 60 min and 10 min, respectively (Figures S4 and S5).

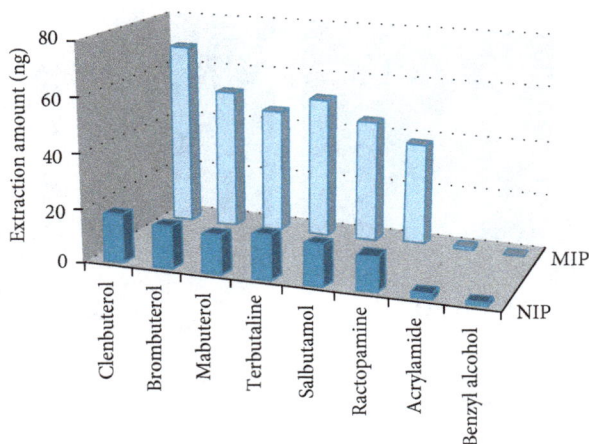

FIGURE 5: Extraction amount of clenbuterol and its structural analogues and reference compounds with MIP and NIP.

TABLE 1: Linear dynamic range, correlation coefficients, LODs and LOQs, average recoveries, and interday and intraday precisions achieved for the six β-agonists.

Compounds	Linear range (μg/L)	R^2	LOD[a] (μg/L)	LOQ[b] (μg/L)	Recovery (%) Spiked level (μg/kg)	Pork	Liver	Interday assay[c] (RSD %, $n = 5$)	Intraday assay[c] (RSD %, $n = 5$)
Clenbuterol	0.5–35	0.9991	0.05	0.10	1.0	85.0	87.2	3.9	4.2
					5.0	86.5	97.9		
Salbutamol	1–35	0.9981	0.10	0.25	1.0	85.3	84.5	4.5	5.1
					5.0	76.5	88.2		
Ractopamine	1–35	0.9952	0.15	0.35	1.0	75.8	80.2	5.1	5.3
					5.0	83.6	72.6		
Mabuterol	1–35	0.9978	0.10	0.25	1.0	81.2	80.8	4.4	2.6
					5.0	85.3	84.9		
Brombuterol	1–35	0.9936	0.10	0.25	1.0	84.5	82.6	4.6	3.8
					5.0	85.8	85.6		
Terbutaline	1–35	0.9957	0.15	0.30	1.0	84.3	83.1	4.9	4.1
					5.0	85.4	85.2		

[a]S/N = 3; [b]S/N = 10; [c]assays at 5.0 μg/kg level.

Therefore, 60 min and 10 min were adopted for extraction and desorption procedures.

The stirring speed varying from 0 to 800 rpm was investigated. The results indicate that the extraction amounts increased with the stirring speed and the maximum extraction amount was obtained at 300 rpm. Beyond 300 rpm, no obvious increase was attained. Thus, 300 rpm was chosen as the optimal stirring speed. In addition, the change in the extraction temperature exerted no obvious influence on the extraction amounts. For convenience, the adsorption experiments were carried out at room temperature.

3.5. Durability Investigation. To investigate the stability of the coating, the MIP-coated stir bar was immersed into conventional organic solvents, such as toluene, acetonitrile, methanol, n-hexane, ethanol, chloroform, ethyl acetate, dimethylsulfoxide, and tetrahydrofuran. After stirring for 24 h, no decomposition and exfoliation were obtained, indicating that the coatings of MIP-coated stir bar possessed excellent antisolvent fastness properties. The MIP-coated stir

bar could be reused after the analyte was removed by methanol-acetic acid (9 : 1, v/v). As the extraction was performed in a round-bottom flask instead of a flat-bottom bottle, the wearing of the stir bar during the stirring has been dramatically reduced. The coating was kept intact, and no decrease in the extraction efficiency was obtained after extraction for 60 times.

3.6. Method Validation. Assay of clenbuterol and its structural analogues by MIP-coated SBSE coupled with HPLC-MS/MS was performed. The blank pork and liver samples were spiked with six β-agonists and taken for analysis. The linear dynamic ranges, correlation coefficients, recoveries, LODs, LOQs, and reproducibility under the optimized experimental conditions are listed in Table 1. The linear range was determined to be 0.50–35 μg/kg for clenbuterol and 1.0–35 μg/kg for ractopamine, mabuterol, brombuterol, or terbutaline. Good linearity was obtained for six β-agonists with correlation coefficients (R^2) higher than 0.994. The limits of detection (LODs, signal-to-noise ratio

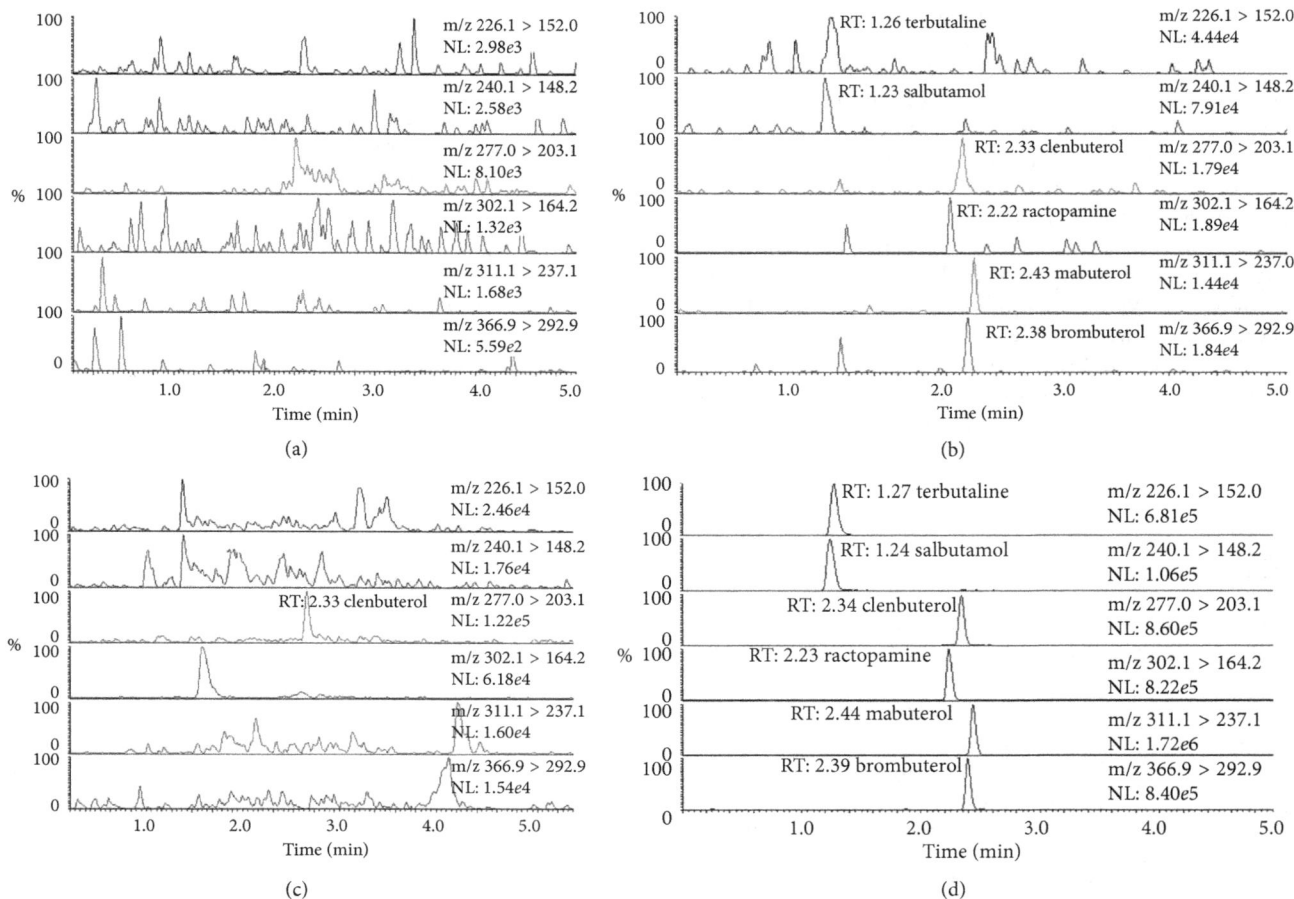

FIGURE 6: Multiple reaction monitoring (MRM) chromatograms showing the transitions corresponding to the quantifier ions of six β-agonists. (a) The blank pork sample; (b) the pork sample spiked with each β-agonist at LOD level; (c) the real pork sample; (d) the pork sample spiked with each β-agonist at 5.0 μg/kg.

TABLE 2: Comparison of the proposed method based on MIP-SBSE with other analytical methods for the determination of β-agonists in pork sample.

Sample pretreatment	Analytical method	Target analyte number	Matrix	Recovery (%)	Precision RSD (%)	LOD (μg/kg)	References
MIP-SBSE	HPLC	3	Pork	83.7–92.3	2.9–8.1	0.10–0.21	[40]
SPE	GC-MS	4	Pork	75.5–89.6	2.5–4.2	0.4–0.7	[41]
SPE	UPLC-MS/MS	4	Pork	65.5–109.1	6.6–15.4	0.1	[42]
QuEChERS	LC-MS/MS	10	Meat	73.7–103.5	2.7–15.3	0.2–0.9	[43]
SPE	HPLC-MS/MS	3	Pork	79.25–112.82	7.14%–17.80	0.13–0.15	[44]
MIP-SBSE	HPLC-MS/MS	6	Pork	75.8–97.9	2.6–5.3	0.05–0.15	This work

equal to 3) and limits of quantification (LOQs, signal-to-noise ratio equal to 10) were in the ranges of 0.05–0.15 μg/kg and 0.10–0.30 μg/kg, respectively. The recoveries for the spiked pork and liver samples ranged from 75.8% to 97.9%. The RSD for intraday and interday assays of six β-agonists varied from 2.6 to 5.3% ($n = 5$).

3.7. Application of the MIP-Coated Stir Bar.
To validate the selectivity of the method for assay of complex sample matrix, the blank pork samples were spiked with six β-agonists at

LOD level and 5.0 μg/kg. Furthermore, the real sample collected from the local market was analyzed by the proposed method. The multiple reaction monitoring (MRM) chromatograms showing the transitions corresponding to the quantifier ions of six β-agonists are depicted in Figure 6. β-Agonists could not be quantitatively assayed in the blank pork sample (Figure 6(a)). However, the six β-agonists were accurately analyzed based on MIP-SBSE (Figure 6(b)). Using the MIP-coated stir bar, the matrix interference has been largely eliminated. As shown in Figure 6(c), the peak at the retention time of 2.33 min was detected in the channel of m/z

277.0/203.1, suggesting that clenbuterol might exist in the pork sample. The extracts were subjected to further analysis using accurate mass AB Sciex Triple Quad 5500 LC/MS/MS in MS/MS/MS scan mode, and the content of clenbuterol in the positive pork sample was determined to be 1.31 μg/kg. The analysis of the pork sample spiked with each β-agonist at 5.0 μg/kg is shown in Figure 6(d). The appearance time for the six β-agonists was within 3 min, and the total analysis time for each specimen was only 5 min. The performance comparison of the proposed method based on MIP-SBSE and other analytical methods for the determination of β-agonists is shown in Table 2. Our work was advantageous in terms of detection throughput, sensitivity, robustness, and reproducibility. The proposed method has been successfully applied to the determination of trace β-agonists in foodstuff of animal origin with improved enrichment ratio of target analytes and reduced sample matrix effect.

4. Conclusions

A new clenbuterol MIP-coated stir bar for the assay of β-agonists with similar structures was constructed. The new coatings possessed homogeneous and porous morphology and exhibited excellent extraction capability and selectivity for six β-agonists. The MIP-coated SBSE coupled with HPLC-MS/MS has been successfully applied to the analysis of pork and liver samples spiked with β-agonists, and the recoveries ranged from 75.8 to 97.9%. In comparison with other analytical methods, our work possesses significant advantages of high throughput, high sensitivity, high efficiency, and low detection limit. The approach described herein may be extended to the preparation of a series of selective MIP-coated stir bars with appropriate template molecules for complex sample analysis.

Conflicts of Interest

The authors declare that they have no conflicts of interest.

Acknowledgments

The authors are grateful for the financial support from the National Natural Science Foundation of China (21375150) and the Project of Innovation-driven Plan in Administration of Quality and Technology Supervision of Hunan Province (2015KYJH07).

References

[1] P. Guggenbuhl, "Evaluation of β2-adrenergic agonists repartitioning effects in the rat by a non-destructive method," *Journal of Animal Physiology and Animal Nutrition*, vol. 75, no. 1–5, pp. 31–39, 2010.

[2] B. J. Johnson, S. B. Smith, and K. Y. Chung, "Historical overview of the effect of beta-adrenergic agonists on beef cattle production," *Asian-Australasian Journal of Animal Sciences*, vol. 27, no. 5, pp. 757–766, 2014.

[3] G. Brambilla, T. Cenci, F. Franconi et al., "Clinical and pharmacological profile in a clenbuterol epidemic poisoning of contaminated beef meat in Italy," *Toxicology Letters*, vol. 114, no. 1–3, pp. 47–53, 2000.

[4] J. Barbosa, C. Cruz, J. Martins et al., "Food poisoning by clenbuterol in Portugal," *Food Additives and Contaminants*, vol. 22, no. 6, pp. 563–566, 2005.

[5] L. Beucher, G. Dervilly-Pinel, S. Prévost, F. Monteau, and B. Le Bizec, "Determination of a large set of beta-adrenergic agonists in animal matrices based on ion mobility and mass separations," *Analytical Chemistry*, vol. 87, no. 18, pp. 9234–9242, 2015.

[6] A. D. Cooper and M. J. Shepherd, "Evaluation of a novel immunoaffinity phase for the purification of cattle liver extracts prior to high-performance liquid chromatographic determination of β-agonists," *Food and Agricultural Immunology*, vol. 8, no. 3, pp. 205–213, 1996.

[7] D. Jiang, B. Cao, M. Wang et al., "Development of a highly sensitive and specific monoclonal antibody based enzyme-linked immunosorbent assay for the detection of a new beta-agonist, phenylethanolamine A, in food samples," *Journal of the Science of Food and Agriculture*, vol. 97, no. 3, pp. 1001–1009, 2017.

[8] H. Liu, N. Gan, Y. Chen et al., "Novel method for the rapid and specific extraction of multiple β2-agonist residues in food by tailor-made Monolith-MIPs extraction disks and detection by gas chromatography with mass spectrometry," *Journal of Separation Science*, vol. 39, no. 18, pp. 3578–3585, 2016.

[9] M. Z. Zhang, M.-Z. Wang, Z.-L. Chen et al., "Development of a colloidal gold-based lateral-flow immunoassay for the rapid simultaneous detection of clenbuterol and ractopamine in swine urine," *Analytical and Bioanalytical Chemistry*, vol. 395, no. 8, pp. 2591–2599, 2009.

[10] Y. Zhu, S. Xie, D. Chen et al., "Targeted analysis and determination of beta-agonists, hormones, glucocorticoid and psychiatric drugs in feed by liquid chromatography with electrospray ionization tandem mass spectrometry," *Journal of Separation Science*, vol. 39, no. 13, pp. 2584–2594, 2016.

[11] H. Du, Y. Chu, H. Yang et al., "Sensitive and specific detection of a new β-agonist brombuterol in tissue and feed samples by a competitive polyclonal antibody based ELISA," *Analytical Methods*, vol. 8, no. 17, pp. 3578–3586, 2016.

[12] H. Liu, X. Lin, T. Lin, Y. Zhang, Y. Luo, and Q. Li, "Magnetic molecularly imprinted polymers for the determination of β-agonist residues in milk by ultra high performance liquid chromatography with tandem mass spectrometry," *Journal of Separation Science*, vol. 39, no. 18, pp. 3594–3601, 2016.

[13] Z. Zhang, H. Yan, F. Cui et al., "Analysis of multiple β-agonist and β-blocker residues in porcine muscle using improved QuEChERS Method and UHPLC-LTQ Qrbitrap mass spectrometry," *Food Analytical Methods*, vol. 9, no. 4, pp. 915–924, 2016.

[14] X. Wang, Y. Liu, Y. Su et al., "High-throughput screening and confirmation of 22 banned veterinary drugs in feedstuffs using LC-MS/MS and high-resolution Orbitrap mass spectrometry," *Journal of Agricultural and Food Chemistry*, vol. 62, no. 2, pp. 516–527, 2014.

[15] L. Couchman and P. E. Morgan, "LC-MS in analytical toxicology: some practical considerations," *Biomedical Chromatography*, vol. 25, no. 1-2, pp. 100–123, 2011.

[16] M. Nakamura, "Analyses of benzodiazepines and their metabolites in various biological matrices by LC-MS(/MS)," *Biomedical Chromatography*, vol. 25, no. 12, pp. 1283–1307, 2011.

[17] M. K. Henze, G. Opfermann, H. Spahn-Langguth, and W. Schänzer, "Screening of β-2 agonists and confirmation of fenoterol, orciprenaline, reproterol and terbutaline with gas chromatography-mass spectrometry as tetrahydroisoquinoline derivatives," *Journal of Chromatography B: Biomedical Sciences and Applications*, vol. 751, no. 1, pp. 93–105, 2001.

[18] K. Sharafi, N. Fattahi, A. Hossein Mahvi, M. Pirsaheb, N. Azizzadeh, and M. Noori, "Trace analysis of some organophosphorus pesticides in rice samples using ultrasound-assisted dispersive liquid-liquid microextraction and high-performance liquid chromatography," *Journal of Separation Science*, vol. 38, no. 6, pp. 1010–1016, 2015.

[19] L. Hermida, R. Rodríguez, L. Lazo et al., "A recombinant envelope protein from Dengue virus purified by IMAC is bioequivalent with its immune-affinity chromatography purified counterpart," *Journal of Biotechnology*, vol. 94, no. 2, pp. 213–216, 2002.

[20] Y. Cai, G. Jiang, J. Liu, and Q. Zhou, "Multiwalled carbon nanotubes as a solid-phase extraction adsorbent for the determination of bisphenol A, 4-n-nonylphenol, and 4-tert-octylphenol," *Analytical Chemistry*, vol. 75, no. 10, pp. 2517–2521, 2003.

[21] M. E. Lindsey, T. M. Meyer, and E. M. Thurman, "Analysis of trace levels of sulfonamide and tetracycline antimicrobials in groundwater and surface water using solid-phase extraction and liquid chromatography/mass spectrometry," *Analytical Chemistry*, vol. 73, no. 19, pp. 4640–4646, 2001.

[22] W. M. Mullett, A. Paul Martin, and J. Pawliszyn, "In-tube molecularly imprinted polymer solid-phase microextraction for the selective determination of propranolol," *Analytical Chemistry*, vol. 73, no. 11, pp. 2383–2389, 2001.

[23] G. Ouyang, D. Vuckovic, and J. Pawliszyn, "Nondestructive sampling of living systems using in vivo solid-phase microextraction," *Chemical Reviews*, vol. 111, no. 4, pp. 2784–2814, 2011.

[24] E. Baltussen, P. Sandra, F. David, and C. Cramers, "Stir bar sorptive extraction (SBSE), a novel extraction technique for aqueous samples: theory and principles," *Journal of Microcolumn Separations*, vol. 11, no. 10, pp. 737–747, 2015.

[25] M. Kawaguchi, R. Ito, K. Saito, and H. Nakazawa, "Novel stir bar sorptive extraction methods for environmental and biomedical analysis," *Journal of Pharmaceutical and Biomedical Analysis*, vol. 40, no. 3, pp. 500–508, 2006.

[26] Y. Xia, J. E. McGuffey, S. Bhattacharyya et al., "Analysis of the tobacco-specific nitrosamine 4-(methylnitrosamino)-1-(3-pyridyl)-1-butanol in urine by extraction on a molecularly imprinted polymer column and liquid chromatography/atmospheric pressure ionization tandem mass spectrometry," *Analytical Chemistry*, vol. 77, no. 23, pp. 7639–7645, 2005.

[27] N. Masqué, R. M. Marcé, F. Borrull, P. A. G. Cormack, and D. C. Sherrington, "Synthesis and evaluation of a molecularly imprinted polymer for selective on-line solid-phase extraction of 4-nitrophenol from environmental water," *Analytical Chemistry*, vol. 72, no. 17, pp. 4122–4126, 2000.

[28] M. Lachová, J. Lehotay, I. Skacani, and C. Jozef, "Study of selectivity of molecularly imprinted polymers prepared under different conditions," *Journal of Chromatographic Science*, vol. 48, no. 5, pp. 395–398, 2010.

[29] F. G. Tamayo, E. Turiel, and A. Martin-Esteban, "Molecularly imprinted polymers for solid-phase extraction and solid-phase microextraction: recent developments and future trends," *Journal of Chromatography A*, vol. 1152, no. 1-2, pp. 32–40, 2007.

[30] C. Chen, L. Yang, and Z. Jie, "Trace bensulfuron-methyl analysis in tap water, soil, and soybean samples by a combination of molecularly imprinted stir bar sorption extraction and HPLC-UV," *Journal of Applied Polymer Science*, vol. 122, no. 2, pp. 1198–1205, 2011.

[31] M. Díazálvarez, E. Turiel, and A. Martínesteban, "Molecularly imprinted polymer monolith containing magnetic nanoparticles for the stir-bar sorptive extraction of triazines from environmental soil samples," *Journal of Chromatography A*, vol. 1469, pp. 1–7, 2016.

[32] B. B. Prasad, A. Srivastava, and M. P. Tiwari, "Highly selective and sensitive analysis of dopamine by molecularly imprinted stir bar sorptive extraction technique coupled with complementary molecularly imprinted polymer sensor," *Journal of Colloid and Interface Science*, vol. 396, pp. 234–241, 2013.

[33] X. Wu, J. Liu, J. Wu et al., "Molecular imprinting-based micro-stir bar sorptive extraction for specific analysis of glibenclamide in herbal dietary supplements," *Journal of Separation Science*, vol. 35, no. 24, pp. 3593–3599, 2012.

[34] Q. Zhong, Y. Hu, Y. Hu, and G. Li, "Online desorption of molecularly imprinted stir bar sorptive extraction coupled to high performance liquid chromatography for the trace analysis of triazines in rice," *Journal of Separation Science*, vol. 35, no. 23, pp. 3396–3402, 2012.

[35] L. Zhu, G. Xu, F. Wei, J. Yang, and Q. Hu, "Determination of melamine in powdered milk by molecularly imprinted stir bar sorptive extraction coupled with HPLC," *Journal of Colloid and Interface Science*, vol. 454, pp. 8–13, 2015.

[36] X. Zhu, J. Cai, J. Yang, Q. Su, and Y. Gao, "Films coated with molecular imprinted polymers for the selective stir bar sorption extraction of monocrotophos," *Journal of Chromatography A*, vol. 1131, no. 1-2, pp. 37–44, 2006.

[37] Z. Xu, C. Song, Y. Hu, and G. Li, "Molecularly imprinted stir bar sorptive extraction coupled with high performance liquid chromatography for trace analysis of sulfa drugs in complex samples," *Talanta*, vol. 85, no. 1, pp. 97–103, 2011.

[38] Z. Xu, Z. Yang, and Z. Liu, "Development of dual-templates molecularly imprinted stir bar sorptive extraction and its application for the analysis of environmental estrogens in water and plastic samples," *Journal of Chromatography A*, vol. 1358, pp. 52–59, 2014.

[39] R. Liu, X. Li, Y. Li, P. Jin, W. Qin, and J. Qi, "Effective removal of rhodamine B from contaminated water using non-covalent imprinted microspheres designed by computational approach," *Biosensors and Bioelectronics*, vol. 25, no. 3, pp. 629–634, 2010.

[40] Z. Xu, Y. Hu, Y. Hu, and G. Li, "Investigation of ractopamine molecularly imprinted stir bar sorptive extraction and its application for trace analysis of β2-agonists in complex samples," *Journal of Chromatography A*, vol. 1217, no. 22, pp. 3612–3618, 2010.

[41] Q. S. Wang, W. M. Sun, and J. R. Liu, "Determination of four clenbuterol residues in pork by solid phase extraction and gas

chromatography-mass spectrometry," *Occupation and Health*, vol. 31, no. 2, pp. 181–183, 2015.

[42] T. Z. Wang, "Simultaneous determination of 4 kinds of β-agonist residues in pork by ultra performance liquid chromatography-electrospray tandem mass spectrometry," *Journal of Food Safety and Quality*, vol. 7, no. 6, pp. 2483–2489, 2016.

[43] L. Xiong, Y.-Q. Gao, W.-H. Li, X.-L. Yang, and S. P. Shimo, "Simple and sensitive monitoring of β2-agonist residues in meat by liquid chromatography-tandem mass spectrometry using a QuEChERS with preconcentration as the sample treatment," *Meat Science*, vol. 105, pp. 96–107, 2015.

[44] R. Y. Zhang, "Determination of 3 kinds of β-agonists residues in pork by high performance liquid chromatography-tandem mass spectrometry," *Journal of Food Safety and Quality*, vol. 8, no. 10, pp. 3831–3836, 2017.

16

Chemical Differentiation and Quantitative Analysis of Different Types of *Panax* Genus Stem-Leaf based on a UPLC-Q-Exactive Orbitrap/MS Combined with Multivariate Statistical Analysis Approach

Lele Li ⓘ, Yang Wang, Yang Xiu ⓘ, and Shuying Liu ⓘ

Jilin Ginseng Academy, Changchun University of Chinese Medicine, Changchun 130117, China

Correspondence should be addressed to Yang Xiu; ys830805@sina.com and Shuying Liu; syliu@ciac.ac.cn

Academic Editor: Antony C. Calokerinos

Two quantitative methods (−ESI full scan and −ESI PRM MS) were developed to analyze ginsenosides in ginseng stem-leaf by using UPLC-Q-Exactive Orbitrap/MS. By means of −ESI PRM MS method, the contents of eighteen ginsenosides in Asian ginseng stem-leaf (ASGSL) and American ginseng stem-leaf (AMGSL) were analyzed. The principal component analysis (PCA) model was built to discriminate Asian ginseng stem-leaf (ASGSL) from American ginseng stem-leaf (AMGSL) based on −ESI PRM MS data, and six ginsenosides (F11, Rf, R2, F1, Rb1, and Rb3) were obtained as the markers. To further explore the differences between cultivated ginseng stem-leaf and forest ginseng stem-leaf, the partial least squares-discriminant analysis (PLS-DA) model was built based on −ESI full scan data. And twenty-six markers were selected to discriminate cultivated ginseng stem-leaf (CGSL) from forest ginseng stem-leaf (FGSL). This study provides reliable and effective methods to quantify and discriminate among different types of ginseng stem-leaf in the commercial market.

1. Introduction

Asian ginseng (*Panax ginseng* C. A. Meyer) and American ginseng (*P. quinquefolium* L.) are two different medicinal herbs highly valued all over the world [1–4]. Asian ginseng (ASG) has been used to cure various diseases for thousands of years in Asian countries [5]. Recently, American ginseng (AMG) has also been well known in Asian countries because of its therapeutic functions [2]. These two related and similar-looking functional herbs, however, show certain differences in their biological activities and pharmacological effects. The root of *Panax* genus is the most commonly used part of the plant. And the stem-leaf is attracting more and more attention because of its similar pharmacological activities to the root [6, 7]. The quality and chemical composition of ginseng vary widely depending on the species, variety, geographical origin, cultivation method, environment, and harvesting time [8]. According to the cultivation

environment, ASG could be divided into cultivated ginseng (CG) and forest ginseng (FG). The former is cultivated in artificial conditions and grown for 3–7 years, while the latter is produced by sowing seeds in forest and grown in natural environment for over 10 years. In the commercial market, the specifications and grades of commercial products are numerous, and the prices fluctuate widely with the types of ginseng. Since there is no practical criterion for differentiating different CG, FG, and AMG, it is valuable to find analytical markers and provide methods for discriminating different types of ginseng. Much effort has been devoted to the quantitative and qualitative analysis of ginseng roots which are cultivated in different areas and ages [9]. However, little is known about the difference of stem-leaf in chemical compositions between different types of *Panax* genus.

Ginsenosides are the major pharmacologically active constituents of *Panax* genus [10, 11]. They are considered to be responsible for the activities of antioxidant, anti-inflammatory,

TABLE 1: Chemical information of the eighteen ginsenoside standards.

Compounds	Formula	Accurate mass	Retention time (min)	Concentration (μg/mL)
Noto R1	$C_{47}H_{80}O_{18}$	932.5345	4.89	288
Noto R2	$C_{41}H_{70}O_{13}$	770.4816	16.54	222
Rb1	$C_{54}H_{92}O_{23}$	1108.6029	18.61	108
Rb2	$C_{53}H_{90}O_{22}$	1078.5924	19.92	162
Rb3	$C_{53}H_{90}O_{22}$	1078.5924	20.20	240
Rc	$C_{53}H_{90}O_{22}$	1078.5924	19.21	72
Rd	$C_{48}H_{82}O_{18}$	946.5501	21.20	372
Re	$C_{48}H_{82}O_{18}$	946.5501	8.29	222
Rf	$C_{42}H_{72}O_{14}$	800.4922	15.70	68
Rg1	$C_{42}H_{72}O_{14}$	800.4922	7.41	252
Rg2	$C_{42}H_{72}O_{13}$	784.4973	17.60	264
Rg3	$C_{42}H_{72}O_{13}$	784.4973	25.98	4
Rh1	$C_{36}H_{62}O_{9}$	638.4394	17.37	114
Rh2	$C_{36}H_{62}O_{8}$	622.4445	27.02	276
Ro	$C_{48}H_{76}O_{19}$	956.4981	19.54	180
F1	$C_{36}H_{62}O_{9}$	638.4394	19.50	150
F2	$C_{42}H_{72}O_{13}$	784.4973	25.30	120
F11	$C_{42}H_{72}O_{14}$	800.4922	15.57	72

TABLE 2: The information about growth years and collecting location for all *Panax* genus stem-leaf samples.

Number	Growth years	Collecting location	Number	Growth years	Collecting location
AMGSL-1	3	Suihua city, Heilongjiang province	CGSL-7	4	Suihua city, Heilongjiang province
AMGSL-2	4	Suihua city, Heilongjiang province	CGSL-8	4	Hunchun city, Jilin province
AMGSL-3	3	Jiaohe city, Jilin province	CGSL-9	5	Hunchun city, Jilin province
AMGSL-4	5	Jiaohe city, Jilin province	CGSL-10	5	Hunchun city, Jilin province
AMGSL-5	4	Antu county, Jilin province	CGSL-11	5	Wangqing county, Jilin province
AMGSL-6	3	Jiaohe city, Jilin province	FGSL-1	16	Huadian city, Jilin province
AMGSL-7	4	Jiaohe city, Jilin province	FGSL-2	14	Huadian city, Jilin province
AMGSL-8	3	Suihua city, Heilongjiang province	FGSL-3	11	Panshi city, Jilin province
AMGSL-9	4	Antu county, Jilin province	FGSL-4	12	Huadian city, Jilin province
AMGSL-10	5	Panshi city, Jilin province	FGSL-5	15	Huadian city, Jilin province
CGSL-1	3	Suihua city, Heilongjiang province	FGSL-6	16	Suihua city, Heilongjiang province
CGSL-2	4	Suihua city, Heilongjiang province	FGSL-7	14	Huadian city, Jilin province
CGSL-3	3	Suihua city, Heilongjiang province	FGSL-8	13	Panshi citiy, Jilin province
CGSL-4	3	Suihua city, Heilongjiang province	FGSL-9	10	Panshi city, Jilin province
CGSL-5	3	Tieli county, Heilongjiang province	FGSL-10	10	Panshi city, Jilin province
CGSL-6	4	Hunchun city, Jilin province	FGSL-11	11	Dongning county, Heilongjiang province

antiapoptotic, and immunostimulant properties of ginseng [12, 13]. More than 70 ginsenosides have been isolated and identified in *Panax* genus [14]. Few analytical methods, including high-performance liquid chromatography (HPLC) [15], have been developed to exclusively determine ginsenosides in stem-leaf. However, in most of these methods, ginsenosides in minor or trace amounts cannot be detected due to the limited resolution and sensitivity of the detectors. Comparing with HPLC, LC-electrospray-mass spectrometry (ESI-MS) can achieve much higher sensitivity and selectivity and quantify chemical markers in the complex matrixes with only a small amount of sample. Q-Exactive Orbitrap mass spectrometry, with extremely high resolution, sensitivity, and mass accuracy, is a powerful technique for the detection of ginsenosides, especially for those in minor amounts [16, 17]. It provides multiple scan modes, including ESI parallel reaction monitoring (PRM) and ESI full scan MS, which are efficient for the

quantification of ginsenosides [18]. Compared with the full scan mode, the PRM mode shows better sensitivity and selectivity without establishing the baseline chromatographic separation of target analytes, but the PRM mode only can be carried out by using standards to achieve accurate quantification. Ginsenosides could be detected in both positive and negative ion modes [19]. Their ionization efficiency and limit of detection are different in each mode. Therefore, it is necessary to evaluate the quantitative performance using PRM in both positive and negative ion modes.

LC-ESI-MS combined with multivariate statistical analysis has been widely applied for the systematic identification and quantification of all metabolites in a given organism or biological sample in the study of metabolomics [9, 20]. The enhanced resolution provided by mass spectrometry, along with powerful chemometric software, allows the simultaneous determination and comparison of

TABLE 3: Results of optimization for product ions and normalized collision energy.

Compounds	Negative			Positive		
	Precursor ion	Product ion	NCE (%)	Precursor ion	Product ion	NCE (%)
Noto R1	977.53	931.53	15	955.52	775.46	30
Noto R2	815.48	769.48	15	793.45	335.10	35
Rb1	1107.60	945.54	25	1131.59	365.11	30
Rb2	1123.59	1077.58	10	1101.58	335.10	35
Rb3	1123.59	1077.58	10	1101.58	335.10	35
Rc	1123.59	1077.58	10	1101.58	335.10	35
Rd	991.55	945.54	15	969.54	789.48	30
Re	991.55	945.54	15	969.54	789.48	30
Rf	845.49	475.38	30	823.48	365.11	40
Rg1	845.49	799.48	15	823.48	643.42	25
Rg2	829.50	783.49	15	807.49	349.11	35
Rg3	829.50	783.49	20	807.49	365.11	35
Rh1	683.44	637.43	15	661.43	203.05	35
Rh2	667.44	621.44	15	645.43	203.05	35
Ro	955.49	793.44	30	979.49	641.40	25
F1	683.44	637.43	15	661.43	203.05	30
F2	829.50	783.49	20	807.49	627.42	30
F11	845.49	653.43	30	823.48	497.36	40

thousands of chemical entities [21, 22]. It also has been used to differentiate ginsengs which were cultivated in different areas [23]. In the present study, a UPLC-Q-Exactive Orbitrap/MS method was developed to accurately quantify and identify the ginsenosides extracted from forest ginseng stem-leaf (FGSL), cultivated ginseng stem-leaf (CGSL), and American ginseng stem-leaf (AMGSL).

Both −ESI PRM MS and +ESI PRM MS were validated [24], then the −ESI PRM MS was chosen to quantify the ginsenosides in FGSL, CGSL, and AMGSL. The multivariate statistical analysis was further employed to differentiate the Panax genus stem-leaf of various types and also to discover the chemical markers in terms of the detected ginsenosides.

2. Materials and Methods

2.1. Chemicals and Materials. HPLC-grade methanol (MeOH), acetonitrile (ACN), and formic acid were obtained from TEDIA (Fairfield, OH, USA). Ultrapure water was filtered through a Milli-Q system (Millipore, Billerica, MA, USA). Other reagents were analytical grade. Notoginsenoside R1, notoginsenoside R2, ginsenoside Rb1, ginsenoside Rb2, ginsenoside Rb3, ginsenoside Rc, ginsenoside Rd, ginsenoside Re, ginsenoside Rf, ginsenoside Rg1, ginsenoside Rg2, ginsenoside Rg3, ginsenoside Rh1, ginsenoside Rh2, ginsenoside Ro, ginsenoside F1, ginsenoside F2, and pseudoginsenoside F11 standards (they were represented by R1, R2, Rb1, Rb2, Rb3, Rc, Rd, Re, Rf, Rg1, Rg2, Rg3, Rh1, Rh2, Ro, F1, F2, and F11, resp. in the rest sections) were purchased from Nanjing Zelang Co. (purity ≥ 98%, Nanjing, China). Their chemical information is shown in Table 1. AMGSL (10 samples), CGSL (11 samples), and FGSL (11 samples) were collected from Jilin and Heilongjiang provinces. And detailed information of them is shown in Table 2.

2.2. Sample Preparation. A certain amount of Noto R1, Noto R2, Rb1, Rb2, Rb3, Rc, Rd, Re, Rf, Rg1, Rg2, Rg3, Rh1, Rh2, Ro, F1, F2, and F11 were dissolved in methanol, respectively, to get eighteen standard solutions. They were mixed and diluted by 70% methanol to get stock solutions and concentrations of each standard in stock solutions are shown in Table 1. Then the stock solutions were diluted to be 1.0, 3.3, 10.0, 33.3, 100.0, 333.3, 1000.0, 3333.3, 10,000.0, 33,333.3, 100,000.0, and 333,333.3 fold dilutions by 70% methanol, respectively, for method validation. All the prepared solutions were stored at 4°C.

The stem-leaf of Panax genus samples were dried and pulverized to powder and then passed through a 10-mesh sieve. The obtained powder was weighed (0.1 g) and extracted with 5 mL of 70% methanol in an ultrasonic waterbath for 60 min. The extract was filtered through a syringe filter (0.22 μm) and injected directly into the UPLC system.

2.3. Method Validation. An external calibration method was used for the quantitative analysis. And one of the CGSL samples was used in method validation. The linear calibration curves were constructed by plotting the concentrations of six to eleven mixed standard solutions against the corresponding peak areas. The limit of detection (LOD) and limit of quantification (LOQ) were measured with the signal-to-noise ratio of 3 and 10, respectively. The precision was performed by analysis of the standard solution six times and the results were expressed as the relative standard deviation (RSD). For repeatability test, six independent sample solutions were prepared using the procedures described in the last section. The recovery of this method was achieved using the standard addition method. Different concentration levels of the standard solutions were added to the sample six times. The average recoveries were determined by the following formula:

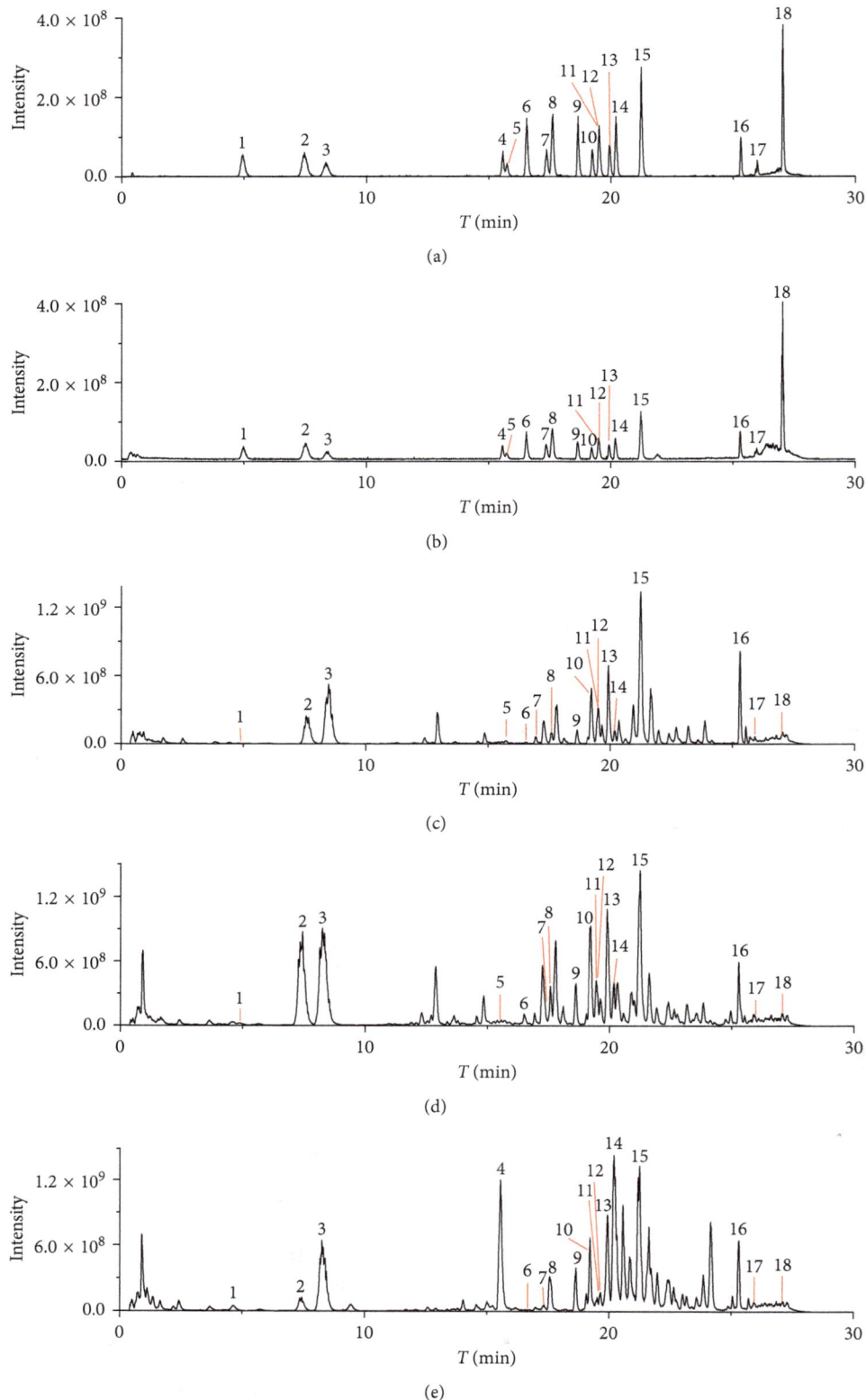

FIGURE 1: Total ion current chromatograms of eighteen ginsenoside standards based on (a) −ESI full scan and (b) +ESI full scan MS; Total ion current chromatograms of (c) FGLS, (d) CGLS and (e) AMGLS based on −ESI full scan MS (1 notoginsenoside R1; 2 ginsenoside Rg1; 3 ginsenoside Re; 4 pseudoginsenoside F11; 5 ginsenoside Rf; 6 notoginsenoside R2; 7 ginsenoside Rh1; 8 ginsenoside Rg2; 9 ginsenoside Rb1; 10 ginsenoside Rc; 11 ginsenoside F1; 12 ginsenoside Ro; 13 ginsenoside Rb2; 14 ginsenoside Rb3; 15 ginsenoside Rd; 16 ginsenoside F2; 17 ginsenoside Rg3; 18 ginsenoside Rh2).

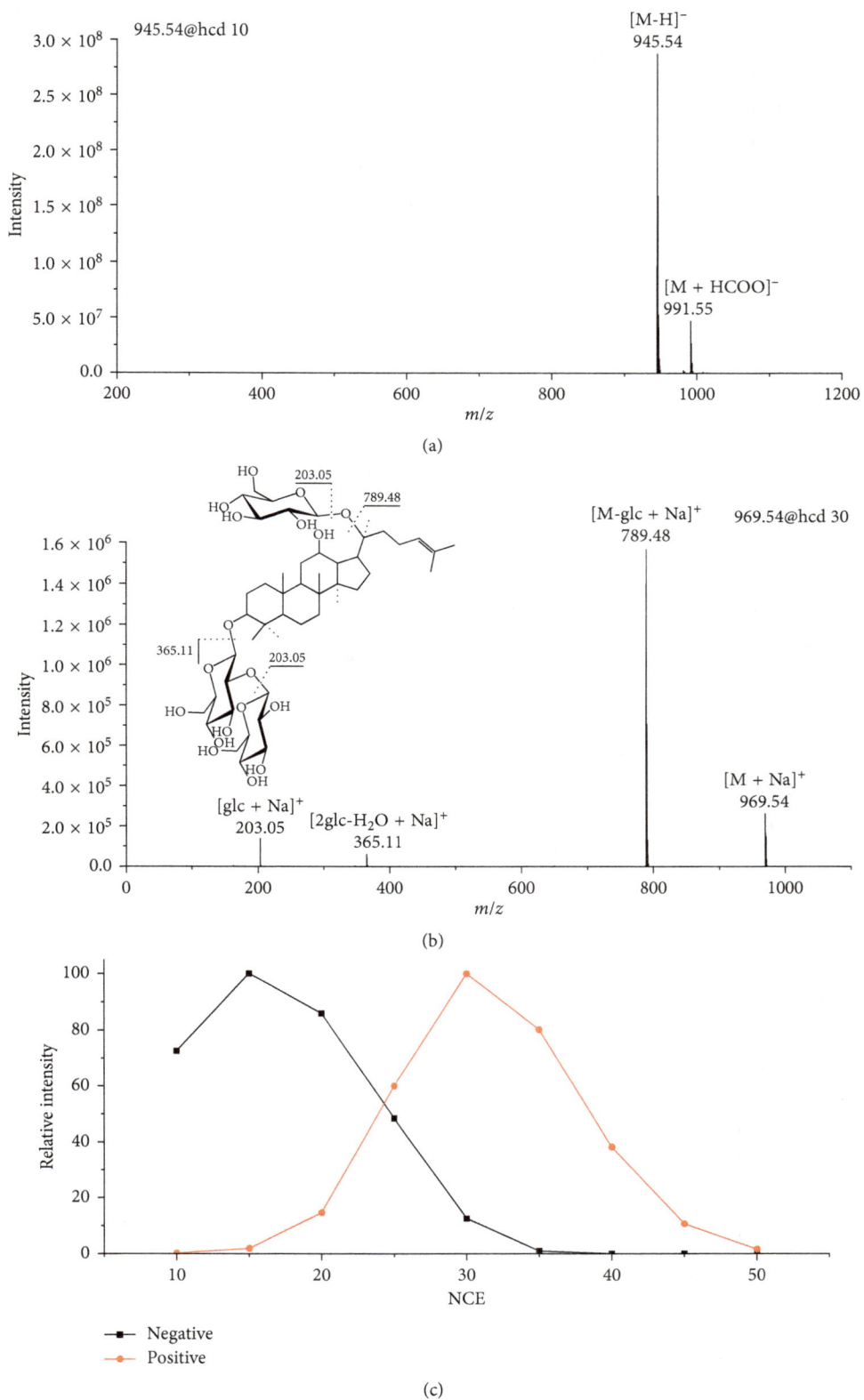

FIGURE 2: MS^2 spectra of (a) [M + HCOO]$^-$ and (b) [M + Na]$^+$ for Rd. (c) The signal intensity of product ions for Rd with respect to the NCE (glc represents glucose).

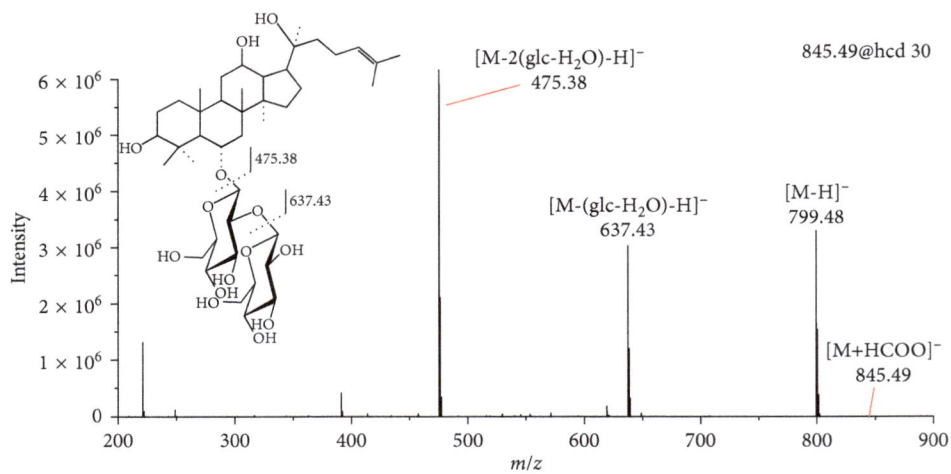

(a)

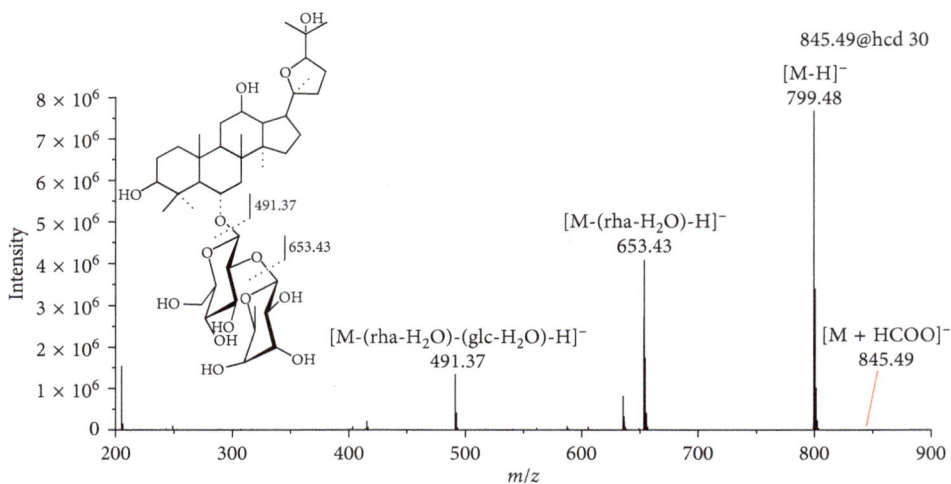

(b)

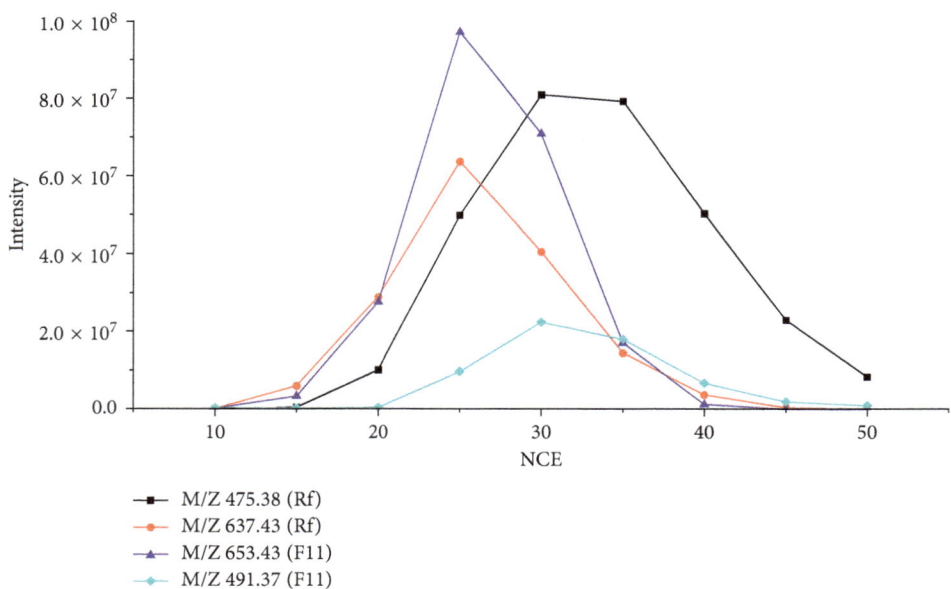

(c)

FIGURE 3: Continued.

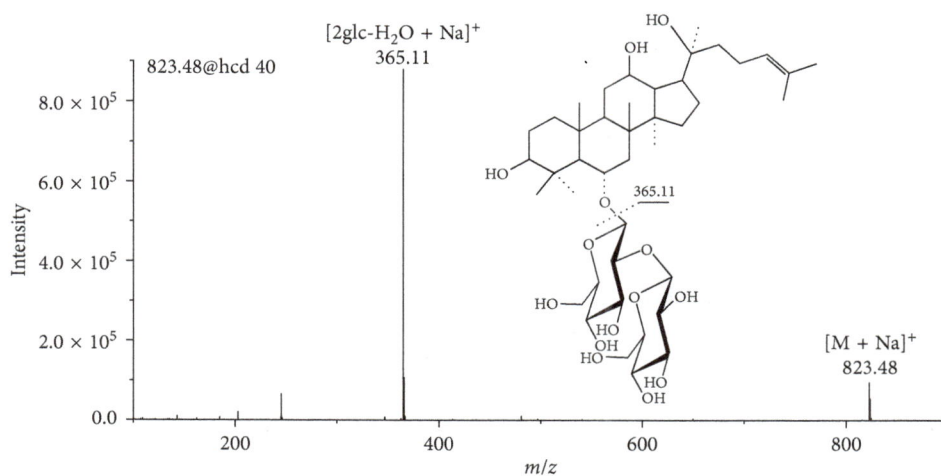

(d)

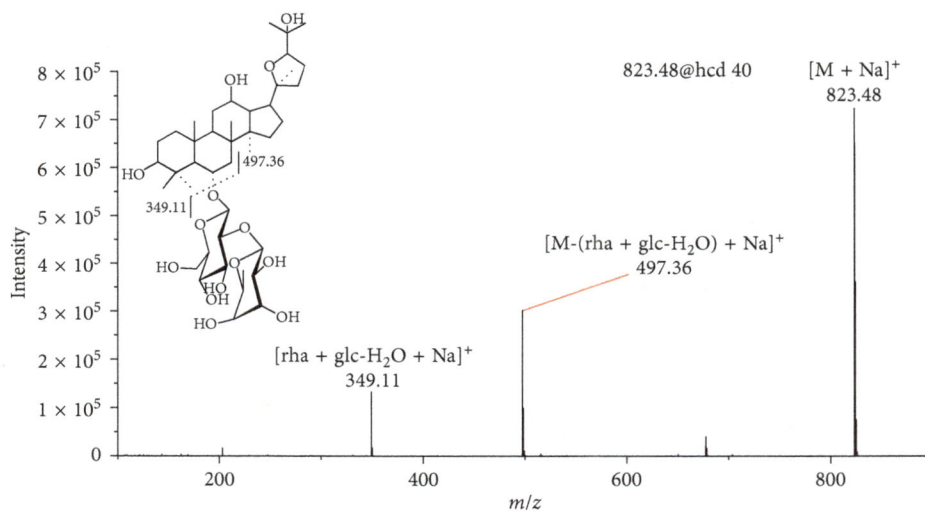

(e)

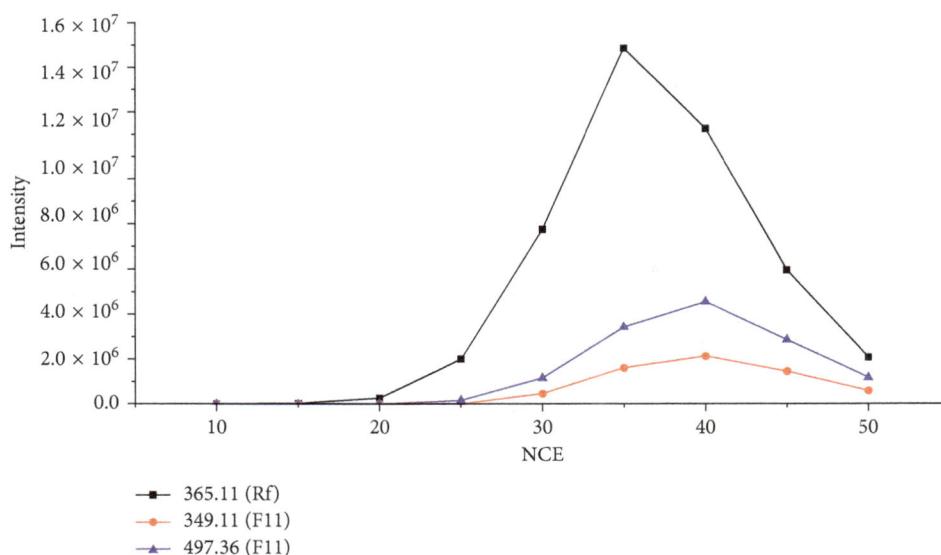

(f)

FIGURE 3: MS^2 spectra of [M + HCOO]$^-$ for (a) Rf and (b) F11 in the negative mode; MS^2 spectra of [M + Na]$^+$ for (d) Rf and (e) F11 in the positive mode. The signal intensity of product ions for Rf and F11 with respect to the NCE in the negative (c) and positive (f) modes (glc represents glucose and ara represents arabinose).

$$\text{Recovery}(\%) = \frac{(\text{observed amount} - \text{original amount})}{\text{spiked amount}} \times 100\%,$$

$$\text{RSD}(\%) = \left(\frac{\text{SD}}{\text{mean}}\right) \times 100\%.$$

(1)

2.4. UPLC-Q-Exactive Orbitrap/MS Analysis. Chromatographic separation was performed on an Ultimate 3000 system (Dionex, Sunnyvale, CA, USA) coupled with Golden C18 column (2.1 × 50 mm, 1.9 µm; Thermo Fisher, USA). The column temperature was maintained at 30°C, and the mobile phases A and B were water with 0.1% formic acid and acetonitrile, respectively. The gradient elution was programmed to get the ginsenoside profile, and the proportion of acetonitrile (B) was increased from 19% to 19% (0–5 min), 19–28% (5–12 min), 28–40% (12–22 min), 40–85% (22–24 min), and finally adjusted from 85% to 19% (24–25 min) and maintained at 19% for five minutes. The injection volume was 5 µL and the flow rate was 0.3 mL/min.

Mass spectrometric detection was carried out on a Q-Exactive Orbitrap/MS (Thermo, San Jose, CA, USA) equipped with an electrospray ionization (ESI) ion source operated in the positive or negative ion mode. The parameters of ion source were set as follows: sheath gas flow of 40 Arb, aux gas flow of 12 Arb, and sweep gas flow of 1 Arb. The S-Lens RF level was 55%. Capillary voltage was set to +3.5 kV or −3.5 kV with a capillary temperature of 333°C and an aux gas heater temperature of 317°C. Full scan MS data were acquired at the centroid mode from m/z 200 to 2000, 70,000 resolution, automatic gain control (AGC) target of 1e6, and maximum injection time (IT) of 100 ms. The mass range of MS^2 spectra varies depending on the precursor ions. The precursor and product ions used for the quantification in the PRM mode are listed in Table 3, together with the corresponding normalized collision energy (NCE). PRM MS data were acquired at the centroid mode, 17,500 resolution, AGC target of 1e5, and maximum IT of 50 ms.

2.5. Multivariate Statistical Analysis. For the −ESI PRM MS, the contents of eighteen ginsenosides in ginseng stem-leaf were used as a dataset containing sample code, ginsenoside, and content. For the −ESI full scan MS, the raw data of samples were processed by the Sieve software (version 2.1, Thermo, San Jose, CA, USA), which could detect the mass, retention time, and intensity of the peaks in each chromatogram. The maximum retention time shift was set at 0.25 min and the m/z width was 10 ppm to align the features. The base peak minimum intensity and background were set at 10^5 and 3, respectively. After being aligned, the intensity of each ion was normalized by the total ion intensity of each chromatogram. The resultant dataset, containing m/z value @ retention time, the normalized intensity and the sample code, was used to perform the

TABLE 4: Precision and repeatability for the eighteen ginsenosides in ginseng stem-leaf.

Compounds	RSD (%) precision ($n = 6$)		RSD (%) repeatability ($n = 6$)	
	−ESI PRM	+ESI PRM	−ESI PRM	+ESI PRM
R1	2.30	2.34	5.63	2.94
R2	4.31	2.93	7.28	2.73
Rb1	2.44	3.93	4.14	3.31
Rb2	2.89	4.47	4.54	2.90
Rb3	4.50	4.67	3.52	4.05
Rc	3.60	2.42	4.05	2.18
Rd	3.21	1.83	7.14	7.28
Re	2.47	2.04	7.85	5.88
Rf	2.76	2.24	8.56	2.46
Rg1	3.26	2.15	4.98	6.41
Rg2	3.40	1.38	9.30	3.04
Rg3	3.76	2.40	2.89	10.05
Rh1	1.93	2.14	7.32	8.33
Rh2	3.07	2.50	4.11	7.62
Ro	1.18	1.17	4.85	5.04
F1	3.22	2.05	2.43	4.40
F2	2.93	4.07	8.26	7.35
F11	3.02	3.33	4.36	6.24

TABLE 5: LOD and LOQ for the eighteen ginsenosides in ginseng stem-leaf.

Compounds	LOD (µg/mL)		LOQ (µg/mL)	
	−ESI PRM	+ESI PRM	−ESI PRM	+ESI PRM
R1	0.00288	0.0288	0.0096	0.096
R2	0.0222	0.0222	0.074	0.074
Rb1	0.0036	0.0108	0.0108	0.036
Rb2	0.00054	0.0162	0.00162	0.054
Rb3	0.0024	0.024	0.008	0.08
Rc	0.00024	0.0072	0.00072	0.024
Rd	<0.00124	0.0372	0.00124	0.124
Re	0.00074	0.0222	0.00222	0.074
Rf	0.000228	0.0228	0.000684	0.0684
Rg1	0.00252	0.0252	0.0084	0.084
Rg2	0.00088	0.088	0.00264	0.264
Rg3	0.000134	0.00402	0.000402	0.0134
Rh1	0.0038	0.114	0.0114	0.38
Rh2	0.0276	0.0092	0.092	0.0276
Ro	0.0006	0.006	0.0018	0.018
F1	0.0015	0.015	0.015	0.05
F2	0.012	0.004	0.04	0.012
F11	0.00072	0.024	0.0024	0.072

multivariate statistical analysis. Then, both the datasets were saved as .csv files and imported into SIMCA-P software 11.5 (Umetrics, Umea, Sweden) to conduct the multivariate statistical analysis including principal component analysis (PCA) and partial least squares-discriminant analysis (PLS-DA). In the PLS-DA model, ions with variable importance in projection (VIP) 1 and VIP 2 values larger than 1 were highlighted and were further filtered by Student's t-test (SPSS19.0, Chicago, IL, USA). The components with $p < 0.05$ were considered significant and were selected as analytical markers.

TABLE 6: Calibration curve for the eighteen ginsenosides in ginseng stem-leaf.

Compounds	−ESI PRM			+ESI PRM		
	Calibration curve	Linear range (μg/mL)	r	Calibration curve	Linear range (μg/mL)	r
R1	$Y = 3.27018 \times 10^7 X + 1.27657 \times 10^7$	0.0096–288	0.9994	$Y = 7.12403 \times 10^5 X + 1.14396 \times 10^6$	0.096–28.8	0.9892
R2	$Y = 2.87993 \times 10^7 X + 5.33922 \times 10^7$	0.074–74	0.9942	$Y = 1.82435 \times 10^5 X + 2.53254 \times 10^5$	0.074–22.2	0.9897
Rb1	$Y = 4.44403 \times 10^6 X + 1.70415 \times 10^6$	0.0108–36	0.9985	$Y = 9.55059 \times 10^5 X + 1.91469 \times 10^6$	0.036–36	0.9850
Rb2	$Y = 2.47272 \times 10^7 X + 1.29975 \times 10^7$	0.00162–54	0.9978	$Y = 5.82544 \times 10^5 X + 1.65322 \times 10^6$	0.054–54	0.9846
Rb3	$Y = 2.15503 \times 10^7 X + 3.80369 \times 10^7$	0.008–80	0.9914	$Y = 2.89207 \times 10^5 X + 1.49713 \times 10^6$	0.08–80	0.9818
Rc	$Y = 4.72793 \times 10^7 X + 1.12854 \times 10^7$	0.00072–24	0.9979	$Y = 8.45948 \times 10^5 X + 1.38423 \times 10^6$	0.024–24	0.9804
Rd	$Y = 3.37678 \times 10^7 X + 1.08663 \times 10^8$	0.00124–124	0.9815	$Y = 3.53253 \times 10^5 X + 3.07447 \times 10^6$	0.124–124	0.9807
Re	$Y = 3.19849 \times 10^7 X + 2.13175 \times 10^7$	0.00222–222	0.9970	$Y = 4.69691 \times 10^5 X + 2.40931 \times 10^6$	0.074–74	0.9796
Rf	$Y = 8.08178 \times 10^7 X + 9.99727 \times 10^6$	0.000684–22.8	0.9993	$Y = 4.29504 \times 10^5 X + 5.30340 \times 10^5$	0.0684–22.8	0.9906
Rg1	$Y = 8.38404 \times 10^6 X + 3.00715 \times 10^7$	0.0084–252	0.9958	$Y = 1.15403 \times 10^6 X + 2.47 \times 10^6$	0.084–25.2	0.9771
Rg2	$Y = 4.17409 \times 10^7 X + 1.14333 \times 10^7$	0.00264–26.4	0.9974	$Y = 5.8844 \times 10^4 X + 4.43358 \times 10^5$	0.264–88	0.9879
Rg3	$Y = 3.47553 \times 10^8 X + 6.73022 \times 10^6$	0.000402–1.34	0.9964	$Y = 1.58307 \times 10^6 X + 1.31307 \times 10^5$	0.0134–1.34	0.9890
Rh1	$Y = 3.4999 \times 10^6 X + 2.45245 \times 10^6$	0.0114–38	0.9950	$Y = 4.7729 \times 10^4 X + 4.6489 \times 10^4$	0.38–11.4	0.9878
Rh2	$Y = 1.3957 \times 10^6 X + 1.25926 \times 10^6$	0.092–27.6	0.9937	$Y = 1.45527 \times 10^6 X + 1.52536 \times 10^6$	0.092–27.6	0.9937
Ro	$Y = 5.2621 \times 10^6 X + 5.94325 \times 10^6$	0.0018–60	0.9925	$Y = 1.82466 \times 10^6 X + 2.10478 \times 10^6$	0.018–60	0.9819
F1	$Y = 4.03707 \times 10^6 X + 1.93111 \times 10^6$	0.015–50	0.9981	$Y = 1.06804 \times 10^6 X + 2.54794 \times 10^5$	0.05–15	0.9982
F2	$Y = 3.14728 \times 10^6 X + 152{,}642$	0.04–12	0.9998	$Y = 1.48171 \times 10^6 X + 2.40890 \times 10^5$	0.04–12	0.9878
F11	$Y = 8.90659 \times 10^5 X + 20{,}891$	0.0024–24	1.0000	$Y = 1.93487 \times 10^6 X + 6.14331 \times 10^5$	0.072–24	0.9752

3. Results and Discussion

3.1. Optimization of LC-MS Conditions. As shown in Figures 1(a) and 1(b), the total ion chromatograms (TIC) of eighteen ginsenoside standards were obtained in −ESI full scan and +ESI full scan MS. Most of the standards were separated distinctly within 30 min, with an exception of Rf and F11. Rf and F11 were structural isomers and the characteristic components in Asian ginseng and American ginseng, respectively. Theoretically, they cannot be present in one ginseng sample. Although they exhibited a poor chromatographic separation in the full scan mode, their quantification could be achieved in PRM mode, in which they were separated by their distinct ion pairs.

The ion pairs used for the quantitative analysis of ginsenosides were firstly optimized. Generally, the base peak in the full scan spectra (usually the adduct ion) was taken as the precursor ion, while the base peak in the MS^2 spectra was selected as the product ion. Taking Rd as an example, the [M + HCOO]$^-$ ion at m/z 991.55 was the main adduct ion of Rd in the −ESI full scan MS and was chosen as the precursor ion. Its corresponding MS^2 spectrum is shown in Figure 2(a). The [M-H]$^-$ ion at m/z 945.54 was the base peak in the MS^2 spectrum and was selected as the quantitative product ion. Similarly, the [M+Na]$^+$ at m/z 969.54 in the +ESI full scan MS and [M-glc + Na]$^+$ at m/z 789.48 ions in the MS^2 spectrum were chosen as the precursor and product ions, respectively. The NCE which ranged from 10% to 50% was further optimized to get the maximum intensity of the product ions. As shown in Figure 2(c), the intensity of the negative ion at m/z 945.54 and positive ion at m/z 789.48 was maximized at NCE of 15% and 30%, respectively.

To perform the quantification of isomeric ginsenosides Rf and F11, their PRM conditions were optimized. The precursor ions of Rf and F11 were the [M + HCOO]$^-$ ion at m/z 845.49 in the −ESI full scan MS and the [M + Na]$^+$ ion at m/z 823.48 in the +ESI full scan MS. As shown in Figures 3(a) and 3(b), Rf and F11 had the same [M-H]$^-$ ion at m/z 799.48

TABLE 7: Recovery for the eighteen ginsenosides in ginseng stem-leaf.

Compounds	−ESI PRM		+ESI PRM	
	Recovery (%)	RSD (%) ($n = 6$)	Recovery (%)	RSD (%) ($n = 6$)
R1	118.32	7.43	103.29	2.03
R2	102.29	5.68	131.23	7.58
Rb1	80.77	6.86	75.38	7.35
Rb2	87.66	3.74	81.24	2.34
Rb3	119.55	5.47	130.08	2.91
Rc	97.41	9.06	84.07	4.05
Rd	111.12	6.75	116.12	3.18
Re	93.25	7.22	78.65	4.32
Rf	76.68	6.51	79.45	5.33
Rg1	95.33	7.75	85.91	4.10
Rg2	90.21	8.09	137.71	4.36
Rg3	101.49	4.73	129.48	10.60
Rh1	112.92	6.18	130.61	4.93
Rh2	103.84	7.90	76.49	2.35
Ro	73.42	6.69	65.42	6.69
F1	80.70	8.16	99.57	4.55
F2	97.93	6.39	78.70	1.80
F11	107.52	4.39	121.05	6.24

TABLE 8: Contents of the eighteen ginsenosides in ginseng stem-leaf.

Content (mg/g)	R1	R2	Rb1	Rb2	Rb3	Rc	Rd	Re	Rg1
FGSL	0.101 ± 0.026	0.078 ± 0.027	0.233 ± 0.063	2.792 ± 0.497	0.289 ± 0.069	1.098 ± 0.184	3.737 ± 0.585	7.394 ± 1.601	2.405 ± 0.593
CGSL	0.214 ± 0.114	0.104 ± 0.075	0.368 ± 0.126	2.635 ± 0.938	0.346 ± 0.192	1.221 ± 0.460	3.390 ± 0.985	8.286 ± 2.364	3.996 ± 1.739
AMGSL	0.391 ± 0.128	0.008 ± 0.002	0.702 ± 0.240	3.543 ± 0.638	4.977 ± 0.481	0.933 ± 0.207	3.424 ± 0.379	8.162 ± 1.635	1.644 ± 0.691

Content (mg/g)	Rg2	Rg3	Rh1	Rh2	Ro	F1	F2	Rf	F11
FGSL	0.396 ± 0.130	0.007 ± 0.002	0.031 ± 0.016	0.003 ± 0.002	0.140 ± 0.070	4.409 ± 0.961	1.067 ± 0.786	0.148 ± 0.050	0.000
CGSL	0.571 ± 0.275	0.006 ± 0.003	0.061 ± 0.049	0.004 ± 0.004	0.087 ± 0.066	4.770 ± 2.490	1.062 ± 0.825	0.160 ± 0.065	0.000
AMGSL	0.899 ± 0.277	0.011 ± 0.003	0.093 ± 0.049	0.005 ± 0.003	0.066 ± 0.088	0.079 ± 0.015	0.921 ± 0.518	0.000	3.271 ± 0.464

and distinct product ions at m/z (475.38, 637.43) and (491.37, 653.43), respectively, which were obtained by the successive losses of the two saccharide moieties. These distinct ions could be used as the characteristic product ions for their differentiation and quantification. Therefore, the ion at m/z 475.38 and 653.43 were selected as the product ions of Rf and F11, respectively, as their intensities were higher than the counterparts. In addition, the NCE was optimized to be 30% and 25% for Rf and F11 (Figure 3(c)). In the positive ion mode, the fragment ions were also unambiguously identified, as shown in Figures 3(d) and 3(e). All the optimized PRM conditions are shown in Table 3.

3.2. Validation of the Method. As shown in Table 4, the precisions for all the eighteen ginsenosides from two scan modes were in the range from 1.17% to 4.67%, while the repeatability ranged from 2.18% to 10.05%. The results showed that the developed UPLC-Q-Exactive Orbitrap/MS

method had a good precision and repeatability for quantifying ginsenosides in all the tested modes.

The LOD and LOQ of eighteen ginsenosides are shown in Table 5. In total, the lower LOD and LOQ were obtained in −ESI PRM MS instead of +ESI PRM MS. In addition, the −ESI PRM MS has the lower LOD and LOQ less than 0.0276 and 0.092 μg/mL for all the ginsenosides.

As shown in Table 6, all the compounds showed good linearity ($r \geq 0.99$) when the method was operated in the negative ion mode. The concentration range could reach up to 3 orders of magnitude, indicating the developed method a good capability to quantify the ginsenosides. Only a few ginsenosides showed good linearity in +ESI PRM MS. It is observed that the linearity of ginsenosides detected in the negative ion mode was better than that in the positive ion mode.

The recoveries detected in both the modes were in the range from 65.42% to 137.71% with RSD less than 10.60%, as shown in Table 7.

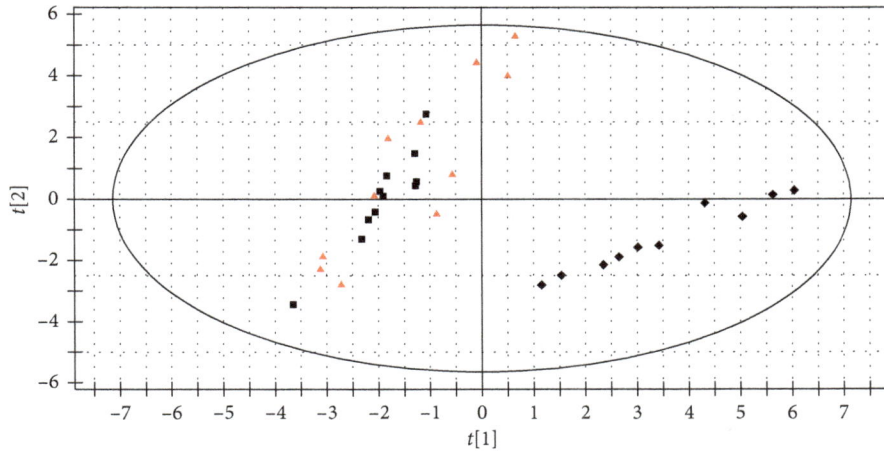

(a)

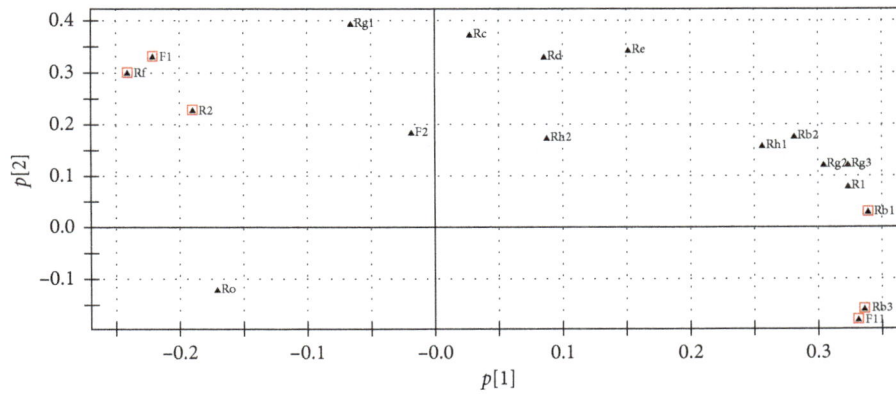

(b)

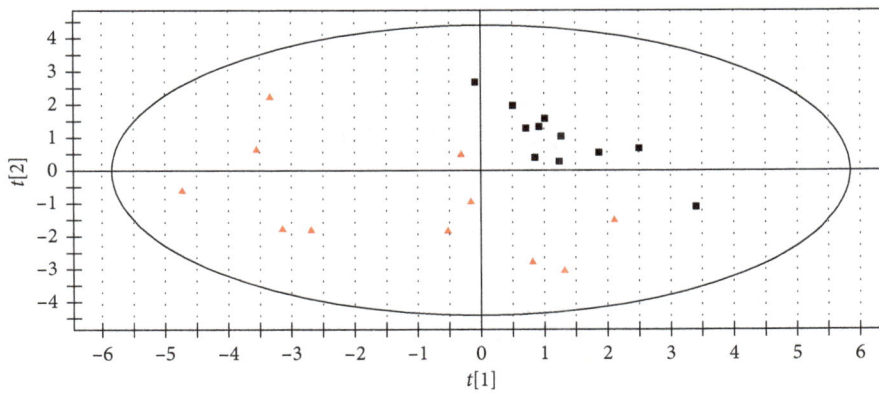

(c)

Figure 4: Continued.

(d)

(e)

(f)

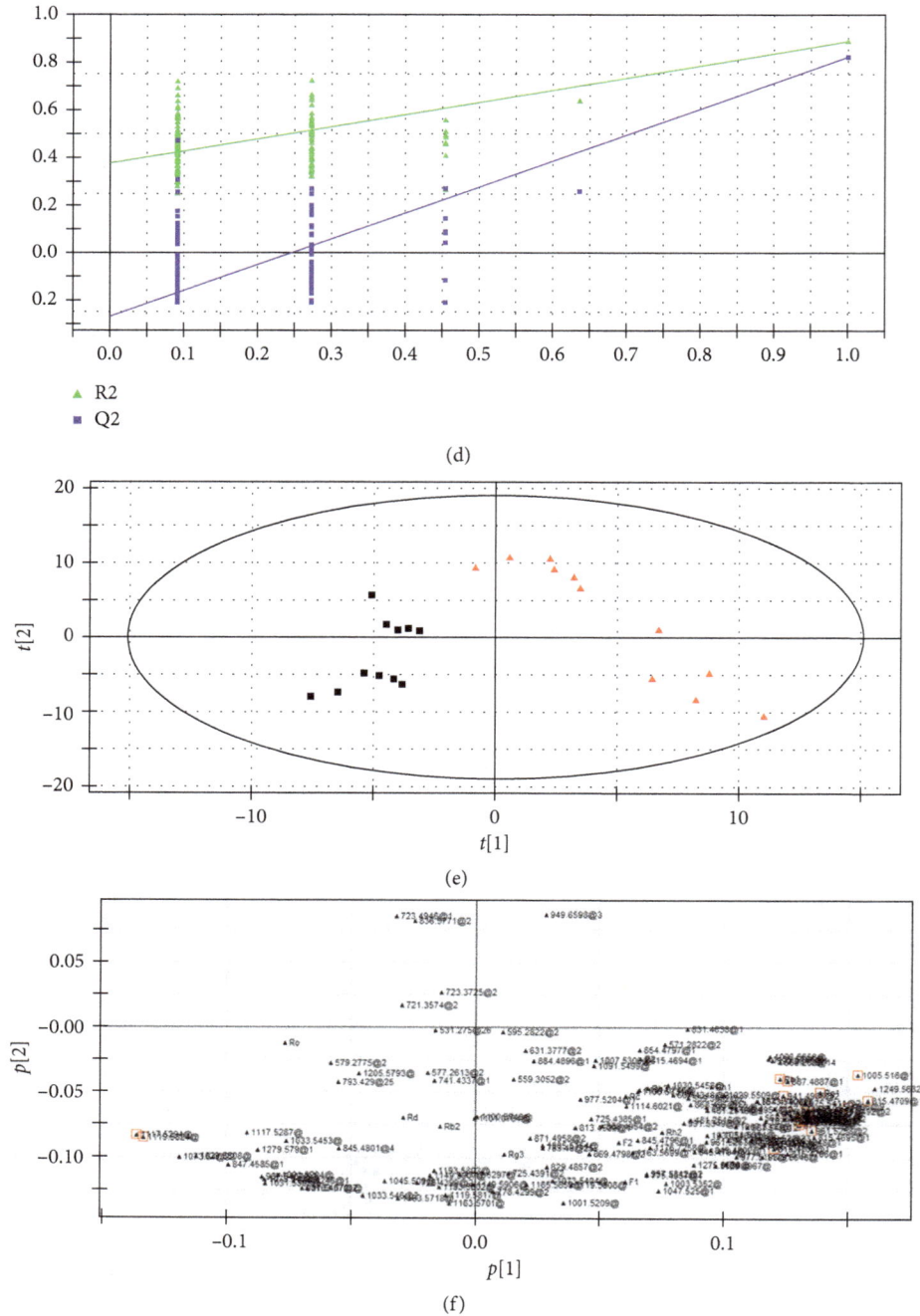

FIGURE 4: (a) Score plot and (b) loading plot of PCA model based on data of eighteen ginsenoside contents in CGLS, FGLS, and AMGLS. (c) Score plot of PLS-DA model based on data of eighteen ginsenoside contents in CGLS and FGLS. (d) Permutation test, (e) score plot, and (f) loading plot of PLS-DA model based on −ESI full scan MS data (▲, CGLS; ■, FGLS; ◆, AMGLS).

The validation results indicated that the −ESI PRM MS was better than +ESI PRM MS in LOD, LOQ, linearity, concentration range, and mean recovery. Therefore, the described −ESI PRM MS method was subsequently applied to the analysis of all the samples. The developed UPLC-Q-Exactive Orbitrap/MS methods showed lower LOD and LOQ, larger concentration range than HPLC methods [18, 25].

3.3. Quantification Analysis Based on −ESI PRM MS Data. The TIC plots of the extracted AMGSL, CGSL, and FGSL are shown in Figures 1(c)–1(e). The differences between AMGSL and Asian Ginseng stem-leaf (ASGSL) could be clearly observed, as Rf and F11 were only present in ASGSL and AMGSL, respectively. The contents of eighteen ginsenosides in every type of samples were detected using the developed PRM/− method, and the quantitative results

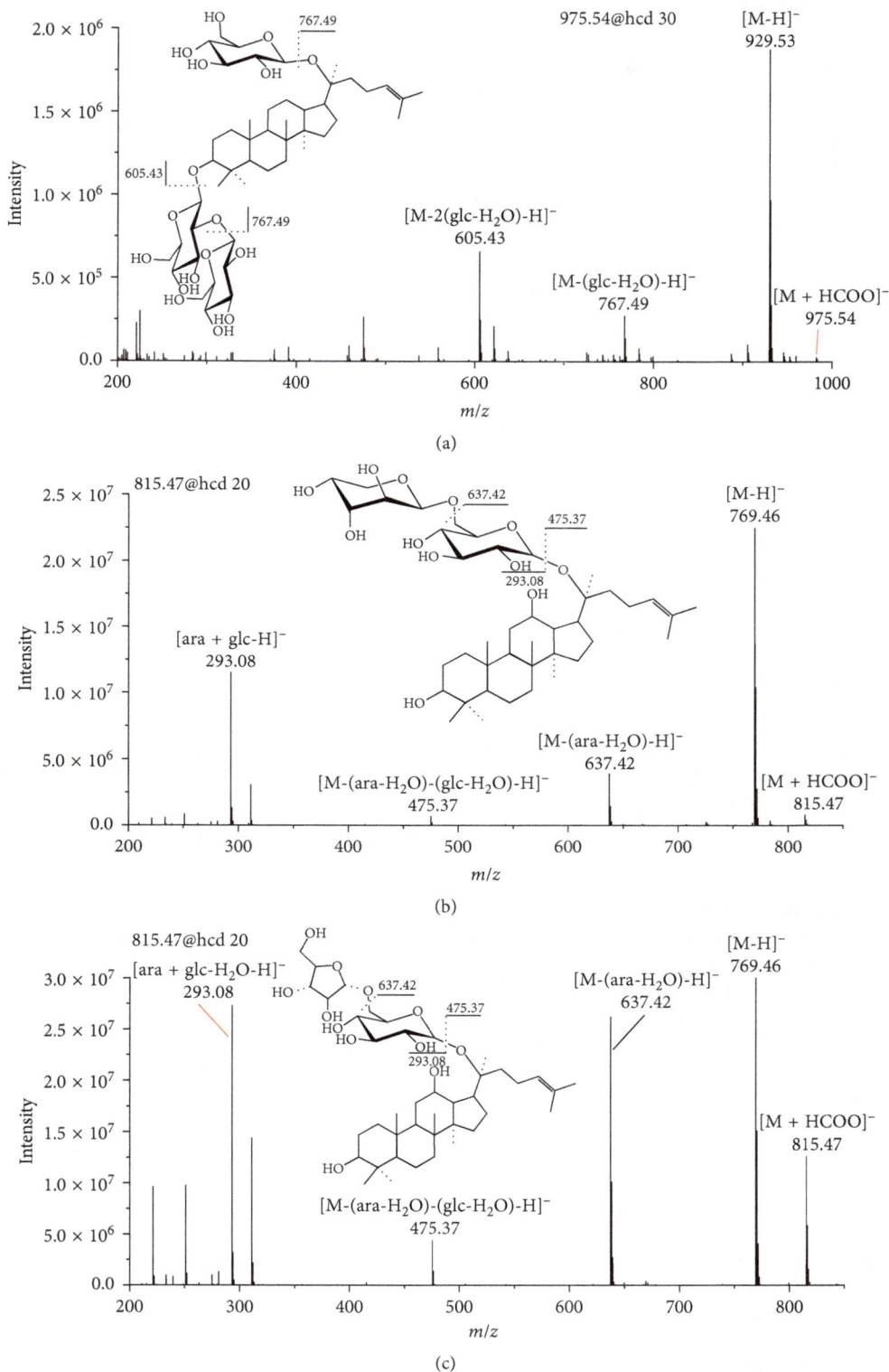

FIGURE 5: MS^2 spectra of [M + HCOO]$^-$ for (a) Vina-ginsenoside R3, (b) ginsenoside F3, and (c) ginsenoside F5 in the negative mode. (The glc represent glucose and the ara represent arabinose.)

are shown in Table 8. Excepting Rf and F11, AMGSL and ASGSL also had large differences in the contents of R2, Rb3, and F1. AMGSL contained less R2, F1 but more Rb3 than ASGSL. For CGSL and FGSL, it was found that the contents of eighteen ginsenosides were very similar. Therefore, it is difficult to discriminate CGSL from FGSL just using the data

TABLE 9: Characterization of analysis markers in CGSL and FGSL based on −ESI full scan MS data.

Variable	Theoretical (*m/z*)	Measured (*m/z*)	Mass error (ppm)	Compounds	Elemental composition	Adduct ions
1117.53@22.49	1117.5284	1117.5286	0.21	—	$C_{49}H_{84}O_{25}$	$[M + HCOO]^-$
1119.58@22.31	1119.5804	1119.5806	0.18	—	$C_{50}H_{90}O_{24}$	$[M + HCOO]^-$
1005.52@12.54	1005.5123	1005.5150	2.78	—	$C_{44}H_{80}O_{22}$	$[M + HCOO]^-$
Rb1	1107.5957	1107.5918	−3.52	Rb1	$C_{54}H_{92}O_{23}$	$[M-H]^-$
Rg1	845.4904	845.4794	−3.54	Rg1	$C_{42}H_{72}O_{14}$	$[M + HCOO]^-$
R1	977.5327	977.5301	−2.66	R1	$C_{47}H_{80}O_{18}$	$[M + HCOO]^-$
975.54@20.84	975.5534	975.5498	−3.69	Vina-ginsenoside R3	$C_{48}H_{82}O_{17}$	$[M + HCOO]^-$
887.49@12.82	887.4857	887.4889	3.78	—	$C_{40}H_{74}O_{18}$	$[M + HCOO]^-$
941.50@25.29	941.4963	941.4987	2.56	—	$C_{44}H_{78}O_{21}$	$[M-H]^-$
1033.55@15.42	1033.5436	1033.5457	2.1	—	$C_{46}H_{84}O_{22}$	$[M + HCOO]^-$
699.42@15.01	699.4266	699.4235	−4.78	—	$C_{43}H_{58}O_5$	$[M + HCOO]^-$
1017.51@25.97	1017.5123	1017.5147	2.44	—	$C_{45}H_{80}O_{22}$	$[M + HCOO]^-$
815.47@17.26	815.4798	815.4695	−4.05	Ginsenoside F5	$C_{41}H_{70}O_{13}$	$[M + HCOO]^-$
939.48@25.88	939.4806	939.4844	4.01	—	$C_{44}H_{76}O_{21}$	$[M-H]^-$
1149.56@25.57	1149.5546	1149.5550	0.37	—	$C_{50}H_{88}O_{26}$	$[M + HCOO]^-$
975.51@22.56	975.5018	975.5056	4.12	—	$C_{43}H_{78}O_{21}$	$[M + HCOO]^-$
991.50@15.87	991.4967	991.4990	2.45	—	$C_{48}H_{82}O_{18}$	$[M + HCOO]^-$
957.50@25.45	957.4912	957.4942	3.29	—	$C_{43}H_{76}O_{20}$	$[M + HCOO]^-$
971.51@25.5	971.5068	971.5085	1.78	—	$C_{44}H_{78}O_{20}$	$[M + HCOO]^-$
957.50@24.95	957.4912	957.4941	3.18	—	$C_{43}H_{76}O_{20}$	$[M + HCOO]^-$
961.53@22.27	961.5225	961.5259	3.71	—	$C_{43}H_{80}O_{20}$	$[M + HCOO]^-$
1107.54@20.3	1107.5440	1107.5445	0.43	—	$C_{49}H_{88}O_{27}$	$[M + HCOO]^-$
1093.57@14.55	1093.5648	1093.5688	3.69	—	$C_{49}H_{90}O_{26}$	$[M + HCOO]^-$
973.49@17.4	973.4861	973.4883	2.36	—	$C_{43}H_{76}O_{21}$	$[M + HCOO]^-$
1089.53@24.31	1089.5335	1089.5349	1.32	—	$C_{49}H_{86}O_{26}$	$[M + HCOO]^-$
815.47@18.11	815.4798	815.4777	−2.58	Ginsenoside F3	$C_{41}H_{70}O_{13}$	$[M + HCOO]^-$

of contents of eighteen ginsenosides. To discriminate among different ginseng stem-leaves and find analytical markers, the multivariate statistical analysis was performed in the next section.

3.4. Multivariate Statistical Analysis. The dataset of the ginsenoside contents was subjected to the multivariate statistical analysis to differentiate the samples from different origins and to discover the chemical markers. Firstly, a PCA mode, an unsupervised pattern recognition technique, was established with $R^2(X)$ (cum) = 0.850 and Q^2 (cum) = 0.561. As shown in Figure 4(a), the samples are separated distinctly into two groups. CGSL and FGSL were clustered together, suggesting their similar content of ginsenosides. And the clear separation between AMGSL and ASGSL was achieved, indicating the significant difference between these two species. To reveal the ginsenosides that contribute most to the separations of samples from different origins, the loading plot was used to select the chemical markers. As shown in Figure 4(b), the ginsenosides far from the origin were considered to contribute most to the separation and were selected as chemical marker candidates. These ginsenosides were further filtered by Student's *t*-test. The variables with contents which are statistically significant different ($p < 0.05$) between two groups were selected as the analytical markers. Finally, six ginsenosides were obtained as the markers and marked with red boxes in the loading plot. They were F11, Rf, R2, F1, Rb1, and Rb3 respectively.

To further explore the differences of targeted ginsenosides present in ASGSL samples, the PLS-DA model [26] was established by setting CGSL as group I and FGSL as group II. The PLS-DA scores plot is established with the model parameters of $R^2(Y) = 0.742$ and Q^2 (cum) = 0.615 and shown in Figure 4(c). A clear separation is still not observed, since the limited ginsenosides could not provide enough characteristic information for their differentiation.

Therefore, the PLS-DA model was established based on the data obtained in −ESI full scan MS method was used to separate CGSL and FGSL. As shown in Figure 4(e), clear separations are observed with $R^2(Y) = 0.890$ and Q^2 (cum) = 0.824, suggesting good fitness and prediction ability of the established PLS-DA model. Permutation test ($n = 200$) was further performed to validate the model. No overfitting was found because the permutated R2 and Q2 values on the left are lower than the original point on the right (Figure 4(d)). These results indicated that the established PLS-DA mode has the high goodness of fit and predictability. Then loading plot (Figure 4(f)) and VIP were used to reveal the potential analytical markers that contribute most to the separation between CGSL and FGSL. After *t*-test, twenty-six chemical components showed statistical differences between groups and considered to be the chemical markers for the separation of ASGSL and marked with red boxes in the loading plot (Figure 4(f)).

Excepting Rb1, Rg1, and R1, the identification of rest analytical markers was conducted by means of accurate mass-to-charge ratio and MS^2 information obtained by MS. The marker 975.54@20.84 was assigned to Vina-ginsenoside R3 [27] whose MS/MS spectrum is shown in Figure 5(a). The $[M + HCOO]^-$ ion of Vina-ginsenoside R3 at *m/z* 975.54 showed the characteristic product ions at *m/z* 767.49 and

605.43. The markers 815.47@18.11 and 815.47@17.26 were assigned to ginsenoside F3 and ginsenoside F5, respectively. The fragmentation schemes for the product ions of ginsenoside F3 and ginsenoside F5 are shown in Figures 5(b) and 5(c). The chemical information of analytical markers is shown in Table 9. All the contents of Rg1, Rb1, R1, Vinaginsenoside R3, ginsenoside F3, and ginsenoside F5 in the FGSL were less than in the CGSL. The reason may be related to cultivation.

4. Conclusions

By using UPLC-Q-Exactive Orbitrap/MS, a rapid, simple, and reliable method to simultaneously determine eighteen ginsenosides (R1, R2, Rb1, Rb2, Rb3, Rc, Rd, Re, Rf, Rg1, Rg2, Rg3, Rh1, Rh2, Ro, F1, F2, and F11) in *Panax* genus stem-leaf was first developed and validated. This method provides an excellent quantitative tool for analysis of ginsenosides in stem-leaf due to its high capacity, high sensitivity, and high selectivity. F11, Rf, R2, F1, Rb1, and Rb3 are analytical markers which could be used to identified ASGSL and AMGLS. Rf was found only in ASGSL, and F11 was found exclusively in AMGLS, which were similar to the difference between Asian ginseng root and American ginseng root [28]. Furthermore, our results suggest that the approach in the present study could be effectively applied to discriminate CGSL from FGSL. Based on UPLC-Q-Exactive Orbitrap/MS combined with multivariate statistical analysis, a reliable and effective approach aimed to discriminate among different types of *Panax* genus stem-leaf has been successfully developed.

Conflicts of Interest

The authors declare that they have no conflicts of interest.

Authors' Contributions

Both Professor Shuying Liu and Yang Xiu played key roles in experiment design and data analysis.

Acknowledgments

The authors are grateful to the financial support by the National Natural Science Foundation of China (21475012) and the Science and Technology Development Plan Project of Jilin Province (20160520123JH and 20160309002YY).

References

[1] Y. Chen, Z. Zhao, H. Chen, Y. Tao, M. Qin, and Z. Liang, "Chemical differentiation and quality evaluation of commercial Asian and American ginsengs based on a UHPLC-QTOF/MS/MS metabolomics approach," *Phytochemical Analysis*, vol. 26, no. 2, pp. 145–160, 2015.

[2] S. Cui, J. Wu, J. Wang, and X. Wang, "Discrimination of American ginseng and Asian ginseng using electronic nose and gas chromatography–mass spectrometry coupled with chemometrics," *Journal of Ginseng Research*, vol. 41, no. 1, pp. 85–95, 2017.

[3] T. T. Tu, N. Sharma, E. J. Shin et al., "Treatment with mountain-cultivated ginseng alleviates trimethyltin-induced cognitive impairments in mice via IL-6-dependent JAK2/STAT3/ERK signaling," *Planta Medica*, vol. 83, no. 17, pp. 1342–1350, 2017.

[4] J. Zhang, X. Y. Howard, W. Gul, D. S. Pasco, M. A. Elsohly, and N. D. Pugh, "In vitro bioassays for standardization of immune enhancing dietary supplements," *Planta Medica*, vol. 82, no. 5, 2016.

[5] Z. Liu, J. Xia, C. Z. Wang et al., "Remarkable impact of acidic ginsenosides and organic acids on ginsenoside transformation from fresh ginseng to red ginseng," *Journal of Agricultural and Food Chemistry*, vol. 64, no. 26, pp. 5389–5399, 2016.

[6] B. Yan, G. Wang, A. Jiye et al., "Construction of the fingerprints of ginseng stem and leaf saponin reference substances and spiked plasma sample by LC-ESI/MS and its application to analyzing the compounds absorbed into blood after oral administration of ginseng stem and leaf saponin in rat," *Biological & Pharmaceutical Bulletin*, vol. 30, no. 9, pp. 1657–1662, 2007.

[7] H. Wang, D. Peng, and J. Xie, "Ginseng leaf-stem: bioactive constituents and pharmacological functions," *Chinese Medicine*, vol. 4, no. 1, p. 20, 2009.

[8] I. H. Cho, H. J. Lee, and Y. S. Kim, "Differences in the volatile compositions of ginseng species (Panax sp.)," *Journal of Agricultural and Food Chemistry*, vol. 60, no. 31, pp. 7616–7622, 2012.

[9] B. Bradbury and P. Saunders, "Discrimination of white ginseng origins using multivariate statistical analysis of data sets," *Journal of Ginseng Research*, vol. 38, pp. 187–193, 2014.

[10] D. Huang, Y. Li, M. Zhang et al., "Tartaric acid induced conversion of protopanaxadiol to ginsenosides Rg3 and Rg5 and their in situ recoveries by integrated expanded-bed adsorption chromatography," *Journal of Separation Science*, vol. 39, no. 15, pp. 2995–3001, 2016.

[11] X. Huang, Y. Liu, Y. Zhang et al., "Multicomponent assessment and ginsenoside conversions of *Panax quinquefolium* L. roots before and after steaming by HPLC-MS n," *Journal of Ginseng Research*, 2017, In press.

[12] L. Jia, Y. Zhao, and X. J. Liang, "Current evaluation of the millennium phytomedicine–ginseng (II): collected chemical entities, modern pharmacology, and clinical applications emanated from traditional Chinese medicine," *Current Medicinal Chemistry*, vol. 16, no. 22, pp. 2924–2942, 2009.

[13] W. Wu, F. Song, D. Guo et al., "Mass spectrometry-based approach in ginseng research: a promising way to metabolomics," *Current Analytical Chemistry*, vol. 8, no. 1, pp. 43–66, 2012.

[14] P. Xia, Z. Bai, T. Liang et al., "High-performance liquid chromatography based chemical fingerprint analysis and chemometric approaches for the identification and distinction of three endangered *Panax* plants in Southeast Asia," *Journal of Separation Science*, vol. 39, no. 20, pp. 3880–3888, 2016.

[15] J. Xie, "American ginseng leaf: ginsenoside analysis and hypoglycemic activity," *Pharmacological Research*, vol. 49, no. 2, pp. 113–117, 2004.

[16] Q. Li, X. Liang, L. Zhao et al., "UPLC-Q-exactive orbitrap/MS-based lipidomics approach to characterize lipid extracts from bee pollen and their in vitro anti-inflammatory properties," *Journal of Agricultural and Food Chemistry*, vol. 65, no. 32, pp. 6848–6860, 2017.

[17] J. B. Renaud, L. Sabourin, E. Topp, and M. W. Sumarah, "Spectral counting approach to measure selectivity of high-

resolution LC-MS methods for environmental analysis," *Analytical Chemistry*, vol. 89, no. 5, pp. 2747–2754, 2017.

[18] F. Liu, N. Ma, C. He et al., "Qualitative and quantitative analysis of the saponins in *Panax notoginseng* leaves using ultra performance liquid chromatography coupled with time-of-flight tandem mass spectrometry and high performance liquid chromatography coupled with UV detector," *Journal of Ginseng Research*, vol. 42, no. 2, pp. 149–157, 2017.

[19] L. W. Qi, H. Y. Wang, H. Zhang, C. Z. Wang, P. Li, and C. S. Yuan, "Diagnostic ion filtering to characterize ginseng saponins by rapid liquid chromatography with time-of-flight mass spectrometry," *Journal of Chromatography A*, vol. 1230, pp. 93–99, 2012.

[20] W. Wu, L. Sun, Z. Zhang, Y. Guo, and S. Liu, "Profiling and multivariate statistical analysis of *Panax ginseng* based on ultra-high-performance liquid chromatography coupled with quadrupole-time-of-flight mass spectrometry," *Journal of Pharmaceutical and Biomedical Analysis*, vol. 107, pp. 141–150, 2015.

[21] R. Goodacre, "Metabolomics shows the way to new discoveries," *Genome Biology*, vol. 6, no. 11, pp. 1-2, 2005.

[22] J. R. Idle and F. J. Gonzalez, "Metabolomics," *Cell Metabolism*, vol. 6, no. 5, pp. 348–351, 2007.

[23] D. Y. Lee, J. K. Kim, S. Shrestha et al., "Quality evaluation of *Panax ginseng* roots using a rapid resolution LC-QTOF/MS-based metabolomics approach," *Molecules*, vol. 18, no. 12, pp. 14849–14861, 2013.

[24] N. Guo, K. Ablajan, B. Fan, H. Yan, Y. Yu, and D. Dou, "Simultaneous determination of seven ginsenosides in Du Shen Tang decoction by rapid resolution liquid chromatography (RRLC) coupled with tandem mass spectrometry," *Food Chemistry*, vol. 141, no. 4, pp. 4046–4050, 2013.

[25] Q. Mao, L. I. Yi, L. I. Song-Lin, J. Yang, P. H. Zhang, and Q. Wang, "Chemical profiles and anticancer effects of saponin fractions of different polarity from the leaves of *Panax notoginseng*," *Chinese Journal of Natural Medicines*, vol. 12, no. 1, pp. 30–37, 2014.

[26] L. Li, S. Liu, L. Ma, Y. Guo, Y. Wang, and S. Liu, "Application of direct analysis in real time-orbitrap mass spectrometry combined with multivariate data analysis for rapid quality assessment of Yuanhu Zhitong Tablet," *International Journal of Mass Spectrometry*, vol. 421, pp. 33–39, 2017.

[27] N. M. Duc, R. Kasai, K. Ohtani et al., "Saponins from Vietnamese ginseng, *Panax vietnamensis* Ha et Grushv. collected in central Vietnam. II," *Chemical and Pharmaceutical Bulletin*, vol. 42, no. 3, pp. 115–122, 1994.

[28] T. W. D. Chan, P. P. H. But, S. W. Cheng, I. M. Y. Kwok, F. W. Lau, and H. X. Xu, "Differentiation and authentication of *Panax ginseng*, *Panax quinquefolius*, and ginseng products by using HPLC/MS," *Analytical Chemistry*, vol. 72, no. 6, pp. 1281–1287, 2000.

17

Development of Micellar HPLC-UV Method for Determination of Pharmaceuticals in Water Samples

Danielle Cristina da Silva[1] and Cláudio Celestino Oliveira ⓘ[2]

[1]*Universidade Tecnológica Federal do Paraná, Campus Dois Vizinhos, Estrada para Boa Esperança, Km 04 85660-000 Dois Vizinhos, PR, Brazil*
[2]*Departamento de Química, Universidade Estadual de Maringá, Avenida Colombo, 5790 87020-900 Maringá, PR, Brazil*

Correspondence should be addressed to Cláudio Celestino Oliveira; ccoliveira@uem.br

Academic Editor: Antonio V. Herrera-Herrera

Method for extraction and determination of amoxicillin, caffeine, ciprofloxacin, norfloxacin, tetracycline, diclofenac, ibuprofen, nimesulide, levonorgestrel, and 17α-ethynylestradiol exploiting micellar liquid chromatography with PDA detector and solid-phase extraction was proposed. The usage of toxic solvents was low; the chromatographic separation of the medicaments was performed using a C18 column and mobile phases A and B containing 15.0% (v/v) ethanol, 3.0% (m/v) sodium dodecyl sulfate (SDS), and 0.02 mol·L^{-1} phosphate at pHs 7.0 and 8.0, respectively. The method is simple, selective, and fast, and the analytes were separated in 23.0 min. For extraction, 1000 mL of sample containing 2.0% (v/v) ethanol and 0.002 mol·L^{-1} citric acid at pH 2.50 was loaded through a 1000 mg of C18 cartridge. The analytes were eluted using 3.0 mL of ethanol, which were evaporated and redissolved in 0.5 mL of mobile phase. Concentration factors better than 1200, except amoxicillin (224), were obtained. The analytical curves were linear (R^2 better than 0.992); LOD and LOQ ($n = 10$) presented values in the range of 0.019–0.247 and 0.058–0.752 mg·L^{-1}, respectively. Recoveries of 99% were obtained, and the results are in agreement with those obtained by the comparative methods.

1. Introduction

For many years, the analysis of environmental contaminants was done to determine the presence of pesticides, air pollutants, petroleum residues, and many other substances designated as conventional pollutants [1, 2]. Recently, a large number of compounds not regulated by countries legislation have been identified as potential pollutants, which have been designated as emerging contaminants [3]. The pharmaceuticals and their metabolites belong to this new class of pollutants, although their presence in the environment is not new [4, 5].

Many tons of these drugs are produced annually and used in human and veterinary medicine. Generally, the exact amount of pharmaceuticals produced is not published in the literature [6], but it is known that Brazil, United States, France, and Germany belong to the group of the world's largest consumers of these drugs [7]. Thus, the monitoring of these compounds in the environment has gained great interest

as they are often found in effluents from wastewater treatment plants and natural waters [4, 8, 9].

In Brazil, the population self-medication is common because part of the population does not have access to adequate medical care and has easy access to medicines due to the high number of drugstores, including those with unethical business practices [8, 10]. After the administration, many pharmaceuticals are transformed into one or more metabolites and excreted in the urine and feces, causing serious problems to the environment [4, 8, 10]. It is known that the rampant use of pharmaceuticals such as antibiotics can make some microorganisms resistant to these drugs as some bacteria have the ability to modify their genetic material [6, 11, 12].

Furthermore, some drugs are used in animal treatment in the rearing of livestock, pigs, and chickens, and the waste generated from these activities has become a major source of environmental contamination due to its use as fertilizer in farmland. Thus, the pharmaceutical compounds are not metabolized and their metabolites can pass into ground-water and eventually to watercourses such as rivers and lakes, affecting aquatic life [6, 8, 12]. Another source of environmental contamination by these drugs is associated with the disposal of waste from pharmaceutical companies and hospitals in landfills that can contaminate underground waters [4, 6].

The presence of hormones in the environment has been indicated as responsible for causing endocrine disturbances in human and animal organisms and endocrine disruptors [13, 14]. There is evidence that reproductive system of certain terrestrial and aquatic organisms is affected by estrogen, resulting in the development of abnormalities and reproductive impairment in exposed organisms, even when these drugs are present at low concentrations [6]. Caffeine is a natural stimulant and the most widely consumed psychoactive drug in the world as it is present in soft drinks, coffee, tea, cocoa, and chocolate and is used concomitantly with various medications as a stimulant [15]. Due to its widespread consumption, caffeine has been used as a potential indicator of anthropogenic pollution of surface water resulted from human activity [16, 17].

Due to the growing concern of the presence of antibiotics, anti-inflammatories and hormones in water intended for public supply a number of methodologies devoted to identify and quantify these compounds in various samples that have been proposed. The most common are those involving separation techniques such as gas and liquid chromatography and capillary electrophoresis coupled with several detectors such as MS, UV-Vis, FID, ECD, and others [18–27] associated with SPE with solvents such as formic acid/water [20], methanol/methyl tertiary-butyl ether [21], n-hexane/ethyl acetate/methanol [22], or more modern methods such as ultrasonic-assisted extraction/centrifugation and purification with SPE [18] and the pressurized liquid extraction followed by extract purification using SPE [19].

As it is not an easy task the development of extraction and determination methods to chemical compounds with very different characteristics, few methods permit the determination of the analytes present in this paper in a single run. Thus, environmental researchers and laboratories dedicated to the analysis of drugs should use several methods for the determination of drugs from different classes in environmental samples, which increase the cost and time of the analysis. The exception is the method based on MS detector that can furnish adequate results to water analysis of emergent pollutants, but it should be considered that a lot of laboratories dedicated to routine water analysis did not have mass spectrometer, mainly due to the cost to acquire the equipment or lack of skilled professionals capable of properly using the equipment.

Thus, to fill this gap, in the present work is proposed a robust, simple, and green HPLC-UV method associated with SPE extraction for the determination of antibiotics (amoxicillin, ciprofloxacin, norfloxacin, and tetracycline), anti-inflammatories (diclofenac, ibuprofen, and nimesulide), hormones (17α-ethynylestradiol and levonorgestrel), and caffeine in water samples, providing the laboratories dedicated to water analysis with a tool able to determine the

more representative pharmaceuticals from classes of antibiotics, anti-inflammatories, and hormones used in Brazil in a single chromatographic run and using a simple HPLC-UV; a common equipment in a number of laboratories devoted to contaminant analysis, contributing to environmental researchers involving emerging pollutants. The task can be done exploiting the micellar chromatography as the surfactant can modify the C18 and the aqueous phase, increasing the possibilities of interactions of the analytes with both phases, permitting the separation of substances with different polarities in the same chromatographic run. The selected compounds are considered toxic, are among the most used in Brazil, present biological activity, and their presence in the aquatic environment has already been demonstrated, which justify their determination.

2. Experimental

2.1. Reagents. All reagents and solvents were of analytical grade (purity higher than 98%); the aqueous solutions were prepared using deionized water and were ultrasonically degassed and vacuum-filtered through a cellulose acetate membrane of $0.45\,\mu m$ before chromatographic use.

Solutions containing ethanol or n-propanol or n-butanol of HPLC-grade (Tedia, Fairfield, USA), anhydrous dibasic sodium phosphate (Na_2HPO_4), SDS, or CTAB (cetyl trimethyl ammonium bromide), or SDBS (sodium dodecyl benzene sulfonate), or Triton X-100 (polyoxyethylene (9-10) p-phenol tertoctyl) (both from Sigma-Aldrich, St. Louis, USA) at different concentrations and pH values were tested as mobile phase. The main mobile phases containing 3.0% (m/v) SDS, $20.0\,mmol\cdot L^{-1}$ Na_2HPO_4, and 15.0% (v/v) ethanol at pH 7.0 (phase A) and pH 8.0 (phase B) were prepared by dissolving 3.046 g of SDS, 0.284 g of Na_2HPO_4, and 15.0 mL of ethanol at approximately 70.0 mL of water and leaving the solution under stirring until complete homogenization. Later, the pH was adjusted by adding HCl or NaOH $1.0\,mol\cdot L^{-1}$, and the final volume was made up to 100.0 mL in a volumetric flask. All mobile phases used in this work were prepared following the same procedures and just modifying the percentages of reagents or changing the organic solvent to n-propanol or n-butanol or the surfactant.

Amoxicillin, ciprofloxacin, norfloxacin, tetracycline, diclofenac, ibuprofen, nimesulide, 17α-ethynylestradiol, levonorgestrel, and caffeine were supplied by Sigma-Aldrich (St. Louis, USA), and $20.0\,mg\cdot L^{-1}$ stock solutions were prepared by dissolving 1.0 mg of each drug in the mobile phase A, the solution was submitted to sonication for 3.0 min to ensure complete dissolution of analytes, and the volume was completed to 50.0 mL in a volumetric flask. The solutions were stored in amber flasks, protected from light, and were frozen. Just before analysis, the solutions were left to attain thermal equilibrium with the room temperature.

To verify the stability of each drug in the mobile phase and the wavelength to monitor them, 50 mL of $10.0\,mg\cdot L^{-1}$ of each drug standard solution was prepared in the mobile phase A and divided into two portions of 25.0 mL. The first set was frozen and protected from light, and the second set was kept at 25°C and immediately used to obtain the UV-Vis spectra (200 to 600 nm). The procedure was repeated each 60 min for 3 h during the first day and then the solution was stored in the refrigerator at 10°C. Later, the solutions (two sets) were analyzed every day during 4 days.

2.2. Sampling, Sample Treatments, and Optimization of the SPE Procedure. A total of 7 river water samples were collected from rivers flowing in the center of Maringá (Paraná, Brazil) in zones with different population densities and industrial activities. Samples 1–3 were collected from the south of Maringá in Moscados stream (ca. 2.9 km from its source in the Inga Park), Cleópatra stream (ca. 2.3 km from its source, located inside the Pioneiros Forest Park), and Borba Gato stream (ca. of 3.3 km from its source located in the Horto Florestal Park), respectively. Samples 4–6 were collected in the north of Maringá in Mandacarú stream (ca. 2.0 km from the source), Morangueiro stream (ca. 4.0 km from the source), and Maringá stream (ca. 2.0 km from the search), respectively; sample 7 was tap water collected in the analytical chemistry laboratory at Maringá State University.

Before sample collection, each bottle was prerinsed with the sample for three times. The samples were sent in boxes packed with ice to the laboratory at Maringá State University. Immediately upon reception, the samples were vacuum filtered through Millipore 0.45 μm membrane; then, to 1.0 L of each sample was added 2.0% (v/v) ethanol and 2.0 mmol$\cdot$L^{-1} citric acid, and the pH was adjusted to 2.5 with HCl 1.0 mol$\cdot$L^{-1}.

For sampling, 4 L of each water sample was collected according to the standard protocol established by the Water Resources Company Management [28], which considers sampling timing, sampling point, sampling tools and containers, sampling operation, field records, labeling, transport, and storage of samples. Immediately upon reception, the samples were vacuum filtered through Millipore 0.45 μm membrane and to 1.0 L of each sample was added 2.0% (v/v) ethanol and 2.0 mmol$\cdot$L^{-1} citric acid, and the pH was adjusted to 2.5 with HCl 1.0 mol$\cdot$L^{-1}. Later, the samples were submitted to the following extraction procedure: before the extraction, the C18 stationary phase was preconditioned by passing 5 mL of methanol, 5 mL of pure water, and 5 mL of an aqueous solution containing 2.0% (v/v) ethanol and 2.0 mmol$\cdot$L^{-1} citric acid at pH 2.5; then, 1000 mL of each sample was passed through the cartridge at a flow rate of 7.0 mL$\cdot$min^{-1}. The solid phase was dried under vacuum for 30 s, and the analytes were eluted using 3.0 mL of ethanol at 1.0 mL$\cdot$min^{-1}. The sample was filtered through a teflon membrane, evaporated, and redissolved in 0.5 mL with the mobile phase A and injected into the chromatograph. To obtain the extraction procedure able to be applied to the ten analytes, the following variables were tested: pH (2.0–8.7), sample (1.0–10.0 mL$\cdot$min^{-1}) and eluent (0.5–3.0 mL$\cdot$min^{-1}) flow rates, sample (25.0–1000 mL) and eluent (1.0–10 mL) volumes, amount of stationary phase (500 and 1000 mg), and chemical nature of the eluent (methanol, ethanol, acetone, and acetonitrile).

2.3. Equipment and Separation Conditions. The chromatographic separation was performed using an HPLC from Thermo Electron Corporation (Waltham, USA) containing quaternary pump model Surveyor LC Plus, manual injector valve of 20 μL Rheodyne Model 8096, UV-Vis photodiode array detector model Surveyor PDA with quartz cell with optical path of 5.0 cm, ChromQuest software version 4.2 (Macherey-Nagel, Germany) for acquisition and signal recording and a column RP-18 ODS off base 250 × 4.6 mm (id) equipped with a guard column RP-18 ODS 10 × 4.0 mm (id) both with particles of 5-micron pore size of 100 Å and carbon content of 15.5%.

Before the first and after the last injection of the day, the column was cleaned with ultrapure water for 30.0 min at a flow rate of 0.5 mL$\cdot$min^{-1}. The initial conditioning of the stationary phase was performed by passing mobile phase A through the column for 20.0 min at a flow rate of 1.0 mL$\cdot$min^{-1}. After standard/sample injection (20.0 μL), the separation process was carried out as follows: 0.0–2.0 min with 100.0% of phase A with the instantaneous change to the 100.0% of phase B and keeping it until 25 min always at a flow rate of 1.0 mL$\cdot$min^{-1}. The temperature was fixed at 25°C, and the analytes were monitored at 220 (amoxicillin, norfloxacin, tetracycline, diclofenac, ibuprofen, nimesulide, 17α-ethynylestradiol and caffeine), 240 (levonorgestrel), and 280 nm (ciprofloxacin), simultaneously. After each analysis, the column was reconditioned for 10.0 min using phase A at a flow rate of 1.0 mL$\cdot$min^{-1}.

2.4. Optimization of the Chromatographic Method. Preliminary experiment was done to identify the main variables affecting the chromatographic separation such as pH (5.0, 6.0, 7.0, and 8.0), percentage of the organic modifier in the mobile phase (15.0, 20.0, and 25.0% (v/v)), nature of the organic modifier (n-butanol, n-propanol, and ethanol), SDS concentration (2.5, 3.0, and 3.5% (m/v)), and flow rate (0.8, 1.0, 1.2, and 1.5 mL$\cdot$min^{-1}). Therefore, a factorial design experiment 2^3 was carried out in duplicate in order to optimize the separation conditions and the contribution of the variables (Na$_2$PO$_4$, ethanol, and SDS concentrations) in the chromatographic separation (Table 1).

2.5. Method Calibration, Characterization, and Sample Analysis. For quantitative determination of pharmaceuticals and caffeine in the water samples, calibration curves were plotted using the peak area (y) versus concentration of the analytes in the following ranges: amoxicillin, ciprofloxacin, diclofenac, ibuprofen, tetracycline, levonorgestrel, nimesulide, and norfloxacin from 0.1 up to 25.0 mg$\cdot$L^{-1}; 17α-ethynylestradiol from 0.5 up to 25.0 mg$\cdot$L^{-1}; and caffeine from 0.08 up to 25.0 mg$\cdot$L^{-1}. The calibration curves were used to determine the analyte concentrations in the samples and in the blank with and without spiking.

The LOD and LOQ ($n = 10$) were estimated using the signal-to-noise ratio of 3.3 and 10.0, respectively [9], and 7 river water samples were analyzed by the proposed method and the obtained results were compared to those obtained by the six comparative HPLC-DAD methods involving SPE

TABLE 1: Factorial experiment 2^3 to optimize the separation of the analytes.

Experiment	[Ethanol], % (m/v)	[SDS], % (m/v)	[Phosphate], mmol·L^{-1}
1	12	2.5	20
2	15	2.5	20
3	12	3.0	20
4	15	3.0	20
5	12	2.5	30
6	15	2.5	30
7	12	3.0	30
8	15	3.0	30

[18, 24, 29, 30], stir-bar sorptive and liquid desorption [25], and liquid-liquid extraction [31] methods to the following analytes: tetracycline [18]; ibuprofen, diclofenac, and nimesulide [24]; levonorgestrel and 17α-ethynylestradiol [25]; caffeine [29]; ciprofloxacin and norfloxacin [30]; and amoxicillin [31].

Sample recovery tests were done spiking the samples with the following: 5.0 μg·L^{-1} of ibuprofen, 17α-ethynylestradiol, diclofenac, nimesulide, levonorgestrel, ciprofloxacin, and norfloxacin; 10 μg·L^{-1} of tetracycline and caffeine; and 60 μg·L^{-1} of amoxicillin. After extraction procedure, the extracts were evaporated completely and stored in the refrigerator; just before analysis, the samples were redissolved in 0.5 mL of the mobile phase A. The spiked samples were analyzed ($n = 5$) by the proposed and by comparative methods during five consecutive days ($n = 5$).

3. Results and Discussion

3.1. Spectra and Stability of the Analytes. The UV-Vis spectra of analytes showed electromagnetic radiation absorption in the region between 200 and 600 nm with higher intensity in the UV region. It is observed that signal overlaps, which makes it difficult for the simultaneous determination of the analytes without prior separation. It was decided to monitor the analytical signals at 220, 240, and 280 nm, because the molar absorptivity is high to the majority of the analytes and the mobile phase absorption is low; the exceptions are levonorgestrel ($\lambda_{max} = 240$ nm) and ciprofloxacin ($\lambda_{max} = 280$ nm) that presented low molar absorptivity at 220 nm. For this reason, the signals can be monitored at 220 nm by laboratories that do not have chromatograph with PDA detector, but the sensitivity to levonorgestrel and ciprofloxacin will be poor.

The UV-Vis spectra also indicated that the analytes were stable in the mobile phase A for a long period, except tetracycline that presented degradation of 4.1, 8.5, 13.4, and 16.0% after 24, 48, 72, and 96 h, respectively. When the samples were kept frozen and in the absence of light before analysis, the tetracycline degradation decreased to lower than 2.0%, indicating that the standards and samples should be kept under these conditions until analysis.

3.2. Extraction of the Analytes. Initially, it was decided to use the C18 phase, control the pH, and use methanol as eluent,

as 10 mL of this solvent permitted to get the elution of ten analytes from the cartridge.

The sample pH was varied from 2.0 up to 8.5, and pH 2.5 was chosen as the better condition due to the optimum extraction percentage for most of the analytes. Amoxicillin, high polar molecule, presented low affinity by the solid phase, and its recuperation was always lower than 51%, whereas ibuprofen, diclofenac, and nimesulide presented high interaction with the solid phase and could only be eluted around pH 2.0 with recoveries of 83.6, 59.0, and 75.5%, respectively. Ciprofloxacin, norfloxacin, and tetracycline have two positive charges at pH 2.5, and the increase in the pH reduced their recuperation from 92.0, 89.0, and 100.0% for pH 2.5 to 36.3, 19.4, and 42.2% for pH 8.5, respectively; and caffeine, 17α-ethynylestradiol, and levonorgestrel did not suffer a significant change in their interaction with the solid phase with the pH and presented recuperation percentages of 94.7, 92.0, and 80.0%, respectively.

A flow rate of 7.0 mL·min^{-1} was chosen as the better compromise between time and efficiency of extraction; for this condition, recoveries around 95% were obtained for the analytes; the exception was amoxicillin (50%). Variations in the citric acid concentration (2.0, 3.0, and 5.0 mmol·L^{-1}) did not change the extraction efficiencies; thus, 2.0 mmol·L^{-1} of citric acid was maintained.

The loading sample volume should be as high as possible to get high concentration factors. When solutions of the analytes in the concentrations of 1.0, 0.5, 0.25, 0.05, and 0.025 mg·L^{-1} were associated with extraction sample volumes of 25, 50, 100, 500, and 1000 mL (to keeping the final concentration after elution in 2.5 mg·L^{-1}), it was noted an improvement in the extraction with the increase of sample volume; however, for the more polar analytes (amoxicillin, tetracycline, caffeine, and diclofenac), a reduction in the recovery percentages was observed. For sample volume of 500 mL, the best recoveries for most analytes were obtained with values higher than 85%, except for amoxicillin (15.3%); however, in order to obtain high concentration factor to levonorgestrel and 17α-ethynylestradiol hormones, that could be present at very low concentrations in the samples, the condition of 1000 mL was selected, even with a reduction in the extraction efficiency to amoxicillin, tetracycline, caffeine, and diclofenac.

The amount of solid phase in the cartridge was also studied, and it was noted that reducing the amount of solid phase led to lower extraction efficiency to tetracycline and amoxicillin. Thus, the cartridge with 1000 mg of C18 was selected.

In order to achieve the highest possible concentration factor with minimum use of organic solvents (methanol, ethanol, acetone, and acetonitrile), the volume of the solvent was varied (1.5, 3.0, 5.0, 7.0, and 10.0 mL) and it was not observed significant variations in the analytes recoveries when the eluent was methanol or ethanol and the eluent volume was varied from 3.0 to 10.0 mL; thus, 3.0 mL of ethanol was elected as the better condition due to its low toxicity. To acetone and acetonitrile, the recoveries were low to all analytes, mainly to tetracycline, norfloxacin, and ethynylestradiol. The eluent flow rate was varied, and it was observed a decrease in the extraction efficiencies for all the analytes to flow rates higher than 1.0 mL·min^{-1}; then, the eluent flow rate was fixed at 1.0 mL·min^{-1}.

Thus, the final extraction conditions for simultaneous extraction and concentration of the ten analytes were as follows: 1000 mL of the sample containing 2.0% (v/v) ethanol, 2.0 mmol·L^{-1} citric acid at pH 2.50, C18 cartridge with 1000 mg of the solid phase, and flow rate of 7.0 mL·min^{-1}. After analytes' retention, the adsorbent was dried under vacuum 30 s and the analytes were eluted with only 3.0 mL of ethanol at a flow rate of 1.0 mL·min^{-1}. The solvent was evaporated, and the samples were redissolved in 0.5 mL of a solution with 15.0% (v/v) ethanol, 3.0% (m/v) SDS, and 20.0 mmol·L^{-1} phosphate buffer at pH 7.0 (mobile phase A). Under these conditions, it was yielded recovery percentage values of higher than 95%, except to tetracycline (64.3%) and caffeine (66.0%), and concentration factors higher than 1200, except to amoxicillin (224).

3.3. Chromatographic Separation

3.3.1. Preliminary Tests. The developed chromatographic method should be as green as possible, and taking into account the different polarities of the analytes, it was decided to exploit the micellar chromatography [32, 33] due to the possibility to get the solubilization of organic compounds of low polarity in the aqueous medium and, at the same time, change the polarity of the C18 stationary phase with the surfactants. For the task, SDS, CTAB, SDBS, and Triton X-100 surfactants were tested and the anionic surfactant SDS was chosen because of its low critical micellar concentration (cmc of 0.0082 mol·L^{-1} in water), which permits to use lower surfactant concentration. Furthermore, SDS presents low light absorption and scattering effect in the UV region, where the analytes should be monitored, allowing to get lower baseline signals and better sensitivity.

Low toxic organic modifier solvent to assist the solubilization of the ten analytes in the micellar medium was chosen after solubility tests with several proportions of ethanol or n-propanol or n-butanol with water and SDS. It was observed that solutions containing 15% of ethanol or butanol or n-propanol together with at least 3.0% SDS and 20 mmol·L^{-1} of phosphate buffer at pH 7.0 were able to solubilize all the analytes as well as elute them from the C18 column (Figure 1). The differences in the chromatograms for the different organic modifiers can be attributed to the different polarities of each organic solvent aliphatic chain, which explains the highest retention time obtained when ethanol was used as an organic modifier (Figure 1). Due to this effect, ethanol was considered the better organic modifier because it improved the symmetry and chromatographic resolution of some peaks (except nimesulide, ciprofloxacin, and norfloxacin) and is low toxic than n-butanol and n-propanol. In addition, the use of ethanol permitted the identification of the tetracycline peak, which could not be done when it was used n-propanol or n-butanol.

3.3.2. pH Effect in the Chromatographic Separation. pH is an important variable to be studied when the analytes have ionizable groups, as variations in pH can promote changes in

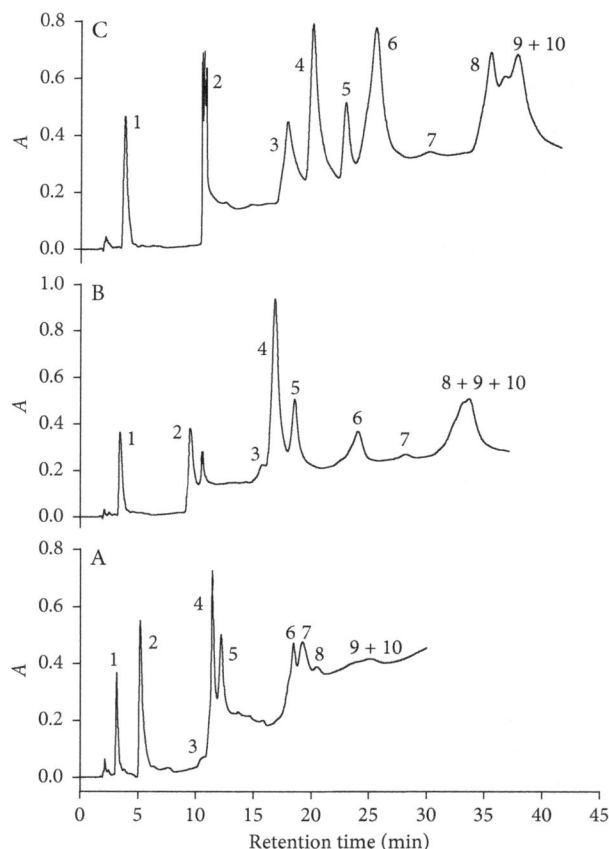

FIGURE 1: Influence of the nature of the organic modifier in the analyte separation. Solvents: (a) n-butanol, (b) n-propanol, and (c) ethanol. Data were obtained for standard solution of amoxicillin (1), caffeine (2), tetracycline (3), diclofenac (4), ibuprofen (5), 17α-ethynylestradiol (6), levonorgestrel (7), nimesulide (8), ciprofloxacin (9), and norfloxacin (10) in the concentration of 20.0 mg·L^{-1}, with a injected volume of 20 μL, at a flow rate of 1.0 mL·min^{-1}, at 25°C, and detection at 220 nm. Phase A with 3.0% (v/v) of solvent and 0.3% (m/v) SDS and phase B with 15% (v/v) of solvent and 3.0% (m/v) SDS and both with 20 mmol·L^{-1} of phosphate at pH 7.0. The mobile phase was changed from 100% of phase A to 100% of phase B in 30 min. A in the y-axis is the absorbance.

the solubility and in the ionic interactions of analytes with the micellar medium and with the stationary phase. When the pH was varied from 6.0 up to 8.0 (Figure 2), caffeine (peak 2, pKa 2.19), 17α-ethynylestradiol (peak 6, pKa 9.44), and levonorgestrel (peak 7, pKa 1.05) did not show significant variation in their retention time as these compounds are neutral in that pH range; the different retention time periods were attributed to their different hydrophobicities. On the other hand, the density of negative charges increased with pH to diclofenac (peak 4, pKa 5.35), ibuprofen (peak 5, pKa 5.82), and nimesulide (peak 8, pKa 7.15), causing large reduction in their retention time due to the high affinity of the charged analytes by the micellar aqueous mobile phase and their repulsion by the negatively charged groups of the surfactants adsorbed on the stationary phase.

Tetracycline (pKa's 3.94, 7.62, and 9.19), ciprofloxacin (pKa's 3.32, 7.12, and 8.42), and norfloxacin (pKa's 3.38, 7.16,

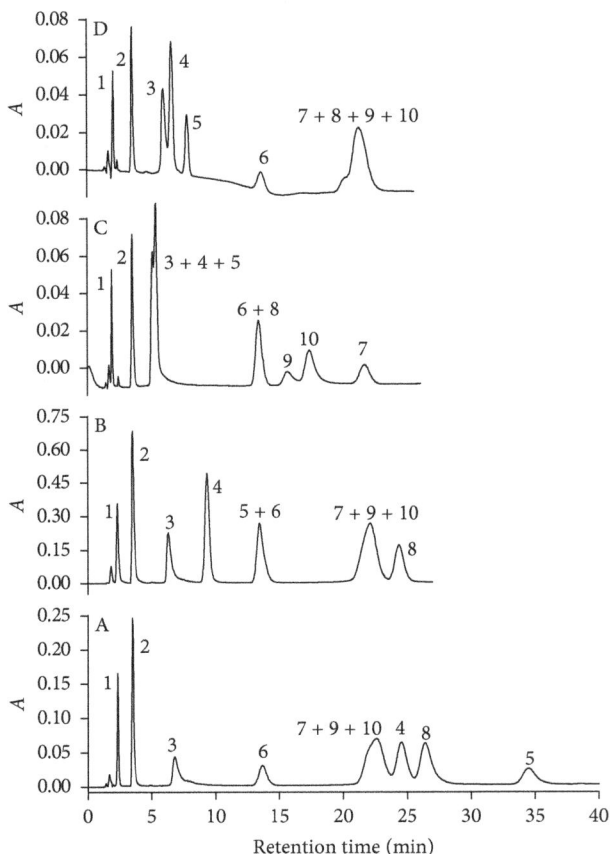

FIGURE 2: Effect of pH on the analyte separation. pH: (a) 6.0; (b) 7.0; (c) 8.0, and (d) 50% of phase A (pH 7.0) and 50% of phase B (pH 8.0). Data were obtained for standard solution of amoxicillin (1), caffeine (2), tetracycline (3), diclofenac (4), ibuprofen (5), 17α-ethynylestradiol (6), levonorgestrel (7), nimesulide (8), ciprofloxacin (9), and norfloxacin (10) in the concentration of 20.0 mg·L^{-1}, with an injected volume of 20 μL, at a flow rate of 1.0 mL·min^{-1}, at 25°C, and detection at 220 nm. Mobile phase contains 15.0% (v/v/) ethanol, 3.0% (m/v) SDS, and 20.0 mmol·L^{-1} of phosphate and isocratic elution. A in the y-axis is the absorbance.

and 8.45) have positive charge densities in the tested pH interval and they are the most soluble molecules in aqueous medium, but their electrostatic interaction with the negative part of the surfactant chain attached to C18 phase increased their retention time to values to those obtained with the hydrophobic compounds. The effect also caused peak broadening, and at pH 5.0, it was not possible to identify the tetracycline peak due to excessive molecule retention. Amoxicillin (pKa's 3.01, 7.32, and 9.70, pI 5.20) is also an amphoteric molecule and with the pH reduction from 7.0 to 6.0 (Figures 2(a) and 2(b)), its retention time increased because the molecule almost reached the neutrality and became more hydrophobic.

In none of the tested pH conditions was possible to obtain the chromatographic separation of the ten analytes with resolutions adequate to their analytical determination. Due to the different analyte characteristics, they could be divided into two groups: those with symmetric and thin peaks and presenting low retention time and those presenting broad

peaks, high retention time, and low efficiency and chromatographic resolution.

Then, it was decided to do a mobile phase pH gradient starting with 100% of phase A changing gradually (7.0, 5.0, and 2.0 min) to 100% of phase B (same composition of phase A at pH 8.0). The phase with pH 6.0 was not chosen because some analytes were highly retained in the stationary phase under this condition (Figure 2).

It was observed that as sooner B phase was introduced, lower analytes' retention time and better analytes separation were obtained, especially to nimesulide (peak 8), ciprofloxacin (peak 9), norfloxacin (peak 10), and levonorgestrel (peak 7) (Figure 3). The chromatographic resolution between diclofenac (peak 4) and ibuprofen (5) worsened, but without compromising their quantification. Thus, it was concluded that the better pH gradient would be 100% of phase A until 2.0 min with a sudden change to 100% of phase B 100%. Under this condition, it was possible to get the separation of the ten compounds with adequate chromatographic resolution in only 23.0 min (Figure 3(d)).

3.3.3. Variables' Interactions.

The preliminary experiments demonstrated that the use of ethanol as an organic modifier and gradient of phases A and B were adequate to achieve the analyte separation; thus, a factorial design 2^3 (Table 1) with SDS, ethanol, and phosphate concentrations as variables in both phases was carried out to verify the main and the interaction effects of variables and their importance in the analytes separation.

Reducing ethanol concentration in the mobile phase from 15.0% to 12.0% (v/v) promotes an increase in the analyte retention time and in time of analysis (experiments 4 and 3, Figure 4). The effect can be explained by the reduction of nonpolar characteristic of the mobile phase, which is difficult for the elution of the low polar analytes from the C18 phase, which led to an overlap between tetracycline (peak 3) and diclofenac (peak 4), nimesulide (peak 8) and ciprofloxacin (peak 9), and norfloxacin (peak 10) and levonorgestrel (peak 7), decreasing their chromatographic resolutions (Table 2).

The reduction in SDS concentration from 3.0% to 2.5% (m/v), experiments 4 and 2 (Figure 4), also increased the retention time for peaks 3, 4, 6, 9, 10, and 7. The reduction in the number of micelles in the mobile phase increased the hydrophobic interaction of 17α-ethynylestradiol (peak 6) and levonorgestrel (peak 7) by C18 phase, causing the coelution of 17α-ethynylestradiol (peak 6) with nimesulide (peak 8). The compounds 3, 4, 9, 10, and 7 increased their affinities by the stationary phase due to electrostatic interactions between analytes, partially positive charged, and the negative ionized sulfonic acid groups adsorbed on C18 phase. This factor improved the separation between nimesulide (peak 8) and ciprofloxacin (peak 9) and ibuprofen (peak 9) and norfloxacin (peak 10) (Table 2).

Analyzing the results furnished by experiments 4 and 8, it was noted that the increase in the phosphate concentration (from 20 to 30 mmol·L^{-1}), and therefore Na$^+$ concentration, reduced the affinity of ibuprofen (peak 5), nimesulide (peak 8),

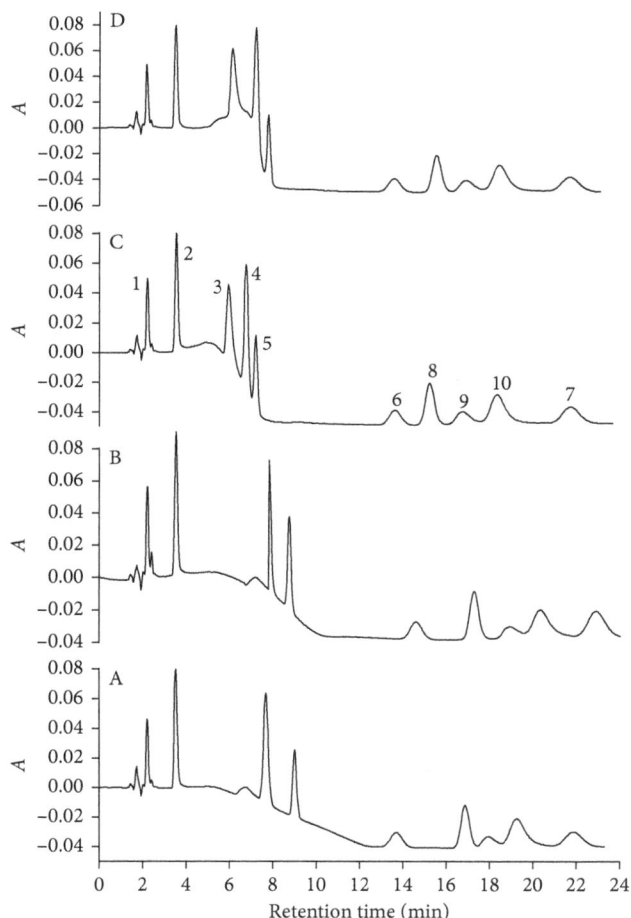

FIGURE 3: Effect of the pH gradient on the analyte separation. Gradients: (a) 7.0 min, (b) 5.0 min, (c) 2.0 min, and (d) isocratic until 2.0 min. Data were obtained for standard solution of amoxicillin (1), caffeine (2), tetracycline (3), diclofenac (4), ibuprofen (5), 17α-ethynylestradiol (6), levonorgestrel (7), nimesulide (8), ciprofloxacin (9), and norfloxacin (10) in the concentration of 20.0 mg·L^{-1}, with an injected volume of 20 μL, at a flow rate of 1.0 mL·min^{-1}, at 25°C, and detection at 220 nm. Mobile phase contains 15.0% (v/v) ethanol, 3.0% (m/v) SDS, and 20.0 mmol·L^{-1} of phosphate at pH 7.0 (phase A) and pH 8.0 (phase B). A in the y-axis is the absorbance.

norfloxacin (peak 10), and ciprofloxacin (peak 9) by the stationary phase due to two reasons. First, because ion-exchange competition between Na$^+$ and the positively charged analytes by the ionized sulfonic acid groups absorbed on C18 phase and, second, due to the partial stabilization of the analytes charged by the phosphate ions. Thus, the chromatographic resolutions between diclofenac (peak 4) and ibuprofen (peak 5) and 17α-ethynylestradiol (peak 6) and nimesulide (peak 8) decreased from 1.68 and 2.10 to 0.51 and 1.42, respectively (Table 2). The variations in ethanol and SDS concentration practically did not affect the amoxicillin peak, whereas caffeine experienced a slight increase in its retention time when ethanol concentration was decreased.

The association of lower percentages of SDS and ethanol led to higher time of analysis (30 min), and despite an improvement in the resolution between peaks 6 and 10, the

resolution between diclofenac (peak 4) and ibuprofen (peak 5) decreased (Figure 4, experiment 1, Table 2). Furthermore, when these factors were associated with higher phosphate concentrations (Figure 4, experiment 5), it was observed the coelution of diclofenac (peak 4) and ibuprofen (peak 5) and 17α-ethynylestradiol (peak 6) and nimesulide (peak 8).

From the main effects (Table 3), it was possible to note that ethanol factor improved the chromatographic resolution between the analytes, especially between peaks 4 and 5, the exception was between peaks 6 and 8. The SDS factor improved the resolution between peaks 6 and 8 and worsened it to peaks 8 and 9. The phosphate factor reduced the resolution between peaks 6 and 8 and increased it to peaks 3 and 4. SDS promoted the higher increase in the time of analysis followed by ethanol, whereas the phosphate concentration presented a contrary effect and induced the reduction in the time of analysis.

The second-order (AB, AC, and BC) and third-order (ABC) interaction effects indicated intense variable interactions (Table 3), showing that the variables could not be studied in an independent form. The ethanol-SDS interaction was significant to increase the resolution between peaks 3 and 4 that was not critical but decreased the chromatographic resolution between peaks 4, 5, 6, and 8 and increased the time of analysis. The ethanol-phosphate interaction was important to increase the resolution between peaks 6 and 8 and contributed to decreasing the resolution between peaks 3 and 4 and 8 and 9. The SDS-phosphate interaction increased the resolution between peaks 8 and 9 and decreased the resolution between peaks 4 and 5 and 6 and 8. The ethanol-SDS-phosphate interaction increased the time of analysis as well as the resolution between peaks 8 and 9 but decreased the resolution between peaks 4 and 5 and 6 and 8 (Table 3).

Considering all variable effects, it was concluded that experiment 4 presented the best characteristics; under this condition, it was possible to separate the ten analytes in only 23.0 min always with chromatographic resolution better than 1.45.

The effect of the flow rate in the analyte separation showed that, for flow rate values higher than 1.0 mL·min^{-1}, the chromatographic resolutions worsened, and the separation between tetracycline and diclofenac, diclofenac and ibuprofen, and ciprofloxacin and norfloxacin was poor, then 1.0 mL·min^{-1} was selected. Thus, the final separation conditions were as follows: mobile phase with 15.0% (v/v) ethanol, 3.0% (m/v) SDS, 20.0 mmol·L^{-1} phosphate buffer at pH 7.0 to phase A and 8.0 to phase B; maintaining 100% of phase A until 2.0 min with abrupt change to 100% of phase B; flow rate of 1.0 mL·min^{-1}; 25°C; injected volume of 20 μL; monitoring the signals at 220, 240, and 280 nm; and time of analysis of 23.0 min (Figure 4, experiment 4).

3.4. Calibration, Characterization, and Sample Analysis. The analytical curves for the ten analytes in the concentration range of 0.10 up to 25.0 mg·L^{-1} for amoxicillin, diclofenac, ibuprofen, levonorgestrel, nimesulide, tetracycline, ciprofloxacin, and norfloxacin; 0.5 up to 25.0 mg·L^{-1} for

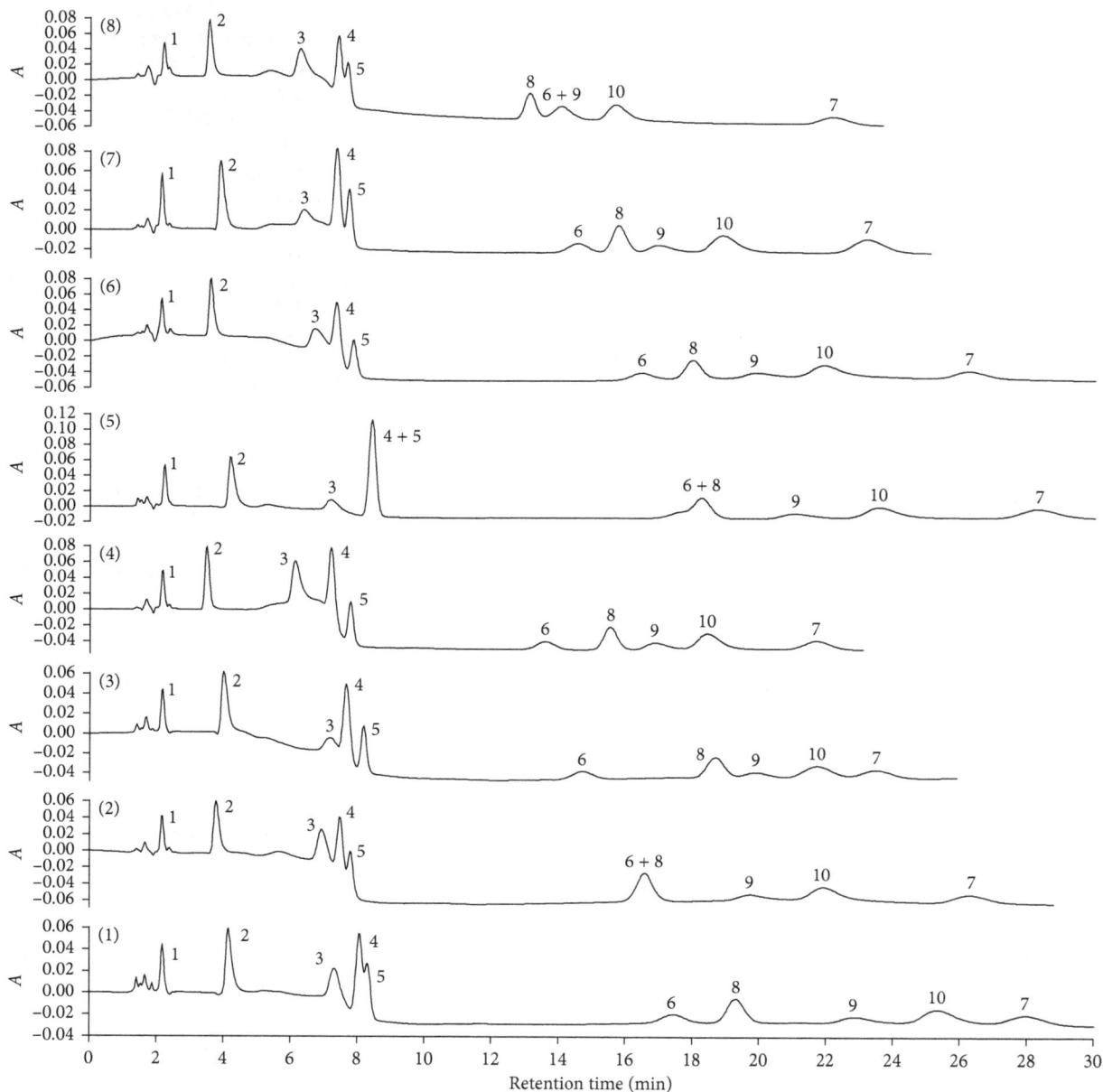

FIGURE 4: Chromatograms obtained from factorial design 2^3. Numbers from (1) to (8) correspond to the experimental condition number. A in the y-axis is the absorbance.

17α-ethynylestradiol; and 0.08 up to 25.0 mg·L^{-1} for caffeine were linear always with R^2 better than 0.992. The limits of detection (LOD) and quantification (LOQ), $n = 10$, considering the signal/noise ratio of 3.30 and 10.0 times, respectively, were estimated between 0.019 and 0.247 mg·L^{-1} and 0.058 and 0.752 mg·L^{-1} for caffeine and 17α-ethynylestradiol, respectively.

When the proposed method was applied to the analysis of the 7 water samples, the standard deviation was ca. 2% and the peaks showed to be free of interferences. The results furnished by the proposed method are in agreement with those obtained by the comparative chromatographic methods (Table 4). Furthermore, the obtained results to sample recovery tests were consistent with high recovery values and low standard deviations (Table 5).

Considering all the obtained results, the proposed method presented analytes average recovery of 99.12% and intraday ($n = 3$) and interday ($n = 5$) precision of 1.11% and 2.30%, respectively, whereas the comparative methods presented analytes average recovery of 98.64% and intraday ($n = 3$) and interday ($n = 5$) precision of 1.34% and 2.97%, respectively, indicating that the proposed method furnished results similar to those obtained by the comparative methods, but with the difference that it was necessary to carry out more than one method for the determination of the ten analytes.

The analysis of river water samples indicated the presence of caffeine in all the samples with concentrations ranging from 0.071 mg·L^{-1} to 1.204 mg·L^{-1}, probably due to

TABLE 2: Chromatographic resolutions obtained in the factorial design.

Factorial design	Rs (1-2)	Rs (2-3)	Rs (3-4)	Rs (4-5)	Rs (5-6)	Rs (6-8)	Rs (8-9)	Rs (9-10)	Rs (10-7)
1	7.17	5.98	1.27	0.53	9.89	1.62	2.75	1.37	1.58
2	6.39	6.60	0.96	0.98	14.11	0.00	3.42	1.40	2.83
3	6.77	5.98	0.85	1.59	9.07	3.66	1.30	1.06	1.24
4	**6.64**	**6.58**	**2.18**	**1.69**	**8.70**	**2.10**	**1.55**	**1.45**	**2.33**
5	7.04	5.33	1.99	0.00	13.44	0.00	2.01	1.42	2.49
6	6.90	5.61	1.06	1.36	10.08	1.30	1.41	1.23	2.81
7	6.99	4.89	1.85	1.08	9.48	1.23	1.26	1.33	2.76
8	5.78	6.26	2.49	0.51	13.53	0.92	1.22	1.47	5.16

Amoxicillin (1), caffeine (2), tetracycline (3), diclofenac (4), ibuprofen (5), 17α-ethynylestradiol (6), levonorgestrel (7), nimesulide (8), ciprofloxacin (9), and norfloxacin (10).

TABLE 3: Main effects, interaction among the variables, and time of analysis.

Factor	Effects					
	Rs (3-4)	Rs (4-5)	Rs (6-8)	Rs (8-9)	Rs (9-10)	Time of analysis
A-ethanol	0.190	0.310	−0.580	0.055	0.100	1.730
B-SDS	0.530	0.460	1.150	−1.070	−0.030	4.630
C-phosphate	0.520	−0.410	−0.950	−0.790	0.037	−0.170
AB	0.800	−0.520	−0.390	0.043	0.170	0.280
AC	−0.320	0.026	0.950	−0.390	−0.120	0.075
BC	0.096	−0.420	−0.820	0.590	0.110	−0.025
ABC	−0.029	−0.370	−0.320	0.240	−0.023	0.370
SD	±0.004	±0.048	±0.130	±0.005	±0.005	±0.048

Tetracycline (3), diclofenac (4), ibuprofen (5), 17α-ethynylestradiol (6), nimesulide (8) ciprofloxacin (9), and norfloxacin (10).

TABLE 4: Analyte determination by the proposed and conventional methods.

Sample	Analyte	Proposed method Concentration	Conventional methods Concentration
Moscados stream (μg·L^{-1})	Caf	0.109 ± 0.002	0.086 ± 0.011
Cleópatra stream (μg·L^{-1})	Caf	1.204 ± 0.034	1.358 ± 0.020
	Tetr	0.188 ± 0.007	0.195 ± 0.009
	Norflo	0.463 ± 0.022	0.427 ± 0.038
Borba Gato stream (μg·L^{-1})	Caf	0.140 ± 0.005	0.136 ± 0.002
Mandacarú stream (μg·L^{-1})	Caf	0.742 ± 0.024	0.766 ± 0.038
	Norflo	0.380 ± 0.016	0.395 ± 0.011
Morangueiro stream (μg·L^{-1})	Caf	0.367 ± 0.020	0.371 ± 0.014
	Cipro	0.295 ± 0.010	0.281 ± 0.019
Maringá stream (μg·L^{-1})	Caf	0.173 ± 0.010	0.190 ± 0.012
Potable water (μg·L^{-1})	Caf	0.071 ± 0.003	0.064 ± 0.007

The conventional methods used were as follows: tetracycline [18]; ibuprofen, diclofenac, and nimesulide [24]; levonorgestrel and 17α-ethynylestradiol [25]; caffeine [29]; ciprofloxacin and norfloxacin [30]; amoxicillin [31].

the human activities. It was verified the presence of antibiotics in three streams: tetracycline and norfloxacin in Cleópatra stream, norfloxacin in the Mandacarú stream, and ciprofloxacin in the Morangueiro stream (Table 4).

The proposed chromatographic and extraction method presented high accuracy and precision for the analysis of water samples, allowing the quantification of the ten analytes with only one extraction and chromatographic method.

4. Conclusions

The proposed extraction and determination method demonstrated to be able to do the extraction and determination of tetracycline, ibuprofen, 17α-ethynylestradiol, caffeine, diclofenac, nimesulide, levonorgestrel, ciprofloxacin, norfloxacin, and amoxicillin using a green mobile phase (15.0% (v/v) ethanol, 3.0% (m/v) SDS, and 20.0 mmol·L^{-1} of phosphate at pHs 7.0 and 8.0) and to extract the analytes in

TABLE 5: Recovery tests for water samples.

Sample	Analyte	Proposed method			Conventional methods		
		Recovery (%)	Precision (RSD, %)		Recovery (%)	Precision (RSD, %)	
		($n = 3$)	Intraday ($n = 3$)	Interday ($n = 5$)	($n = 5$)	Intraday ($n = 3$)	Interday ($n = 5$)
Moscados stream	Caf	99.52	0.94	1.53	98.39	1.21	1.67
	Tetra	98.64	1.33	2.89	96.67	2.04	2.55
	Cipro	98.92	2.45	3.90	96.20	2.83	4.81
	Norflo	99.14	1.09	1.47	98.80	1.55	2.39
	Diclo	99.45	1.77	0.95	99.94	0.90	1.96
	Ibup	98.20	1.61	2.11	98.85	1.32	1.42
	Nime	98.35	2.83	4.54	98.97	2.29	3.11
	Ethynyl	97.21	0.97	3.02	99.91	1.48	2.27
	Norg	99.08	3.14	4.67	99.87	2.11	2.89
	Amox	—	—	—	97.35	1.76	1.53
Cleópatra stream	Caf	97.85	2.37	3.10	98.78	1.05	2.14
	Tetra	99.71	0.59	2.48	99.89	1.49	2.95
	Cipro	98.22	0.83	1.26	98.96	1.68	3.28
	Norflo	101.30	1.11	1.45	100.13	2.27	1.99
	Diclo	99.45	0.76	1.90	98.32	1.84	2.56
	Ibup	99.60	1.28	2.34	98.88	1.16	1.59
	Nime	98.50	0.67	1.19	97.83	2.08	2.31
	Ethynyl	99.31	1.04	1.72	98.36	1.53	2.47
	Norg	98.06	0.95	1.58	98.97	1.87	2.08
	Amox	—	—	—	98.60	1.29	1.84
Borba Gato stream	Caf	98.23	0.61	1.04	98.52	2.08	3.77
	Tetra	99.11	1.58	3.27	99.01	1.73	2.01
	Cipro	97.89	1.13	1.85	99.27	0.99	1.63
	Norflo	98.04	1.20	2.01	99.17	1.62	1.94
	Diclo	99.55	0.44	1.33	98.05	1.31	2.22
	Ibup	99.19	0.82	1.59	98.82	1.47	1.98
	Nime	100.73	0.57	1.98	99.51	0.84	1.72
	Ethynyl	99.82	1.28	2.10	98.46	1.10	2.36
	Norg	98.44	0.49	1.67	99.04	1.26	1.93
	Amox	—	—	—	98.63	0.63	1.69
Mandacarú stream	Caf	100.91	1.23	1.81	98.39	0.55	1.14
	Tetra	97.36	1.05	2.17	98.67	1.62	2.51
	Cipro	99.94	1.41	2.32	97.20	1.28	1.93
	Norflo	99.12	1.95	3.06	98.80	1.99	2.48
	Diclo	101.33	1.03	1.90	99.94	2.11	2.83
	Ibup	99.60	0.55	1.26	98.85	1.33	1.70
	Nime	99.82	0.81	1.78	97.77	0.81	1.66
	Ethynyl	99.01	0.90	1.63	99.91	1.55	3.90
	Norg	100.29	0.78	1.99	99.87	0.90	1.93
	Amox	—	—	—	99.35	1.22	2.07
Morangueiro stream	Caf	99.14	0.32	0.98	98.78	0.95	1.57
	Tetra	98.52	0.70	1.53	99.89	0.49	1.00
	Cipro	98.05	1.51	2.49	98.96	1.35	2.41
	Norflo	98.81	1.63	2.72	100.13	1.21	2.27
	Diclo	99.20	0.76	1.44	98.32	0.88	1.82
	Ibup	100.33	0.52	0.97	98.88	2.04	2.89
	Nime	98.87	1.11	2.06	97.83	1.59	2.70
	Ethynyl	99.00	0.83	1.88	98.33	1.11	1.40
	Norg	100.94	1.47	2.91	98.97	0.97	1.34
	Amox	—	—	—	99.04	0.83	0.90
Maringá stream	Caf	97.99	1.14	2.03	98.52	0.66	1.32
	Tetra	99.10	0.95	1.82	99.01	1.01	1.83
	Cipro	97.66	1.98	3.06	98.27	1.44	1.68
	Norflo	98.34	1.32	2.21	99.17	1.27	2.83
	Diclo	99.62	1.77	3.48	99.05	0.59	1.16
	Ibup	98.41	0.92	1.85	98.82	1.33	2.29
	Nime	99.97	1.06	1.95	98.51	0.78	2.04
	Ethynyl	98.55	1.20	1.79	98.46	1.11	1.98
	Norg	99.67	1.00	1.53	99.04	1.50	2.10
	Amox	—	—	—	98.63	0.95	1.93

TABLE 5: Continued.

| Sample | Analyte | Proposed method | | | Conventional methods | | |
| | | Recovery (%) | Precision (RSD, %) | | Recovery (%) | Precision (RSD, %) | |
		$(n = 3)$	Intraday $(n = 3)$	Interday $(n = 5)$	$(n = 5)$	Intraday $(n = 3)$	Interday $(n = 5)$
	Caf	99.96	0.25	0.81	100.40	2.05	1.19
	Tetra	99.05	2.05	2.97	98.00	1.39	1.72
	Cipro	99.31	0.68	1.14	97.93	1.03	1.58
	Norflo	99.20	0.92	1.67	98.33	0.76	1.47
Tap water	Diclo	100.08	0.74	1.93	99.19	0.62	0.95
	Ibup	99.19	0.33	0.85	100.71	1.28	1.33
	Nime	99.21	0.49	1.36	99.67	0.67	1.59
	Ethynyl	99.89	0.97	2.02	97.59	1.04	1.45
	Norg	98.40	0.81	1.77	97.71	0.95	1.90
	Amox	99.12	1.02	1.84	98.26	1.32	2.34

Data were obtained for water samples spiked with $5.0\,\mu g \cdot L^{-1}$ of ibuprofen, 17α-ethynylestradiol, diclofenac, nimesulide, levonorgestrel, ciprofloxacin, and norfloxacin; $10\,\mu g \cdot L^{-1}$ of caffeine; and $60\,\mu g \cdot L^{-1}$ of amoxicillin. The conventional methods were the same as used in Table 4.

only one procedure using a C18 cartridge and ethanol as eluent. The method was selective, robust, simple, fast, and inexpensive, presented adequate LOD and LOQ, and it can be an important tool for laboratories dedicated to the analysis of emerging pollutants that do not have MS detector.

Conflicts of Interest

The authors declare that they have no conflicts of interest.

Acknowledgments

The authors are grateful to CAPES, CNPq, FINEP, and Fundação Araucária for financial support.

References

[1] S. Konar and S. Mullick, "Problems of safe disposal of petroleum products, detergents, heavy metals and pesticides to protect aquatic life," *Science of the Total Environment*, vol. 134, pp. 989–1000, 1993.

[2] E. Razo-Flores and P. Olguín-Lora, "Biotreatment of water pollutants from the petroleum industry," *Studies in Surface Science and Catalysis*, vol. 151, pp. 513–536, 2004.

[3] C. Mahugo-Santana, Z. Sosa-Ferrera, M. E. Torres-Padron, and J. J. Santana-Rodriguez, "Application of new approaches to liquid-phase microextraction for the determination of emerging pollutants," *TrAC Trends in Analytical Chemistry*, vol. 30, no. 5, pp. 731–748, 2011.

[4] P. Verlicchi, A. Galletti, M. Petrovic, and D. Barcelo, "Hospital effluents as a source of emerging pollutants: an overview of micropollutants and sustainable treatment options," *Journal of Hydrology*, vol. 389, no. 3-4, pp. 416–428, 2010.

[5] H. Wei, J. J. Sun, Y. M. Wang, X. Li, and G. N. Chen, "Rapid hydrolysis and electrochemical detection of trace carbofuran at a disposable heated screen-printed carbon electrode," *Analyst*, vol. 133, no. 11, pp. 1619–1624, 2008.

[6] D. M. Bila and M. Dezotti, "Fármacos no meio ambiente," *Química Nova*, vol. 26, no. 4, pp. 523–530, 2003.

[7] M. Stumpf, T. A. Torres, R. D. Wilken, S. V. Rodrigues, and W. Baumann, "Polar drug residues in sewage and natural waters in the state of Rio de Janeiro," *Science of the total environment*, vol. 225, no. 1-2, pp. 135–141, 1999.

[8] M. Seifrtova, L. Novakova, C. Lino, A. Pena, and P. Solich, "An overview of analytical methodologies for the determination of antibiotics in environmental waters," *Analytica Chimica Acta*, vol. 649, no. 2, pp. 158–179, 2009.

[9] L. Araujo, J. Wild, N. Villa, N. Camargo, D. Cubillan, and A. Pietro, "Determination of anti-inflammatory drugs in water samples, by in situ derivatization, solid phase microextraction and gas chromatography–mass spectrometry," *Talanta*, vol. 75, no. 1, pp. 111–115, 2008.

[10] Z. Klemenc-Ketis, Z. Hladnik, and J. Kersnik, "A cross sectional study of sex differences in self-medication practices among university students in Slovenia," *Collegium Antropologicum*, vol. 35, no. 2, pp. 329–334, 2011.

[11] S. Babic, D. Asperger, D. Mutavdzic, A. J. M. Horvat, and M. Kastelan-Macan, "Solid phase extraction and HPLC determination of veterinary pharmaceuticals in wastewater," *Talanta*, vol. 70, no. 4, pp. 732–738, 2006.

[12] R. Hirsch, T. Ternes, K. Jaberer, and K. L. Kratz, "Occurrence of antibiotics in the aquatic environment," *Science of the Total environment*, vol. 225, no. 1-2, pp. 109–118, 1999.

[13] F. F. Sodre, C. C. Montagner, M. A. F. Locatelli, and W. F. Jardim, "Ocorrência de interferentes endócrinos e produtos farmacêuticos em águas superficiais da região de Campinas (SP, Brazil)," *Journal of the Brazilian Society of Ecotoxicology*, vol. 2, no. 2, pp. 187–196, 2007.

[14] T. A. Torres, "Analytical methods for the determination of pharmaceuticals in aqueous environmental samples," *TrAC Trends in Analytical Chemistry*, vol. 20, no. 8, pp. 419–434, 2001.

[15] R. Siegener and R. F. Chen, "Caffeine in Boston Harbor seawater," *Marine Pollution Bulletin*, vol. 44, no. 5, pp. 383–387, 2002.

[16] P. L. Fernandez, M. J. Martin, A. G. Gonzales, and F. Pablos, "HPLC determination of catechins and caffeine in tea. Differentiation of green, black and instant teas," *The Analyst*, vol. 125, no. 3, pp. 421–425, 2000.

[17] P. R. Gardinali and X. Zhao, "Trace determination of caffeine in surface water samples by liquid chromatography–atmospheric pressure chemical ionization–mass spectrometry (LC-APCI-MS)," *Environment International*, vol. 28, no. 6, pp. 521–528, 2002.

[18] A. R. Shalaby, N. A. Salama, S. H. Abou-Raya, W. H. Emam, and F. M. Mehaya, "Validation of HPLC method for determination of tetracycline residues in chicken meat and liver," *Food Chemistry*, vol. 124, no. 4, pp. 1660–1666, 2011.

[19] M. Lillenberg, S. Yurchenko, K. Kipper et al., "Simultaneous determination of fluoroquinolones, sulfonamides and tetracyclines in sewage sludge by pressurized liquid extraction and liquid chromatography electrospray ionization-mass spectrometry," *Journal of Chromatography A*, vol. 1216, no. 32, pp. 5949–5954, 2009.

[20] Q. T. Dinh, F. Alliot, E. Moreau-Guigon, J. Eurin, M. Chevreuil, and P. Labadie, "Measurement of trace levels of antibiotics in river water using on-line enrichment and triple-quadrupole LC–MS/MS," *Talanta*, vol. 85, no. 3, pp. 1238–1245, 2011.

[21] E. S. Elmolla and M. Chaudhuri, "Degradation of amoxicillin, ampicillin and cloxacillin antibiotics in aqueous solution by the UV/ZnO photocatalytic process," *Journal of Hazardous Materials*, vol. 173, no. 1–3, pp. 445–449, 2010.

[22] L. Tong, P. Li, Y. X. Wang, and K. Z. Zhu, "Analysis of veterinary antibiotic residues in swine wastewater and environmental water samples using optimized SPE-LC/MS/MS," *Chemosphere*, vol. 74, no. 8, pp. 1090–1097, 2009.

[23] S. Wigel, R. Kallenborn, and H. Juhnerfuss, "Simultaneous solid-phase extraction of acidic, neutral and basic pharmaceuticals from aqueous samples at ambient (neutral) pH and their determination by gas chromatography–mass spectrometry," *Journal of Chromatography A*, vol. 1023, no. 2, pp. 183–195, 2004.

[24] A. Panusa, G. Multari, G. Incarnato, and L. Galiardi, "High-performance liquid chromatography analysis of anti-inflammatory pharmaceuticals with ultraviolet and electrospray-mass spectrometry detection in suspected counterfeit homeopathic medicinal products," *Journal of Pharmaceutical and Biomedical Analysis*, vol. 43, no. 4, pp. 1221–1227, 2007.

[25] C. Almeida and J. M. F. Nogueira, "Determination of steroid sex hormones in water and urine matrices by stir bar sorptive extraction and liquid chromatography with diode array detection," *Journal of Pharmaceutical and Biomedical Analysis*, vol. 41, no. 4, pp. 1303–1311, 2006.

[26] Y. C. Chen and C. E. Lin, "Migration behavior and separation of tetracycline antibiotics by micellar electrokinetic chromatography," *Journal of chromatography A*, vol. 802, no. 1, pp. 95–105, 1998.

[27] M. Hernandez, F. Borrull, and M. Calull, "Determination of amoxicillin in plasma samples by capillary electrophoresis," *Journal of chromatography B*, vol. 731, no. 2, pp. 309–315, 1999.

[28] COGERH, *Recomendações e Cuidados na Coleta de Amostras de Água*, Companhia de Gestão e Dos Recursos Hídricos, Fortaleza, CE, Brazil, 2001.

[29] M. J. Martin, F. Pablos, and A. G. Gonzalez, "Simultaneous determination of caffeine and non-steroidal anti-inflammatory drugs in pharmaceutical formulations and blood plasma by reversed-phase HPLC from linear gradient elution," *Talanta*, vol. 49, no. 2, pp. 453–459, 1999.

[30] P. G. Gigosos, P. R. Revesado, O. Cadahia et al., "Determination of quinolones in animal tissues and eggs by high-performance liquid chromatography with photodiode-array detection," *Journal of chromatography A*, vol. 871, no. 1-2, pp. 31–36, 2000.

[31] Q. Pei, G. P. Yang, Z. J. Li, X. D. Peng, J. H. Fan, and Z. Q. Liu, "Simultaneous analysis of amoxicillin and sulbactam in human plasma by HPLC-DAD for assessment of bioequivalence," *Journal of chromatography B*, vol. 879, no. 21, pp. 2000–2004, 2011.

[32] E. C. Vidotti, W. F. Costa, and C. C. Oliveira, "Development of a green chromatographic method for determination of colorants in food samples," *Talanta*, vol. 68, no. 3, pp. 516–521, 2006.

[33] V. Kienen, W. F. Costa, J. V. Visentainer, N. E. de Souza, and C. C. Oliveira, "Development of a green chromatographic method for determination of fat-soluble vitamins in food and pharmaceutical supplement," *Talanta*, vol. 75, no. 1, pp. 141–146, 2008.

18

Development and Validation of Gas Chromatography-Triple Quadrupole Mass Spectrometric Method for Quantitative Determination of Regulated Plasticizers in Medical Infusion Sets

So Hyeon Jeon,[1,2] Yong Pyo Kim,[3] Younglim Kho,[4] Jeoung Hwa Shin,[5] Won Hyun Ji,[6] and Yun Gyong Ahn ⓘ[1]

[1]Western Seoul Center, Korea Basic Science Institute, Seoul 03759, Republic of Korea
[2]Department of Environmental Science and Engineering, Ewha Womans University, Seoul 03759, Republic of Korea
[3]Department of Chemical Engineering and Material Science, Ewha Womans University, Seoul 03760, Republic of Korea
[4]Department of Health, Environment & Safety, Eulji University, Seongnam 13135, Republic of Korea
[5]Seoul Center, Korea Basic Science Institute, Seoul 02841, Republic of Korea
[6]Institute of Mine Reclamation Technology, Mine Reclamation Corporation, Wonju 26464, Republic of Korea

Correspondence should be addressed to Yun Gyong Ahn; ygahn@kbsi.re.kr

Academic Editor: Paolo Montuori

A method for the quantitative determination of dibutyl phthalate (DBP), benzyl butyl phthalate (BBP), bis(2-ethylhexyl) adipate (DEHA), bis(2-ethylhexyl) phthalate (DEHP), di-n-octyl phthalate (DNOP), dioctyl terephthalate (DOTP), diisononyl phthalate (DINP), and diisodecyl phthalate (DIDP) in medical infusion sets was developed and validated using gas chromatography coupled with triple quadrupole mass spectrometry (GC-MS/MS) in the multiple reaction monitoring (MRM) mode. Solvent extraction with polymer dissolution for sample preparation was employed prior to GC-MS/MS analysis. Average recoveries of the eight target analytes are typically in the range of 91.8–122% with the relative standard deviations of 1.8–17.8%. The limits of quantification (LOQs) of the analytical method were in the ranges of 54.1 to 76.3 ng/g. Analysis using GC-MS/MS provided reliable performance, as well as higher sensitivity and selectivity than GC-MS analysis, especially for the presence of minority plasticizers at different concentrations.

1. Introduction

Phthalates as a class of synthetic chemicals are widely used in a variety of consumer products including medical devices, toys for children, food wrapper, building materials, automotive parts, and so on [1–3]. Since they may cause harm to human health by altering endocrine function or through other biological mechanisms [4], environmental monitoring of phthalates has been accomplished from various media (air, soil, food, water, etc.) [5–9]. Most medical devices are made of flexible polyvinylchloride (PVC), a produced synthetic plastic polymer due to its numerous benefits, which include chemical stability, biocompatibility, clarity and transparency, flexibility, durability, chemical and mechanical resistance, sterilizability, and low cost [10].

Since plastics are mainly used as plasticizers to soften PVC, phthalates are abundant in PVC-based medical devices, and they may enter into contact with the patients through leaching out into infused solutions [11, 12]. The larger molecular weight phthalates—di(2-ethylhexyl) phthalate (DEHP), di-n-butyl phthalate (DBP), and diisononyl phthalate (DINP)—are suspected carcinogens, toxic to liver, kidneys [13], and reproductive organs [14]. Benzyl butyl phthalate (BBP), DBP, and DEHP are weakly estrogenic [15]. There have been reports of extraction of phthalates such as BBP, DBP, and DEHP in dialysis tubing and infusion bags [16, 17]. Consequently, the European Chemicals Agency (ECHA) recommended that several compounds of very high concern should not be used without specific authorization. Furthermore, according to South Korea Ministry of Food

TABLE 1: Summary of chemical information and instrument conditions of target analytes.

Compounds	Abbreviation	CAS number	Formula	M.W. (g/mol)	R.T. (min)	Precursor ion	Product ion	Collision energy (eV)
Dibutyl phthalate	DBP	84-74-2	$C_{16}H_{22}O_4$	278.35	10.126	149	121	12
Benzyl butyl phthalate	BBP	85-68-7	$C_{19}H_{20}O_4$	312.37	13.734	149	121	11
Bis(2-ethylhexyl) adipate	DEHA	103-23-1	$C_{22}H_{42}O_4$	370.57	14.032	129	101	2
Bis(2-ethylhexyl) phthalate	DEHP	117-81-7	$C_{24}H_{38}O_4$	390.56	15.251	149	121	13
Di-n-octyl phthalate	DNOP	117-84-0	$C_{24}H_{38}O_4$	390.56	16.642	279	149	3
Dioctyl terephthalate	DOTP	6422-86-2	$C_{24}H_{38}O_4$	390.56	16.716	261	149	8
Diisononyl phthalate	DINP	20548-62-3	$C_{26}H_{42}O_4$	418.61	17.482	293	149	3
Diisodecyl phthalate	DIDP	26761-40-0	$C_{28}H_{46}O_4$	446.67	18.877	307	149	4

and Drug Safety (MFDS) regulation, phthalates within intravascular administration products including DEHP, DBP, and BBP are no longer being used in the production of medical infusion sets since July 2015.

Due to the expanding use of phthalates and assimilated analytes in medical devices, the analytical methods for plasticizers from infusion sets are increasing in order to create plans for safety. Several analytical methods capable of detecting and quantifying the alternative plasticizers in medical devices have been developed and validated by means of separative and nonseparative methods up to date [18]. The separative methods have been adopted with gas chromatography, supercritical fluid chromatography [19], and liquid chromatography combined with various detectors such as mass spectrometer, flame ionization detector, evaporative light scattering detector, or UV techniques. Nonseparative methods have been performed on a nuclear magnetic resonance and Fourier transform infrared spectrometry [20, 21]. Among these methods, gas chromatography coupled with mass spectrometry was the most specific and sensitive method and suggested to be a suitable method to perform a regulatory control. Furthermore, application of GC-triple quadrupole mass spectrometry (GC-MS/MS) is recently increasing in terms of quantitative confirmation in various matrices [22].

For the sample preparation, an easy and inexpensive technique is polymer dissolution which is well known as a solid-liquid extraction [23]. The whole polymer is first dissolved in a solvent-like tetrahydrofuran (THF) or dimethylacetamide [24, 25]. After then, when solvent is added to the dissolution for solvent extraction, PVC-based polymers are removed by precipitation. There are already two publications by Gimeno et al. and Bourdeaux et al. dealing with the GC-MS analysis of phthalates and assimilated analytes in medical devices [26, 27]. But the viewpoint of the detection level and target analytes is clearly different from our study. It is important to monitor the presence of the minority plasticizers, as well as the majority, because their migration and toxic potential can be extremely different depending on their chemical properties. Furthermore, alternative plasticizers and mixtures have been increasingly developed to provide flexibility for medical devices in these days. Also, their concentration levels in the samples can be quite varied, but they should be evaluated depending on the type of phthalates and assimilated analytes. Accordingly, the

identification and quantification of toxic plasticizers in medical devices, especially the tubings of medical infusion sets, are needed since they are able to leach from infusion sets and result in human exposure directly.

The aim of this study was to evaluate the capabilities of GC-MS/MS for the quantification of phthalates and assimilated analytes from infusion sets. Analysis using GC-MS with a single quadrupole mass analyzer has been typically used in most cases for the determination of phthalates and assimilated analytes in medical devices. However, this method requires a clean matrix to avoid the interference of unwanted ions [28]. By conducting interlaboratory collaborative studies, it was found that there was a lack of accuracy and precision for quantitative GC/MS analysis due to insufficient cleanup procedures. The advantages of the proposed method rely also on the fact that there is no need for complete chromatographic separation due to the MRM mode. As for the sample preparation according to polymer dissolution, GC-MS/MS provides sensitive, efficient, and reliable results. This method would be a useful tool to control the safety of intravascular administration set and assess unintended harmful substances, especially for the presence of both majority and minority plasticizers at the level of various concentrations.

2. Materials and Methods

2.1. Chemicals and Reagents. Standard solutions of eight analytes (BBP, DBP, DEHA, DEHP, DNOP, DOTP, DINP, and DIDP; see Table 1 for their full chemical names and information) for GC-MS/MS analysis were purchased from Sigma-Aldrich (St. Louis, MO, USA), and stock solution was at a concentration of 2000 mg/L in hexane. Internal standards of benzyl benzoate (BB) were purchased from Sigma-Aldrich (St. Louis, MO, USA) and used at a concentration of 100 mg/L in hexane as a stock solution. Organic solvents (hexane and THF) of the GC analysis grade were purchased from Burdick & Jackson (Philipsburg, NJ, USA).

2.2. Preparation of Samples. 50 mg of small pieces from the tubings in medical infusion sets in 5 mL of THF was sonicated at room temperature for 30 min in a glass tube. Subsequently, 10 mL of hexane was added into the glass tube, and the solution was vortexed for the precipitating polymer matrix. After precipitating the polymer matrix, the solution was left as

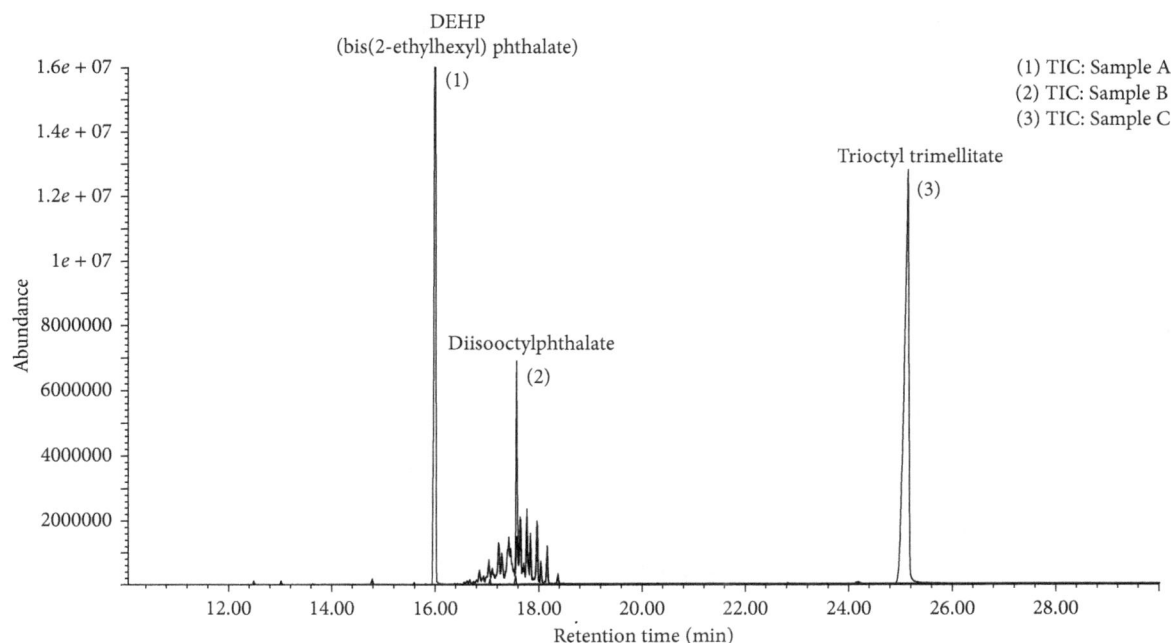

FIGURE 1: Total ion chromatograms of the hexane extracts after polymer dissolution in tubings of three domestic medical infusion sets. In the TICs, majority plasticizers for each sample are represented in a chromatogram with the three main peaks, each identified by different colors.

it was for 10 min. 0.5 mL of supernatant extracted samples was transferred into a 2 mL of vial while 100 ng of internal standard was added. A total volume of 1 mL was homogenized and filtered by an Acrodisc 0.2 μm GHP syringe filter.

2.3. Analysis Using GC-MS/MS. Analysis was performed by an Agilent 7890B gas chromatograph, equipped with a 7010 mass selective detector triple quadrupole mass spectrometer system (Palo Alto, CA, USA). Chromatographic separation was achieved using a DB-5MS UI (5% diphenyl-95% dimethyl siloxane phase, 30 m × 0.25 mm I. D; 0.25 μm film thickness) from a J&W Scientific (Santa Clara, CA, USA) capillary column. The temperature of injector was 300°C. One microliter of each extract was injected in the split mode (2 : 1). Helium as carrier gas (99.999%) flow was 1 mL/min. The GC oven temperature program was as follows. The initial temperature of 150°C was held for 3 min after injection before it was increased up to 300°C at 10°C/min held for 12 min. Nitrogen (99.999%) was used as collision gas. The running time was 30 min, divided into seven segments of time for each selected product ion to increase sensitivity and selectivity. The transfer line and ion source temperature were set at 250 and 230°C, respectively. The mass spectrometer is tuned on electron impact ionization (EI) at 70 eV in the multiple reaction monitoring (MRM) mode.

3. Results and Discussion

3.1. Sample Preparation. Previous studies and conventional extraction techniques, including Soxhlet, solvent extraction, or a method that firstly dissolves the whole polymer and then

separates the plasticizers from the PVC by precipitation, are suggested. Due to the large volume of solvent needed and the long procedure times, solvent extraction after the polymer dissolution at room temperature has been used as an alternative efficient and simple technique instead of Soxhlet extraction [10]. In this study, 50 mg of the cutting infusion tube was dissolved in 5 mL of THF and sonicated for 30 min. The mixed solution was left for 10 min or more, and then the polymer was precipitated due to difference in polarity between THF and hexane. According to the solvent used for the phthalate extraction, an organic solvent such as dichloromethane [29] or acetone [30], hexane, and acetonitrile [31] has been used. The selection of different solvents for extraction can have significant effects on the discrimination of target analytes from the sample matrix, as well as the extraction recoveries [32]. Although the advantage of this sample preparation is simple without purification, the matrix effect based on the different organic plastic additives in infusion sets can lead to false results of target analytes [33]. As an extraction solvent after the polymer dissolution, hexane was chosen because the partitioning of the extract with hexane was able to minimize matrix interferences compared with other solvents. Figure 1 shows the total ion chromatograms of the hexane extracts after polymer dissolution in three domestic infusion sets using a GC-MS system. Sample A clearly showed a large amount of DEHP released from the sample suspected to be the PVC sets. Although the use of DEHP in medical tubing has been restricted, it is still used in medical devices within the medical industrial field. In the case of samples B and C, diisooctylphthalate (DIOP) and trioctyl trimellitate (TOTM) were mainly detected, respectively. These chemicals could be used as alternatives to regulated analytes in domestic medical devices, with their unwanted ions interfering with the detection of the target analytes. For spiking experiments, polyurethane (PU) sample containing none of the eight target analytes was used to evaluate the suitability of the analytical method.

3.2. GC-MS Analysis. A gas chromatography coupled with single quadrupole mass spectrometer with selected ion monitoring (SIM) mode has been frequently used to evaluate the analytical performance and validation for quantitative analysis in the clinical research field owing to high sensitivity and the ability to achieve low limits of detection [34]. Based on the test method of the standard operating procedure for the determination of phthalates in PVC products by the U.S. Consumer Product Safety Commission's (CPSC) testing laboratory (LSC) [35], three laboratories were involved in an interlaboratory collaborative study to test the practicability of the modified CPSC method for the quantitative determination of phthalates and assimilated analytes in infusion tubing samples. Accuracy, precision, and linearity as evaluation parameters between laboratories were performed using the sample preparation procedures (Section 2.2) with the same type of commercial GC-MS instrument. All the investigated calibration curves were obtained by the acceptable correlation coefficients ($R^2 > 0.99$) at five concentrations ranging from 0.05 to 5 μg/g for DBP, BBP, DEHA, DEHP, DNOP, and DOTP and from 0.15 to 15 μg/g for DINP and DIDP, respectively. However, it was found that there was a lack of accuracy and precision from the results of the spike recovery comparison of collaborative studies. The accuracy and precision were assessed by recovery experiments between laboratories using triplicate target analyte free samples spiked with three different concentration points (low, middle, and high) in the calibration range compared with the pure authentic standards. Over 60% of the 144 individual results reported fell outside of the acceptable limits for recoveries (normally ranging from 70 to 130%) and precision (<20% RSD). These results indicated that the analytical method of extraction solvent after the polymer dissolution in the medical infusion sets followed by GC-MS was unsuitable for the determination of trace level concentration of phthalates and assimilated analytes. As mentioned by the U.S. Consumer Product Safety Commission (CPSC), it was intended to address the determination of concentrations of more than 0.1 percentage per plasticized component part of toys for children or child care article in order to protect children from hazard [35]. Therefore, triple quad GC-MS system allows the simplification of sample preparation while maintaining high selectivity and sensitivity when targeting trace level of analytes in complex sample extracts [36].

3.3. Optimization of the GC-MS/MS Conditions. For GC-MS/MS analysis using the triple quadrupole, full-scan spectra were obtained to select the precursor ions of each target analyte. Figure 2 shows the full-scan MS/MS data of each analyte to optimized MRM transitions. A mass-to-charge ratio of 149 or 129 was selected as precursor ions

of DBP, BBP, DEHA, and DEHP due to their highest mass intensity. Mass-to-charge ratios of 279, 261, 293, and 307 were selected as precursor ions of DNOP, DOTP, DINP, and DIDP generally considered as confirmation ions for increasing selectivity. Collision-induced dissociation (CID) for selected precursor ions using 99.999% nitrogen was acquired to find the appropriate collision energy of each analyte. To determine collision energies (CEs) for both the quantifying and qualifying MRM ion transitions, they were varied between 2 and 20 eV and optimized as shown in Table 1. This shows that the precursor ions and dominant fragmented ions of phthalates in EI ionization were typically m/z 149, corresponding to the protonated phthalic anhydride ion $[C_8H_5O_3]^+$ [37]. The useful dissociation pathway for phthalates was provided by the relatively low collision energy of the molecular ion. Figure 3 shows the mass spectral fragmentation pathway of DINP in the MRM mode as an example. The molecular ion loses the alkyl fragment accompanied by two hydrogen migrations. Since the molecular ion peak for phthalates with long chain alkyl groups is usually weak, it was not always present in the mass spectra. But the ions loosing alkyl group $[M-R]^+$ and those with oxygen $[M-OR]^+$ fragments can be a secondary form of identification [38]. In the MRM mode, m/z 293 cleaved again to m/z 149, losing the remaining alkyl fragment as shown in Figure 3. The main product ion from m/z 149 is m/z 121, resulting from fragmentation with loss of the aldehyde group. A similar fragmentation pattern was observed in the other analytes except different alternative plasticizers such as bis(2-ethylhexyl) adipate as shown in Figure 2.

To verify the presence of target analytes in matrix blank samples, 5 μg/g (the concentration of solid by weight) was spiked in the matrix sample and the same experimental process was conducted. The spiked and matrix blank sample were analyzed by the same instrument condition. Figure 4 shows MRM chromatograms of matrix blank sample (a) and those spiked with target analytes (5 μg/g) (b). As a result, target analytes were not observed in the matrix blank sample, and each peak of target analytes were separated clearly in the MRM mode.

3.4. Analytical Performance. Method validation was performed by spike recovery experiments based on the optimized analytical methods. In this study, benzyl benzoate was used as an internal standard (IS) for target analytes. The calibration curves were generated in the range of 5–500 ng/g using a least-square linear regression analysis. The correlation coefficients were all greater than 0.999 compared with results obtained by GC-MS. The limit of quantification (LOQ) for each analyte was determined in accordance with ICH guidelines [39]. The calculated LOQs ranged from 54.1 to 76.3 ng/g, and they were approximately 20 times lower compared with the results from the latest study which was performed to determine LOQ by the GC/MS method [18]. To validate the accuracy and precision of the analytical method, three replicate analyses of matrix blank samples spiked with the known amounts of the analytes at low and middle concentration levels (17 and 100 ng/g) were prepared

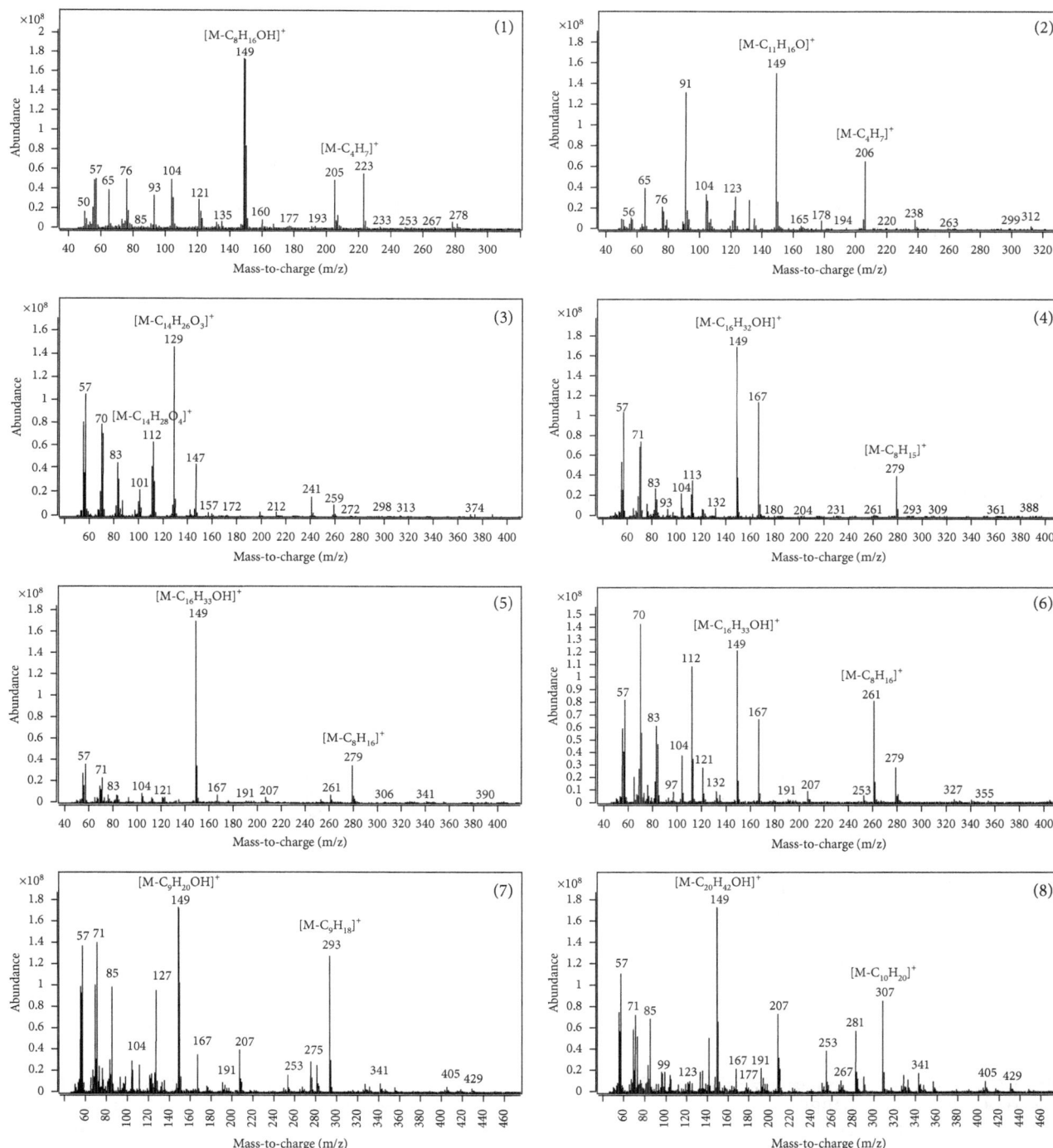

FIGURE 2: Full-scan MS/MS data of each target analyte to optimized MRM transitions. Each individual parent ion scan experiment is represented as follows: (1) dibutyl phthalate (DBP); (2) benzyl butyl phthalate (BBP); (3) bis(2-ethylhexyl) adipate (DEHA); (4) bis(2-ethylhexyl) phthalate (DEHP); (5) di-n-octyl phthalate (DNOP); (6) dioctyl terephthalate (DOTP); (7) diisononyl phthalate (DINP); (8) diisodecyl phthalate (DIDP).

and determined by the analytical procedure. The accuracy and precision were assessed by the average recovery and percentage relative standard deviation (%RSD) of three results at each concentration, as shown in Table 2. The average recoveries of eight analytes ranged from 91.8% to 122%, with RSDs ranging from 1.8% to 17.8%. The linearity, LOQ,

accuracy, and precision were summarized in Table 2. The developed method was applied to determine the regulated plasticizers in domestic infusion sets. As detected infusion sets containing DEHP (5.4 μg/g) and DOTP (153.7 μg/g), the chromatogram obtained by the GC-MS SIM mode (a) in comparison with the corresponding chromatogram obtained

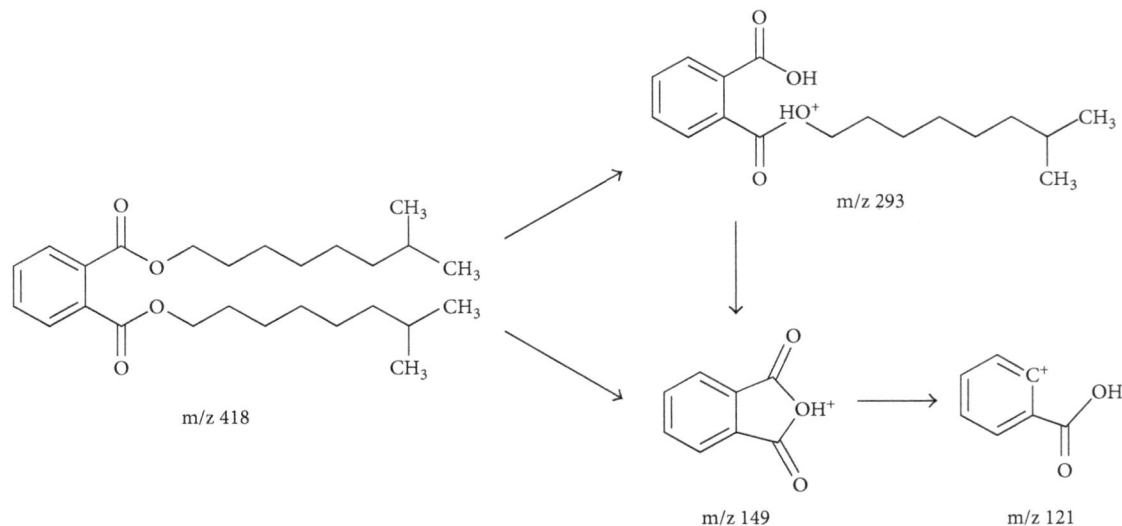

FIGURE 3: Fragmentation pathway of diisononyl phthalate (DINP) in the MRM mode. A similar fragmentation pattern is found for the other target phthalates. The dominant fragmented ion of m/z 149 in phthalates is able to distinguish among different alternative plasticizers, such as bis(2-ethylhexyl) adipate (DEHA).

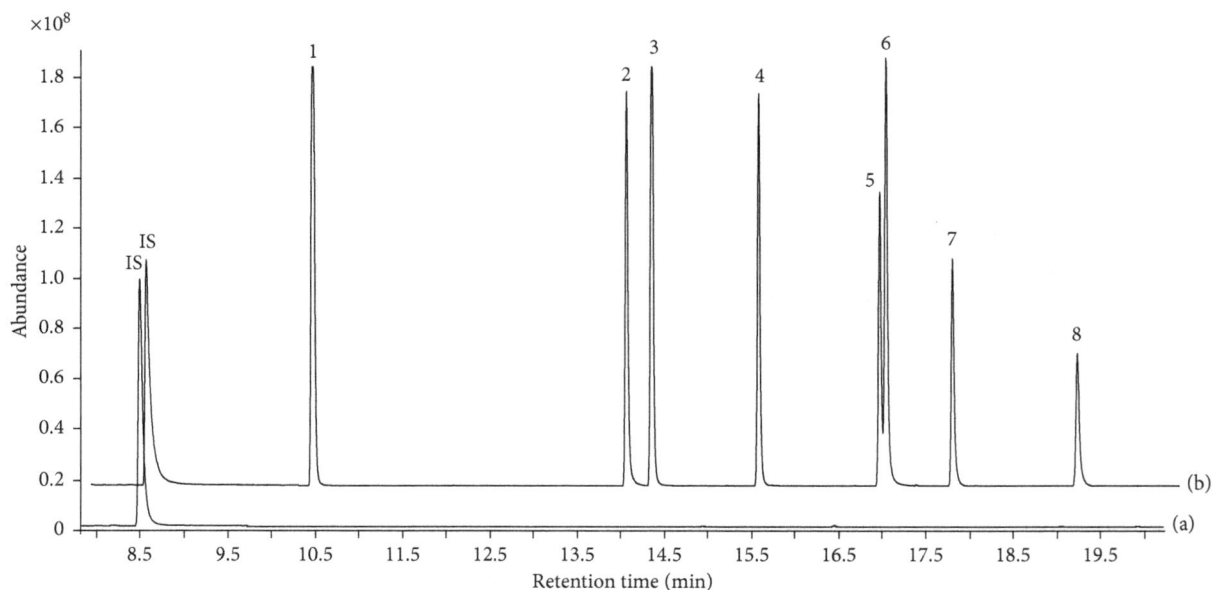

FIGURE 4: GC-MS/MS MRM chromatograms of matrix blank sample (a) and sample spiked with analytes at 5 μg/g (b). Peak identities are as follows: (1) dibutyl phthalate (DBP); (2) benzyl butyl phthalate (BBP); (3) bis(2-ethylhexyl) adipate (DEHA); (4) bis(2-ethylhexyl) phthalate (DEHP); (5) di-n-octyl phthalate (DNOP); (6) dioctyl terephthalate (DOTP); (7) diisononyl phthalate (DINP); (8) diisodecyl phthalate (DIDP); IS, benzyl benzoate (BB).

by the GC-MS/MS MRM mode (b) are shown in Figure 5. Due to the presence of relatively high concentrations of DEHP and DOTP, the calibrations with ten times lower concentration of internal standard, but using the same analyte concentration range, were adopted, and their concentrations in the sample were multiplied by the dilution factor. Upon comparing the chromatograms, the superiority of MRM was obvious. In the MRM mode, the baseline noise of the chromatogram was reduced and distinct peaks appeared, whereas the interferences observed in the SIM mode even look like a false-positive DNOP peak. Apparently, MRM gave higher sensitivity and selectivity than the SIM mode for the determination of target analytes in medical infusion sets.

4. Conclusion

The suitability of solvent extraction with the polymer dissolution method followed by the GC-MS/MS MRM mode for the determination of eight regulated plasticizers

TABLE 2: Linearity, LOQ, accuracy, and precision of target analytes in tubings of medical infusion sets.

Compound	R^{2*}	LOQ (ng/g)[†]	17 ng/g[‡]		100 ng/g[‡]	
			Accuracy (%)	Precision (%RSD)	Accuracy (%)	Precision (%RSD)
DBP	0.9996	59.5	122	1.8	100	11.4
BBP	0.9996	58.6	104	5.8	103	13.4
DEHA	0.9994	56.9	95.9	4.1	91.8	12.4
DEHP	0.9991	76.3	113	2.0	99.7	13.2
DNOP	0.9993	54.1	116	13.5	98.0	16.0
DOTP	0.9997	75.5	98.8	5.4	118	11.5
DINP	0.9991	72.4	121	17.8	109	9.8
DIDP	0.9994	64.1	109	11.0	92.6	9.2

[*]The square of the correlation coefficient; [†]limit of quantitation (LOQ) refers to the lowest concentrations that can be quantified with adequate accuracy and precision; [‡]the concentrations of the analytes in the spiked samples.

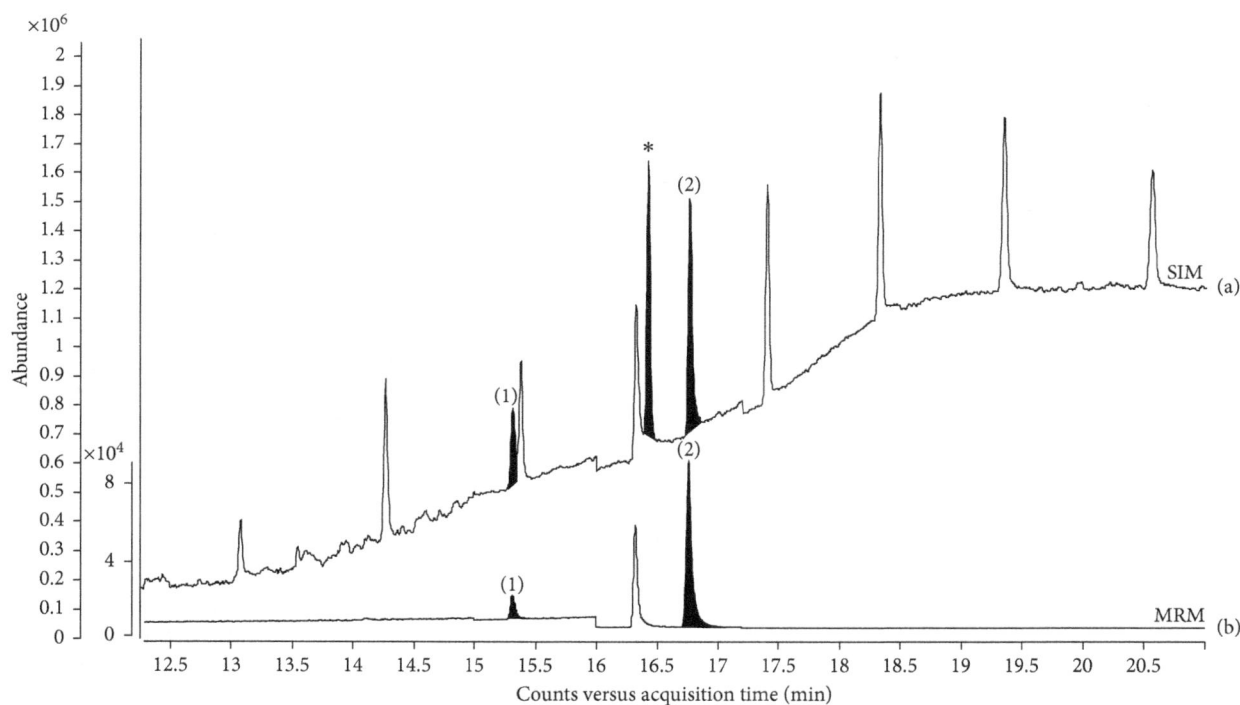

FIGURE 5: GC-MS SIM chromatogram (a) in comparison with the corresponding GC-MS/MS MRM chromatogram (b) in a tubing of medical infusion set containing bis(2-ethylhexyl) phthalate (DEHP) (1) and dioctyl terephthalate (DOTP) (2). SIM chromatogram shows an inconsistent high noise level and interfering peaks, and false-positive identification of di-n-octyl phthalate (DNOP) is represented by an asterisk in SIM chromatogram, but elimination is clear in MRM chromatogram.

(DBP, BBP, DEHA, DEHP, DNOP, DOTP, DINP, and DIDP) in the tubing of medical infusion sets was described. The validated method was successfully used to analyze the samples of medical infusion sets, maintaining high sensitivity and selectivity. The LOQ of this study using the GC-MS/MS MRM mode was lower than those of other studies using conventional GC-MS. The proposed method could be a useful tool to control the safety of intravascular administration sets and to assess unintended harmful substances, especially plasticizers. Furthermore, it will be helpful for biomonitoring as a methodology for tracking the fate of these chemicals associated with human exposures through direct contact and use because the analytical method enables the trace level determination of target analytes.

Conflicts of Interest

The authors declare that they have no conflicts of interest.

Acknowledgments

This research was supported by the Bio-synergy Research Project (NRF-2017M3A9C4065961) of the Ministry of Science, ICT, and Future Planning through the National

Research Foundation and Korea Basic Science Institute grant (C37705).

References

[1] A. L. Andrady and M. A. Neal, "Applications and societal benefits of plastics," *Philosophical Transactions of the Royal Society of London B: Biological Sciences*, vol. 364, no. 1526, pp. 1977–1984, 2009.

[2] G. Ginsberg, J. Ginsberg, and B. Foos, "Approaches to children's exposure assessment: case study with diethylhexylphthalate (DEHP)," *International Journal of Environmental Research and Public Health*, vol. 13, no. 7, p. 670, 2016.

[3] A. Abaamrane, S. Qourzal, M. El Ouardi et al., "Modeling of photocatalytic mineralization of phthalic acid in TiO_2 suspension using response surface methodology (RSM)," *Desalination and Water Treatment*, vol. 53, no. 1, pp. 249–256, 2015.

[4] J. D. Meeker, S. Sathyanarayana, and S. H. Swan, "Phthalates and other additives in plastics: human exposure and associated health outcomes," *Philosophical Transactions of the Royal Society of London B: Biological Sciences*, vol. 364, no. 1526, pp. 2097–2113, 2009.

[5] P. Mikula, Z. Svobodova, and M. Smutna, "Phthalates: toxicology and food safety-a review," *Czech Journal of Food Sciences*, vol. 23, no. 6, pp. 217–223, 2005.

[6] J. Adibi, R. Whyatt, D. Camann, D. Peki, W. Jedrychowski, and F. Perera, "Phthalate diester levels in personal air samples during pregnancy in two urban populations," *Proceedings of Indoor Air*, vol. 4, pp. 177–182, 2002.

[7] B. C. Tran, M. J. Teil, M. Blanchard, F. Alliot, and M. Chevreuil, "Fate of phthalates and BPA in agricultural and non-agricultural soils of the Paris area (France)," *Environmental Science and Pollution Research*, vol. 22, no. 14, pp. 11118–11126, 2015.

[8] S. E. Serrano, J. Braun, L. Trasande, R. Dills, and S. Sathyanarayana, "Phthalates and diet: a review of the food monitoring and epidemiology data," *Environmental Health*, vol. 13, no. 1, p. 43, 2014.

[9] M. F. Zaater, Y. R. Tahboub, and A. N. Al Sayyed, "Determination of phthalates in Jordanian bottled water using GC–MS and HPLC-UV: environmental study," *Journal of Chromatographic Science*, vol. 52, no. 5, pp. 447–452, 2013.

[10] L. Bernard, B. Décaudin, M. Lecoeur et al., "Analytical methods for the determination of DEHP plasticizer alternatives present in medical devices: a review," *Talanta*, vol. 129, pp. 39–54, 2014.

[11] M. Veiga, D. Bohrer, P. C. Nascimento, A. G. Ramirez, L. M. Carvalho, and R. Binotto, "Migration of phthalate-based plasticizers from PVC and non-PVC containers and medical devices," *Journal of the Brazilian Chemical Society*, vol. 23, no. 1, pp. 72–77, 2012.

[12] L. Bernard, R. Cueff, C. Breysse, B. Décaudin, V. Sautou, and A. S. Group, "Migrability of PVC plasticizers from medical devices into a simulant of infused solutions," *International Journal of Pharmaceutics*, vol. 485, no. 1-2, pp. 341–347, 2015.

[13] A. Gomez-Hens and M. P. Aguilar-Caballos, "Social and economic interest in the control of phthalic acid esters," *TrAC Trends in Analytical Chemistry*, vol. 22, no. 11, pp. 847–857, 2003.

[14] S. H. Swan, K. M. Main, F. Liu et al., "Decrease in anogenital distance among male infants with prenatal phthalate exposure," *Environmental Health Perspectives*, vol. 113, no. 8, pp. 1056–1061, 2005.

[15] S. Keresztes, E. Tatár, Z. Czégény, G. Záray, and V. G. Mihucz, "Study on the leaching of phthalates from polyethylene terephthalate bottles into mineral water," *Science of the Total Environment*, vol. 458–460, pp. 451–458, 2013.

[16] H. G. Wahl, A. Hoffmann, H.-U. Häring, and H. M. Liebich, "Identification of plasticizers in medical products by a combined direct thermodesorption–cooled injection system and gas chromatography–mass spectrometry," *Jounal of Chromatography A*, vol. 847, no. 1-2, pp. 1–7, 1999.

[17] I. S. Kostić, T. D. Anđelković, D. H. Anđelković, T. P. Cvetković, and D. D. Pavlović, "Determination of di(2-ethylhexyl) phthalate in plastic medical devices," *Hemijska Industrija*, vol. 70, no. 2, pp. 159–164, 2016.

[18] L. Bernard, D. Bourdeaux, B. Pereira et al., "Analysis of plasticizers in PVC medical devices: Performance comparison of eight analytical methods," *Talanta*, vol. 162, pp. 604–611, 2017.

[19] R. Z. Al Bakain, Y. Al-Degs, B. Andri, D. Thiébaut, J. Vial, and I. Rivals, "Supercritical fluid chromatography of drugs: parallel factor analysis for column testing in a wide range of operational conditions," *Journal of Analytical Methods in Chemistry*, vol. 2017, Article ID 5340601, 13 pages, 2017.

[20] A. P. Tüzüm Demir and S. Ulutan, "Migration of phthalate and non-phthalate plasticizers out of plasticized PVC films into air," *Journal of Applied Polymer Science*, vol. 128, no. 3, pp. 1948–1961, 2013.

[21] Q. Du, L. Shen, L. Xiu, G. Jerz, and P. Winterhalter, "Di-2-ethylhexyl phthalate in the fruits of Benincasa hispida," *Food Additives and Contaminants*, vol. 23, no. 6, pp. 552–555, 2006.

[22] L. Zhang, C. Shang, and C. Sun, "Simultaneous determination of 17 phthalate esters in Shengmaiyin by gas chromatography-triple quadrupole mass spectrometry," *Chinese Journal of Chromatography*, vol. 32, no. 6, pp. 653–657, 2014.

[23] T. Niino, T. Asakura, T. Ishibashi et al., "A simple and reproducible testing method for dialkyl phthalate migration from polyvinyl chloride products into saliva simulant," *Journal of the Food Hygienic Society of Japan*, vol. 44, no. 1, pp. 13–18, 2003.

[24] Q. Wang and B. K. Storm, "Separation and analysis of low molecular weight plasticizers in poly (vinyl chloride) tubes," *Polymer Testing*, vol. 24, no. 3, pp. 290–300, 2005.

[25] S. Genay, C. Luciani, B. Décaudin et al., "Experimental study on infusion devices containing polyvinyl chloride: to what extent are they di(2-ethylhexyl) phthalate-free?," *International Journal of Pharmaceutics*, vol. 412, no. 1-2, pp. 47–51, 2011.

[26] P. Gimeno, S. Thomas, C. Bousquet et al., "Identification and quantification of 14 phthalates and 5 non-phthalate plasticizers in PVC medical devices by GC-MS," *Journal of Chromatography B*, vol. 949-950, pp. 99–108, 2014.

[27] D. Bourdeaux, M. Yessaad, P. Chennell et al., "Analysis of PVC plasticizers in medical devices and infused solutions by GC-MS," *Journal of Pharmaceutical and Biomedical Analysis*, vol. 118, pp. 206–213, 2016.

[28] S. H. Jeon, J. H. Shin, Y. P. Kim, and Y. G. Ahn, "Determination of volatile alkylpyrazines in microbial samples using gas chromatography-mass spectrometry coupled with head space-solid phase microextraction," *Journal of Analytical Science and Technology*, vol. 7, no. 1, p. 16, 2016.

[29] J. Möller, E. Strömberg, and S. Karlsson, "Comparison of extraction methods for sampling of low molecular compounds in polymers degraded during recycling," *European Polymer Journal*, vol. 44, no. 6, pp. 1583–1593, 2008.

[30] K.-C. Ting, M. Gill, and O. Garbin, "GC/MS screening method for phthalate esters in children's toys," *Journal of AOAC International*, vol. 92, no. 3, pp. 951–958, 2009.

[31] C. Liao, P. Yang, Z. Xie et al., "Application of GC-triple quadrupole MS in the quantitative confirmation of polycyclic aromatic hydrocarbons and phthalic acid esters in soil," *Journal of Chromatographic Science*, vol. 48, no. 3, pp. 161–166, 2010.

[32] Z. Guo, S. Wang, D. Wei et al., "Development and application of a method for analysis of phthalates in ham sausages by solid-phase extraction and gas chromatography–mass spectrometry," *Meat Science*, vol. 84, no. 3, pp. 484–490, 2010.

[33] K. Jaworek and M. Czaplicka, "Determination of phthalates in polymer materials-comparison of GC/MS and GC/ECD methods," *Polimeros*, vol. 23, no. 6, pp. 718–724, 2013.

[34] X. Meng, X. Zhao, S. Wang et al., "Simultaneous determination of volatile constituents from Acorus tatarinowii schott in rat plasma by gas chromatography-mass spectrometry with selective ion monitoring and application in pharmacokinetic study," *Journal of Analytical Methods in Chemistry*, vol. 2013, Article ID 949830, 7 pages, 2013.

[35] United States Consumer Product Safety Commission, *Test Method: CPSC-CH-C1001-09.3 Standard Operating Procedure for Determination of Phthalates*, Consumer Products Safety Commission, Directorate for Laboratory Sciences, Gaithersburg, MD, USA, 2010.

[36] G. T. Deepa, M. Chetti, M. C. Khetagoudar, P. T. Goroji, and D. Bilehal, "Application of GC-MS/MS for pesticide residue analysis in pomegranate fruits," 2010.

[37] Y. A. Jeilani, B. H. Cardelino, and V. M. Ibeanusi, "Density functional theory and mass spectrometry of phthalate fragmentations mechanisms: modeling hyperconjugated carbocation and radical cation complexes with neutral molecules," *Journal of the American Society for Mass Spectrometry*, vol. 22, no. 11, pp. 1999–2010, 2011.

[38] A. O. Earls, I. P. Axford, and J. H. Braybrook, "Gas chromatography-mass spectrometry determination of the migration of phthalate plasticisers from polyvinyl chloride toys and childcare articles," *Journal of Chromatography A*, vol. 983, no. 1-2, pp. 237–247, 2003.

[39] ICH Harmonized Tripartite Guideline, *Validation of Analytical Procedures: Text and Methodology*, Q2(R1), 2005.

19

Comparison between Different Extraction Methods for Determination of Primary Aromatic Amines in Food Simulant

Morteza Shahrestani,[1,2] **Mohammad Saber Tehrani** [ID][1,5] **Shahram Shoeibi,**[2,3]
Parviz Aberoomand Azar,[1] **and Syed Waqif Husain**[1,4]

[1]*Department of Analytical Chemistry, Faculty of Basic Science, Science and Research Branch, Islamic Azad University, Tehran, Iran*
[2]*Food and Drug Laboratories Research Center (FDLRC), Iran Food and Drug Administration (IFDA), MOH, Tehran, Iran*
[3]*Department of Food Chemistry, Food and Drug Laboratories Research Center (FDLRC), Iran Food and Drug Administration (IFDA), MOH, Tehran, Iran*
[4]*Department of Chemistry, Faculty of Science, Science and Research Branch, Islamic Azad University, Tehran, Iran*
[5]*Department of Analytical Chemistry, Faculty of Basic Sciences, Azad University, Sciences and Researches Branch, P.O. Box 14515-775, Poonak-Hesarak, Tehran, Iran*

Correspondence should be addressed to Mohammad Saber Tehrani; drmsabertehrani@yahoo.com

Academic Editor: Krishna K. Verma

The primary aromatic amines (PAAs) are food contaminants which may exist in packaged food. Polyurethane (PU) adhesives which are used in flexible packaging are the main source of PAAs. It is the unreacted diisocyanates which in fact migrate to foodstuff and then hydrolyze to PAAs. These PAAs include toluenediamines (TDAs) and methylenedianilines (MDAs), and the selected PAAs were 2,4-TDA, 2,6-TDA, 4,4′-MDA, 2,4′-MDA, and 2,2′-MDA. PAAs have genotoxic, carcinogenic, and allergenic effects. In this study, extraction methods were applied on a 3% acetic acid as food simulant which was spiked with the PAAs under study. Extraction methods were liquid-liquid extraction (LLE), dispersive liquid-liquid microextraction (DLLME), and solid-phase extraction (SPE) with C18 ec (octadecyl), HR-P (styrene/divinylbenzene), and SCX (strong cationic exchange) cartridges. Extracted samples were detected and analyzed by HPLC-UV. In comparison between methods, recovery rate of SCX cartridge showed the best adsorption, up to 91% for polar PAAs (TDAs and MDAs). The interested PAAs are polar and relatively soluble in water, so a cartridge with cationic exchange properties has the best absorption and consequently the best recoveries.

1. Introduction

Primary aromatic amines (PAAs) are food contaminants. Migration of PAAs into food is through colored plastics, printed paper, cooking utensils, and flexible packaging. One of the main sources of PAAs is polyurethane (PU) adhesives which are used extensively in lamination of multilayer films. PU adhesives might contain unreacted aromatic diisocyanates coming from the imperfect polymerization process of polyurethane; after packaging, water present in foods hydrolyzes residual aromatic diisocyanates, thus leading to PAA formation [1].

Toluene diisocyanate (TDI) and methylene diphenyl isocyanate (MDI) are used in the production of PU adhesives.

TDIs are often a mixture (80 : 20) of the two isomers: 2,4-TDI and 2,6-TDI, while MDIs consist of a mixture of higher oligomer homologues, 4,4′-MDI (40–50%), 2,4′-MDI (2.5–4.0%), and 2,2′-MDI (0.1–0.2%) [2].

PAAs have genotoxic, carcinogenic, and allergenic effects. Epidemiological studies of the 4,4′-MDA indicate a risk of bladder cancer in humans, and the 2,4-TDA and 4,4′-MDA are listed as possible human carcinogens [3].

In the European Union (EU) Regulation (10/2011), plastic material and articles shall not release PAAs in detectable quantity (DL = 0.01 mg/kg) in food or food simulants. Chromatography method is recommended for individual identification and quantification of migrated PAAs [4].

The extraction/preconcentration of PAAs is required in order to reach the detection limit, prior to analysis. Sample preparation is commonly done for LLE or SPE method [5]. Each one of these two general approaches has advantages and disadvantages; also there has been a new development in the two methods in order to save time, labor, chemicals, and other materials used. This process has led to the creation of new methods such as solid-phase microextraction (SPME), liquid-phase microextraction (LPME), and dispersive liquid-liquid microextraction (DLLME) [6].

DLLME technique has been developed and often used for the determination of organic compound in water samples. DLLME is a very simple and rapid extraction method which is used for preconcentration of organic and inorganic compounds from aqueous samples. This method is based on the fast injection of the appropriate mixture of extraction (high-density solvent) and dispersive solvents (polar) into the aqueous solution to form a cloudy ternary component solvent. Extraction process is being completed with centrifugation step, and the enriched analyte is collected in the shape of droplets in the sediment phase [7–9].

Several methods are provided in the literatures for determination of TDIs and MDIs or TDAs and MDAs in the urine, blood, PU foam, water, and waste water samples. These methods are often used with or without hydrolysis of samples, extraction/preconcentration in case of liquid-liquid extraction (LLE) or solid-phase extraction (SPE), and with or without derivatization for GC or HPLC analysis. The selection of extraction methods depends on the type and condition of the sample [10–14].

In this study, two different methods of liquid-phase extraction (LLE, DLLME) and solid-phase extraction (Chromabond C18 ec, Chromabond HR-P, and SCX) were investigated for the extraction of interested PAAs, which were spiked in a 3% acetic acid as food simulant [15]. Use of 3% acetic acid which is accepted by the FDA offers the best condition for simulation and a worst-case scenario for this study, and also it is easy to evaluate [16, 17]. The selection of aromatic amines was based on the type of diisocyanates used to produce PU adhesives. The polyurethane adhesives are often produced from toluene diisocyanates (2,4-TDI and 2,6-TDI) or methylene diphenyl isocyanate (4,4′-MDI); thus, the PAAs presented are 2,4-TDA, 2,6-TDA, 4,4′-MDA, 2,4′-MDA, and 2,2′-MDA, and the relative recoveries of the PAAs of interest are determined in the 3% acetic acid matrix.

2. Materials and Methods

2.1. Standards and Reagents. The reagents used in this work were all of analytical grade with no further purification. All the reagents were obtained from Merck KGaA, Darmstadt, Germany.

2,6-Toluenediamine, 2,4-toluenediamine, and 4,4′-methylenedianiline (purity > 98%) were obtained from Sigma-Aldrich, and 2,4′-methylenedianiline and 2,2′-methylenedianiline (purity > 95%) were from Angene.

2.2. Apparatus and Conditions. HPLC system was an Agilent 1200, equipped with a quaternary pump (G1311A), a column

thermostat (G1316A), a degasser unit (G1322A) an autosampler (G1329), and a diode array detector (G1315D). The HPLC system was controlled, and data were analyzed by a computer equipped with LC software (Agilent Chem Station).

The chromatographic conditions were as follows: HPLC analysis was performed with methanol (solvent A) and an ammonium acetate buffer 10 mM (solvent B) as mobile phase. Separation with solvent programming was accomplished on a Phenomenex C18 column (250 × 4.6 mm i.d., 5 μm particle sizes); flow rate was 1 ml·min^{-1}, detection wavelength was 235 nm, and temperature was adjusted to 25°C. The gradient elution process was performed as follows in order to achieve the optimum separation: solvent A was kept at 15% and solvent B was kept at 85% for 2 min, solvent A percentage was increased to 65% along a linear gradient curve for 24 min, and then, solvent A concentration was again increased from 65% to 100% along a linear gradient curve for further 2 min and kept steady at 100% for another 2 min and then was decreased to 15% in the same manner in 1 min and continued for a further 2 min, giving the total run time of 30 min.

2.3. Standard Solutions and Sample Solutions. The individual stock solutions of each standard with concentration of 100 mg·l^{-1} in methanol were prepared. These solutions were kept in the darkness and under refrigeration (4°C) for up to six months. Mixed intermediate standard solutions in methanol were prepared by dissolving appropriate amounts of each individual solution to yield concentrations of 10 mg·l^{-1}. Calibration solutions in methanol/sodium citrate solution 0.1 M (25/75) were prepared daily in the range of 50–800 ng·ml^{-1}.

The sample solutions were prepared from standard mix solutions 10 mg·l^{-1} with three concentrations 7.5, 15, and 30 ng·ml^{-1} in 3% acetic acid (w/v).

2.4. Extraction Methods

2.4.1. Liquid-Liquid Extraction Method. For the extraction of PAAs by solvents, the optimized method was applied [12, 13] and modified as follows: an aliquot sample solution of 8 ml was transferred to a test tube, the pH was adjusted to above 10 with sodium hydroxide, and the solution was saturated with sodium chloride. The sample was extracted three times with 5 mL portion of mixed dichloromethane/petroleum benzene (3:1) and followed by twice with 5 mL of mixed dichloromethane/methyl tert-butyl ether (1:1). The organic solvent fractions were combined and evaporated to dryness by a stream of nitrogen gas. Finally, the extract was dissolved in 0.8 mL of methanol/sodium citrate solution 0.1 M (25/75) and filtered through a 0.22 μm filter for HPLC analysis.

2.4.2. Dispersive Liquid-Liquid Microextraction Method. The method used in DLLME was obtained from Wang et al. [8] and Zhou et al. [9], which was modified by the description given below.

A volume of 5 mL sample solution was placed in a conical bottom tube, 1 g sodium chloride was added, and the pH was adjusted to 12 using a 5 N NaOH solution.

TABLE 1: Linearity of PAAs in 3% acetic acid.

Compounds	Regression equation	Linear range (mgl^{-1})	Correlation coefficient (R^2)
2,6-TDA	$Y = 0.438X - 2.152$	0.05–0.8	0.9996
2,4-TDA	$Y = 0.356X + 2.6455$	0.05–0.8	0.9987
4,4'-MDA	$Y = 0.5104X - 6.0336$	0.05–0.8	0.9998
2,4'-MDA	$Y = 0.4868X - 4.2338$	0.05–0.8	0.9999
2,2'-MDA	$Y = 0.3734X - 2.3448$	0.05–0.8	0.9999

TABLE 2: The recovery of PAAs in 3% acetic acid.

Compounds	Amount spiked ($\mu g \cdot l^{-1}$)	Recovery (%)	RSD	Amount spiked ($\mu g \cdot l^{-1}$)	Recovery (%)	RSD	Amount spiked ($\mu g \cdot l^{-1}$)	Recovery (%)	RSD
First day									
2,6-TDA	7.5	90.26	2	15	93.74	1.4	30	96.46	1.3
2,4-TDA	7.5	74.4	3.3	15	87.77	3.3	30	92.92	2.6
4,4'-MDA	7.5	83.94	2.6	15	95.28	0.95	30	96.33	0.95
2,4'-MDA	7.5	95	1.6	15	94.76	1.4	30	96.86	1.4
2,2'-MDA	7.5	81.58	2.69	15	93.58	2.71	30	94.79	1.1
Second day									
2,6-TDA	7.5	90.77	2.6	15	94.04	1.4	30	96.94	1.3
2,4-TDA	7.5	74.90	4.2	15	91.97	4.1	30	93.9	1.1
4,4'-MDA	7.5	89.26	2.7	15	94.53	1.3	30	97.03	0.43
2,4'-MDA	7.5	96.76	1.8	15	97.08	1.1	30	97.15	0.96
2,2'-MDA	7.5	85.02	2.18	15	93.16	2.07	30	94.93	4.6
Third day									
2,6-TDA	7.5	89.76	2.4	15	94.90	1.5	30	95.80	1.5
2,4-TDA	7.5	77.9	4.8	15	91.21	2.2	30	92.9	1.8
4,4'-MDA	7.5	88.13	0.94	15	94.79	2.5	30	96.29	0.4
2,4'-MDA	7.5	94.94	2.17	15	97.17	2.2	30	98.05	1.1
2,2'-MDA	7.5	85.38	3.37	15	94.82	4.6	30	95.44	0.61

A mixture of 500 μL acetonitrile and 90 μL toluene was rapidly injected into the sample using a syringe, and a cloudy solution was obtained. The solution thus prepared was then shaken for 2 min and was then centrifuged at 6000 rpm for 5 min. The dispersed fine droplets of toluene were sedimented at the bottom of the test tube. The upper layer was removed, and the residual phase was blown to almost dryness with low-pressure nitrogen gas. Finally, the extract was dissolved in 0.5 mL of methanol/sodium citrate solution 0.1 M (25/75) and filtered through a 0.22 μm filter for HPLC analysis.

2.4.3. SPE Chromabond® C18 ec.
SPE cartridges were conditioned based on the method explained by Oostdyk et al. [5] as the procedure with 5 ml methanol and 5 ml distillated water. Then, 50 ml sample solution was adjusted to pH 10 with 10 M sodium hydroxide solution and was passed slowly through the cartridges. The cartridges were eluted three times with 1 ml of ethyl acetate. Eluents were evaporated in a stream of nitrogen. The remaining residue was dissolved in 5 mL of methanol/sodium citrate solution 0.1 M (25/75) and collected into a 5 ml graduated tube. The solutions were filtered through a 0.22 μm filter for HPLC analysis.

2.4.4. SPE Chromabond HR-P.
SPE cartridges were conditioned as advised by Macherey-Nagel Company [11], with

2 × 5 ml methanol, 2 × 5 ml acetonitrile, and 2 × 5 ml 10^{-5} M sodium hydroxide solution. Then, 50 ml sample solution was adjusted to pH 9 with 10 M sodium hydroxide solution and was passed slowly through the cartridges. The cartridges were eluted by 3 × 1.5 ml methanol/acetonitrile (1 : 1). Eluents were evaporated in a stream of nitrogen. The remaining residue was dissolved in 5 mL of methanol/sodium citrate solution 0.1 M (25/75) and collected into a 5 ml graduated tube. The solutions were filtered through a 0.22 μm filter for HPLC analysis.

2.4.5. SPE SCX.
Aznar et al. [16] applied SCX cartridges for PAAs, and the method was modified as below: cartridges were conditioned with 2 × 3 ml methanol and 2 × 3 ml 3% acetic acid (w/v). Then, 50 ml sample solution was passed slowly through the cartridges. The cartridges were eluted with 5 × 1 ml methanol/sodium citrate solution 0.1 M (25/75) and collected into a 5 ml graduated tube. The solutions were filtered through a 0.22 μm filter for HPLC analysis.

2.5. Method Validation of SPE SCX

2.5.1. Linearity.
Linearity was evaluated for mix of five PAAs with concentrations of 50, 100, 200, 400, and 800 μgl^{-1} in

TABLE 3: Intra- and interday precision of determination of PAAs in 3% acetic acid.

Compounds	Amount spiked (μgl^{-1})	Precision (intraday) RSD	Precision (interday) RSD
2,6-TDA	75	0.115	0.146
	150	0.115	0.278
	300	0.2	0.239
2,4-TDA	75	0.115	0.188
	150	0.173	0.161
	300	0.152	0.215
4,4'-MDA	75	0.058	0.139
	150	0.152	0.273
	300	0.1	0.256
2,4'-MDA	75	0.115	0.115
	150	0.404	0.292
	300	0.058	0.153
2,2'-MDA	75	0.1	0.086
	150	0.152	0.165
	300	0.1	0.1

methanol/sodium citrate solution 0.1 M (25/75) with three replications ($n = 3$) (Table 1).

There was a good linearity in the range of 0.05–0.8 mgl^{-1} for all five target compounds.

2.5.2. Accuracy. In order to investigate the recoveries of the method including SPE procedures, blank food simulants (3% acetic acid) spiked with five PAAs at three different concentrations 7.5, 15, and 30 μgl^{-1} were performed with three replications ($n = 3$). The recoveries are presented for the five aromatic amines (Table 2).

These data confirm not only the accuracy of the method, but also the integrity of the SPE procedure.

2.5.3. Precision. Precision calculation was done based on the repeatability criterion in one day (intraday) and in 3 consecutive days (interday).

Intraday precision was evaluated with spiked blank food simulant (3% acetic acid) at three different concentrations 7.5, 15, and 30 μgl^{-1} including SPE procedures. The RSD of intraday was obtained (Table 3).

Interday precision was evaluated with spiked blank food simulant (3% acetic acid) at three different concentrations 7.5, 15, and 30 μgl^{-1} including SPE procedures on three different days. The RSD of interday was obtained (Table 3).

2.5.4. Sensitivity. Relative standard deviation (RSD) of the method was performed with spiked blank food simulant (3% acetic acid) by five PAAs at concentration 7.5 μgl^{-1}, including the SPE procedure. LOD and LOQ of the method were calculated (Table 4).

3. Results and Discussion

The migration of primary aromatic amines (PAAs) from flexible food packaging represents a serious risk to public health as these compounds are potentially carcinogenic substances. The source of PAAs is from the residues of aromatic

TABLE 4: LOD and LOQ of the method of determination of PAAs in 3% acetic acid.

Compounds	Limit of detection (μgl^{-1})	Limit of quantitation (μgl^{-1})	RSD
2,6-TDA	1.88	5.71	0.250
2,4-TDA	1.86	5.64	0.201
4,4'-MDA	1.44	4.37	0.223
2,4'-MDA	1.36	4.13	0.201
2,2'-MDA	2	6.08	0.227

diisocyanates (2,4-TDI, 2,6-TDI, and 4,4'-MDI) arising from incomplete curing of the main polyurethane (PU) adhesive, and also other aromatic diisocyanates such as 2,4'-MDI and 2,2'-MDI are present in adhesive inappropriately.

The five PAAs (2,4-TDA, 2,6-TD, 4,4'-MDA, 2,4'-MDA, and 2,2'-MDA) used in this study were selected on the basis of their origin of PU adhesives used in food packaging. Toxicity of PAAs was evaluated and classified into three groups based on their levels of toxicity with 2,4-TDA, 2,6-TDA, and 4,4'-MDA being in the high toxicity class [18]. According to the EU Regulation (EU 10/2011), detection limit of the released PAAs must be below 10 μg/kg of food or food simulant. The method was developed and designed based on the worst-case scenario for the migration of PAAs from packaging material into a food simulant which in this study was 3% acetic acid. The aim of the research was to design and select an extraction method, liquid and/or solid phase, of analysis for these PAAs (TDAs and MDAs) with high sensitivity and also to make possible to understand the origin of the PAAs detected. The method was designed for 3% acetic acid (as food simulant) in purified water, since this simulant was considered the most restrictive, that means it uses the worst-case scenario for the migration of TDAs and MDAs from food packaging. The direct analysis of the simulant by HPLC-UV did not provide enough sensitivity, as the sample amount injected into the system was very low. For this reason, the LLE and SPE experiments were needed prior to the HPLC detection.

TDAs and MDAs have two functional groups (-NH$_2$) that are polar and partly soluble in water; as a result, the

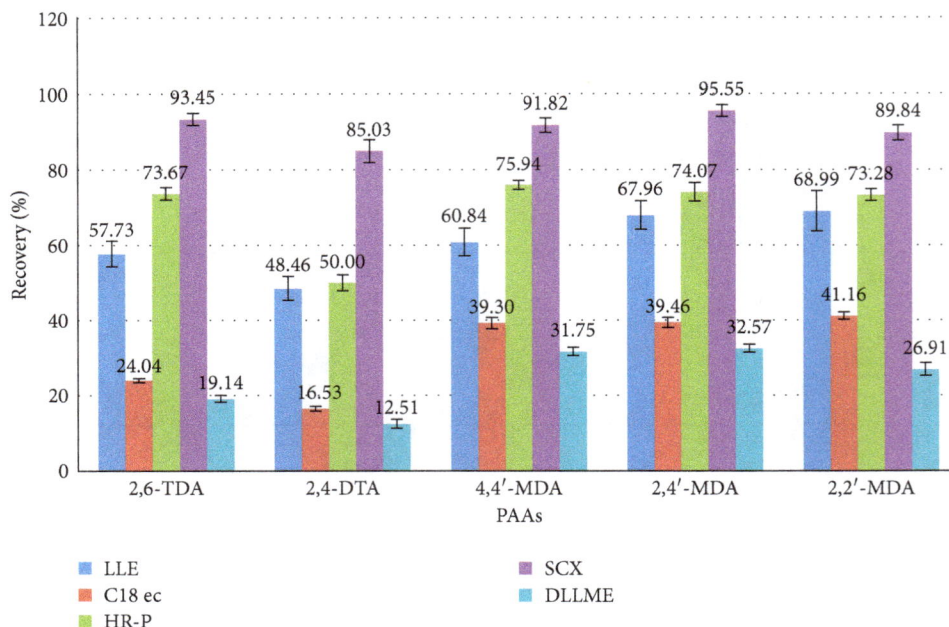

FIGURE 1: Compressive of recoveries of extraction methods: LLE, DLLME, SPE C18 ec, SPE HR-P, and SPE SCX for 5 primary aromatic amines.

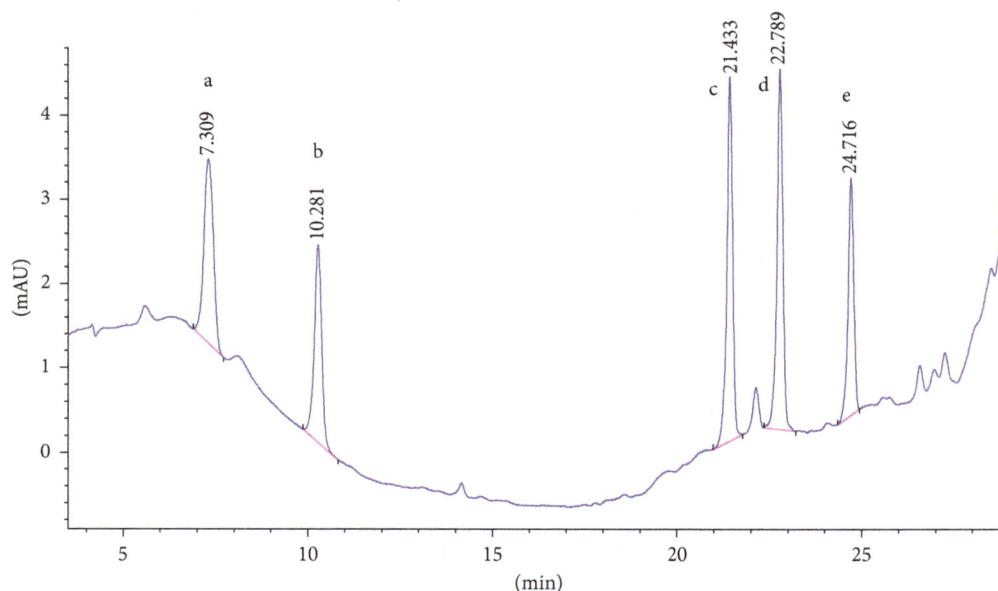

FIGURE 2: Chromatogram of the standard solution (100 ppb). Peak identified as a: 2,6-TDA, b: 2,4-TDA, c: 4,4'-MDA, d: 2,4'-MDA, and e: 2,2'-MDA.

extraction of these compounds from aqueous solution is difficult [11].

TDAs (single ring) in comparison with MDAs (two rings) have more polarity and are more soluble in water. Therefore, the mean recoveries for the extraction of TDA derivatives (2,4-TDA and 2,6-TDA) should be less than that of MDA derivatives (4,4'-MDA, 2,4'-MDA, and 2,2'-MDA).

In liquid extraction methods, in general data obtained for mean recoveries of extraction for both LLE and DLLME were not very satisfactory. The mean recoveries of LLE is better than those of DLLME, because the number of extraction steps could be increased in the LLE method, while there is just one time dispersion in the DLLME method. The mean recoveries for the extraction of MDAs were greater than that of TDAs in both methods. DLLME were used in extraction of aromatic amines in aqueous matrix with acceptable recoveries [8, 9], but in our research, the interested aromatic amines have two functional group ($-NH_2$) which are more soluble in aqueous solutions, and consequently, recoveries are lower than LLE and other methods.

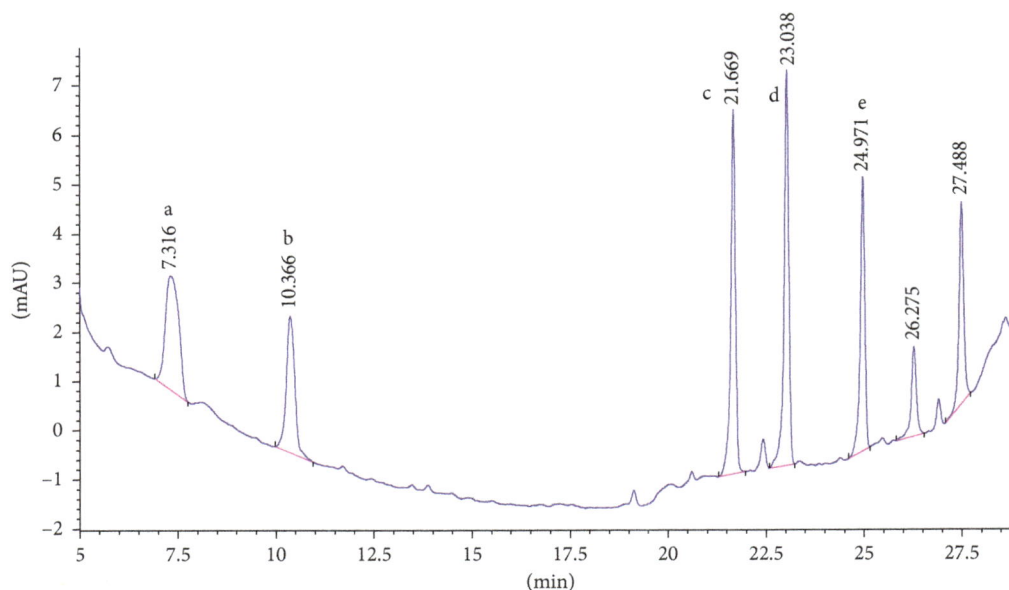

FIGURE 3: Chromatogram of extracted sample spiked (15 ppb) in 3% acetic acid with SPE SCX. Peak identified as a: 2,6-TDA, b: 2,4-TDA, c: 4,4′-MDA, d: 2,4′-MDA, and e: 2,2′-MDA.

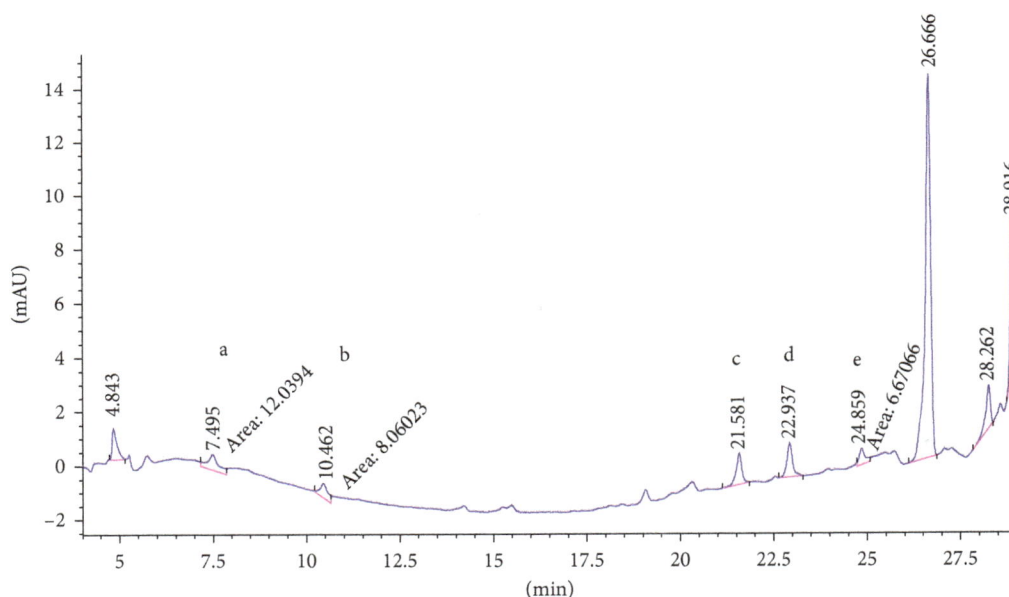

FIGURE 4: Chromatogram of extracted sample spiked (15 ppb) in 3% acetic acid with LLDME. Peak identified as a: 2,6-TDA, b: 2,4-TDA, c: 4,4′-MDA, d: 2,4′-MDA, and e: 2,2′-MDA.

In SPE methods, the mean recoveries of TDAs and MDAs were in the order of SCX > HR-P > C18 ec. Octadecyl sorbents (nonpolar) are not suitable for TDAs and MDAs in the C18 cartridge. In HR-P columns, the cross-linking in SDVB leads to better adsorption than C18 for the polar PAAs. The mean recoveries of the five extraction methods were compared, and the SPE method with SCX cartridge was the best (more than 90%). The result of all applied methods are compared and shown in Figure 1.

SCX cartridge with the mechanism of action being based on ion exchange could properly adsorb polar PAAs and so has the best mean recovery among the studied cartridges. Figure 2 shows the chromatogram of the standard solution of five PAAs ($100 \, \text{ng·ml}^{-1}$). The chromatogram of $15 \, \text{ng·ml}^{-1}$ spiked food simulant, in which the final concentration of the extracted and preconcentrated solution by SCX cartridge is $150 \, \text{ng·ml}^{-1}$, is shown in Figure 3. Also the chromatograms of each method are compared for the performance of the extraction and shown in Figures 4–7.

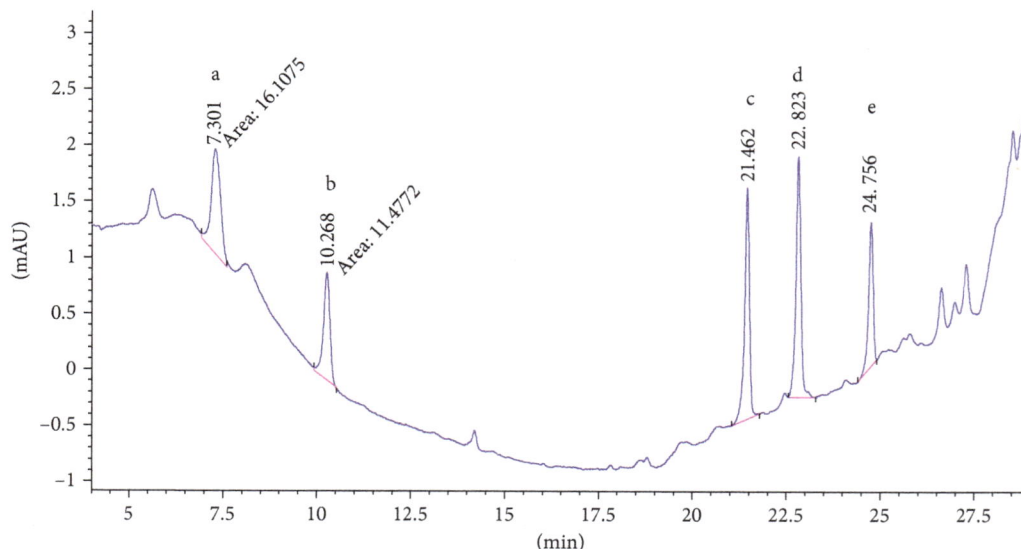

FIGURE 5: Chromatogram of extracted sample spiked (15 ppb) in 3% acetic acid with SPE C18. Peak identified as a: 2,6-TDA, b: 2,4-TDA, c: 4,4′-MDA, d: 2,4′-MDA, and e: 2,2′-MDA.

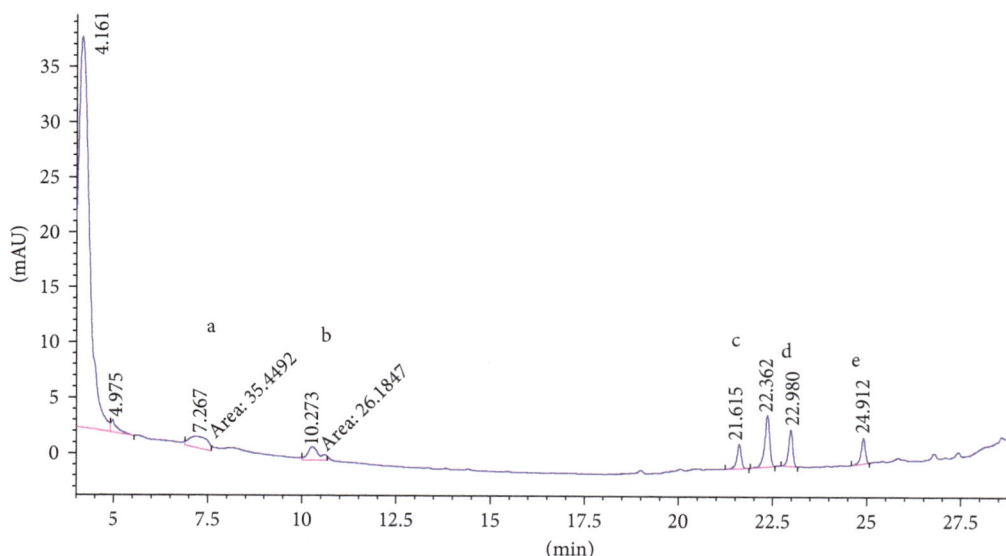

FIGURE 6: Chromatogram of extracted sample spiked (15 ppb) in 3% acetic acid with SPE HR-P. Peak identified as a: 2,6-TDA, b: 2,4-TDA, c: 4,4′-MDA, d: 2,4′-MDA, and e: 2,2′-MDA.

HPLC method combined with SCX cartridge was validated and developed for the separation and quantitation of five primary aromatic amines in 3% acetic acid as food simulant. After optimizing all separation parameters, the good separation of five PAAs in food simulant was feasible within 30 min. Additionally, the method was validated for linearity, LOD, precision, and intra- and interday variation.

4. Conclusions

It is important to determine and control the migration of contaminants from materials in contact with food into the food. Packaging materials provided this route for migration of contaminants such as PAAs from PU adhesives which are used in laminated multilayer films; some PAAs can migrate to the food. Analysis of PAAs is very valuable in the European regulations, and for this purpose, EU has established regulations for PAAs. Simultaneous extraction and analysis of PAAs (2,4-TDA, 2,6-TDA, 4,4′ MDA, 2,4′ MDA, and 2,2′ MDA) lead to the process of food safety assessment and maintenance, also a means to monitor the use of authorized adhesives and prevention of unauthorized ones being used. Solid-phase extraction SCX is introduced as a suitable method for extraction of polar primary aromatic amines (PAAs) which might have migrated from the laminated multilayer films of packaging materials into 3% acetic acid as food simulant.

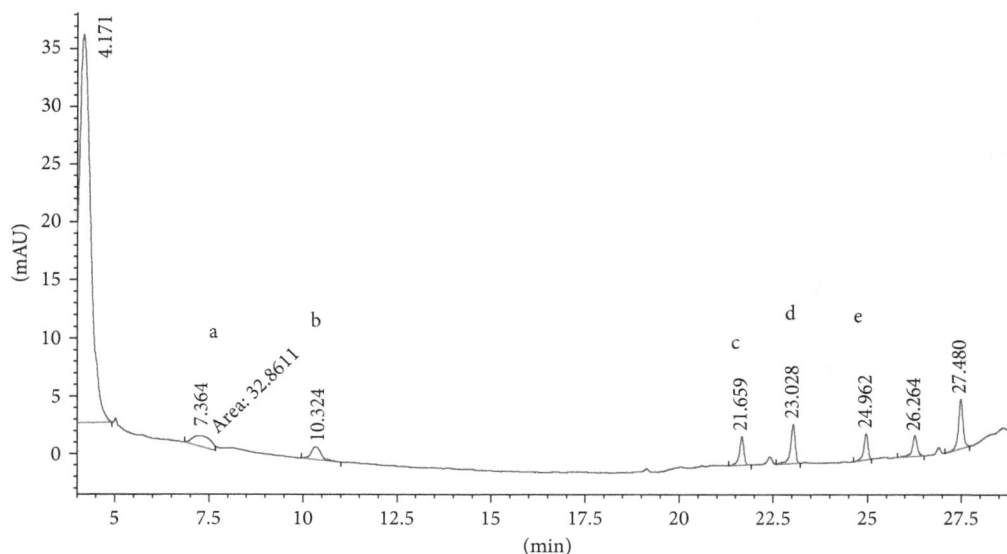

FIGURE 7: Chromatogram of extracted sample spiked (15 ppb) in 3% acetic acid with LLE. Peak identified as a: 2,6-TDA, b: 2,4-TDA, c: 4,4'-MDA, d: 2,4'-MDA, and e: 2,2'-MDA.

Abbreviations

DLLME:　Dispersive liquid-liquid microextraction
HPLC-UV:　HPLC-ultraviolet detector
LLE:　Liquid-liquid extraction
MDA:　Methylenedianiline
MDI:　Methylene diphenyl isocyanate
PAAs:　Primary aromatic amines
PU:　Polyurethane
SPE:　Solid-phase extraction
TDA:　Toluenediamine
TDI:　Toluene diisocyanate.

Conflicts of Interest

The authors declare that they have no conflicts of interest.

Acknowledgments

The authors thank the scientific experts of the Food and Drug Laboratories Research Center (FDLRC) for their kind cooperation during the study, and special thanks are due to the Henkel Corporation for their professional and skillful technical assistance.

References

[1] M. Mattarozzi, F. Lambertini, M. Suman, and M. Careri, "Liquid chromatography–full scan-high resolution mass spectrometry-based method towards the comprehensive analysis of migration of primary aromatic amines from food packaging," *Journal of Chromatography A*, vol. 1320, pp. 96–102, 2013.

[2] C. J. Sennbro, C. H. Lindh, H. Tinnerberg et al., "Development, validation and characterization of an analytical method for the quantification of hydrolysable urinary metabolites and plasma protein adducts of 2,4- and 2,6-toluene diisocyanate,

1,5-naphthalene diisocyanate and 4,4'-methylenediphenyl diisocyanate," *Biomarkers*, vol. 8, no. 3-4, pp. 204–217, 2003.

[3] X. Trier, B. Okholm, A. Foverskov, M. L. Binderup, and J. H. Petersen, "Primary aromatic amines (PAAs) in black nylon and other food-contact materials," *Food Additives and Contaminants*, vol. 27, no. 90, pp. 1325–1335, 2010.

[4] O. Yavuz, S. Valzacchi, E. Hoekstra, and C. Simoneau, "Determination of primary aromatic amines in cold water extract of coloured paper napkin samples by liquid chromatography-tandem mass Spectrometry," *Food Additives and Contaminants*, vol. 33, no. 6, pp. 1072–1079, 2016.

[5] T. S. Oostdyk, R. L. Grob, J. L. Snyder, and M. E. McNally, "Solid-phase extraction of primary aromatic amines from aqueous samples; comparison with liquid-liquid extraction techniques," *Journal of Environmental Science and Health*, vol. 29, no. 8, pp. 1607–1628, 1994.

[6] L. Xu, C. Basheer, and H. K. Lee, "Review chemical reactions in liquid-phase microextraction," *Journal of Chromatography A*, vol. 1216, no. 4, pp. 701–707, 2009.

[7] H. Deng, F. Yang, Z. Li et al., "Rapid determination of 9 aromatic amines in mainstream cigarette smoke by modified dispersive liquid-liquid microextraction and ultra-performance convergence chromatography tandem mass spectrometry," *Journal of Chromatography A*, vol. 1507, pp. 37–44, 2017.

[8] X. Wang, L. Fu, G. Wei et al., "Determination of four aromatic amines in water samples using dispersive liquid–liquid microextraction combined with HPLC," *Journal of Separation Science*, vol. 31, no. 16-17, pp. 2932–2938, 2008.

[9] Q. Zhou, L. Pang, G. Xie et al., "Dispersive liquid phase microextraction of aromatic amines in environmental water samples," *International Journal of Environmental Analytical Chemistry*, vol. 90, no. 14, pp. 1099–1107, 2010.

[10] T. Zimmermann, W. J. Ensinger, and T. C. Schmidt, "In situ derivatization/solid-phase micro extraction: determination of polar aromatic amines," *Analytical Chemistry*, vol. 76, no. 4, pp. 1028–1038, 2004.

[11] T. C. Schmidt, M. Leβ, R. Haas, E. V. Löw, and K. Steinbach, "Determination of aromatic amines in ground and waste water by two new derivatization methods," *International Journal of Environmental Analytical Chemistry*, vol. 74, no. 1, pp. 25–41, 2011.

[12] R. C. Snyder and C. V. Breder, "High performance liquid chromatographic determination of 2,4 and 2,6-toluenediamine in aqueous extracts," *Journal of Chromatography*, vol. 236, no. 2, pp. 429–440, 1982.

[13] P. Carbonnelle, S. Boukortt, D. Lison, and J. P. Buchet, "Determination of toluenediamines in urine of workers occupationally exposed to isocyanates by high-performance liquid chromatography," *Analyst*, vol. 121, no. 5, pp. 663–669, 1996.

[14] C. Brede, I. Skjevrak, and H. Herikstad, "Determination of primary aromatic amines in water food simulant using solid-phase analytical derivatization followed by gas chromatography coupled with mass spectrometry," *Journal of Chromatography A*, vol. 983, no. 1-2, pp. 35–42, 2003.

[15] C. Molins-Legua and P. Campins-Falco, "Solid phase extraction of amines," *Analytica Chimica Acta*, vol. 546, no. 2, pp. 206–220, 2005.

[16] M. Aznar, E. Canellas, and C. Nerín, "Quantitative determination of 22 primary aromatic amines by cation-exchange solid-phase extraction and liquid chromatography–mass spectrometry," *Journal of Chromatography A*, vol. 1216, no. 27, pp. 5176–5181, 2009.

[17] G. Kollbach, *Smart Cure Technology from Henkel Achieves Cure and Regulatory Compliance in 3 Days*, Flexible Packaging Adhesives, Henkel Corporation, Düsseldorf, Germany, 2009.

[18] G. M. Cramer and R. A. Ford, "Estimation of toxic hazard," *Food and Cosmetics Toxicology*, vol. 16, no. 3, pp. 255–276, 1987.

20

Quality Analysis of Long dan Xie gan Pill by a Combination of Fingerprint and Multicomponent Quantification with Chemometrics Analysis

Jing Liu ⊕,[1] Hui Liu,[1,2] Zhong Dai ⊕,[1] and Shuangcheng Ma ⊕[1]

[1]National Institutes for Food and Drug Control, Beijing 100050, China
[2]Institute of Food and Drug, Yanbian Korean Autonomous Prefecture, Jilin Province 133002, China

Correspondence should be addressed to Zhong Dai; daizhong@nifdc.org.cn and Shuangcheng Ma; masc@nifdc.org.cn

Jing Liu and Hui Liu contributed equally to this work.

Academic Editor: Ricardo Jorgensen Cassella

Long dan Xie gan pill is a traditional complex compound preparation with a long history for treatment of diseases, including hepatocolic hygropyrexia, dizziness, tinnitus, and deafness. Quality of products from different manufacturers may be varied. Since the current standard could not control the quality of products in a comprehensive and effective way, this study aimed at establishing a practical and convenient approach for holistic quality control of the preparation. This study included both qualitative and quantitative works to get information on the overall composition and main components, respectively. As a result, HPLC fingerprint (UV 240 nm) similarities of all fifty samples were in the range of 0.65~0.99. Results indicated that there was a difference among products from different manufacturers. Additionally, ten characteristic peaks of the fingerprint were tentatively identified by LC-MS. Further chemometrics analysis was utilized to evaluate the products from different manufacturers. At the same time, the HPLC (UV 285 nm) multicomponent quantification result showed that contents of gentiopicrin, baicalin, baicalein, and wogonin were in the range of 0.61–5.40, 1.96–5.33, 0.10–3.40, and 0.046–1.16 mg·g^{-1}, respectively. Data analysis verified the main different component of baicalein from the fingerprint statistical analysis. It is worth mentioning that the qualitative fingerprint and quantitative multicomponent determination were simultaneously accomplished by HPLC-DAD with dual channels. The study provided sound basis for improving quality control standards. This study also provided practical strategy for overall quality control of traditional Chinese medicines.

1. Introduction

Long dan Xie gan pill is prepared from ten species of crude drugs including *Gentianae radix* et rhizoma, *Scutellariae radix*, and *Akebiae caulis* (Mutong) in Chinese Pharmacopoeia (2015 Edition, Volume I) [1]. It is widely used for the treatment of diseases, including hepatocolic hygropyrexia, dizziness, tinnitus and deafness, hypochondriac pain [1]. The preparation has attracted widespread attention, since it caused aristolochic acid nephropathy (AAN) [2, 3]. At that time, the prescription collected the crude drug of Caulis aristolochiae manshuriensis (Guanmutong) instead of *Akebiae caulis* [4–6]. Because of the serious adverse effect of aristolochic acids, the medicinal standard of Caulis aristolochiae manshuriensis was abolished and

replaced by *Akebiae caulis* without containing such toxic constituents since 2003.

As one commonly used Chinese patent medicine (CPM) with a long history, Long dan Xie gan pill has about 200 manufacturers. Therefore, the quality of products from different manufacturers may be varied. Since quality is directly related to drug safety and efficacy, it is very important to evaluate the holistic quality of the products. Researchers have been working on the essential quality control and evaluation methods for years [2]. For most CPMs, the effective components are not clear and the consistency of product quality is a key indicator of quality product evaluation. It is worth mentioning that fingerprint is an internationally recognized effective method because it could reflect the overall quality information [2–10]. However, fingerprint is usually used for qualitative consistency evaluation. And the quantification could be achieved by applying multicomponent determination [10–13]. In recent years, more and more chromatographic and spectroscopic methods including LC, LC-MS^n, and quantitative nuclear magnetic resonance (QNMR) are applied for the aforementioned qualitative and quantitative work [14–21]. Among these methods, high-performance liquid chromatography (HPLC) is still the main method deployed in quality control of traditional Chinese medicines (TCMs) because of its advantages, including good repeatability, wide application, and high efficiency.

In this study, the qualitative and quantitative consistency information of Long dan Xie gan pill samples was achieved at the same time by a combination of fingerprint with multicomponent quantification by HPLC-diode array detector (DAD) with dual channels (UV 240 nm and 285 nm). Additionally, further deep mining of the data by chemometrics analysis helped to evaluate the differences of products in a more comprehensive and effective way. The results indicated that the established method could comprehensively analyze the product quality. This strategy could provide a practical approach for the holistic quality control of TCM.

2. Materials and Methods

2.1. Chemicals and Reagents. Gentiopicrin (97.6%, batch no. 110770-201716), baicalin (93.5%, batch no. 110715-201720), baicalein (98.5%, batch no. 111595-201607), and wogonin (100%, batch no. 111514-201605) were from the National Institutes for Food and Drug Control, Beijing, China. Methanol (analytical reagent) was from National Drug Chemical Reagents Co. Ltd. Acetonitrile (chromatographic pure) and formic acid (mass spectrometry reagent) were from Thermo Fischer Scientific. The water was of ultrahigh purity.

2.2. Materials. Fifty batches of Long dan Xie gan pill samples were from 8 manufacturers (A~H). All the samples involve two dosage forms including water-bindered pills (WBP) and big candied pills (BCP). The detailed information is listed as follows (Table 1).

2.3. Instrumentations. HPLC analysis was performed on a Waters 2690 HPLC instrument (Waters, Milford, USA), equipped with a DAD, an autosampler and a column heater. METTLER XS105 electronic analytical balance (Mettler-toledo, Zurich, Switzerland), Milli-Q water purification system (Milli-pore, Burlington, USA), and KQ-300DA numerical control ultrasound cleaning instrument (Kunshan Ultrasonic Instruments Co. Ltd., Kunshan, China) were used. Chemometrics analysis was achieved by ChemPattern software (Chenmind Technologies Co., Ltd., Beijing, China).

TABLE 1: Sample information.

No.	Manufacturers	Batch number	Dosage
1		B16099	
2		B17001	
3		B17045	WBP
4		B16033	
5	A	B16040	
6		A17057	
7		A16065	
8		A16119	BCP
9		A17004	
10		A17104	
11		20170502	
12		20170504	
13	B	20170503	WBP
14		171002	
15		171003	
16		1704058	
17		1801024	
18	C	1801027	WBP
19		1801026	
20		1711016	
21		20180336	
22		20180335	
23		20180334	WBP
24		20180452	
25	D	20180451	
26		20170621	
27		20180419	
28		20170620	BCP
29		20170504	
30		20170308	
31		180203	
32		180201	
33	E	180204	WBP
34		180205	
35		180202	
36		171102	
37		171002	
38	F	170902	WBP
39		171101	
40		171001	
41		1801001	
42		1802005	
43	G	1801003	WBP
44		1801004	
45		1802006	
46		180102	
47		180101	
48	H	171001	BCP
49		171002	
50		160501	

2.4. Preparation of Standard Solutions. Standard stock solutions of baicalin (0.1 mg·mL^{-1}) were prepared by dissolving suitable amounts of reference substance in methanol for fingerprint establishment.

Standard stock mixed solution of gentiopicrin (0.4972 mg·mL^{-1}), baicalin (0.5268 mg·mL^{-1}), baicalein (0.4688 mg·mL^{-1}), and wogonin (0.5080 mg·mL^{-1}) was prepared by dissolving suitable amounts of each reference substance in methanol for multicomponents assay.

2.5. Preparation of Sample Solutions.

For Long dan Xie gan pill (WBP) (6 g per small bag), 5 bags were mixed and pulverized to powder. For Long dan Xie gan pill (BCP) (6 g per pill), 5 pills were cut into small pieces. Then, 2 g were weighed accurately and put into a 50 mL plug conical bottle. Twenty-five mL methanol (for WBP samples) and 25 mL 80% methanol-water solution (for BCP samples) were added precisely and weighed, respectively. After extracting by ultrasonication (power: 300 W; frequency: 40 kHz) for 30 min, the extract was cooled down and then made up for lost weight by adding methanol (for WBP samples) or 80% methanol (for BCP samples). The continuous filtrate was taken and then filtered by 0.22 μm microporous filter membrane.

2.6. HPLC-DAD Chromatographic Condition.

Column: Phenomenex Gemini C18 (4.60 × 250 mm, 5 μm); mobile phase: gradient elution with acetonitrile- (A-)0.1% formic acid-water solution (B) (0–5 min, 7%A–10%A; 5–11 min, 10%A–15%A; 11–15 min, 15%A–20%A; 15–32 min, 20% A–30%A; 32–54 min, 30%A–55%A; 54–60 min, 55%A–80% A; and 60–68 min, 7%A); flow rate: 1.0 mL·min^{-1}; column temperature: 30°C; injection volume: 10 μL; detection wavelength: UV 240 nm for fingerprint and UV 285 nm for multicomponent determination. The typical chromatograms were shown as Figures 1 and 2, respectively.

2.7. Mass Spectrometry Condition.

MS analysis was performed on an Agilent1260-6410B LC-MS couplet system equipped with Agilent Mass Hunter ChemStation (Agilent, Santa Clara, USA). The mass spectrometry settings were as follows: split ratio = 1 : 9; desolvation temperature: 350°C; desolvation air flow N$_2$: 540 L·h^{-1}; nebulizer pressure: 30 psi; and capillary: 4000 V. Both positive and negative modes were performed with a scan range of m/z 50–1200.

3. Results and Discussion

3.1. Results of Fingerprint.

For fingerprint study, all the samples were prepared and analyzed according to conditions under 2.5 and 2.6. Baicalin (t_R = 34.02 min) was taken as the reference peak. Relative retention times (RRTs) and relative peak areas (RPAs) of the characteristic peaks were calculated for method validation.

3.1.1. Instrument Precision.

The same sample solution (no. 21) was injected for six consecutive times. The result showed that the RSDs of RRTs and RPAs were in the range of 0.073%–2.0% and 0.26–1.74%, respectively. It showed that the precision of the instrument was good.

3.1.2. Repeatability.

The same batch sample (no. 21) was taken and prepared for six independent sample solutions for analysis. The result showed that the RSDs of RRT and RPA were in the range of 0.010%–0.44% and 0.51–2.24%, respectively. It indicated that method repeatability was good.

3.1.3. Stability.

The same sample solution (no. 21) was injected at 0, 4, 8, 12, 16, 20, and 24 h at room temperature. The result showed that the RSDs of RRT and RPA were in the range of 0.039%–0.81% and 0.71–4.58%, respectively. It demonstrated the sample solution was stable within 24 h.

3.1.4. Establishment of Fingerprint.

After fifty batches of the sample solutions were analyzed, their chromatograms (UV 240 nm) were recorded (Figure 3) and imported to ChemPattern software. All variables were used as the common peak screening condition. And Gauss curve simulation method was applied to generate the common mode with 16 characteristic peaks (Figure 4). All sample chromatograms were analyzed by comparison with the common mode.

3.1.5. Identification and Attribution of Characteristic Peaks.

The sample solution was analyzed according to the conditions under Sections 2.6 and 2.7. By combination with the chromatographic behavior of the components, ten characteristic peaks were identified by comparing with reference standards. Also the main origins of the peaks were attributed (Table 2), and they were mainly from five species of crude drugs in the prescription.

3.1.6. Statistical Analysis

(1) Similarity Analysis. The above HPLC (UV 240 nm) fingerprint common mode was taken as a reference. During the analysis, the included angle cosine method was used to calculate the similarity of each sample (Figure 5). Finally, the similarities of all samples were in the range of 0.65~0.99. Among them, similarities of seven batches of samples were lower than 0.8, including all samples from enterprise B. The result indicated that there was a certain difference among the overall product quality of these samples from others. Also it was clear that the uniformity of most BCP samples (no. 6~10 and 26~30) was not good as the WBP ones.

(2) Principle Component Analysis. Principle component analysis (PCA) was carried out after standardization of all sample data (Figures 6 and 7). The contribution rates of the first and second principle component (PC1) were 45.92% and 39.22%, respectively. And the total contribution rate of 85.14% showed that it could reflect the differences between samples in a more comprehensive way. PCA scatter plot (Figure 6) displayed that samples from each enterprise basically could be grouped into a class. It showed that samples from enterprise B deviated far away from others. The principle component load diagram (Figure 7) gave the proportion of each chromatographic peak in the principal component. And the greater the distance from X = 0

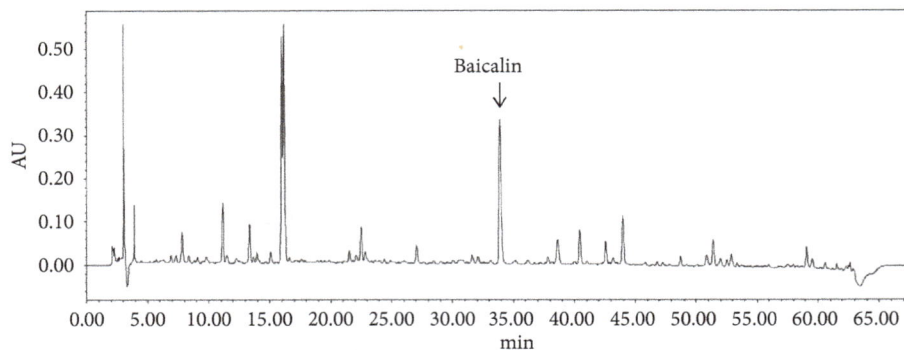

FIGURE 1: HPLC (UV 240 nm) chromatogram of the typical sample.

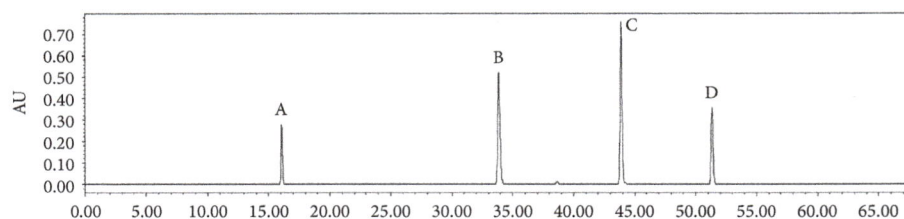

(a)

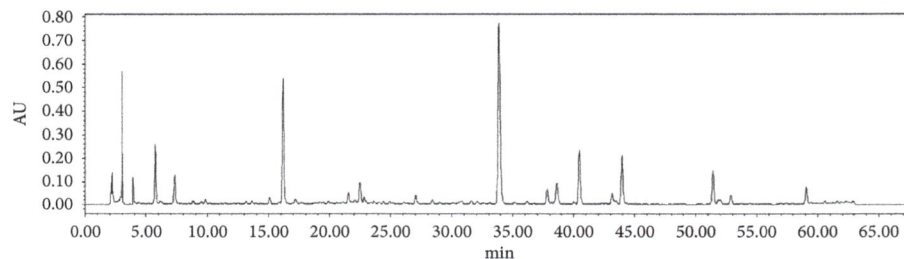

(b)

FIGURE 2: HPLC (UV 285 nm) chromatograms of mixed standard solution (a) (A, Gentiopicrin; B, Baicalin; C, Baicalein; and D, Wogonin) and typical sample (b).

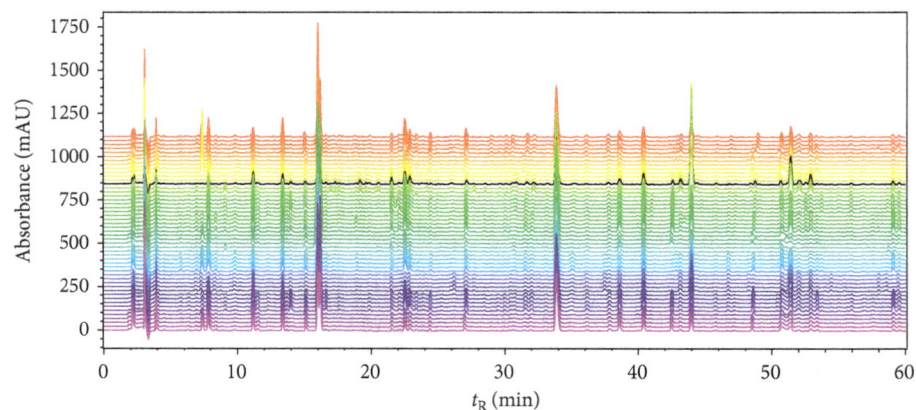

FIGURE 3: HPLC chromatograms (UV240 nm) of Long dan Xie gan pill samples.

longitudinal axis, the greater the contribution to PC1, such as gentiopicrin and baicalein. Likewise, the greater the distance from $Y = 0$ transverse axis, the greater the contribution to PC2, such as baicalein and geniposide. The result displayed that samples from enterprise B separated with others along both PC1 and PC2. Therefore, the major

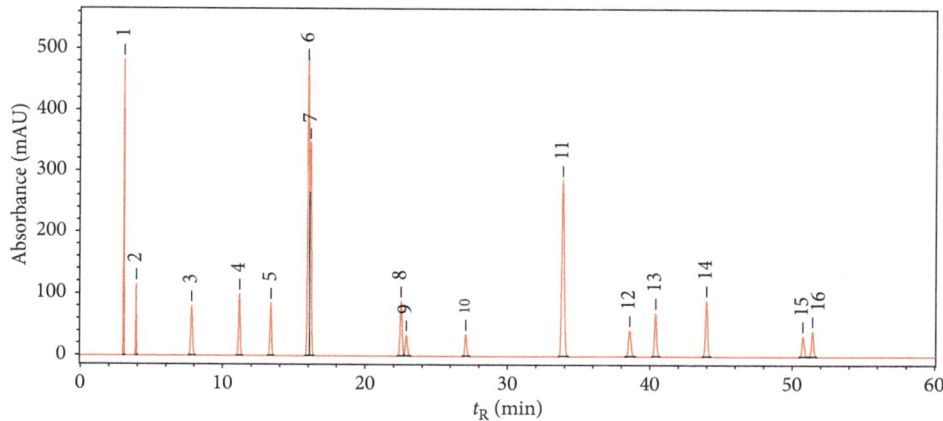

FIGURE 4: HPLC (UV240 nm) common mode of Long dan Xie gan pill samples.

TABLE 2: Identification of characteristic peaks of Long dan Xie gan pill.

Peak no.	Compounds	t_R (min)	Molecular formula	Molecular weight	Quasimolecular ions	Origin
3	Geniposidic acid	7.83	$C_{16}H_{22}O_{10}$	374.34	372.9 $[M-H]^-$	*Gardeniae fructus*
6	Geniposide	15.07	$C_{17}H_{24}O_{10}$	388.37	433.0 $[M+HCOO]^-$	*Gardeniae fructus*
7	Gentiopicrin	16.18	$C_{16}H_{20}O_9$	356.32	400.9 $[M+HCOO]^-$	*Gentianae radix* et rhizoma
8	Verbascoside	22.86	$C_{29}H_{36}O_{15}$	624.59	623.0 $[M-H]^-$	Plantaginis semen/rehmanniae radix
10	Baicalin	33.88	$C_{21}H_{18}O_{11}$	446.36	445.0 $[M-H]^-$	*Scutellariae radix*
11	Isomer of wogonoside	38.62	$C_{22}H_{20}O_{11}$	460.39	459.0 $[M-H]^-$	*Scutellariae radix*
12	Wogonoside	40.44	$C_{22}H_{20}O_{11}$	460.39	459.0 $[M-H]^-$	*Scutellariae radix*
14	Baicalein	43.82	$C_{15}H_{10}O_5$	270.24	268.9 $[M-H]^-$	*Scutellariae radix*
15	Isomer of wogonin	51.40	$C_{16}H_{12}O_5$	284.26	282.9 $[M-H]^-$	*Scutellariae radix*
16	Wogonin	51.99	$C_{16}H_{12}O_5$	284.26	282.9 $[M-H]^-$	*Scutellariae radix*

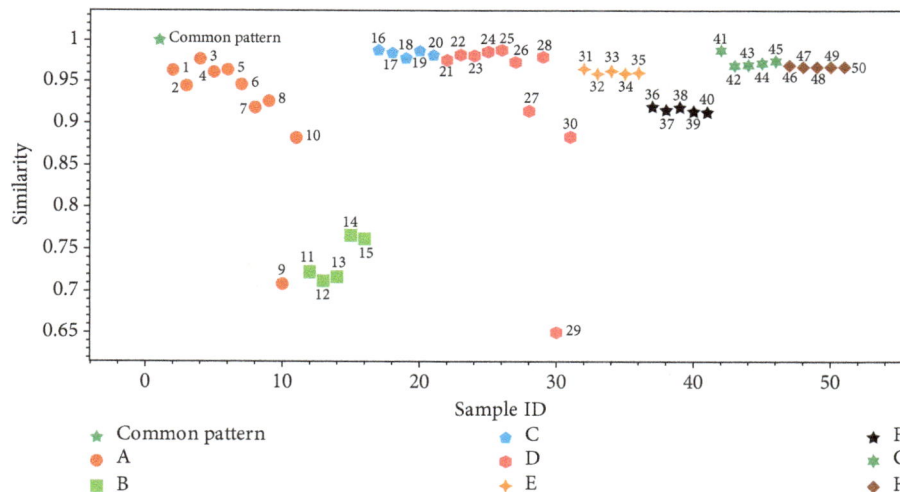

FIGURE 5: Similarity analysis results of Long dan Xie gan pill samples.

contributions to PC1 and PC2 were their main differential components from others. Because of the location at the position of PC1 > 0 and PC2 < 0, contents of such compounds were in positive correlation with PC1 and in negative correlation with PC2. Therefore, the contents of baicalein and wogonin were higher in these samples. Along the PC1, the concentration of baicalein distinguished the samples of

enterprise B, whose values were higher than those presented for the others (Figure 6). On the contrary, along with the PC2, the levels of geniposide and gentiopicrin showed that there was a tendency for the separation of the products of enterprises D and G from others. One sample of enterprise A was grouped with samples of enterprises D and G, which presented higher levels for these compounds.

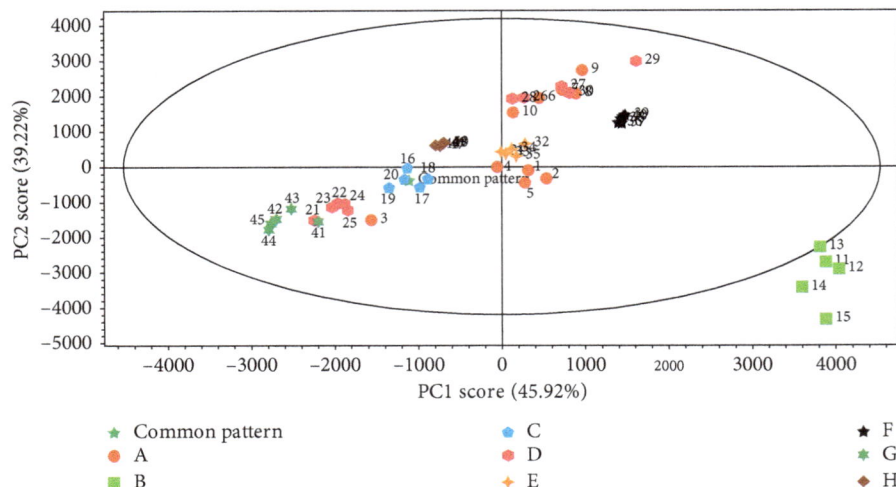

FIGURE 6: PCA score scatter plot of Long dan Xie gan pill samples.

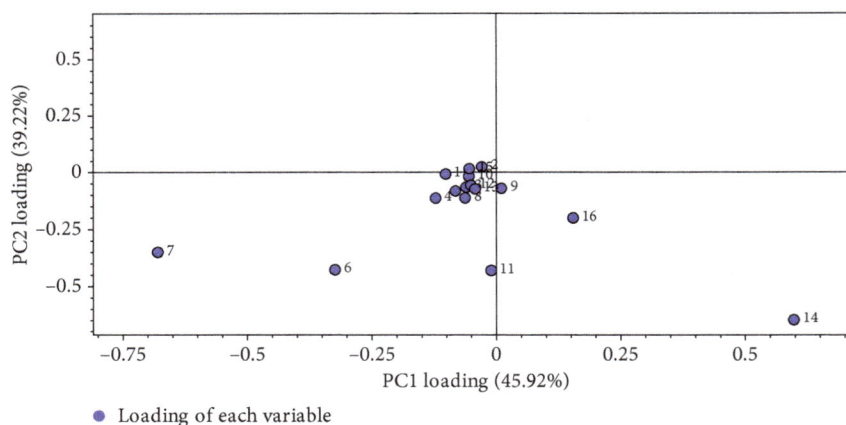

FIGURE 7: PCA loading scatter plot of Long dan Xie gan pill samples (1–16 represents the number of chromatographic peaks in the common mode).

(3) Cluster Analysis. Hierarchical cluster analysis (HCA) is a conventional cluster analysis method. It is a detection tool that clearly reveals the natural grouping of data. The block distance was selected for distance calculation and HCA (Figure 8) was performed by error square sum method. Similar to the result of PCA, except that some samples from manufacturer D are not distinguished from those of A, other samples from different enterprises could basically distinguish. Additionally, samples from B were relatively most far away from others. It indicated there existed some differences of these samples.

3.2. Results of Multicomponent Quantification

3.2.1. Linearity, LOD, and LOQ. Working standard solutions containing gentiopicrin, baicalin, baicalein, and wogonin were prepared by diluting the stock mixed solution with methanol to a series of proper concentrations. Then, they were injected and analyzed. The results of regression equations, linearity, determination coefficient, and limits of detection and quantification of the method are presented in

Table 3. The linear range varied from 3.25 to 492.56 μg mL^{-1}, in accordance with the analyte. All analytes presented a determination coefficient (R^2) of the 0.9999, which allows the method to be considered linear. The limits of detection (LOD) and quantification (LOQ) were calculated according to guidelines for validation of analytical methods for pharmaceutical quality standards [22].

3.2.2. Instrument Precision. The same sample solution (no. 21) was injected for six consecutive times and analyzed. The RSDs of peak areas for gentiopicrin, baicalin, baicalein, and wogonin were 0.63%, 0.29%, 0.41%, and 0.15%, respectively. It indicated that the precision of the instrument was in accordance with the requirement in guidelines for validation of analytical methods for pharmaceutical quality standards [22].

3.2.3. Repeatability. The same batch of sample (no. 21) was taken and prepared for six independent sample solutions. Then, they were analyzed according to conditions under 2.6.

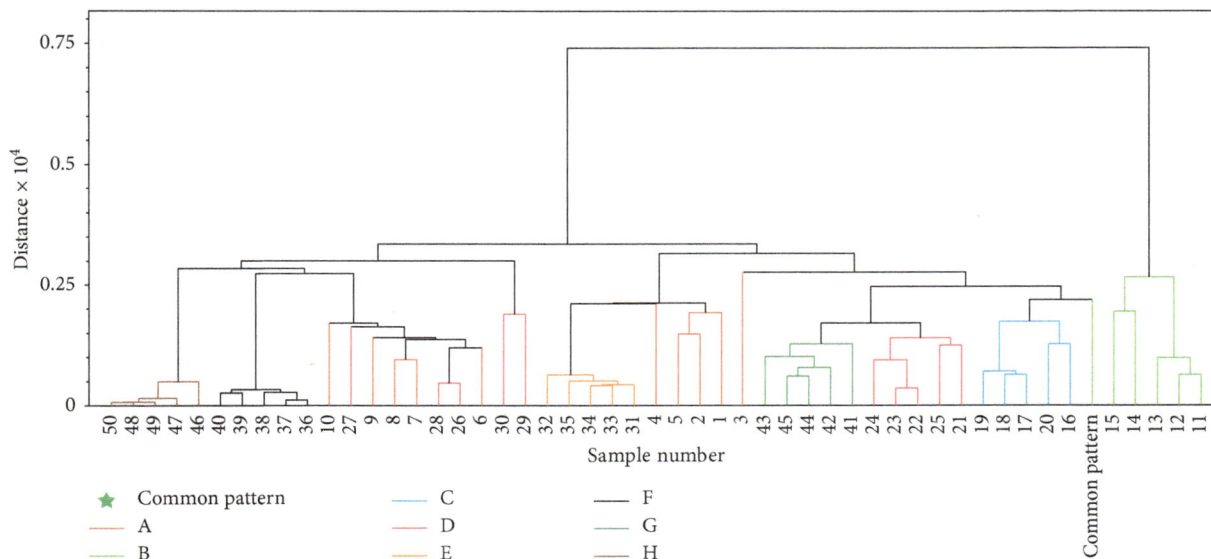

FIGURE 8: Dendrograms of hierarchical cluster analysis of Long dan Xie gan pill samples.

TABLE 3: Regression equations, linearity, determination coefficient, and limits of detection and quantification of the method.

Components	Regression equations	Linear range (μg·mL^{-1})	R^2	LOD (ng)	LOQ (ng)
Gentiopicrin	$y = 11524\,x + 12562$	7.76~485.27	0.9999	0.837	2.790
Baicalin	$y = 32266\,x + 4256.2$	7.88~492.56	0.9999	0.431	1.077
Baicalein	$y = 43732\,x - 23130$	7.39~461.77	0.9999	0.439	1.025
Wogonin	$y = 47996\,x + 23628$	3.25~203.20	0.9999	0.190	0.474

The average contents of gentiopicrin, baicalin, baicalein, and wogonin were 4.77, 3.84, 0.62, and 0.40 mg·g^{-1}, respectively. And the RSDs were 0.49%, 1.11%, 0.40%, and 0.56%, respectively. It indicated that method repeatability was in accordance with the requirement in guidelines for validation of analytical methods for pharmaceutical quality standards [22].

3.2.4. Stability. The same sample solution (no. 21) was injected at 0, 4, 8, 12, 18, and 24 h at room temperature. The RSDs of contents for gentiopicrin, baicalin, baicalein, and wogonin were 1.88%, 1.54%, 1.70%, and 3.22%, respectively. It indicated that the sample solution was stable within 24 h.

3.2.5. Recovery. The recovery experiment was performed by adding a known amount of individual reference standards into a certain amount of sample (no. 21).

Six separate samples of 1 g (contents of gentiopicrin, baicalin, baicalein, and wogonin were 4.77, 3.84, 0.62, and 0.40 mg·g^{-1}, respectively) were weighed accurately. And 25 mL of mixed reference standard solution

(concentrations of gentiopicrin, baicalin, baicalein, and wogonin were 0.1861, 0.1505, 0.04856, and 0.0300 mg·mL^{-1}, respectively) was added separately and prepared. The results (Table 4) showed that the average recoveries of four components ranged from 97.71% to 100.59% with RSDs in the range of 0.72%–1.29%, which indicated that the method was accurate.

3.2.6. Sample Analysis. Fifty batches of sample solutions were prepared and analyzed. The results (Table 5) displayed that the contents of gentiopicrin, baicalin, baicalein, and wogonin were in the range of 0.61–5.40, 1.96–5.33, 0.10–3.40, and 0.046–1.16 mg·g^{-1}, respectively. It was easily to find the differences among samples from different enterprises by the scatter diagram (Figure 9). It showed that the general content trends of baicalin, baicalein, and wogonin were basically similar. Among them, the contents of baicalein and wogonin in samples from B were apparently higher than others; especially, the content of baicalein was much higher. The determination result was in accordance with the abovementioned PCA analysis result.

3.3. Optimization of Experimental Conditions

3.3.1. Investigation of Extraction Methods. The extraction method was optimized in order to make the fingerprint reflect the chemical composition information as much as possible. For both dosage forms of samples, different extraction solvent (80% methanol and 50% methanol-water), and extraction mode and time (ultrasonic extraction for 30 min, 45 min, and 60 min) were investigated. The result showed that extraction time had little effect on both dosage forms. For WBP samples, the chromatogram could reflect rich chemical information with good separation of peaks with methanol extraction for 30 min. While for BCP

TABLE 4: Recovery results of four components in Long dan Xie gan pill samples.

Components	No.	Sampling amount (g)	Sample content (mg)	Added amount (mg)	Detected amount (mg)	Recovery (%)	Average recovery (%)
Gentiopicrin	1	1.0005	4.772	4.653	9.371	98.83	
	2	1.0017	4.778	4.653	9.432	100.02	
	3	1.0006	4.773	4.653	9.431	100.11	100.59% (RSD
	4	1.0047	4.792	4.653	9.473	100.60	1.22%)
	5	1.0030	4.784	4.653	9.528	101.96	
	6	1.0026	4.782	4.653	9.529	102.02	
Baicalin	1	1.0005	3.842	3.762	7.478	96.65	
	2	1.0017	3.847	3.762	7.537	98.09	
	3	1.0006	3.842	3.762	7.503	97.32	97.96% (RSD
	4	1.0047	3.858	3.762	7.583	99.02	0.87%)
	5	1.0030	3.852	3.762	7.543	98.11	
	6	1.0026	3.850	3.762	7.554	98.56	
Baicalein	1	1.0005	0.620	1.214	1.807	97.78	
	2	1.0017	0.621	1.214	1.802	97.28	
	3	1.0006	0.620	1.214	1.793	96.62	97.71% (RSD
	4	1.0047	0.623	1.214	1.809	97.69	0.72%)
	5	1.0030	0.622	1.214	1.816	98.35	
	6	1.0026	0.622	1.214	1.818	98.52	
Wogonin	1	1.0005	0.400	0.750	1.141	98.80	
	2	1.0017	0.401	0.750	1.146	99.33	
	3	1.0006	0.400	0.750	1.141	98.80	99.98% (RSD
	4	1.0047	0.402	0.750	1.153	100.13	1.29%)
	5	1.0030	0.401	0.750	1.156	100.67	
	6	1.0026	0.401	0.750	1.167	102.13	

TABLE 5: Contents of four components in Long dan Xie gan pills ($mg \cdot g^{-1}$).

No.	Manufacturers (dosage)	Gentiopicrin	Baicalin	Baicalein	Wogonin
1		1.579	3.191	0.778	0.433
2		1.391	2.912	0.992	0.449
3	A (WBP)	3.295	3.102	0.741	0.429
4		1.801	2.582	0.718	0.381
5		1.884	3.006	0.976	0.526
6		1.510	2.133	0.330	0.198
7		0.979	2.173	0.298	0.130
8	A (BCP)	0.816	2.269	0.337	0.176
9		0.746	2.148	0.176	0.103
10		1.777	2.409	0.362	0.176
11		1.019	4.153	3.049	1.129
12		1.013	4.394	3.141	1.158
13	B (WBP)	0.963	3.763	3.280	1.080
14		0.860	4.396	3.133	0.994
15		0.826	5.330	3.405	1.084
16		2.801	3.460	0.295	0.232
17		2.442	3.359	0.502	0.300
18	C (WBP)	2.138	3.353	0.423	0.274
19		2.822	3.363	0.411	0.270
20		2.724	3.889	0.294	0.216
21		4.753	3.747	0.622	0.406
22		4.166	3.123	0.570	0.367
23	D (WBP)	4.344	3.118	0.590	0.381
24		4.028	3.047	0.585	0.385
25		4.084	3.642	0.589	0.390
26		1.562	2.412	0.203	0.110
27		1.157	2.371	0.235	0.125
28	D (BCP)	1.688	2.337	0.189	0.102
29		0.606	2.879	0.100	0.046
30		0.670	3.170	0.111	0.052

Permissions

List of Contributors

Yun Gyong Ahn, So Hyeon Jeon and Geum-Sook Hwang
Western Seoul Center, Korea Basic Science Institute, Seoul 03759, Republic of Korea

Hyung Bae Lim
Air Quality Research Division, National Institute of Environmental Research, Incheon 22689, Republic of Korea

Na Rae Choi and Ji Yi Lee
Department of Environmental Science and Engineering, Ewha Womans University, Seoul 03759, Republic of Korea

Yong Pyo Kim
Department of Chemical Engineering and Material Science, Ewha Womans University, Seoul 03760, Republic of Korea

Yi Peng, Minghui Dong, Jing Zou and Zhihui Liu
Department of Pharmacy, Nanjing University of Traditional Chinese Medicine Affiliated Hospital, Nanjing 210029, China

Luisa F. Angeles and Diana S. Aga
Department of Chemistry, e State University of New York, Buffalo, NY 14260, USA

Naz Hasan Huda, Bhawna Gauri, Heather A. E. Benson and Yan Chen
School of Pharmacy and Biomedical Sciences, Curtin Health Innovation Research Institute, Curtin University, Perth, WA 6845, Australia

Xu Wang
The First Clinical Medical College, Nanjing University of Chinese Medicine, Nanjing 210023, China

Xin Shao
The First Clinical Medical College, Nanjing University of Chinese Medicine, Nanjing 210023, China
Department of Endocrinology, Nanjing Hospital of Traditional Chinese Medicine, Nanjing 210001, China

Jie Zhao
Pharmaceutical Animal Experimental Center, China Pharmaceutical University, Nanjing 210009, China

Yi Tao
School of Pharmacy, Nanjing University of Chinese Medicine, Nanjing 210023, China

Li-hua Chen, Yao Wu, Yong-mei Guan, Chen Jin and Wei-feng Zhu
Key Laboratory of Modern Preparation of TCM, Ministry of Education, Jiangxi University of Traditional Chinese Medicine, No. 18 Yun Wan Road, Nanchang 330004, China

Ming Yang
Jiangxi Sinopharm Co. Ltd., No. 888 National Medicine Road, Nanchang 330004, China

Iñaki Elorduy, Nieves Durana, José Antonio García, María Carmen Gómez and Lucio Alonso
Chemical and Environmental Engineering Department, School of Engineering, University of the Basque Country, Alameda de Urquijo s/n, 48013 Bilbao, Spain

Mohammed A. Meetani and Anas A. Alaidaros
Chemistry Department, United Arab Emirates University, Al-Ain, UAE

Rashed Alremeithi
Chemistry Department, United Arab Emirates University, Al-Ain, UAE
General Department of Forensic Science and Criminology, Dubai Police, Dubai, UAE

Adnan Lanjawi and Khalid Alsumaiti
General Department of Forensic Science and Criminology, Dubai Police, Dubai, UAE

Vinit V. Gholap and Matthew S. Halquist
Department of Pharmaceutics, School of Pharmacy, Virginia Commonwealth University, Richmond, VA 23298, USA

Leon Kosmider
Department of Pharmaceutics, School of Pharmacy, Virginia Commonwealth University, Richmond, VA 23298, USA
Center for the Study of Tobacco Products, Virginia Commonwealth University, Richmond, VA 23298, USA

Wei Liu and Zeshu Hu
School of Resources and Environmental Engineering, Wuhan University of Technology, No. 122 Luoshi Road, Wuhan 430070, China

Ji Quan
School of Management, Wuhan University of Technology, No. 122 Luoshi Road, Wuhan 430070, China

Chunhui Cao, Zhongping Li, Liwu Li and Li Du
Key Laboratory of Petroleum Resources, Gansu Province/Key Laboratory of Petroleum Resources Research, Institute of Geology and Geophysics, Chinese Academy of Sciences, Lanzhou 730000, China

Rosa del Carmen Lopez-Sanchez, Victor Javier Lara-Diaz, Alejandro Aranda-Gutierrez, Jorge A. Martinez-Cardona and Jose A. Hernandez
Tecnologico de Monterrey, Escuela de Medicina y Ciencias de la Salud, Ave. Morones Prieto 3000, 64710 Monterrey, NL, Mexico

Sriram Valavala, Nareshvarma Seelam, Subbaiah Tondepu and V. Shanmukha Kumar Jagarlapudi
Department of Chemistry, K L University, Green Fields, Vaddeswaram, Guntur 522502, Andhra Pradesh, India

Vivekanandan Sundarmurthy
Research and Development, Bluefish Pharmaceuticals Private Limited, Bangalore 560115, Karnataka, India

Ji Hyun Jeong, Seon Yu Lee, Bo Na Kim, Guk Yeo Lee and Seong Ho Ham
National Development Institute of Korean Medicine, Udae land gil 288, Jangheung-gun, Jeollanam-do 59338, Republic of Korea

Jianxiu Wang, Shuyun Shi and Shengqiang Hu
College of Chemistry and Chemical Engineering, Central South University, Changsha 410083, China

Jiwang Tang
College of Chemistry and Chemical Engineering, Central South University, Changsha 410083, China
Hunan Testing Institute Product and Commodity Supervison, Changsha 410007, China

Liejiang Yuan
Hunan Testing Institute Product and Commodity Supervison, Changsha 410007, China

Lele Li, Yang Wang, Yang Xiu and Shuying Liu
Jilin Ginseng Academy, Changchun University of Chinese Medicine, Changchun 130117, China

Danielle Cristina da Silva
Universidade Tecnológica Federal do Paraná, Campus Dois Vizinhos, Estrada para Boa Esperança, Km 04 85660-000 Dois Vizinhos, PR, Brazil

Cláudio Celestino Oliveira
Departamento de Química, Universidade Estadual de Maringá, Avenida Colombo, 5790 87020-900 Maringá, PR, Brazil

Yun Gyong Ahn
Western Seoul Center, Korea Basic Science Institute, Seoul 03759, Republic of Korea

So Hyeon Jeon
Western Seoul Center, Korea Basic Science Institute, Seoul 03759, Republic of Korea
Department of Environmental Science and Engineering, Ewha Womans University, Seoul 03759, Republic of Korea

Yong Pyo Kim
Department of Chemical Engineering and Material Science, Ewha Womans University, Seoul 03760, Republic of Korea

Younglim Kho
Department of Health, Environment & Safety, Eulji University, Seongnam 13135, Republic of Korea

Jeoung Hwa Shin
Seoul Center, Korea Basic Science Institute, Seoul 02841, Republic of Korea

Won Hyun Ji
Institute of Mine Reclamation Technology, Mine Reclamation Corporation, Wonju 26464, Republic of Korea

Parviz Aberoomand Azar
Department of Analytical Chemistry, Faculty of Basic Science, Science and Research Branch, Islamic Azad University, Tehran, Iran

Morteza Shahrestani
Department of Analytical Chemistry, Faculty of Basic Science, Science and Research Branch, Islamic Azad University, Tehran, Iran
Food and Drug Laboratories Research Center (FDLRC), Iran Food and Drug Administration (IFDA), MOH, Tehran, Iran

Syed Waqif Husain
Department of Analytical Chemistry, Faculty of Basic Science, Science and Research Branch, Islamic Azad University, Tehran, Iran
Department of Chemistry, Faculty of Science, Science and Research Branch, Islamic Azad University, Tehran, Iran

Mohammad Saber Tehrani
Department of Analytical Chemistry, Faculty of Basic Science, Science and Research Branch, Islamic Azad University, Tehran, Iran

Department of Analytical Chemistry, Faculty of Basic Sciences, Azad University, Sciences and Researches Branch, Poonak-Hesarak, Tehran, Iran

Shahram Shoeibi
Food and Drug Laboratories Research Center (FDLRC), Iran Food and Drug Administration (IFDA), MOH, Tehran, Iran
Department of Food Chemistry, Food and Drug Laboratories Research Center (FDLRC), Iran Food and Drug Administration (IFDA), MOH, Tehran, Iran

Jing Liu, Zhong Dai and Shuangcheng Ma
National Institutes for Food and Drug Control, Beijing 100050, China

Hui Liu
National Institutes for Food and Drug Control, Beijing 100050, China
Institute of Food and Drug, Yanbian Korean Autonomous Prefecture, Jilin Province 133002, China

Index